AF393622

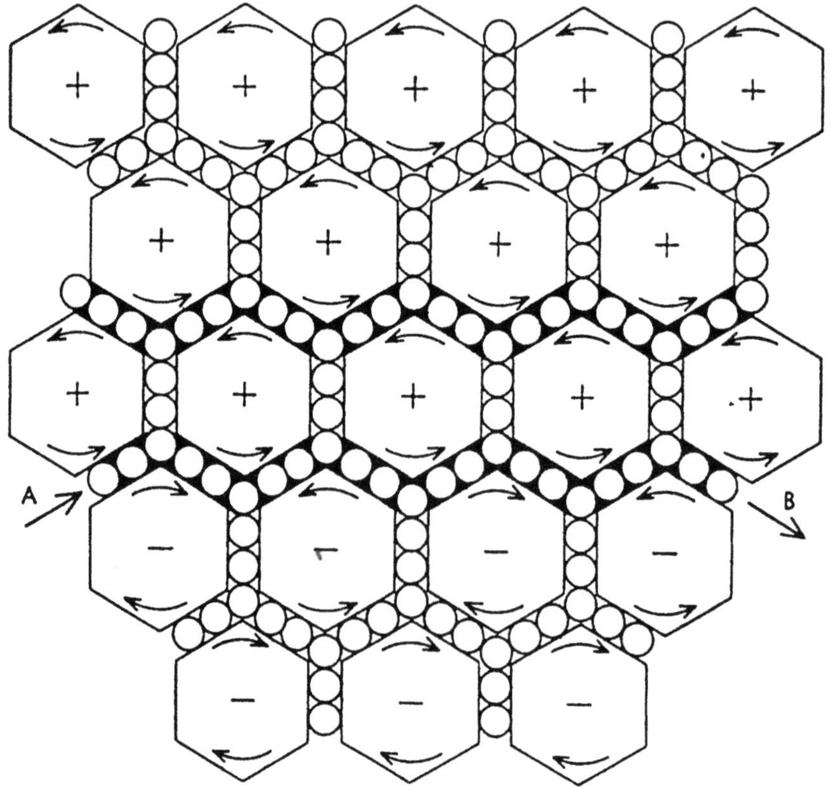

With his Theory of Relativity, Albert Einstein in 1905 put an end to all mechanical ether interpretations of electromagnetic phenomena, such as the ether model shown above. In it, the magnetic field was imagined as a system of molecular vortices rotating around the field lines, with 'ball bearings' between vortices consisting of charge particles. The velocity of rotation is to be proportional to the field strength, and when neighboring vortices rotate with differing velocities, the charge particles get displaced. This model was the basis for the derivation of the Maxwell equations. "I never satisfy myself unless I can make a mechanical model of a thing ... that is why I cannot get the electromagnetic theory ..." (Lord Kelvin, 1884).

Roman U. Sexl
Helmuth K. Urbantke

Relativity, Groups, Particles

Special Relativity and Relativistic
Symmetry in Field and Particle
Physics

Revised and translated
from the German
by H. K. Urbantke

SpringerWienNewYork

Dr. Roman U. Sexl †
Dr. Helmuth K. Urbantke
Institut für Theoretische Physik
Universität Wien, Vienna, Austria

This edition succeeds the third, revised German-language edition, *Relativität, Gruppen, Teilchen,*
© 1992 Springer-Verlag/Wien

Camera-ready copies provided by the author
Printed in Austria by Novographic Druck G.m.b.H., A-1230 Wien
Printed on acid-free and chlorine-free bleached paper
SPIN 10756865

With 56 figures and a frontispiece

Library of Congress Cataloging-in-Publication Data

Sexl, Roman Ulrich.
    [Relativität, Gruppen, Teilchen. English]
    Relativity, groups, particles : special relativity and relativistic symmetry in field and particle
physics / Roman U. Sexl, Helmuth K. Urbantke ; revised and translated from the German by
Helmuth K. Urbantke. – Rev. ed.
       p. cm.
    This edition succeeds the third, revised German-language edition, Relativität, Gruppen,
Teilchen, c1992 Springer-Verlag/Wien – T.p. verso.
    Includes bibliographical references and index.
    ISBN 3211834435 (alk. paper)
    1. Relativity (Physics) 2. Field theory (Physics) 3. Representations of groups. 4. Particles
(Nuclear physics) I. Urbantke, Helmuth Kurt. II. Title.

QC173.65.S4813 2000
530.11 – dc21
                                                00-063782

ISBN 3-211-83443-5 Springer-Verlag Wien New York

Preface and Introduction

Like many textbooks, the present one is the outgrowth of lecture courses, mainly given at the University of Vienna, Austria; on the occasion of the English edition, it may be mentioned that our first such lecture course was delivered by my late co-author, Roman U. Sexl, during the fall and winter term 1967-68 in the USA—more precisely, at the University of Georgia (Athens). Since then, Particle Physics has seen spectacular revolutions; but its relativistic symmetry has never been shaken. On the other hand, new technological developments have enabled applications like the GPS (Global Positioning System) that, in a sense, brought Relativity to the domain of everyday use.

The purpose of the lecture courses, and thus of the book, is to fill a gap that the authors feel exists between the way Relativity is presented in introductory courses on mechanics and/or electrodynamics on the one hand and the way relativistic symmetry is presented in particle physics and field theory courses on the other. The reason for the gap is a natural one: too many other themes have to be addressed in the introductory courses, and too many applications are impatiently waiting for their presentation in the particle and field theory courses.

In this text we try to bridge this gap, and guide the reader (him *and* her, we hope) to more abstract points of view concerning space-time geometry and symmetry wherever they are useful. At the same time, the reader is introduced to the world of groups and their realizations, particularly Lie groups and Lie algebras. Much of this material could have been omitted given a severe restriction to the groups actually to be dealt with, but a slight broadening was intentional. However, we stress that we certainly do not see the need of entering the realm of the simple Lie algebras of rank greater than one, which would be necessary for the discussion of the inner symmetries of particle physics. Naturally, mathematical developments tend to occupy a large amount of space here, but we hope that the gradual transition from the explicit component-matrix format to the more abstract version of linear algebra will, in the end, work against loss of sight of the basic concepts. Motivation and heuristic considerations are in the foreground, and our presentation will essentially remain at the heuristic level whenever functional analysis would be needed to cope with the infinite-dimensional spaces that occur. Also, the precise definition of manifolds is not given, although we try to give the reader at least a vague impression of group manifolds, covering spaces, fiber bundles, etc., since these objects are there and should be named for ease of addressing. Moreover, all these concepts pervade modern theoretical physics in many other places. For their precise definition, the reader is referred to suitable mathematical textbooks, some of which we quote. However, basic group theory and abstract (multi)linear algebra are summarized in two of the appendices.

At this point, we may list things the reader should be acquainted with. On the mathematical side, these include linear algebra (first only in three but later in arbitrary dimensions), multivariable calculus, and a rudimentary knowledge of the Dirac delta function; the basic definitions from group theory are useful to be known already as well. On the physics side, they include the basic concepts of theoretical mechanics, electrodynamics, and quantum theory (on a level that assumes multivariable calculus); thus, e.g., small parts of the well-known books by Goldstein, Jackson, and Schiff will suffice. Enough experimental background is assumed, so that our only very occasional mention of experiments suffices to assure the reader that we are indeed talking about physics rather than pure mathematics.

Throughout this book, particularly so in its first half, we have interpolated historical remarks: if short enough, they appear in small print paragraphs interspersed in the main text; if longer, they take the form of whole sections (namely, sects. 1.6 and 2.11, written together with R. Mansouri, now at Sharif University of Technology, Teheran, who also contributed to sect. 10.3). Similarly, mathematical asides of interest or of relevance in later sections may appear in small print paragraphs. These may be omitted on a first reading of the section they appear in, but must sometimes be (re)turned to on studying later sections. (In other words, there is no strict separation in the book enabling a "track one" and a "track two" reading.) In any case, they are hoped to whet the reader's appetite and to allow looking at some of the features of Relativity from a "higher" point of view.

The table of contents gives a general overview of our subject matter, so here we make only a few general remarks on how the development proceeds. Chapter 1 gives a "derivation" of the Lorentz transformation starting from the usual "axioms" (which are not to be understood in the sense of logicians). The role of group structure should already be apparent in this stage, even if that term is introduced only later. The role of the rotation group of Euclidean 3-space is very much in the foreground here, which is perhaps somewhat unusual. Chapter 2 discusses standard elementary consequences of the Lorentz transformation, including Thomas rotation. The sections on superluminal phenomena and non-Einstein synchronized reference frames may appear somewhat outside the canonical textbook content. Chapters 3, 4, and 5 are standard, but the latter includes, in a semi-historical section, the history of 'classical electron theory' and the role played by relativistic covariance in the later developments of that theory.

With chapter 6, we enter the group-representation part of the book, and a reader who knows standard relativistic mechanics and electrodynamics might well begin with this chapter, perhaps first reading sections 1.5, 2.9, 2.10 and the introduction to chapter 3. Chapter 6 includes an investigation of the structure of the Lorentz group (its quasidirect product structure in particular, since that is closely related to our initial derivation of the group) as well as the basic definitions and theorems from the theory of representations. All of these are well-illustrated with reference to material in previous chapters.

Chapter 7 is preparatory to chapters 8 and 9; in particular, section 7.10 on multivalued representations may be helpful to some readers. In chapter 8 on the finite-dimensional representations of the Lorentz group, we hope we have made clear the often-confused role played by the use of complex numbers in this context; we explain complex structure, real structure, complexification, realification and the job they do for us here.

Chapter 9 first discusses the representation theoretic aspect of covariant wave equations; after a general discussion of relativistic symmetry in quantum mechanics, it then introduces the well-known Wigner classification. The mention of helicity as a 'topological quantum number' is perhaps not frequently encountered in other texts.

Chapter 10, on conservation laws associated with relativistic space-time symmetry, can be read almost independently of the preceding ones. Section 10.3 shows an application of a phenomenologically constructed energy-momentum tensor.

We have already commented on two of the appendices (A and B); Appendix C continues an already quite lengthy appendix to section 9.1 on Dirac spinors: both are intended to encourage an essentially basis-free attitude towards the 'gamma' matrices, such as would be required when going to the curved space-time of General Relativity. Appendix D tries to give a modest introduction to relativistic covariance in Quantum Field Theory.

There are exercises to most sections; in the later chapters, many of them ask the reader to provide proofs, following given hints, for theorems of a general nature that were quoted and applied in the main text. Essentially, these exercises intend to further the reader's intuition about linear spaces.

Thanks are due to many persons who contributed in one way or another to the previous (German) editions: their names are listed there. Added here to that list must be my colleague Helmut Kühnelt, who tried (essentially in vain) to educate me in LaTeX and, in any case, helped me, as also did Ulrich Kiermayr, to overcome many difficulties. Of course, the responsibility for any imperfections in typesetting, as well as for infelicities of language and content, is entirely with me. Every new edition gives opportunity not only to eliminate mistakes in the previous one but also to create new ones. At least, a reasonable balance is hoped for. I will be grateful to anybody bringing mistakes and ambiguous or cryptic formulations to my attention, which in our electronic age should be easy using urbantke@galileo.thp.univie.ac.at; I plan to make the collection of corrections so obtained available via link on the homepage of my institution, http://www.thp.univie.ac.at/, in due time, so that even readers of this present edition may profit from such activity.

Our big hope is that the present edition contribute to an increase of joy in physics by widening more people's scope for "seeing" symmetry in nature! Naturally, this edition is dedicated to the memory of my former co-author, teacher and friend,

ROMAN ULRICH SEXL

whose untimely and tragic death, now 14 years ago, prevents him from greeting the new millennium.

Vienna, August 2000 Helmuth K. Urbantke

Contents

1 The Lorentz Transformation

Traditionally, two postulates are put at the beginning of Special Relativity, from which all other results can be derived:

A. The Principle of Relativity

B. The constancy of the speed of light

From these principles the Lorentz transformation may be derived in numerous ways, some more and some less elementary, as is done in most presentations of Relativity. Already from 1910 on, authors occasionally pointed out[1] that the principle of relativity alone already determines almost all of the structure, and in particular implies the existence of a (numerically unspecified) invariant speed. This approach does not concentrate on a single Lorentz transformation but works with the totality of all transformations admitted by the principle of relativity. Thus, group theoretical ideas, on which we are going to elaborate in this book, come in implicitly or explicitly right from the beginning. We therefore here set out to derive the Lorentz transformation in a manner that takes into account this central role of principle A, and take B only to decide between the numbers -1, 0, and 1.

To understand the principle of relativity, we have to analyze the concept of 'inertial systems of reference', which we do first.

1.1 Inertial Systems

Consider a number of labs in free flight (Fig. 1.1)—we assume we can neglect their mutual interactions (by gravitation, say). Within each of them, Newton's First Axiom

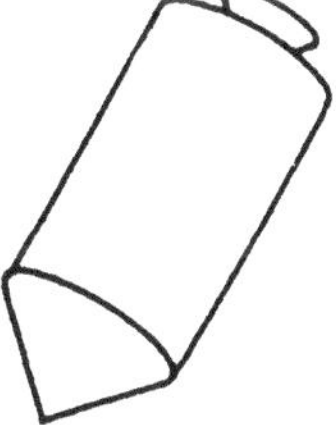

Fig. 1.1. Labs in free flight

(the law of inertia) holds, every body with no forces acting on it remains—as judged from the lab—in a state of rest or of uniform rectilinear motion. Such a lab defines an *inertial system* I. Each (pointlike) *event* may be recorded by noting its coordinates

---

[1] W. v. Ignatowsky, Phys. Z. *11*, 927 (1910); P. Frank, H. Rothe, Ann. Phys. (Leipzig) *34*, 825 (1911); see also G. Süßmann, Z. Naturforsch. *24a*, 495 (1969).

x, y, z with respect to a rectangular Cartesian coordinate system anchored in I together with the reading t of a clock attached to I. We shall term this setup an *inertial reference frame*, and we restrict to positively oriented coordinate axes at the moment. It is useful to consider t, x, y, z as four coordinates $x^i = (x^0, x^1, x^2, x^3) := (t, x, y, z)$. Time thus appears—at first in a purely formal manner—as a fourth ('zeroth') coordinate.

To describe the motion of some point mass with respect to such an inertial system I it is also helpful to use *space-time diagrams*. (For actual drawings we must restrict to less than three space dimensions, however (see Fig. 1.2).) The consecutive positions of the moving point mass in this diagram make up its *world line*. As one easily convinces oneself, for rectilinear uniform motion the world line is straight, and conversely.

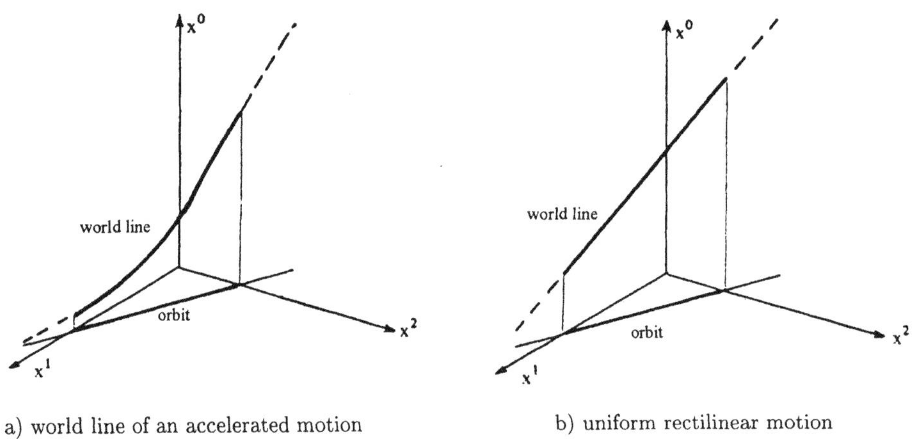

a) world line of an accelerated motion   b) uniform rectilinear motion

Fig. 1.2. Space-time diagrams for the motion of a mass point

Our next task is to find the relation between different inertial frames. If I is inertial, then from experience we know that a reference frame $\bar{\mathrm{I}}$ is again inertial if with respect to I it is

    a. parallely displaced by $\boldsymbol{a}$
    b. rotated by $\boldsymbol{\alpha}$
    c. moving at constant velocity $\mathbf{v}$ (10 parameters)
    d. time delayed by a^0.

Here $\boldsymbol{\alpha}$ is the rotation vector (see later; it may be replaced by any other triple of numbers capable of fixing a rotation), and a^0 is the time lag between the clocks attached to the two systems; parallel displacement and rotation refer to Euclidean Geometry, valid by experience in every inertial system. One does *not*, however, obtain new inertial systems by considering systems accelerated against I. We exclude transformations of units of length and time by assuming—as justified from experience—the existence of measuring rods and clocks insensitive to accelerations, which may be used to gauge all inertial frames (cf. sect. 2.8).

Formally, the relation between inertial frames I, $\bar{\text{I}}$ is described by specifying, for each event x, the relation between its coordinates x^i with respect to I and its coordinates $x^{\bar{i}}$ with respect to $\bar{\text{I}}$. We are thus looking for the transformation

$$x^{\bar{i}} = f^i(x^k). \tag{1.1.1}$$

The possible form of the functions f^i is restricted drastically already by the requirement that both, I and $\bar{\text{I}}$, be inertial: straight world lines with respect to I (describing uniform rectilinear motion) have to be transformed into straight world lines with respect to $\bar{\text{I}}$ by the transformation (1.1.1). It is reasonable also to require that finite coordinate values are always transformed to finite ones: it is then well-known that transformations with these properties are given by affine transformations

$$x^{\bar{i}} = L^i{}_k\, x^k + a^i, \qquad i = 0, 1, 2, 3. \tag{1.1.2}$$

Here we have used Einstein's *summation convention*, according to which for each index occurring twice in a monomial a summation over its range ($k = 0, 1, 2, 3$ in eq. (1.1.2)) is understood. We shall continue to use this convention in the sequel, the range of Latin indices being $\{0, 1, 2, 3\}$ and that of Greek ones $\{1, 2, 3\}$ only. (Observe that many authors do just the opposite!)

If one does not want to make the additional finiteness postulate—since it cannot be ascertained by experiments restricted to a finite domain—there is still the possibility of projective (i.e., fractional linear) transformations, and principle B would have to be used much more extensively. (See Weyl (1923), who also contrasts derivations of the transformation working on arbitrarily small open sets but postulating differentiability conditions with derivations using all space but using not even continuity. We shall not go into these details.)

1.2   The Principle of Relativity

Consider two experiments, set up in exactly the same manner in inertial frames I and $\bar{\text{I}}$, such as measuring the attraction between electron and proton (Fig. 1.3).

Fig. 1.3. Measuring Coulomb's Law in differing frames of reference

It turns out that the result of this experiment, and of every other one set up in identical manner in both systems, is the same for both systems. If all processes of nature satisfying identical initial and boundary conditions in I and $\bar{\mathrm{I}}$ lead to identical results, it should be possible to formulate the basic laws that serve for their description in a manner which assumes the same form in I and $\bar{\mathrm{I}}$ and in any other inertial frame. In other words, we have the postulate that the laws of nature be *covariant* with respect to the set of transformations between inertial frames. This is the *Principle of Relativity*. One can see the reason for this designation by formulating the principle negatively: there is no absolute rest or absolute velocity in some absolute space which could show up, e.g., by the attraction between electron and proton becoming extremal (maximum or minimum) when measured in a reference system at absolute rest. Similarly, there is no distinguished point in space, no distinguished direction, no distinguished instant of time[1]; therefore only *relative* values of velocities, distances, angles, times matter.

The Principle of Relativity also holds in Newtonian mechanics. Its prominent position in the framework of Relativity Theory comes from the fact that at the end of the 19th century doubts were cast on its validity: it seemed possible to unify classical mechanics and electrodynamics only by postulating some absolute space, called 'ether'. Einstein showed in 1905 that the correct way out was not to dismiss the principle of relativity but to change classical mechanics.

1.3 Consequences from the Principle of Relativity

In this section we derive the restrictions on eq. (1.1.2) which follow from the principle of relativity. Since there are no restrictions on the space-time translations a^i, we will consider here only the homogeneous transformations, eq. (1.1.2) with $a^i = 0$, and take up translations only much later. As we have stated, there are no absolute directions and velocities. As a consequence, the relation between I and $\bar{\mathrm{I}}$, and thus the matrix $L^i{}_k$, must be expressible by the *axial vector* $\boldsymbol{\alpha}$ describing the relative angular orientation between their spatial axes, together with the *polar vector* $\mathbf{v}$ of relative velocity.

If there is only a relative rotation between the systems, $L^i{}_k$ has to be formed from the rotation vector $\boldsymbol{\alpha}$ alone. ($\boldsymbol{\alpha}$ is parallel to the axis of rotation; its length is the angle of rotation in radians, restricted by $\alpha := |\boldsymbol{\alpha}| \leq \pi$, and the vector is oriented by the usual right hand rule.) In this case, eq. (1.1.2) has the form

$$x^{\bar{0}} = x^0$$

$$\bar{\mathbf{x}} = \frac{\boldsymbol{\alpha}\,\mathbf{x}}{\alpha^2}\,\boldsymbol{\alpha} + \left(\mathbf{x} - \frac{\boldsymbol{\alpha}\,\mathbf{x}}{\alpha^2}\,\boldsymbol{\alpha}\right)\cos\alpha - \frac{\boldsymbol{\alpha}}{\alpha} \times \mathbf{x}\sin\alpha, \tag{1.3.1}$$

i.e., $L^0{}_0 = 1$, $L^0{}_\alpha = 0 = L^\alpha{}_0$, $L^\mu{}_\nu = R^\mu{}_\nu$, where $R^\mu{}_\nu$ is the proper orthogonal matrix

$$R^\mu{}_\nu := \frac{\alpha^\mu \alpha^\nu}{\alpha^2} + \left(\delta^\mu{}_\nu - \frac{\alpha^\mu \alpha^\nu}{\alpha^2}\right)\cos\alpha + \frac{\sin\alpha}{\alpha}\epsilon^{\mu\nu\lambda}\alpha^\lambda. \tag{1.3.2}$$

However, if the systems differ only by uniform rectilinear relative motion, then only $\mathbf{v}$ is at our disposal for constructing $L^i{}_k$, and the transformation must look like

$$x^{\bar{0}} = a(v)\,x^0 + b(v)\,\mathbf{v}\,\mathbf{x} \tag{1.3.3a}$$

---

[1] This is often formulated as *homogeneity of space, isotropy of space, homogeneity of time*.

$$\bar{\mathbf{x}} = c(v)\,\mathbf{x} + \frac{d(v)}{v^2}\mathbf{v}\,(\mathbf{v}\,\mathbf{x}) + e(v)\,\mathbf{v}\,x^0. \tag{1.3.3b}$$

Here[2] $\mathbf{v} = (v^1, v^2, v^3)^\top$ is the velocity of $\bar{\mathrm{I}}$ as measured in I, and $v := |\mathbf{v}|$. The following arguments lead to eq. (1.3.3): first, $L^0_{\ 0}$, $L^0_{\ \alpha}\,x^\alpha$ have to be scalars, hence $L^0_{\ 0}$ must be a scalar formed from $\mathbf{v}$, $L^0_{\ 0} = a(v)$, and $L^0_{\ \alpha}$ must be a vector formed from $\mathbf{v}$, $L^0_{\ \alpha} = b(v)\,v^\alpha$. Second, $\bar{\mathbf{x}}$ is a polar vector formed from the scalar x^0 and the polar vectors $\mathbf{x}$ and $\mathbf{v}$ such that it is linear in the x^i; the only possibility for this is eq. (1.3.3b). (See Appendix 2 to this section for a formal proof of these physicist's folklore arguments.)

A first restriction for the unknown functions $a(v)$, $b(v)$, $c(v)$, $d(v)$ and $e(v)$ comes from the condition that the origin of $\bar{\mathrm{I}}$ be moving with velocity $\mathbf{v}$ relative to I, which means that $\mathbf{x} = \mathbf{v}x^0$ must imply $\bar{\mathbf{x}} = \mathbf{0}$, and this is the case if

$$c(v) + d(v) + e(v) = 0. \tag{1.3.4}$$

Further conditions for the unknown functions now follow from the principle of relativity. Let us exchange the roles of I and $\bar{\mathrm{I}}$: then I is moving against the latter with velocity $\bar{\mathbf{v}} = -\mathbf{v}$. This statement about the velocity *components*—often called *reciprocity*—is so plausible that for decades nobody thought of deriving it from the principle of relativity explicitly until Berzi and Gorini did so in 1969. (A version of their proof is given as Appendix 1 to this section.) Since the form (1.3.3) of the relation between nonrotated moving inertial systems is universal, we must require that the transformation (1.3.3) have the inverse

$$x^0 = a(\bar{v})\,x^{\bar{0}} + b(\bar{v})\,\bar{\mathbf{v}}\,\bar{\mathbf{x}}$$
$$\mathbf{x} = c(\bar{v})\,\bar{\mathbf{x}} + \frac{d(\bar{v})}{\bar{v}^2}\,\bar{\mathbf{v}}\,(\bar{\mathbf{v}}\,\bar{\mathbf{x}}) + e(\bar{v})\,\bar{\mathbf{v}}\,x^{\bar{0}}, \tag{1.3.5}$$

where $\bar{\mathbf{v}} = -\mathbf{v}$, $\bar{v} = v$. Substituting this and eq. (1.3.5) into eq. (1.3.3), we will obtain an identity only if

$$c^2 = 1, \quad a^2 - ebv^2 = 1, \quad e^2 - ebv^2 = 1, \quad e(a+e) = 0, \quad b(a+e) = 0, \tag{1.3.6}$$

as is best checked by specializing $\mathbf{v} = (v, 0, 0)^\top$.

The value $c = -1$ would correspond to a 180° rotation contained in (1.3.3b) and has to be excluded here. From the third equality of eqs. (1.3.6) we have $e \neq 0$, hence $a + e = 0$ from the fourth. This satisfies the fifth also, and the second and third become equivalent. Thus we have

$$b = \frac{1 - a^2}{av^2}, \quad c = 1, \quad d = a - 1, \quad e = -a. \tag{1.3.7}$$

The only yet unknown function $a(v)$ will finally result from the application of the principle of relativity to three inertial frames I, $\bar{\mathrm{I}}$, $\bar{\bar{\mathrm{I}}}$, where $\bar{\mathrm{I}}$ is moving with $\mathbf{v}$ against

---

[2] In the text, column vectors like $\mathbf{v}$ are written as row vectors with the superscript $\top$ for transposition attached, just to save space.

I and $\bar{\bar{\text{I}}}$ is moving with $\bar{\mathbf{w}}$ against $\bar{\text{I}}$. If here $\mathbf{v}$ and $\bar{\mathbf{w}}$ are proportional, the relation between $\bar{\bar{\text{I}}}$ and I has again to be a pure 'boost' of type (1.3.3) in the same direction. (If they are not proportional, one can form the axial vector $\mathbf{v} \times \bar{\mathbf{w}}$, so that under composition of arbitrary boosts a relative rotation between $\bar{\bar{\text{I}}}$ and I is formally conceivable; indeed the Thomas precession considered in sect. 2.10 is related to this.) Putting $\mathbf{v}$ and $\bar{\mathbf{w}}$ into the 1-directions, the product of the transformations

$$x^{\bar{0}} = a(v)x^0 + \frac{1 - a^2(v)}{v\,a(v)}x^1 \qquad\qquad x^{\bar{\bar{0}}} = a(\bar{w})x^{\bar{0}} + \frac{1 - a^2(\bar{w})}{\bar{w}\,a(\bar{w})}x^{\bar{1}}$$

$$x^{\bar{1}} = a(v)\,x^1 - v\,a(v)\,x^0 \quad\text{and}\quad x^{\bar{\bar{1}}} = a(\bar{w})\,x^{\bar{1}} - \bar{w}\,a(\bar{w})\,x^{\bar{0}} \qquad (1.3.8)$$

$$x^{\bar{2}} = x^2, \quad x^{\bar{3}} = x^3 \qquad\qquad x^{\bar{\bar{2}}} = x^{\bar{2}}, \quad x^{\bar{\bar{3}}} = x^{\bar{3}}$$

must assume the form

$$x^{\bar{\bar{0}}} = a(u)\,x^0 + \frac{1 - a^2(u)}{u\,a(u)}x^1$$

$$x^{\bar{\bar{1}}} = a(u)\,x^1 - u\,a(u)\,x^0 \qquad (1.3.9)$$

$$x^{\bar{\bar{2}}} = x^2, \quad x^{\bar{\bar{3}}} = x^3$$

for some u. Comparing coefficients, we obtain two expressions for $a(u)$; equating them gives

$$\frac{v\,a(v)}{\bar{w}\,a(\bar{w})}(1 - a^2(\bar{w})) = \frac{\bar{w}\,a(\bar{w})}{v\,a(v)}(1 - a^2(v))$$

or

$$\frac{1 - a^2(v)}{v^2\,a^2(v)} = \frac{1 - a^2(\bar{w})}{\bar{w}^2\,a^2(\bar{w})} = K. \qquad (1.3.10)$$

Here K is a constant which is the same for each pair of inertial systems—hence it is universal. Solving eq. (1.3.10) for $a^2(v)$ we obtain

$$a^2(v) = (1 + K\,v^2)^{-1}, \qquad b(v) = K\,a(v),$$

and the relation between I and $\bar{\text{I}}$ thus finally becomes

$$x^{\bar{0}} = a(v)(x^0 + K\,\mathbf{v}\,\mathbf{x})$$

$$\bar{\mathbf{x}} = \mathbf{x} + \frac{a(v) - 1}{v^2}\,\mathbf{v}\,(\mathbf{v}\,\mathbf{x}) - a(v)\,\mathbf{v}\,x^0 \qquad (1.3.11)$$

$$\text{where } a^2(v) := (1 + K\,v^2)^{-1}.$$

We see that the principle of relativity almost completely fixes the transformation, only a universal constant K (and the sign of $a(v)$) remaining undetermined.

If in eq. (1.3.11) we put $K = 0$ and $a = +1$, we obtain the *Galilean boost*

$$x^{\bar{0}} = x^0$$

$$\bar{\mathbf{x}} = \mathbf{x} - \mathbf{v}\,x^0, \qquad\qquad (1.3.12)$$

which underlies Newtonian mechanics (*'Galilean Relativity'*).

When $K \neq 0$, it has dimension (velocity)$^{-2}$, and we can rescale $x^0 \to x^0|K|^{1/2}$, $x^{\bar{0}} \to x^{\bar{0}}|K|^{1/2}$, $\mathbf{v} \to \mathbf{v}|K|^{-1/2}$ to arrive at $K = +1$ or $K = -1$, as announced earlier. The decision for the actual value needs a further empirical fact.

It is interesting at this point to rule out $K = +1$ still on semi-formal grounds. Apart from eq. (1.3.10), we get from eq. (1.3.9) the following:

$$a(u) = a(v)a(\bar{w})(1 - Kv\bar{w}) \tag{1.3.13}$$

$$u = \frac{v + \bar{w}}{1 - Kv\bar{w}}. \tag{1.3.14}$$

(It should be remembered here that in eqs. (1.3.8-10) v and $\bar{w}$ carry a sign, being the components of $\mathbf{v}$ and $\bar{\mathbf{w}}$ which we chose to put into the 1- and $\bar{1}$-directions.) Equation (1.3.14) is a velocity addition theorem, to be discussed more generally in sect. 2.9; the square of eq. (1.3.13) can be checked from eq. (1.3.14) to imply $a^2(u) = (1 + Ku^2)^{-1}$ as we would like to have it by the universality of K. So the independent content of eq. (1.3.13) is just in the coupling of the *signs* involved. Now for $K = -1$ the reality of $a(v)$ requires $|v| < 1$ for all relative velocities between inertial systems, implying $1 - Kv\bar{w} > 0$ (despite v, $\bar{w}$ carrying signs here, as pointed out above), so that $a(v) > 0$ is a choice consistent with eq. (1.3.13) and guaranteeing that there is no (formal) reversal of the sense of time in eq. (1.3.11). The latter agrees with the experience that all clocks run into the future. However, when $K = +1$, this is not so: the reality of $a(v)$ does not restrict the domain of $|v|$, so that even when we take $a(v) > 0$, $a(\bar{w}) > 0$ we may get $a(u) < 0$ from eq. (1.3.13); and there is no restriction on v to save $a > 0$ that would be respected by the composition (1.3.14). (In geometrical language, $K = +1$ gives Euclidean rotations in $(t, \mathbf{x})$-space which may well rotate the t-axis into its negative.)

Appendix 1: Reciprocity of Velocities

As mentioned, the relation $\bar{\mathbf{v}} = -\mathbf{v}$ between the velocity $\mathbf{v}$ of $\bar{I}$ against I and the velocity $\bar{\mathbf{v}}$ of I against $\bar{I}$ may be deduced from the principle of relativity.[1] We first find $\bar{\mathbf{v}}$ by putting $\mathbf{x} = \mathbf{0}$ in eq. (1.3.3): as seen from $\bar{I}$, the origin of I moves with velocity $\bar{\mathbf{v}} = \mathbf{v}e(v)/a(v)$. For the absolute values we will now argue that $\bar{v} = v$. Writing $\bar{v} = f(v)$, the function f enjoys the following properties: $f(0) = 0$, $f(v) > 0$ when $v > 0$; by the principle of relativity, we may exchange the systems to get $v = f(\bar{v}) = f(f(v))$, so f satisfies the functional equation $f \circ f = \mathrm{id}$; it also maps the universal (by the relativity principle) domain of possible relative speeds bijectively (again by the relativity principle) onto itself, so must be strictly monotonic—in fact monotonically increasing by the properties mentioned before. But this condition and the functional equation lead to a contradiction immediately if we assume $f(v) > v$ or $f(v) < v$, so that $f(v) = v$. Hence we get $\bar{\mathbf{v}} = \pm\mathbf{v}$; the formal possibility $\bar{\mathbf{v}} = \mathbf{v}$ (reversal of motion) leads to transformations with time reversal, which we excluded here. – It should be pointed out that the article by Berzi and Gorini also contains numerous references to the literature on the derivation of the Lorentz transformation together with critical remarks.

Appendix 2: Some Orthogonal Concomitants of Vectors

We here give the arguments to support the following 'folklore theorems' that we used in arriving at eq. (1.3.3). By vectors we shall here mean polar vectors, changing components as $\mathbf{x} \mapsto \mathbf{Rx}$ under all (proper and improper) orthogonal transformations R of the reference frame.

i. A scalar depending on a vector—i.e., a function $f(\mathbf{x}) = f(x^1, x^2, x^3)$ of its components $\mathbf{x} = (x^1, x^2, x^3)^\top$ which is independent of the orthonormal frame to which they refer, is a function of its length: $f(\mathbf{x}) = f(\mathbf{Rx})$ for all orthogonal R implies that there is a function g of one variable such that $f(\mathbf{x}) = g(|\mathbf{x}|)$.

---

[1]V. Berzi, V. Gorini, J. Math. Phys. *10*, 1518 (1969); see also their article in Barut (1973), and for later references J. H. Field, Helv. Phys. Acta *70*, 542 (1997).

ii. A vector depending on a vector—whose components $V(x)$ thus transform orthogonally when the x are so transformed and which is called a vectorial concomitant of x—is proportional to x, the factor of proportionality being a function of its length: $V(Rx) = RV(x)$ for all orthogonal R implies $V(x) = g(|x|)x$ for some function g.

iii. A scalar depending on two vectors is a function of their lengths and their scalar product: $f(Rx, Ry) = f(x, y)$ for all orthogonal R implies the existence of a function g of 3 variables such that $f(x, y) = g(|x|, |y|, xy)$.

iv. A vector depending on two vectors—a vectorial concomitant of them—is a linear combination of them with coefficients as in iii: $V(Rx, Ry) = RV(x, y)$ for all orthogonal R implies $V(x, y) = g(\ldots)x + h(\ldots)y$ with g and h as in iii.

For i, ii, and iii proper orthogonality for the R suffices, while for iv all orthogonal R must be admitted for the statement to be true—otherwise the cross product also comes in. Now to prove i, choose a frame whose positive 1-axis is in the direction of the vector, so that $x = (|x|, 0, 0)^\top$. This operation does not change the value of $f(x)$. Define $g(u) = f(u, 0, 0)$: then $f(x) = g(|x|)$. For ii, first consider rotations R about x as an axis: our condition implies that also V is unchanged by them and thus must lie in the axis, i.e., is proportional to x: $V(x) = \lambda(x)x$. Replacing here x by Rx and using the condition on V again, we then see that $\lambda(x)$ satisfies the condition in i. For iii we rotate the frame such that x is in the positive 1-axis and y is in the upper half 1,2-plane: $x = (|x|, 0, 0)^\top$, $y = (xy/|x|, \sqrt{x^2y^2 - (xy)^2}/|x|, 0)^\top$; then $f(x, y) = f(x^1, x^2, x^3, y^1, y^2, y^3) = f(|x|, 0, 0, xy/|x|, \sqrt{x^2y^2 - (xy)^2}/|x|, 0) = g(|x|, |y|, xy)$. For iv, again rotate the frame into the same position and consider the value of V for it. Our frame is determined uniquely up to a reflection in the 1,2-plane, which by our requirement on V must not change it; this is the case iff its 3-component vanishes, making it a linear combination of x and y. This state of affairs then follows to hold in all frames, and as in ii one finds that the coefficients satisfy the condition in iii.

Exercises

1. Prove eq. (1.3.1)!

2. Write eq. (1.3.1) as $\bar{x} = L_R x$, where R is the orthogonal matrix given in eq. (1.3.2), and write eq. (1.3.3) as $\bar{x} = L_v x$; show that $L_{Rv} L_R = L_R L_v$; interpret this as saying that no direction of relative velocity is distinguished.

1.4 Invariance of the Speed of Light. Lorentz Transformation

The yet undetermined constant K has the physical dimension of reciprocal velocity squared. To interpret it we remark that for the transformations (1.3.11), but also for the rotations (1.3.1) as well as for space-time translations we have the fundamental identity

$$(dx^0)^2 + K(d\mathbf{x})^2 \equiv (d\bar{x}^0)^2 + K(d\bar{\mathbf{x}})^2. \tag{1.4.1}$$

As a consequence, for any motion $\mathbf{x} = \mathbf{x}(x^0)$ satisfying $(d\mathbf{x}/dx^0)^2 = -1/K$ in one inertial system the analogous relation is true in any other inertial system. Therefore, $c := 1/\sqrt{-K}$ plays the role of a uniquely determined *invariant speed*. It is an experimental question whether such exists in nature, and if so, what is its value. Numerous well-known experiments[1] show that the speed of propagation of electromagnetic waves

---

[1] They are described, e.g., in French (1971).

in vacuum,

$$c = 2.997925 \times 10^{10}\,\text{cm/sec}, \tag{1.4.2}$$

is independent of the inertial system where it is measured; therefore, K is finite and negative. Note that this state of affairs is *not* covered by the principle of relativity: we are not considering here two emission experiments set up in an identical manner in two different inertial systems but are observing one and the same light wave, emitted by some source somewhere, from different inertial systems. Therefore the invariance of c is sometimes described by saying that it is independent of the motion of the source.

In what follows, we shall most of the time assume performed the rescaling indicated above, and use units where $c = 1$—i.e., speeds are expressed as multiples of c. Then we have

$$K = -1, \quad a(v) = \frac{1}{+\sqrt{1 - v^2}} =: \gamma, \quad \frac{\gamma - 1}{v^2} \equiv \frac{\gamma^2}{\gamma + 1}, \tag{1.4.3}$$

and (1.3.11) becomes the (special) *Lorentz transformation* ('*Lorentz boost*')

$$x^{\bar{0}} = \gamma(x^0 - \mathbf{v}\,\mathbf{x})$$

$$\bar{\mathbf{x}} = \mathbf{x} + \frac{\gamma^2}{\gamma + 1}\mathbf{v}\,(\mathbf{v}\,\mathbf{x}) - \gamma\,\mathbf{v}\,x^0. \tag{1.4.4}$$

In eq. (1.4.3) we expressly chose the positive square root; $a(v)$ negative would correspond to a reversal of the sense of time, uninterpretable with the present meaning of the transformations as being 'passive' (i.e., referring to changes of frames) and thus to be excluded. As was pointed out above, this does not lead to consistency problems.

By composing space-time translations, space rotations and Lorentz boosts in various ways we get more complicated transformations. Homogeneous ones will be called *(general) Lorentz transformations*, inhomogeneous ones will be called *Poincaré transformations*. The kind of relativity realized in Nature, in which transformations between inertial systems are given by them is *Einsteinian Relativity*. What we are going to show in this book are the consequences of this fact for the formulation of physical laws.

One might object that, classically, light is more appropriately described by waves, so that our application of eq. (1.4.1) to light is questionable except in the geometric optics limit. One can replace the argument with one that works with wave motion rather than point particle motion as follows. Let $\bar{\phi}(\bar{t}, \bar{\mathbf{x}})$ be any function; upon substituting the transformations (1.3.1,11) as well as translations, one gets a function $\phi(t, \mathbf{x})$ of the unbarred variables. We then have the identity (exercise)

$$K(\frac{\partial \phi}{\partial t})^2 + (\nabla \phi)^2 \equiv K(\frac{\partial \bar{\phi}}{\partial \bar{t}})^2 + (\bar{\nabla} \bar{\phi})^2. \tag{1.4.5}$$

Consider now a plane wave, described in I by a wave function $\cos \phi$ with phase $\phi = \omega t - \mathbf{k}\mathbf{x} + \delta$, angular frequency ω, wave number vector $\mathbf{k}$ and phase velocity $v_{ph} = \omega/|\mathbf{k}|$, so that $\omega^2 - v_{ph}^2\mathbf{k}^2 = 0$, and similarly described in $\bar{\text{I}}$ by the corresponding barred quantities. Then eq. (1.4.5) says that $-1/K^2$ is the square of an invariant phase velocity.

Exercise

Verify eqs. (1.4.1) and (1.4.5)!
Hint: You can consider pure rotations and pure boosts in 1-direction separately.

1.5   The Line Element

The general Lorentz and Poincaré transformations being much more complicated than eq. (1.3.1) or (1.4.4), we look for yet another possibility of characterizing the transformations between inertial systems. This possibility emerges from comparison with the situation in Galilean Relativity. There we have—see eq. (1.3.12)—:
1. There exists an absolute time t, i.e., in the passage from one inertial system I to another one, $\bar{\text{I}}$, we always have $dx^{\bar{0}} = dx^0 = dt$ *invariant*.
2. The spatial distance between two simultaneous events is independent of the inertial system in which it is measured: $d\mathbf{x}^2 = d\bar{\mathbf{x}}^2$ for $dx^0 = 0$; since an absolute time exists, this simultaneity is then true in all inertial systems: $dx^{\bar{0}} = 0$.
These two properties—the existence of absolute, observer-independent (=invariant) space and time intervals—serve to characterize all transformations of Galilean relativity completely.

In Einsteinian relativity we have $dx^{\bar{0}} \neq dx^0$, as eq. (1.4.4) shows. Therefore, there is *no absolute time*, time and space intervals depend on the observer. Absolute time is relativized here—which led to the designation 'Theory of Relativity'. However, in 1908 Minkowski pointed out that, as already remarked by Poincaré in 1905, all Poincaré transformations may be similarly characterized by an *invariance principle*, namely by the invariance of the four-dimensional *line element ds*,

$$ds^2 := (dx^0)^2 - (d\mathbf{x})^2 \equiv (dx^{\bar{0}})^2 - (d\bar{\mathbf{x}})^2. \tag{1.5.1}$$

Equation (1.5.1) arises from eq. (1.4.1) putting $K = -1$, and assigns a distance to every pair of neighboring events—to the pairs themselves, and not only to their images in some space-time coordinate diagram! Space and time each are no more invariant for themselves, but what remains absolute is *space-time* (the set of all events, also called 'World' by Minkowski) and the distance (1.5.1) defined on it. (More about it will follow in sect. 3.2.)

For the proof of this characterization—which may be omitted until reading sect. 2.10—we have to show that, conversely, all transformations leaving ds^2 invariant are admitted in Einsteinian relativity. The demonstration that these transformations must be linear we postpone to sect. 3.1 where we will develop a formalism which is efficient for this purpose; here we show how every homogeneous transformation of this kind may be decomposed as a product of a rotation (1.3.1) and a boost (1.4.4)—possibly splitting off a space or time reversal. (As mentioned before, the latter must be ignored, however, as long as we consider only the present 'passive' interpretation of the transformations.)

So let $x^{\bar{i}\prime} = L^i{}_k x^k$ or, in matrix notation, $\bar{x}' = L x$, be a linear homogeneous transformation leaving ds^2 invariant. Splitting its matrix L as[1]

$$L = \begin{pmatrix} \gamma & -\mathbf{a}^\top \\ -\mathbf{b} & \mathsf{M} \end{pmatrix} \tag{1.5.2}$$

and inserting $\bar{x}' = L\,x$ into ds^2, we find that γ, $\mathbf{a}$, $\mathbf{b}$, M have to satisfy the relations

$$\gamma^2 - \mathbf{b}^2 = 1 \qquad \mathbf{b}^\top \mathsf{M} = \gamma\,\mathbf{a}^\top \qquad \mathsf{M}^\top \mathsf{M} = \mathbf{a}\,\mathbf{a}^\top + \mathbf{1}$$
$$\left(\Leftrightarrow \mathsf{M}^\top \mathbf{b} = \gamma\,\mathbf{a} \right) \tag{1.5.3}$$

This implies

$$L^{-1} = \begin{pmatrix} \gamma & \mathbf{b}^\top \\ \mathbf{a} & \mathsf{M}^\top \end{pmatrix}, \tag{1.5.4}$$

since the product $L^{-1} L$ gives the 4×4 unit matrix E, by eqs. (1.5.3). From this we also have $L\,L^{-1} = E$ or, after splitting,

$$\gamma^2 - \mathbf{a}^2 = 1 \qquad \mathsf{M}\,\mathbf{a} = \gamma\,\mathbf{b} \qquad \mathsf{M}\,\mathsf{M}^\top = \mathbf{b}\,\mathbf{b}^\top + \mathbf{1}. \tag{1.5.5}$$

Call I and $\bar{\mathrm{I}}'$ the frames to which the coordinates x^i and $x^{i\prime}$ refer, respectively; the former is now assumed to be inertial; we want to show the latter to be inertial also. From the inverse transformation $x = L^{-1}\bar{x}'$ given by eq. (1.5.4) we obtain for the spatial origin $\bar{\mathbf{x}}' = \mathbf{0}$ of $\bar{\mathrm{I}}'$ the relations $x^0 = \gamma\,x^{0\prime}$, $\mathbf{x} = \mathbf{a}\,x^{0\prime}$: this point therefore is moving relative to I with velocity $\mathbf{v} = \mathbf{x}/x^0 = \mathbf{a}/\gamma$, for which from the first of eqs. (1.5.5) we have $|\mathbf{v}| = |\mathbf{a}|\,(1 + \mathbf{a}^2)^{-1/2} < 1$.

If we now write $L_{\mathbf{v}}$ for the matrix of the boost (1.4.4), then $L_{\mathbf{v}}$ leads from I to an inertial frame $\bar{\mathrm{I}}$ which will have the same velocity relative to I as $\bar{\mathrm{I}}'$ has if we put $\mathbf{v} = \mathbf{a}/\gamma$. Then L should differ from $L_{\mathbf{v}}$—and thus $\bar{\mathrm{I}}'$ from $\bar{\mathrm{I}}$—only by a spatial rotation. The matrix $L_{\mathbf{v}}$ becomes

$$L_{\mathbf{v}} := \begin{pmatrix} \gamma & -\gamma\,\mathbf{v}^\top \\ -\gamma\,\mathbf{v} & \mathbf{1} + \dfrac{\gamma^2}{1+\gamma}\,\mathbf{v}\,\mathbf{v}^\top \end{pmatrix} = \begin{pmatrix} \gamma & -\mathbf{a}^\top \\ -\mathbf{a} & \mathbf{1} + \dfrac{\mathbf{a}\,\mathbf{a}^\top}{1+\gamma} \end{pmatrix}, \tag{1.5.6}$$

since γ here and in eq. (1.4.4) has the same meaning, by eq. (1.5.5), if $\gamma > 0$. (If $\gamma < 0$, the transformation L involves a reversal of the sense of time, and as has been said repeatedly, this must be excluded as long as we consider passive transformations only. Formally we can include it by performing a time reversal transformation

$$T := \begin{pmatrix} -1 & \mathbf{0}^\top \\ \mathbf{0} & \mathbf{1} \end{pmatrix} \tag{1.5.7}$$

after $L_{\mathbf{v}}$, in whose definition (1.5.6) γ has to be replaced by $|\gamma|$ and $\mathbf{a}$ by $-\mathbf{a}$.) The relation between $\bar{\mathrm{I}}'$ and $\bar{\mathrm{I}}$ is given by $\bar{x}' = L\,x = L\,L_{\mathbf{v}}^{-1}\bar{x}$, i.e., by the matrix $L\,L_{\mathbf{v}}^{-1}$. Using $L_{\mathbf{v}}^{-1} = L_{-\mathbf{v}}$ and eqs. (1.5.5), matrix multiplication now indeed gives

$$L\,L_{\mathbf{v}}^{-1} = \begin{pmatrix} 1 & \mathbf{0}^\top \\ \mathbf{0} & \mathsf{R} \end{pmatrix} =: L_{\mathsf{R}}, \qquad \text{where} \quad \mathsf{R} := \mathsf{M} - \dfrac{\mathbf{b}\,\mathbf{a}^\top}{1+\gamma}. \tag{1.5.8}$$

---

[1] $\mathbf{a}$, $\mathbf{b}$ are 3-rowed columns, M, R, $\mathbf{1}$ are 3×3 matrices, the superscript $\top$ indicates transposition.

Here the matrix R must be orthogonal, since eq. (1.5.8) shows that $x^{\bar{0}\prime} = x^{\bar{0}}$, and from the invariance of ds^2 under L and $L_{\mathbf{v}}$ then follows $d\bar{\mathbf{x}}'^2 = d\bar{\mathbf{x}}^2$; orthogonality may, however, also be checked directly as $\mathrm{R}^{\mathsf{T}}\,\mathrm{R} = \mathbf{1}$, using eq. (1.5.3). From it we have $(\det \mathrm{R})^2 = 1$, $\det \mathrm{R} = \pm 1$, and for $\det \mathrm{R} = -1$ (improper orthogonal transformation) a space reversal

$$P := \begin{pmatrix} 1 & \mathbf{0}^{\mathsf{T}} \\ \mathbf{0} & -\mathbf{1} \end{pmatrix}, \tag{1.5.9}$$

describing the transition from a right-handed to a left-handed spatial frame, must be performed in $\bar{\mathrm{I}}$ or $\bar{\mathrm{I}}'$ before we can determine the rotation vector $\boldsymbol{\alpha}$ by comparison with eq. (1.3.2) as (Tr indicates the trace of a matrix)

$$1 + 2\cos\alpha = \mathrm{Tr}\,\mathrm{R}, \qquad \alpha^\mu = \frac{1}{2}\,\epsilon^{\mu\nu\lambda}\,R^\nu{}_\lambda\,\frac{\alpha}{\sin\alpha} \tag{1.5.10}$$

for $0 \le \alpha < \pi$, and as eigenvector of R with eigenvalue $+1$ with ambiguous directional sense if $\alpha = \pi$. The announced *Cartan decomposition* thus has been achieved in a *unique* fashion, implying that $\bar{\mathrm{I}}'$ is inertial as well.

To avoid erroneous conclusions it is important to strictly keep track of the frames to which all occurring quantities are referred. To illustrate this point, we read off from eq. (1.5.4) that the components of the relative velocity of I against $\bar{\mathrm{I}}'$ are given by $-\mathbf{b}/\gamma$—and this does *not* contradict the reciprocity discussed in sect. 1.3, since the relation between I and $\bar{\mathrm{I}}'$ contains a rotation. Indeed, from eq. (1.5.5) it follows that

$$\mathrm{R}\,\mathbf{a} \equiv \mathbf{b}; \tag{1.5.11}$$

this says that the same rotation matrix that achieves $\bar{\mathbf{x}}' = \mathrm{R}\,\bar{\mathbf{x}}$ also changes the velocity components $-\mathbf{a}/\gamma$ of I against $\bar{\mathrm{I}}$, which do satisfy reciprocity, to components referring to $\bar{\mathrm{I}}'$, as it should be. Conversely, if we pass from $\bar{\mathrm{I}}'$ to a frame I' by the boost $L_{-\mathbf{b}/\gamma}$, then the latter has the same components $-\mathbf{b}/\gamma$ of relative velocity as does I, and thus should only be rotated against I. By a calculation completely analogous to the one above we indeed have $x' = L_{-\mathbf{b}/\gamma}\,L\,x$, where, by eq. (1.5.4),

$$L_{-\mathbf{b}/\gamma}\,L = \begin{pmatrix} 1 & \mathbf{0}^{\mathsf{T}} \\ \mathbf{0} & \mathrm{R} \end{pmatrix} = L_{\mathrm{R}} \tag{1.5.12}$$

with the same matrix R as in eq. (1.5.8). We therefore have, in the case without time reversal, two decompositions of L, each one unique:

$$L_{\mathrm{R}\mathbf{v}}\,L_{\mathrm{R}} = L = L_{\mathrm{R}}\,L_{\mathbf{v}},$$

$$\mathbf{v} = \mathbf{a}/\gamma, \qquad \mathrm{R} = \mathrm{M} - \frac{\mathbf{b}\,\mathbf{a}^{\mathsf{T}}}{1+\gamma}. \tag{1.5.13}$$

(Observe eq. (1.5.11) and compare to exercise 2 of sect. 1.3.)

As an application, let us investigate the following question. It is obvious that the matrix of a boost (1.5.6) is symmetric. Does the converse hold as well? We have

$$L = L_{\mathrm{R}}\,L_{\mathbf{v}} \Rightarrow L^{\mathsf{T}} = L_{\mathbf{v}}^{\mathsf{T}}\,L_{\mathrm{R}}^{\mathsf{T}} = L_{\mathbf{v}}\,L_{\mathrm{R}^{\mathsf{T}}}\,;$$

if now $L^\top$ is to agree with $L = L_{\mathrm{Rv}}\, L_\mathrm{R}$, uniqueness of the decomposition gives $\mathrm{R}\,\mathbf{v} = \mathbf{v}$, $\mathrm{R} = \mathrm{R}^\top (= \mathrm{R}^{-1})$. If here R is proper-orthogonal, we get from eq. (1.3.2) that $\sin\alpha = 0$, so $\alpha = 0$ or $\alpha = \pi$ and therefore $\mathrm{R} = \mathbf{1}$ or $\mathrm{R} = 2\,\mathbf{n}\,\mathbf{n}^\top - \mathbf{1}$, where $|\mathbf{n}| = 1$. For $\mathbf{v} \neq \mathbf{0}$ we must have $\mathbf{n} = \mathbf{v}/v$, while if $\mathbf{v} = \mathbf{0}$, $\mathbf{n}$ may be an arbitrary unit vector. So we see that apart from boosts (1.4.4) our symmetry condition is also satisfied by 180° rotations, and by products of such rotations with boosts whenever the axis of rotation is in the direction of the relative velocity.

We point out that in eq. (1.5.1) we could have chosen equally well the negative of the expression on the right-hand side for the squared line element. The choice is conventional and varies from one author to another. The convention opposite to eq. (1.5.1) recommends itself if space-time splits are to be performed frequently, since then $dx^0 = 0$ converts eq. (1.5.1) simply into the Euclidean metric (cf. pertaining remarks in sect. 5.9). Our choice of convention offers advantages in connection with the 2-component spinor algebra to be discussed in chap. 8.

We should also mention recent attempts at a *physical* distinction between both possibilities, based on the non-isomorphic Pin groups associated with the two conventions (see the appendix to sect. 9.1 for this concept); this has consequences for Dirac spinor fields if space-time in the large has a non-orientable topological structure deviating from $\mathbf{R}^4$. See S. Carlip, C. DeWitt-Morette, Phys. Rev. Lett. *60*, 1599 (1988), and C. DeWitt-Morette, B. S. DeWitt, Phys. Rev. D *41*, 1901 (1990).

Minkowski's geometric formulation turned out to be extremely useful in the sequel, from the conceptual point of view as well as from the calculational one. We shall see the latter from chap. 3 on. For the former, we remark that only using Minkowski's concept Einstein was able to pass from his 'Principle of Equivalence' to a complete relativistic theory of gravitation—the General Theory of Relativity, as it is called. Historically, it is interesting that Einstein's first reaction to Minkowski's formulation was—as reported by Sommerfeld ("Zum 70. Geburtstag A. Einsteins", Deutsche Beiträge, Bd. III, Nr. 2. München: Nymphenburger Verlagshandlung, 1949)—to say that he would not understand his own theory any more. In fact, it took him almost five years until he made up his mind to use the line element, but then after two more years General Relativity was completed. Without it, again in his own words, that theory would never have got beyond its diapers.

Exercise

Verify, from the definitions (1.5.6,7,8), that $TL_\mathrm{R}T^{-1} = L_\mathrm{R}$ and $TL_\mathbf{v}T^{-1} = L_{-\mathbf{v}}$. Conclude that $TLT^{-1} = L$ iff (=if and only if) L is a pure rotation. Also, instead of this operation of 'conjugation by T', consider the operation of taking the transposed inverse.

1.6 Michelson, Lorentz, Poincaré, Einstein

The approach to Lorentz transformations given here is rather different from the original argumentation of Einstein in 1905. We therefore want to supplement it by a sketch of the historical development, also taking into account the roles of Michelson, Poincaré and Lorentz.

In the 19th century, *ether*, a medium or carrier of electromagnetic waves, was considered as an undoubtable reality (see frontispiece), and a central point of physical research was to measure the motion of the Earth through this medium. Many experiments were devised, and the same number of ad hoc hypotheses had to be made to explain the negative results of all those experiments. Most of them were destined to measure effects up to order v/c, v being the speed of the Earth relative to the ether.

Concerning these, H. A. Lorentz was able, in two basic papers dating from 1892 and 1895, to show that a correctly formulated 'electron theory'—in which Maxwell's equations were supplemented by hypotheses about microscopic charge distributions and their dynamics—would predict a negative result.

In these papers one finds the introduction of a 'local time' $t' = t - (vx)/c^2$ as a purely calculational tool; it had already been used by Voigt in 1887 in an investigation of Doppler's principle. The situation at the turn of the century is described by Lorentz in 1927 (at the "Conference on the Michelson-Morley experiment", held at Mt. Wilson Observatory, published 1928 in Astrophys. J. *68*, 341–402) as follows.

"I remember especially the assembly of the German Society of Natural Sciences in Düsseldorf in 1898, at which numerous German physicists were present, Planck, W. Wien, Drude, and others. We discussed especially the question of first order effects. Some devices with which such an effect might be observed were proposed, but none of these attempts was ever made, as far as I know. The conviction that first-order effects do not exist became by and by too strong. We even got, finally, into the habit of looking at the summary of experimental papers which dealt with such effects. In case the result was properly negative we felt perfectly satisfied."

It therefore became necessary to pass on to effects of order $(v/c)^2$. There were no good theoretical hints for the speed of the Earth relative to the ether, but it was thought that it should be of the order of magnitude of the orbital speed round the sun, so that $(v/c)^2 \approx 10^{-8}$ was very small.

Already in 1882, A. Michelson had proposed and carried out an experiment capable of determining such second-order effects. But in the 1882 experiment as well as in the improved 1887 version, it was impossible to observe the effects of the motion of the Earth through the ether. To explain this negative result, Fitzgerald and Lorentz postulated in 1892 that the length of a body moving through the ether would contract in the direction of motion by a factor $\sqrt{1 - v^2/c^2}$ (Lorentz contraction, see sect. 2.4). There were also other explanations, postulating a dragging of the ether by the earth, but they are contradicted by the aberration effect, and we shall disregard them here. Lorentz was also able to deduce the contraction from the fundamental equations of electrodynamics (see sect. 5.8).

In the following years, a number of papers and books were dedicated to the problem of the motion of the Earth through the ether. For instance, the Lorentz transformation (1.4.4) can be found in Voigt's 1887 paper and in Larmor's (1900) book "Ether and Matter". An essential contribution to the discussion (which is critically analyzed in Whittaker (1960)) is again by Lorentz (reprinted in Lorentz et al. (1952)), who in 1904 proved the covariance of the Maxwell equations under Lorentz transformations, albeit only approximately. This way he was able to explain the negative result of all known experiments, including Michelson's and Morley's.

One further step was made by Poincaré[1] in his paper "Sur la dynamique de l'électron" from July 1905. There he formulates the Principle of Relativity: "It appears that the impossibility to determine motion of the Earth in the ether is a general law of Nature; we are led to assume the validity of this law, which we call the 'postulate

---

[1]H. Poincaré, Rend. Circ. Math. Palermo *21*, 129 (1906); a partial translation into English is given in Kilmister (1970); a translation, with comments, into modern terminology can be found in H. M. Schwartz, Am. J. Phys. *39*, 1287 (1971); *40*, 862 (1972).

of relativity', without any restriction."

In this paper, Poincaré also introduces the concepts of 'Lorentz transformation' and 'Lorentz group', postulating that the laws of Nature must be covariant under Lorentz transformations. However, the role of the formally introduced time coordinate does not become clear and remains undiscussed.

The difficulty presented by the latter is illustrated by the following quotation from Lorentz (paragraph following the one quoted above). "As to the second order effects, the situation was more difficult. The experimental results could be accounted for by transforming the co-ordinates in a certain manner from one system of co-ordinates to another. A transformation of time was also necessary. So I introduced the conception of a local time which is different for different systems of reference which are in motion relative to each other. But I never thought that this had anything to do with real time. This real time for me was still represented by the old classical notion of an absolute time, which is independent of any reference to special frames of co-ordinates. There existed for me only this one true time. I considered my time transformation only as a heuristic working hypothesis. So the theory of relativity is really solely Einstein's work. And there can be no doubt that he would have conceived it even if the work of all his predecessors in the theory of this field had not been done at all. His work is in this respect independent of the previous theories." It may be assumed that Poincaré's point of view was similar—otherwise he would hardly had left unmentioned, in his paper, the most radical and most important step towards the theory of relativity, the elimination of absolute time. As he writes himself, his primary aim was a formal improvement of Lorentz' paper: "The results I achieved coincide, in all their important points, with those of Mr. Lorentz; I was just led to improve on them in some details; the differences, of minor importance, will become clear later." From the point of view of Philosophy of Science, we have in Poincaré's work a partially uninterpreted formalism in which the assignment between theoretical terms and empirical terms is partially absent. (See Leinfellner 1965, p. 107 for this topic.)

It was left to Einstein to derive the Lorentz contraction without any reference to electrodynamics and models of matter. His famous 1905 paper "Zur Elektrodynamik bewegter Körper", reprinted in Lorentz et al. (1952), is highly recommended reading in its original version. The first section carries the title "Definition of Simultaneity" and investigates the concept of simultaneity of distant events (see sect. 2.2 of this book). The next section, entitled "On the relativity of lengths and times", ends with the statement: "We thus see that we must not attribute absolute significance to the concept of simultaneity; rather, two events which are simultaneous as regarded from one system of coordinates are to be conceived of as being not simultaneous if regarded from a system in relative motion with respect to the former system." In the derivation of the Lorentz transformation which follows, Einstein immediately identifies the time coordinates t and $\bar{t}$ with times that are actually measured in the corresponding reference systems (so that an assignment between theoretical and empirical terms is present from the very beginning). In the second chapter of the paper Einstein then shows that the Lorentz transformation, derived from the principles of relativity and of invariance of the speed of light with the help of his analysis of simultaneity, leave the form of Maxwell's equations invariant.

Lorentz (1909; printed version of his 1906 Columbia University Lectures) characterizes the difference in attitude between Einstein and himself as follows: "... the chief difference being that Einstein simply postulates what we have deduced, with some difficulty and not altogether satisfactorily, from the fundamental equations of the electromagnetic field. By doing so, he certainly may take credit for making us see in the negative results of experiments like those of Michelson, Rayleigh and Brace, not a fortuitous compensation of opposing effects, but the manifestation of a general and fundamental principle. Yet, I think, something may also be claimed in favour of the form in which I have presented the theory. I cannot but regard the ether, which can be the seat of an electromagnetic field with its energy and its vibrations, as endowed with a certain degree of substantiality, however different it may be from all ordinary matter. In this line of thought, it seems natural not to assume at starting that it can never make any difference whether a body moves through the ether or not, ..."

This quotation shows that Einstein's theory was not immediately recognized in its full significance but was regarded as a—perhaps somewhat unusual—contribution to the voluminous ether literature rather than essentially marking its end.

From hindsight it should be emphasized that Einstein's way of proceeding separated the problems around 'space–time–relativity' from the problems of 'electron theory': the solution of the latter was effected by quantum theory rather than relativity. In Lorentz' electron theory, the problem of space-time transformations was mixed up with the problem of the dynamics of charged particles, which presents difficulties even today (see sect. 5.10); but also the Zeeman effect, electric conductivity, etc., were to correctly follow from the theory. It became clear only much later how different the theoretical analyses of these subjects had to be.

The above analysis of the contributions by Lorentz, Poincaré, and Einstein is also of interest in view of Whittaker's (1960) historical investigation "A History of the Theories of Aether and Electricity". Chapter 2 of the second volume of this work carries the title "The Relativity Theory of Poincaré and Lorentz". After some extensive valuation of their merits, Whittaker writes: "...In the autumn of the same year [1905], in the same volume of the *Annalen der Physik* as his paper on the Brownian motion, Einstein published a paper which set forth the relativity theory of Poincaré and Lorentz with some amplifications, and which attracted much attention. ..." It is not clear why Whittaker underrated Einstein's merits concerning special relativity in his otherwise excellent book.

There has also been a lot of discussions in recent years about the role played by the Michelson-Morley experiment in Einstein's setting up special relativity. In his original 1905 paper he only refers to "unsuccessful attempts to demonstrate some motion of the Earth relative to the 'light medium'", without, however, singling out any particular experiment. On the other hand, in pedagogically oriented presentations of the theory, one often finds the remark that between the Michelson-Morley experiment and relativity there is a close historical and physical tie. Hardly any textbook on Einstein's theory fails to give a description of the experiment—one sometimes gets the impression that the theory of relativity follows from it.

What then is the historical and physical significance of the Michelson-Morley experiment, as far as relativity is concerned? Is it indeed the 'experimentum crucis' that put an end to the epoch of Newtonian physics and caused a revolution in the physical world view? Einstein himself has given various statements concerning the influence of this experiment on his thoughts while setting up the theory; they are critically analyzed by Holton (1973).

For instance, in 1950 Einstein communicated to Shankland that he had learned about the Michelson-Morley experiment only after the year 1905, from the writings of Lorentz. However, two years later he was not so sure any more about when he had heard about it, saying: "...I was not conscious that it had influenced me directly during the seven years that relativity had been my life. I guess I took it for granted that it was true. ..." In 1954 Einstein wrote to Davenport: "...In my own development, Michelson's result has not had a considerable influence. I do not even remember if I knew of it at all when I wrote my first paper on the subject. ..."

In fact, the experiment was of importance only in the technical discussion of electron theory, being rather different, from that point of view, from other ether drift experiments. But Einstein had left behind the world of concepts of that theory, and in his chain of ideas the Michelson-Morley experiment is only one of many measurements that show the unobservability of the Earth's motion through the ether; whether these experiments were of first or second order in v/c was unimportant in his approach—*all* such experiments were only hints to the nonexistence of the ether.

The distinction between first- and second-order experiments were, however, of fundamental importance for the development of electron theory, as we have explained above. Therefore, also in the years 1895–1905 numerous articles treated the Michelson-Morley experiment and the Lorentz contraction. As one learns from looking into *Physics Abstracts* from those years, people like Abraham, Sommerfeld, Wien, Brillouin, Cohn, Hasenöhrl, Langevin, Kohl, Gans, etc., were engaged in these problems. It seems improbable, therefore, that Einstein's knowledge of the experiment dates from after 1905.

Not only from a historical point of view but also from a logical one, the Michelson-Morley experiment is not the experimentum crucis to distinguish between Newtonian physics and relativity theory: if one were to deduce relativity theory (or something equivalent to it in a sense to be explained in sect. 2.11) by a phenomenological approach, using the experimental evidence alone, two more experiments are necessary, as shown by Robertson[1]: the experiments of Kennedy-Thorndike and of Ives-Stilwell (see also, e.g., Schwartz 1968). These latter experiments together determine the constancy of lengths orthogonal to the direction of relative motion, and time dilation, while Michelson-Morley gives the Lorentz contraction of lengths parallel to the direction of motion and does not suffice, taken alone, to deduce the Lorentz transformation.

Also, the Michelson-Morley experiment did not change the conceptual basis of physics, being satisfactorily explained by the electron theory as supplemented by the Lorentz contraction, for which Lorentz had actually given an explanation within the framework of electron theory already in 1895 (see sect. 5.8). Thus Poincaré and Lorentz, the spearheads of electron theory, did not accept the Einsteinian change in the basic concepts of physics for many years, as witnessed by Lorentz' remarks already quoted, and by a report of Moszkowski (1922) about a talk given by Poincaré on 13 October 1910: "Poincaré talked about the 'new mechanics' ... This revolution, he said, seems to threaten what in science until recently was deemed the safest: the basics of mechanics, as we owe it to Newton's genius. At the moment, this revolution is only a threatening phantom, since it is well possible that sooner or later those well established Newtonian dynamical principles will emerge as the winners. And continuing, he repeated several times that he would become anxious in front of the accumulating hypotheses whose integration into a system seemed difficult, even impossible, to him."

Poincaré did not live long enough to await the arrival of Einstein's General Relativity, but it is

---

[1]H. P. Robertson, Rev. Mod. Phys. *21*, 378 (1949); Robertson and Noonan (1968). "Deduce" is meant here in the sense used in physics, not in the sense of logic—for a discussion of the distinction, see Popper (1982).

interesting to note that Lorentz had no difficulties to accept it immediately and even work on it!

In analyzing the role of the Michelson-Morley experiment, concepts from Kuhn's (1962) "Structure of Scientific Revolutions" are useful. In the course of the development of electron theory as 'normal science', the experiment was indeed an experimentum crucis—necessitating to build in the Lorentz contraction. After this, all experiments had been explained satisfactorily and no change in the usual concepts was necessary.

It was thus Einstein's special relativity theory that brought the 'revolution' by either giving to the concepts space, time, ether, electron a new content or showing them to be irrelevant or assigning them to other branches of physical research. At first, it did not explain more than the old theory, so that ether people were able to retain their views even for decades—perhaps this is the explanation for Whittaker's presentation of the history of relativity. Only gradually did the huge *simplification* brought by the new concepts become clear.

2 Physical Interpretation

As has become apparent in the derivation of the Lorentz transformation, some considerations are facilitated with diagrams in which space and time coordinates are plotted simultaneously. In subsequent investigations of the physical consequences of the Lorentz transformation such diagrams will become indispensable. In particular, the demonstration that some of the apparently paradoxical implications of Special Relativity are actually free from contradiction will be simple to see with space-time diagrams. Their disadvantage is that they are transparent only upon restriction to one space dimension (especially when hand-drawn), so that for practical applications, where almost always all three space dimensions are important and numerical results are being called for, other techniques have to be developed as well (chap. 3). However, it will suffice for the basic questions to be treated in this chapter up to sect. 9 to restrict to one space dimension (coordinate $x = x^1$) and time t.

2.1 Geometric Representation of Lorentz Transformations

Upon restriction to one space dimension, the Lorentz transformation (1.4.4) is

$$\begin{aligned}
\bar{t} &= \gamma(t - v\,x) \\
\bar{x} &= \gamma(x - v\,t).
\end{aligned} \tag{2.1.1}$$

To represent it geometrically in a space-time diagram, we must first determine the relation between the coordinate axes implied by eq. (2.1.1). The $\bar{x}$-axis, given by $\bar{t} = 0$, according to eq. (2.1.1) has the equation $t = v\,x$, and therefore represents, in the (x, t)-diagram, a straight line through the origin with slope $\tan\delta = v$. Similarly, the $\bar{t}$-axis ($\bar{x} = 0$) is given by $x = v\,t$ and hence has slope $\tan\delta' = 1/v$ (Fig. 2.1).

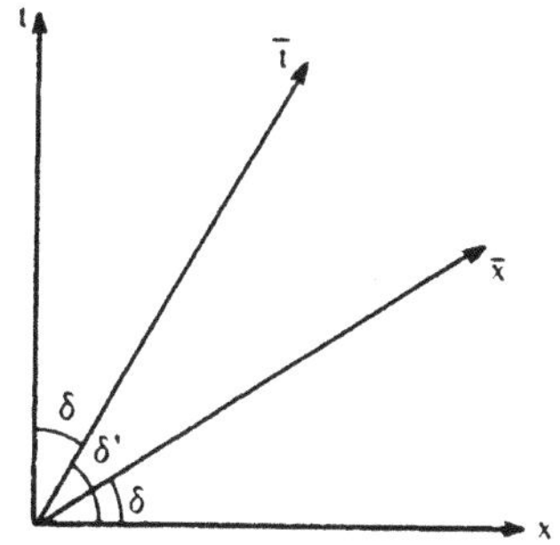

Fig. 2.1. Relation between (t, x) and $(\bar{t}, \bar{x})$

To determine the units on the barred axes we use the identity

$$t^2 - x^2 \equiv \bar{t}^2 - \bar{x}^2 \tag{2.1.2}$$

satisfied by eq. (2.1.1)(cf. eq. (1.3.1)). The unit point on the $\bar{t}$-axis, ($\bar{t} = 1$, $\bar{x} = 0$), therefore satisfies

$$t^2 - x^2 = 1, \tag{2.1.3}$$

and similarly for the unit point ($\bar{t} = 0$, $\bar{x} = 1$) on the $\bar{x}$-axis we have

$$x^2 - t^2 = 1. \tag{2.1.4}$$

The unit points are therefore the intersections of the coordinate axes with the unit hyberbolae (2.1.3,4) as shown in Fig. 2.2. (As may be shown as an exercise, the tangent to these hyperbolae at a unit point is parallel to the other axis—a fact to be observed in qualitative drawings made by hand in order to avoid wrong conclusions.)

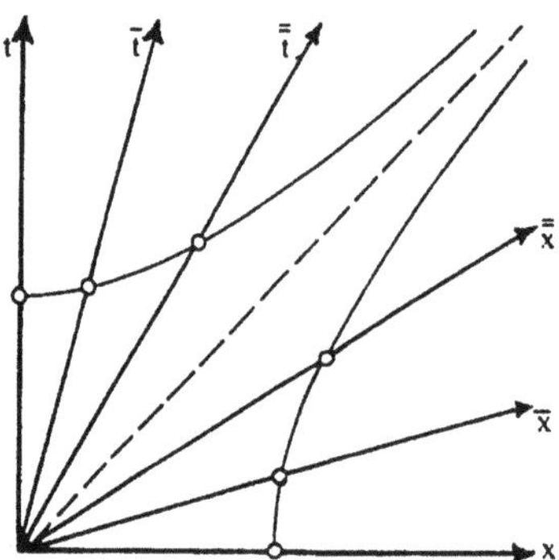

Fig. 2.2. Unit points on the axes

As a further illustration of the transformation we remark that by introducing an imaginary time coordinate $t = i\,x^4$ we may rewrite eq. (2.1.2) as

$$(x^4)^2 + x^2 \equiv (x^{\bar{4}})^2 + \bar{x}^2. \tag{2.1.5}$$

Transformations leaving invariant such a sum of squares are rotations

$$\begin{aligned}
x^{\bar{4}} &= \cos\varphi\, x^4 - \sin\varphi\, x \\
\bar{x} &= \sin\varphi\, x^4 + \cos\varphi\, x.
\end{aligned} \tag{2.1.6}$$

We can therefore regard Lorentz transformations as 'complex rotations'; the transition $x^4 \to -it$ changes the unit circle of ordinary Euclidean geometry (which contains all unit points) into the hyperbolae of Fig. 2.2. To obtain the connection between eqs. (2.1.6) and (2.1.1) we multiply the first of eqs. (2.1.6) by i and put $\alpha := i\varphi$, $\cos\varphi = \cosh\alpha$, $i\sin\varphi = \sinh\alpha$; then

$$\begin{aligned}
\bar{t} &= \cosh\alpha\, t - \sinh\alpha\, x \\
\bar{x} &= -\sinh\alpha\, t + \cosh\alpha\, x.
\end{aligned} \tag{2.1.7}$$

To get real $(\bar{t}, \bar{x})$ when (t, x) is real we must have α real, i.e., φ has to be an imaginary 'angle'. Comparing eqs. (2.1.7) and (2.1.1) we get

$$\cosh \alpha = \gamma, \quad \sinh \alpha = \gamma v, \quad \tanh \alpha = v. \tag{2.1.8}$$

This analogy between Lorentz transformations and ordinary Euclidean rotations is useful to remember; one says that they are different real forms of complex rotations.

Exercise

Prove the property of the tangents to the unit hyperbola quoted in the text.

2.2 Relativity of Simultaneity. Causality

The fundamental difference between the Lorentz transformation and the Galileo transformation emerges when Fig. 2.2 is contrasted with the corresponding diagram for the latter (Fig. 2.3).

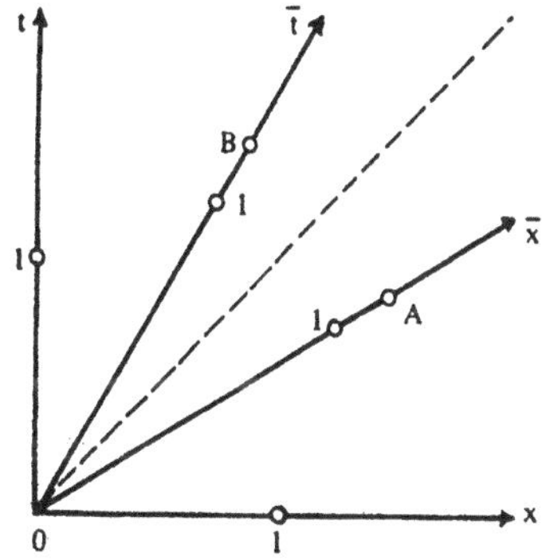
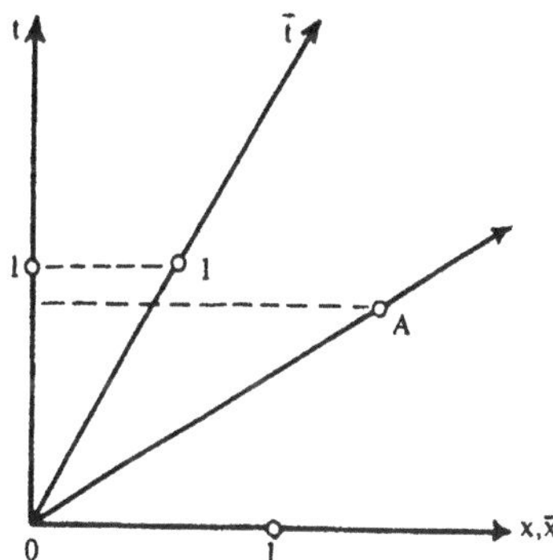

a) Lorentz transformation
$$\bar{x} = \gamma(x - vt)$$
$$\bar{t} = \gamma(t - vx)$$

b) Galileo's transformation
$$\bar{x} = x - vt$$
$$\bar{t} = t$$

Fig. 2.3. Comparing the classical and the relativistic transformation

Galileo's transformation changes only the t-axis while the x-axis remains fixed. Since there is no upper bound on v, one may arrange by a suitable choice of a new inertial system $\bar{\bar{\text{I}}}$ that an arbitrary event A not on the x-axis will lie on the $\bar{\bar{t}}$ axis, so that, relative to $\bar{\bar{\text{I}}}$, A takes place at the same spatial site $\bar{\bar{x}} = 0$ as does the event O. Therefore the spatial distance of nonsimultaneous events in Galilean Relativity— which is at the basis of Newtonian mechanics—depends on the inertial system used and may always be made zero by a suitable choice of that system (*unrestricted relativity of equilocality*). On the contrary, the time difference between arbitrary events in Galilean Relativity is independent of the inertial system and so has here absolute meaning just as does the distance between simultaneous events (*absolute simultaneity*).

The Lorentz transformation leads to a change of both, t- and x-axis. As a consequence, the event A shown in Fig. 2.3, which for I is later than O, comes to lie on the $\bar{x}$-axis and is thus simultaneous with O relative to $\bar{\text{I}}$, both occurring at time $\bar{t} = 0$. This shows that in Einsteinian Relativity *simultaneity of spatially separated events is not an absolute concept* but depends on the inertial system used.

Not every event may be made simultaneous with O by choice of a reference frame, however: eq. (2.1.1) makes no sense when $v = 1$, so the x-axis cannot be moved beyond the pair of lines given by $x^2 = t^2$. Thus, the event B shown in Fig. 2.3a is later than O for all observers (reference frames). Similarly, one cannot have A taking place at the same site as O for any frame, according to Einsteinian Relativity, contrary to what we found in Fig. 2.3b (*restricted relativity of simultaneity and idemlocality*).

The pair of lines $x^2 = t^2$ is called the *light cone* of O, representing the set of all events that can be reached by light rays emanating from O or from which one can reach O on light rays: $x = \pm t$ means motion at the speed of light. The designation 'cone' becomes clear once we add one more space dimension (Fig. 2.4): it describes the history of a spherical wave front contracting towards O and then reexpanding from O with the speed of light according to $x^2 + y^2 = t^2$.

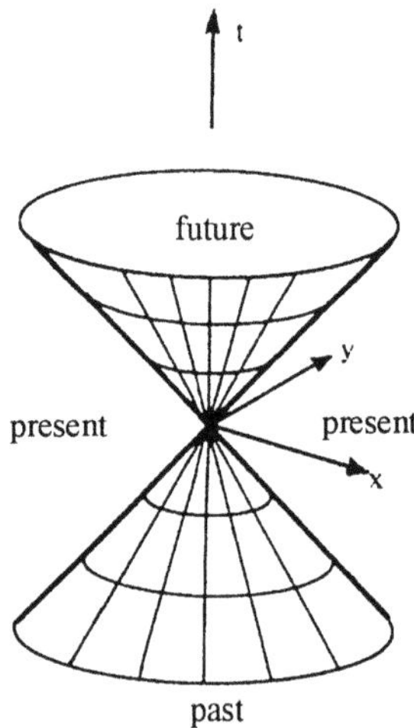

Fig. 2.4. The light cone

The light cones are of fundamental importance for the theory. Every event outside the light cone of O may be made simultaneous with O by a suitable choice of the inertial system, and in this sense belongs to the present of O. Points inside or on the *future light cone* ($t > 0$) belong to the future of O, in that they take place later than O for *all* possible inertial frames. Since there is still a frame-independent distinction between that future excluding or including the light cone itself, one uses the designations *chronological future* or *causal future* of O, respectively. Similarly, the *past light cone* bounds the (chronological or causal) *past* of O. The light cones of the events thus define the *causal structure* of the theory. Events outside the light cone of O can neither have an influence on O nor can they be influenced by O—there is an observer for which such an event is simultaneous with but spatially separated from O. On the other hand, O may influence everything that happens in its causal future and may be influenced by everything that happened in its causal past. (One therefore sometimes encounters the terms future, resp. past 'domain of influence'; a reader uneasy for a certain circularity here is asked to await the next section!)

To illustrate these features in a concrete example, consider the pair annihilation of two electron-positron pairs:

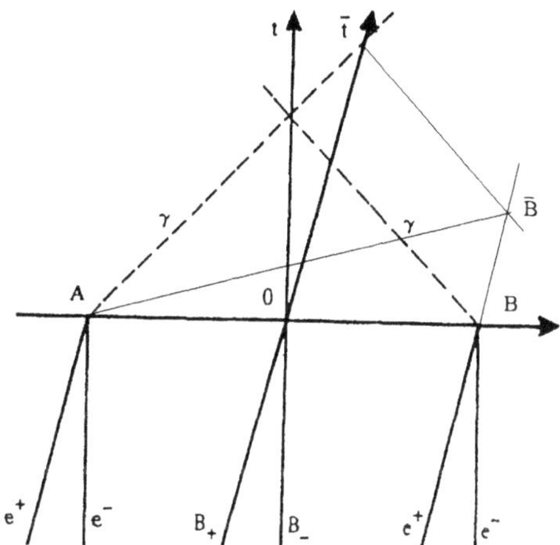

Fig. 2.5. Annihilation of two electron-positron pairs

In Fig. 2.5 the world lines of the particles of both pairs are shown together with the world lines of two observers B_+, B_-. Both electrons e^- are at rest in the (x,t)-system together with B_- midway between them; similarly, the positrons e^+ and B_+ between them are at rest in the $(\bar{x},\bar{t})$-system. At time $t = 0$—i.e., simultaneously in the (x,t)-system—both pairs annihilate in the reaction $e^+ + e^- \rightarrow \gamma + \gamma$ ($\gamma =$ photon = quantum of light; the figure shows only one photon for each reaction, for simplicity). B_- receives both flashes of light precisely at the same time, for him the pairs were annihilating simultaneously indeed. However, B_+ receives the flash from event A much later than the flash from B and thus concludes that B must have happened much earlier than A. Since both observers are on equal footing, absolute simultaneity cannot be defined for A and B.

Note that, conversely, the event $\bar{B}$ shown in the figure would be simultaneous with A as judged by B_+, so that the straight line connecting these two events is parallel to the $\bar{x}$-axis. This gives a much more physical construction of that axis than the one we gave before in Fig. 2.1. The point here is that in Fig. 2.1 we used Euclidean geometry in the (x,t)-space together with the convention

$c = 1$ while the present construction is free from both, using only the affine structure preserved by the linearity of the Lorentz transformation, and light signals. The natural geometry to be used in space-time diagrams that derives from these two features—Minkowski geometry—will be treated in the next chapter. Let us just remark here that looking at space-time diagrams with Euclidean eyes is directly contrary to the idea that all inertial frames are on equal footing: restricting to one space dimension and excluding a space reversal, there is always exactly one such frame whose axes are at right angle in the Euclidean sense, but this system is in no way distinguished physically!

We want to point out here that a completely analogous reasoning led Einstein to Special Relativity in 1905. His starting point was an epistemological analysis of the concept of simultaneity of spatially separated events. In Newtonian mechanics, this concept had never been analyzed but had been considered as being self-evident. Einstein showed the necessity of a definition here. The procedure proposed by him to synchronize two clocks at rest at different sites in an inertial system exactly corresponds to the reasoning given above: he proposed to define as simultaneous two spatially separated events (such as the pointers of two clocks reaching zero positions) just if two light signals emitted by them arrive simultaneously at an observer midway between them (*Einstein synchronization*). Equivalently one could achieve synchrony of the various clocks in one reference frame by (very slowly) transporting a standard clock from place to place.

The version of the Principle of Relativity adopted in chap. 1 already implies such a procedure: arbitrary inertial systems can of course be on equal footing only if the procedure of synchronizing the clocks within each of them does not single out any of them! This is achieved using a procedure completely internal to the system—e.g., by light synchronization or by slow clock transport. Other procedures do not lead to completely equivalent reference frames; however, this does not preclude their use—see sect. 2.11.

2.3 Faster than Light

We have seen already that the Lorentz transformations (2.1.1) make no sense if $v \geq 1$. This implies that the inertial frames admitted in the principle of relativity may be moving relative to each other with speeds $v < 1$ only. Although we have not used any detailed constitutive properties of our reference systems, this will be reinforced in sect. 4.2 by relativistic dynamics in that one cannot accelerate massive objects from $v < 1$ to $v \geq 1$ using only a finite amount of energy.

We can go one step further here and illustrate the difficulties arising if there were signals of any kind which propagate with speed $v > 1$ relative to their source. Consider such kind of signal, with speed $v = \infty$, say, for simplicity: one then could signal into one's own past. Figure 2.6 shows a situation where such a signal is emitted at A and reflected by an observer in relative motion at B, i.e., re-emitted by him with speed ∞ relative to him. (If this were not possible, the two observers would not be on equal footing, violating the principle of relativity!) But this signal enters into the causal past of A, and could, e.g., be received before emission.

It would be a tricky business to avoid paradoxes if this possibility were admitted (imagine the message carried by the signal is 'do not emit'); thus it is easiest to *postulate* that *no signals* (e.g., sound) *exist that can propagate with speeds greater than the speed of light*. Note that this postulate is not implied by Lorentz covariance but is consistent with it: as will be shown more explicitly in sect. 2.9, the domains of subluminal and superluminal speeds are separately Lorentz invariant.

From the postulate that signals should never be superluminal one can derive many consequences for relativistic theories. One is that the classical concept of (accelerat-

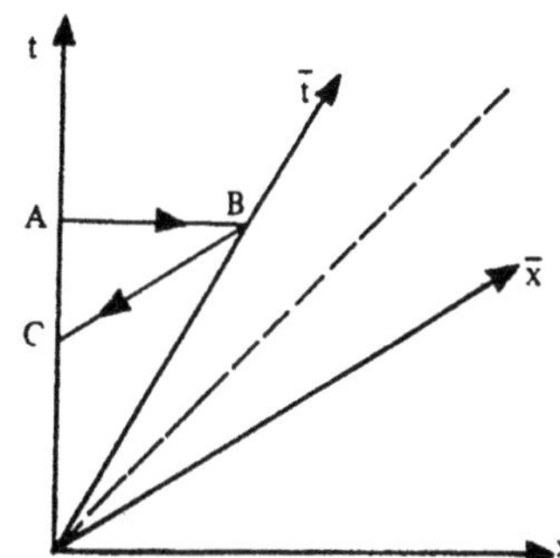

Fig. 2.6. Consequences of infinite signal speeds

ing) *rigid bodies* is *excluded* by it: kicking such a body at one end would cause its other end to move at the same time, according to its definition, resulting in infinite signal speed. (We do not discuss here certain restricted types of accelerated motion which are compatible with rigidity, as they do not occur in signalling attempts.)

The fundamental importance of the postulate of nonexistence of superluminal signals for the theory of relativity makes it necessary to formulate the concept of signal velocity more precisely. For this we first recapitulate the distinction between *phase velocity, group velocity,* and *front velocity* of waves.

Consider a wave $\varphi(x, t)$ propagating in a dispersive medium—i.e., a medium where the index of refraction depends on the wave vector. The phase velocity v_P of the wave $\varphi_k(x, t) = \exp(ikx - i\omega t)$ is defined by $kx - \omega t =: =: k(x - v_P t)$, thus

$$v_P(k) = \omega(k)/k. \tag{2.3.1}$$

However, v_P is irrelevant for the transmission of signals, since the monochromatic wave train $\varphi_k(x, t)$ has infinite length and is unmodulated, carrying no signal. By superposition of waves of various wave numbers—in the simplest case by forming

$$\varphi = \frac{1}{2} \left(\varphi_{k-\Delta k} + \varphi_{k+\Delta k} \right) = \underbrace{\exp(ikx - i\omega t)}_{\text{phase}} \cdot \underbrace{\cos(\Delta k\, x - \Delta\omega\, t)}_{\text{amplitude}} \tag{2.3.2}$$

one obtains wave packets that propagate with group velocity v_G, where ($\Delta k \to 0$)

$$v_G = \frac{d\omega}{dk} . \tag{2.3.3}$$

However, even the group velocity describes propagation of signals correctly only in the simplest cases. There are situations in classical electrodynamics (cf. Jackson 1999; Brillouin 1960) where v_P and/or v_G exceed the speed of light ($v_P > 1$ in wave guides, $v_P > 1, v_G > 1$ in the presence of anomalous dispersion).

In recent years it has been pointed out (R. Landauer, Sci. Am., Aug. 1993) and demonstrated experimentally that in regions of exponential damping one may have superluminal group velocities— let us mention here the work of G. Nimtz and coworkers (e.g., Phys. Lett. A *196*, 154 (1994)) on tunneling in wave guides, of A. W. Steinberg et al. (e.g., Phys. Rev. Lett. *71*, 708 (1993)) on photon tunneling, and of F. Krausz and coworkers (e.g., Phys. Rev. Lett. *73*, 2308 (1994)) on evanescent waves), and recent improvements by D. Mugnai, R. Ruggieri (Phys. Rev. Lett. *84*, 4830 (2000)) and L. I. Wang et al. (Nature *406*, 594 (2000)). In these cases, dispersion is so pronounced that the concept of wave packet becomes rather meaningless, as an initial packet gets completely deformed and unsuitable for perfect signal transmission during the course of propagation due to the vastly differing phase velocities of its various frequency components.

Under such circumstances, only discontinuities in the wave field may be used for signalling (e.g., switching on or off suddenly). Discontinuities propagate at *front velocity*

$$v_F = \lim_{k \to \infty} v_P(k) = \lim_{k \to \infty} \frac{\omega(k)}{k}, \tag{2.3.4}$$

which is also the speed of propagation of a wave front that separates the domains $\varphi \neq 0$ and $\varphi = 0$. Since perfect signals are to be regarded always as a kind of discontinuity—the decision to signal A or non-A at a certain instant should not be recognizable from the wave existing before that instant—signals cannot be transmitted faster than v_F, and this our postulate requires not to exceed 1.

To at least sketch the proof of eq. (2.3.4), we write

$$\varphi(x, t = 0) = \int_{-\infty}^{\infty} f(k)\, e^{-ikx} dk, \tag{2.3.5}$$

where $f(k)$ is required to be analytic except for poles in the upper complex k half plane. Then one may close the path of integration by adding a large semicircle in the lower half plane and obtains, using the residue theorem, $\varphi(x > 0, t = 0) = 0$. Therefore (2.3.5) has the discontinuity necessary for a sharp signal. The time development for this signal is

$$\varphi(x, t) = \int_{-\infty}^{\infty} f(k)\, e^{-i[kx - \omega(k)t]}\, dk. \tag{2.3.6}$$

Again by the residue theorem, this integral vanishes if for $k \to \infty$ we have

$$\lim(kx - \omega(k)t) > 0,$$

since then the path of integration may again be closed in the half plane free of poles. We therefore get

$$\varphi(x, t) = 0 \quad \text{for} \quad x > \left(\lim_{k \to \infty} \frac{\omega(k)}{k}\right) t, \tag{2.3.7}$$

so that the front is indeed propagating at speed v_F.

In electrodynamics as well as in all other sensible field theories one always has $v_F = 1$, since for infinite frequencies (hence infinite photon energies) all influence of the medium upon wave propagation may be neglected.

The details of our proof may, for a special form of the signal, be found in Brillouin (1960) where also a classic paper by Sommerfeld about front velocity is reprinted. It is also shown there that for $v_G < v_F$ only weak forerunners will propagate with front velocity, while the main part of the wave and thus the proper part of the signal propagates with group velocity. For general wave forms, justification of the steps taken above requires the Payley-Wiener theory described, e.g., in Dym and McKean (1972). The question of propagation of discontinuities may also be discussed in the language of configuration space alone, without the Fourier transform, if a dynamical model of the medium is at hand rather than its phenomenological description by the dispersion formula $\omega(k)$, as was pointed out to Sommerfeld in a letter from T. Levi-Cività (included in Brillouin's book). This approach is essentially the determination of the *characteristic hypersurfaces* of the pertinent wave equation; cf. Courant and Hilbert (1962). The dynamical model also shows that signal transmission is accompanied by transmission of energy and thus to energy currents. The latter will be considered in sect. 5.9—see exercise 3 of that section concerning subluminality.

Lack of distinction between phase velocity, group velocity, and front velocity has led to erroneous physical arguments again and again. As an example we mention the speed of sound in nuclear matter. Generally, $v_S^2 = dp/d\rho$ has to be calculated from the equation of state $p(\rho)$; approximate calculations at densities $\rho \approx 10^{15}$ g/cm^3 lead to equations of state which imply $v_S > 1$. Since v_S is a phase velocity, this result does not contradict our postulate. One cannot use the condition $v_S < 1$ directly as a restriction on possible equations of state of nuclear matter to retain only those that satisfy $dp/d\rho < 1$. Arguments of this type have been used frequently in neutron star models, where the equation of state of nuclear matter plays an important role. See the article of Ruffini in DeWitt (1973).

One should note also that the above argument against superluminal signal velocities, Fig. 2.6, is conclusive only under the assumption of free will. Without the latter—which in physics is always assumed to exist—no contradictions (as the one where the observer kills his mother before his conception) result from the possibility of signals running into the past if one suitably restricts the initial conditions and thus avoids certain contrived apparatus; see the discussions on related situations in Hawking and Ellis (1973), p. 189; Terletskii (1968), stressing thermodynamical aspects; H. Schmidt, Found. Phys. *8*, 463 (1978); A. Peres, ibid. *16*, 537 (1986).

The possibility or impossibility of particles moving superluminally (*'tachyons'*) was under frequent discussion for a while, beginning with an article by G. Feinberg (Phys. Rev. *159*, 1089 (1957)), who tried to solve the causality problems described above by reinterpreting the laws of propagation of tachyons (criticized, e.g., in F. Pirani, Phys. Rev. D *1*, 3224 (1970)). Apart from causality questions, it turns out to be impossible to construct a quantum field theory of localizable tachyons, since negative energies occur that upon interaction (e.g., observation) lead to instabilities (see G. Ecker, Ann. Phys. (N.Y.) *58*, 303 (1970)).

From a pro-tachyonic point of view, extensive studies were made by E. Recami, R. Mignani and collaborators. Here one finds even the idea of inertial systems with superluminal relative velocities and a discussion of possible experiments to detect tachyons. See E. Recami and W. A. Rodrigues in Weber and Karade (1985); E. Recami, Riv. Nuovo Cimento *9*, no. 6 (1986); Found. Phys. *17*, 239 (1987).

Worth reading in this connection is again Terletskii (1968), who investigates the problems of an information theoretic and thermodynamic nature relating to tachyons and particles with negative energies.

Questions of superluminal information transfer in connection with the famous quantum paradox of Einstein, Rosen and Podolski (EPR paradox) are discussed in Maudlin (1994) and references given there.

Exercise[1]

In Fig. 2.6, we assumed an infinite signal speed for simplicity. Find, for a given relative velocity v between the two observers, the greatest lower bound (infimum) for the speeds of superluminal signals allowing to influence the past. Invert the relation found. What happens when $v \to 1$?

---

[1]Supplied by P. C. Aichelburg.

2.4   Lorentz Contraction

An extended object—we shall consider in the following a unit measuring rod—is described, in a space-time diagram, by specifying the world lines of its atoms, as indicated in Fig. 2.7, or by specifying the surface of its *'world tube'*. The size of an object is determined by the positions of its atoms at time t, i.e., by the cross section of its world tube with the surface $t = const.$ Because of the relativity of simultaneity, this cross section, and therefore the extension of the object, will depend on the inertial system considered. Figure 2.7 shows a measuring rod with length $\Delta \bar{x} = 1$ in its rest system $\bar{\mathrm{I}}$. The figure shows clearly that the intersection of its world tube with $t = const.$ gives a length $\Delta x < 1$, i.e., the *moving rod is contracted (Lorentz-Fitzgerald contraction)*. This statement holds for the direction of motion, while there is no contraction in the direction orthogonal to it (y-direction in Fig. 2.7b).

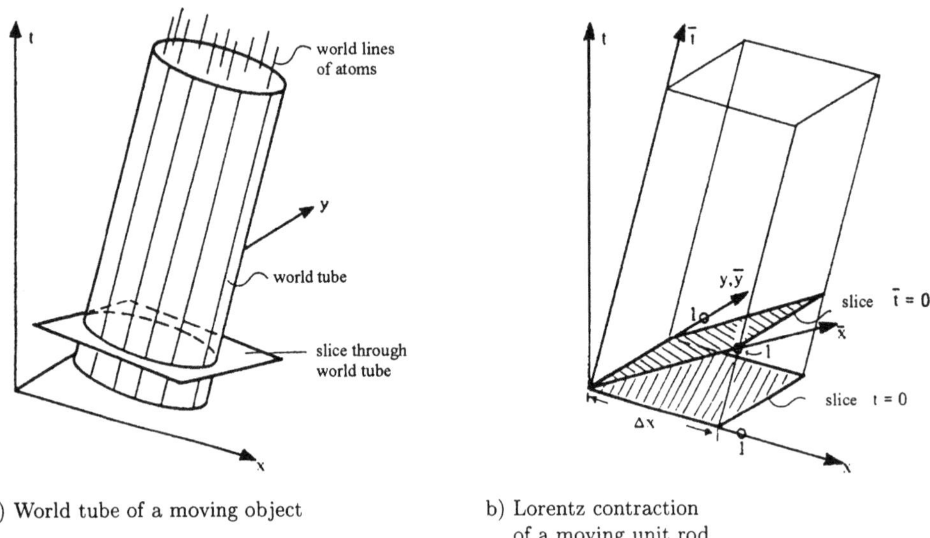

a) World tube of a moving object

b) Lorentz contraction
of a moving unit rod

Fig. 2.7. Lorentz Contraction

We obtain the numerical value of the Lorentz contraction immediately from eq. (2.1.1) by putting $t = 0$ there:

$$\bar{x} = \gamma\, x \quad (t = 0). \tag{2.4.1}$$

Since the rod's length is to be $\Delta \bar{x} = 1$ in its *rest system* $\bar{\mathrm{I}}$, its length in the system I becomes

$$\Delta x = \gamma^{-1} = \sqrt{1 - v^2}. \tag{2.4.2}$$

Observable consequences of the Lorentz contraction will be discussed in sect. 5.8; in the next section we discuss its 'invisibility'.

Exercises

1. Show from Fig. 2.7 and also from eq. (2.1.1) that Lorentz contraction is a reciprocal effect: a rod at rest in the (x, t)-system will be considered shorter by the same factor in the $(\bar{t}, \bar{x})$-system.

2. The length of a thin rod moving relative to I is going to be measured. For this, a number of flash lights go off simultaneously, and the shadow cast by the rod is registered on a photographic plate. Show that the Lorentz contraction is explained by a comoving observer by saying that the flashes do not go off simultaneously.

3. A man carrying horizontally a ladder 2.1 m in length runs into a room 1 m in length at speed $v/c = \sqrt{3}/2$ and closes the door behind him (observe the numbers!).

   a. How is this possible?

   b. How does the man describe the situation?

   c. What happens afterwards?

   d. Draw a number of sections $t = const.$ resp. $\bar{t} = const.$ to describe this story from both points of view.

(This problem was adapted from Rindler (1982), where one finds some more paradoxes involving length contraction. It in fact seems to be the first length contraction paradox in the history of Relativity, published 1960 in the first edition of Rindler's text.)

2.5 Retardation Effects: Invisibility of Length Contraction and Apparent Superluminal Speeds

Up to 1960 most physicists thought that the Lorentz contraction would show up in visual or photographic observation of fast objects. However, in 1959 they were made aware, independently by R. Penrose and J. Terrell, of the fact that the Lorentz contraction cannot be measured this way. Rather, these optical methods register *retarded positions*, where the time it takes for light to propagate from the object to the observer is taken into account—and not the *instantaneous positions* considered in the last section. Let us look at a simple example to illustrate this (Fig. 2.8).

A cube moves past a camera with speed v, and a snapshot is made. This then will involve all light that arrives simultaneously at the camera—and not the light emitted simultaneously from the cube. We investigate the consequences.

Since light from edge A has to overcome a larger distance than light from edge B, edge A will be registered corresponding to an earlier position, farther to the left (A'). We can calculate this effect easily if the camera is very far from the cube (as compared to the cube's extension), since then all path lengths for the light may be taken for parallel rays (the errors being of order $1 - \cos \delta \approx \delta^2/2 \approx 0$). Edge A being a distance ℓ farther away from the camera than B or C, light has to be emitted earlier by $\Delta t = \ell$ from there than from B or C to register at the same time. During this

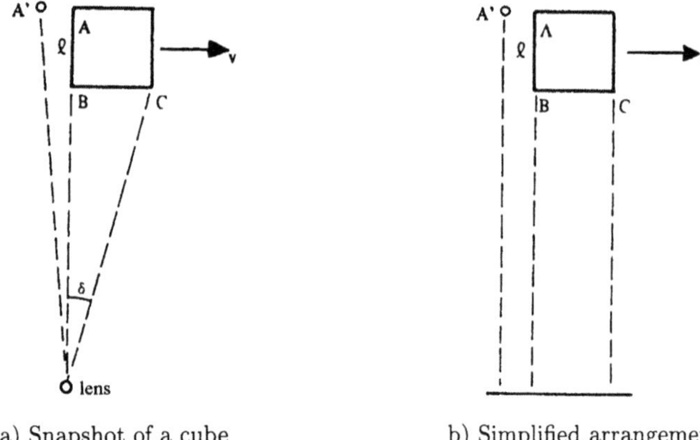

a) Snapshot of a cube   b) Simplified arrangement

Fig. 2.8. Photographing a cube

interval, the cube moves to the right by $\Delta x = v\Delta t = v\ell$, so that $AA' = v\ell$. Thus without Lorentz contraction, the image would look like Fig. 2.9a. Lorentz contraction reduces BC to $\ell \cdot \sqrt{1 - v^2}$, giving Fig. 2.9b, which is just the image of a cube of the same size at rest, which, however, has been *rotated* by the angle $\alpha = \arcsin v$.

a) Image without Lorentz contraction   b) Lorentz contraction *undistorts* the image

Fig. 2.9. Invisibility of Lorentz contraction

Although derived in a special case here, this result holds generally: In photographic images, moving objects far away do not appear contracted but rotated. We shall show this in connection with the aberration of light (which comes in when we take the point of view of the cube) in sect. 4.3.

Another retardation effect turns out to be important in the interpretation of astrophysical phenomena. Consider a spherical shell of gas (many light years in size) surrounding a central object (Fig. 2.10) that emits a flash of intensive radiation. This will cause the gaseous hull to shine for a moment, simultaneously at all its points in its rest system. What are the light phenomena observed by a terrestrial astronomer, if the object is receding from her at speed v (cosmic expansion)?

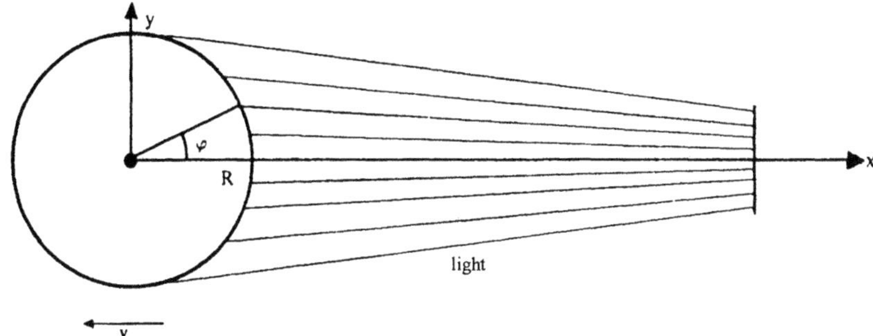

Fig. 2.10. Emission by a gaseous shell

Let us, for simplicity, restrict to two space dimensions and consider a gaseous ring—in some applications this is even more realistic—in the (x, y)-plane of the system I; let its center be at the origin at time $t = 0$, receding with speed v in x-direction from an observer at rest at $(x = D, y = 0)$. If the points of the ring, parametrized as $\bar{x} = R\cos\varphi$, $\bar{y} = R\sin\varphi$ in its comoving system $\bar{\text{I}}$, emit their flashes at comoving time $\bar{t} = 0$, then in the system I of the observer we have

$$
\begin{aligned}
t &= \gamma(\bar{t} - v\,\bar{x}) = -\gamma\,v\,R\cos\varphi \\
x &= \gamma(\bar{x} - v\,\bar{t}) = \gamma\,R\cos\varphi \\
y &= \bar{y} = R\sin\varphi.
\end{aligned}
\tag{2.5.1}
$$

Thus, as seen from I the ring will not shine simultaneously at all points! If it were possible to register the ring by making the (x, y)-plane into a giant photographic plate, its image would not be Lorentz contracted but rather dilated, as from eqs. (2.5.1) we get $y^2 + (x/\gamma)^2 = R^2$, which is the equation of an ellipse with its major semiaxis $\gamma R > R$ in x-direction. This shows that objects shining for a short moment only behave differently compared to objects in snapshots as investigated above.[1]

The emitted light propagates towards the observer; she will receive the flashes coming from the points $\varphi = \pm\varphi_1$ at a time t_1, where

$$
t_1 + \gamma\,v\,R\cos\varphi_1 = [(D - \gamma\,R\cos\varphi_1)^2 + R^2\sin^2\varphi_1]^{1/2} \approx D - \gamma\,R\cos\varphi_1
\tag{2.5.2}
$$

(in the situation considered we have $D \gg R$). Therefore, the observer sees, at time t_1, two shining points at a distance

$$
y_1 = R\sin\varphi_1 = \left[R^2 - (D - t_1)^2\frac{1 - v}{1 + v}\right]^{1/2} =: y_1(t_1)
\tag{2.5.3}
$$

from each other, moving apart with velocity $2dy_1/dt \gg 1$. The observer gets the impression of an object breaking into two parts which initially move apart at a multiple

---

[1]See, e.g., N. C. McGill, Contemp. Phys. *9*, 33 (1968).

of the speed of light, against all predictions of Relativity, then slow down and reverse their motion.

These considerations show how careful one has to be when interpreting optical data. Fake superluminal velocities may arise from retardation effects but may also have other causes without any masses or signals being transferred superluminally.

Radio astronomical observations show that components of the quasars 3C 279 and 3C 273 are moving apart at 6- resp. 8-fold speed of light. This discovery was a great surprise in 1971, and many theories were proposed to remove the apparent contradiction to Relativity. Some authors even considered Relativity to be ruled out by this discovery. The model discussed above is to illustrate that astronomical observations may be explained without exotic hypotheses. An overview about other models and observational facts is found in R. H. Sanders, Nature *248*, 390 (1974).

Exercise

Show that the debris of an exploding mass seem to move apart at superluminal speed if there is a sufficiently large velocity component towards the observer.

2.6   Proper Time and Time Dilation

We now come to interpret the line element ds introduced formally in sect. 1.5. Omitting again the x^2- and x^3-coordinates, we have

$$ds^2 = dt^2 - dx^2 = d\bar{t}^2 - d\bar{x}^2. \tag{2.6.1}$$

Consider the world line of an arbitrarily (but subluminally, as it can be used for signalling and may be at rest in some system) moving mass point (Fig. 2.11). According to sect. 2.3 it has to remain inside the light cone of each of its points. For each such point there is an inertial system $\bar{\mathrm{I}}$ that is instantaneously comoving with the mass, its *instantaneous rest system*. (With three space dimensions, this does not yet specify a *frame* since the directions of the spatial axes would remain unspecified, but this will not concern us here.[1]) The time axis of $\bar{\mathrm{I}}$ is parallel to the tangent of the world line at the point considered. In this rest system—which in general changes from point to point—we have, along the world line, $d\bar{x} = 0$, $ds = d\bar{t}$. Therefore, the line element measures, at each instant, the interval of time shown on a clock carried by the mass point, and is called the element of *proper time*. Since it takes the same value in every inertial system it is the invariant measure (under Poincaré transformations) of length for the world line in the same sense as in Euclidean geometry $d\sigma^2 = dx^2 + dy^2$ measures the length of a curve, invariant under Euclidean motions.

The difference in sign between $ds^2 = dt^2 - dx^2$ and $d\sigma^2 = dx^2 + dy^2$ makes 'arc lengths' of world lines between two fixed points shorter the longer the line looks in the space-time diagram, where 'looking' now refers to Euclidean eyes.

The instantaneous rest system of an accelerating mass point will change all the time; so we will specify the orbit in one single fixed inertial system I as usual by writing $x = x(t)$. Then the velocity with respect to I is $v = dx/dt$ and the proper

---

[1]For a discussion of possible choices see E. G. P. Rowe, Am. J. Phys. *64*, 1184 (1996).

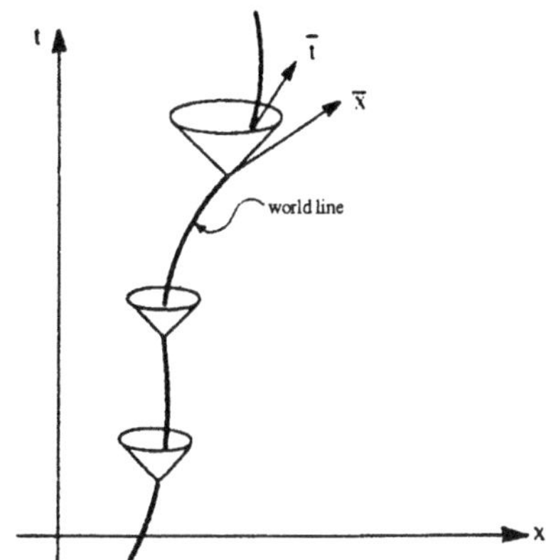

Fig. 2.11. Instantaneous rest system of a mass point

time along the world line is given by $ds^2 = dt^2 - dx^2 = dt^2(1 - v^2)$, so that the time shown by a moving clock is

$$ds = dt\sqrt{1 - v^2} < dt. \tag{2.6.2}$$

Therefore, *moving clocks go slow*. This effect, *time dilation*, may also be read off from the space-time diagram shown in Fig. 2.12. The figure shows two clocks, 1 and 2, at rest at the origins of I and $\bar{\text{I}}$, so that their world lines are just the t- and $\bar{t}$-axis, respectively.

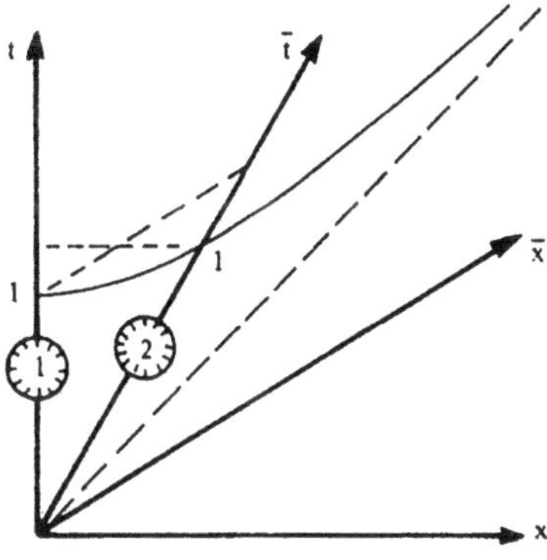

Fig. 2.12. Time dilation

The unit point on the $\bar{t}$-axis corresponds to the event 'clock 2 shows $\bar{t} = 1$'. This event obviously occurs at a time $t > 1$ in I, so that the moving clock goes slow as judged from I. But this is also the case for clock 1 as judged from $\bar{\text{I}}$, where we have, according to eq. (2.6.2),

$$\bar{t} = t\sqrt{1 - v^2} \quad \text{for clock 2, i.e., for } d\bar{x} = 0 \tag{2.6.3}$$

$$t = \bar{t}\sqrt{1 - v^2} \quad \text{for clock 1, i.e., for } dx = 0. \tag{2.6.4}$$

Time dilation is a reciprocal effect: judged from every inertial system, the clocks of any other go slow. Our figure shows that this result is due to the relativity of simultaneity.

When written as eqs. (2.6.3,4), time dilation appears paradoxical if these equations are misinterpreted as formulae for transforming t, $\bar{t}$. Of course this is not legal, the transformation connecting t, $\bar{t}$ is eq. (2.1.1), while eqs. (2.6.3,4) are relations between certain time intervals—not time coordinates—which are defined uniquely by Fig. 2.12. One may perhaps express this even more clearly by writing the latter as

$$\left.\frac{\partial \bar{t}}{\partial t}\right|_{\mathrm{X}} = \gamma, \qquad \left.\frac{\partial \bar{t}}{\partial t}\right|_{\bar{\mathrm{X}}} = \gamma^{-1}. \tag{2.6.5}$$

There is a vast literature on the misunderstanding just mentioned. In particular, in connection with the 'twin paradox'—to be discussed in the next section—numerous articles have appeared. The selected bibliography contained in Marder (1971) contains, e.g., 305 references. Of interest is H. Dingle, who in 1940 published a textbook on Relativity and who writes, in the foreword to its 1961 edition (Dingle 1961): "Since this book was written, reasons have appeared, which to me are conclusive for believing that the theory is no longer tenable". This clearly shows the problems that may be caused by an insufficient mathematical symbolism.

2.7   The Clock or Twin Paradox

The best-known version of the kind of problems alluded to by this title is the twin paradox illustrated in Fig. 2.13.

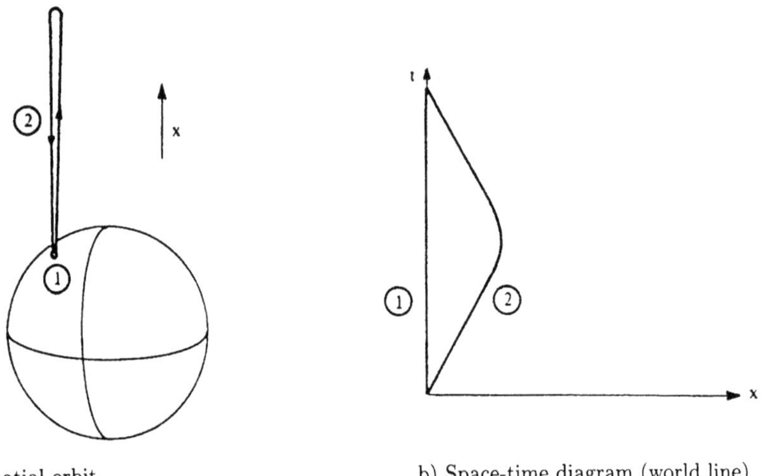

a) Spatial orbit   b) Space-time diagram (world line)

Fig. 2.13. The clock or twin paradox

Twin 1 stays on earth while twin 2 undertakes a space travel at a speed v close to the speed of light, eventually returning to earth. While on earth the time passed is T_1, the time passed for twin 2 should be only, according to eqs. (2.6.2,3),

$$T_2 = T_1\sqrt{1 - v^2}. \tag{2.7.1}$$

The moving twin thus should have aged less than her sister on earth. From the point of view of 2, however, things should be the other way round, since for her it is 1 who was moving all the time.

Let us translate the problem into one more accessible to physical analysis (Fig. 2.13b). Clock 1 is at rest in an inertial system; clock 2 first moves away from it uniformly and rectilinearly, then decelerates and reaccelerates back, finally meeting 1 again. Since 2 is moving at speed v all the time it should, upon return at time $t = T_1$ in the (x, t)-system, show the time $T_2 = T_1\sqrt{1 - v^2}$ only. The argument that leads to the paradox is that one may as well may take the point of view of 2, with respect to which 1 is moving at speed v all the time, so that the relation should be the reversed one, namely $T_1 = T_2\sqrt{1 - v^2}$.

To find the error in this argument we note first that 1 and 2 *by no means enter symmetrically* into the problem, as our space-time diagram Fig. 2.13b shows immediately. Clock 1 is at rest in the inertial system I while clock 2 gets accelerated; in the space-time diagram its world line is not a straight line.

One might think now that the difference between clocks 1 and 2 has to do with the acceleration of 2, in that one here has an influence of acceleration upon clock 2 rather than of velocity. In the next section we shall investigate this more closely; but let us anticipate here that the influence of acceleration may always be eliminated. Also, one may make the period of accelerated motion arbitrarily short as compared with the unaccelerated one, so that its influence as seen from I should be negligible.

Now let us analyze the paradoxical argument more accurately! We saw in the preceding section that world lines that look longer in the space-time diagram have shorter proper time. This shows at once that world line 2 in Fig. 2.13b has shorter proper time.[1] The counter-argument leading to the paradox is the following (Fig. 2.14). If we take the point of view of 2 and draw its world line as the straight line $\bar{x} = 0$, then the world line of 1 will appear curved (Fig. 2.14) and thus longer, corresponding to shorter proper time.[2]

The error in this argument lies in the fact that the coordinate system $(\bar{t}, \bar{x})$ is *curvilinear*, as shown in Fig. 2.13. The $\bar{t}$-coordinate line (i.e., world line 2) is obviously curved (which is a meaningful statement within the affine geometry of space-time diagrams!), corresponding to the fact that a reference system permanently attached to clock 2 is accelerated rather than inertial. It is of course admissible to make use of such a system, just as it is admissible to use curvilinear coordinates—like polar coordinates—in Euclidean geometry. However, just as there, all formulas have to be rewritten to become valid in curvilinear coordinates (noninertial, i.e., accelerated reference systems). So it is perfectly legal to plot, e.g., plane polar coordinates

---

[1] Path 2 looks longer *because* it contains a curved part; however, while this *enables* greater length, the essential contributions to its length nevertheless come from its straight, i.e., unaccelerated parts!

[2] J. Crampin, W. McCrea, D. McNally, Proc. R. Soc. Lond. Ser. A **252**, 156 (1959) give diagrams drawn to scale for some concrete cases.

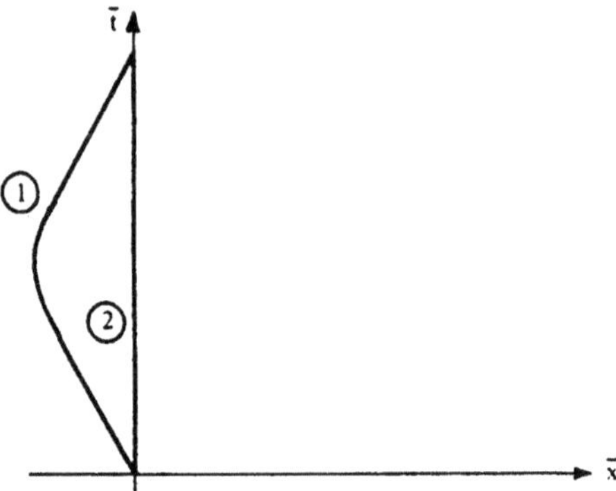

Fig. 2.14. The twin paradox as seen by 2. In this diagram a coordinate system
$(\bar{t}, \bar{x})$ was chosen in which clock 2 is at rest at the origin $\bar{x} = 0$

like Cartesian ones (Fig. 2.15); but the Euclidean distance between (infinitesimally neighboring) points is then not simply given by the formula $d\sigma^2 = dr^2 + d\varphi^2$ but rather has to be calculated from $d\sigma^2 = dr^2 + r^2 d\varphi^2$, which in turn is obtained from the Cartesian version $d\sigma^2 = dx^2 + dy^2$ by the well-known transformation, which is a manifestly nonaffine one. With the clock paradox, we have a completely analogous situation. In the inertial frame I the line element is given by $ds^2 = dt^2 - dx^2$, and this looks alike in all inertial systems; however, when curvilinear coordinates $(\bar{t}, \bar{x})$ are introduced, $ds^2 = d\bar{t}^2 - d\bar{x}^2$ is not valid any more. From a figure like Fig. 2.14 no conclusion may be drawn on proper times. So the error in the argument that puts 1 and 2 on equal footing is to use the formula $ds^2 = dt^2 - dx^2$, which is wrong in accelerated frames and only valid in inertial ones.

It is of course possible to perform the transformation to some accelerated reference

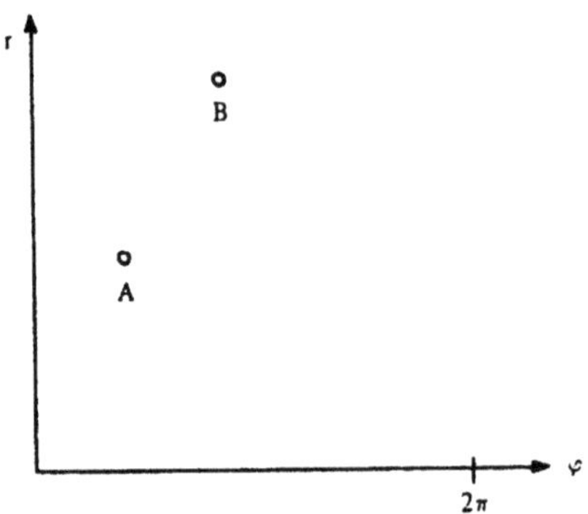

Fig. 2.15. Concerning curvilinear coordinates

system and obtain the correct form of ds^2 there. The general mathematical techniques of handling general coordinate transformations are usually developed in the wider framework of General Relativity; one then can of course show that the result is eq. (2.7.1).

The fact that these techniques in most cases appear only when it comes to General Relativity should not lead one to the erroneous idea that the clock paradox has to do with that theory. It is of course possible to rewrite Special Relativity in accelerated reference systems; it is unnecessary, however, to introduce these slightly more complicated techniques at this stage since we can always refer to an inertial system and get all results more easily.

A very accurate *measurement* of special-relativistic time dilation was achieved in 1968 at CERN in connection with measurements on elementary particles; the dilation factor was $\gamma = 12.1$, the accuracy was 2% (cf. F. M. Farley et al., Nature *217*, 17 (1968)). There were also suggestions to measure the clock effect using atomic clocks (cesium clocks) in earth satellites. While the preparations for this experiment progressed only slowly, the accuracy of Cs-clocks was increased drastically, so that it became possible to measure the effect on ordinary airline flights, as demonstrated by J. Hafele and R. Keating in 1971 with 10% accuracy (cf. Science *177*, 166, 168 (1972); Sexl and Sexl 1975). However, in this experiment also effects of gravitation play a role, which cannot be treated here.

2.8 On the Influence of Acceleration upon Clocks

In the last section we saw that there are effects of velocity on clock rates. We might ask ourselves whether there are similar effects of acceleration, so that eq. (2.6.2) would have to be changed into, say, $ds = \sqrt{1-v^2}\sqrt{1+ba^2}$, where b is a constant and a is the acceleration of the clock. (This kind of an a-dependence was taken as an arbitrary example!) Differentially, such a dependence would mean that the clock rate also depends on the clock's prehistory, i.e., the manner how it reached its state of motion. Without infinitesimals, consider two clocks which are first at rest in the inertial system I, showing the same time and going at the same rate. Assume now that they perform completely different accelerated motions but finally come to rest at the same site in the inertial system $\bar{\text{I}}$ having speed v relative to I. As a slight generalization of the preceding results, we expect them to show different times when they meet (*first clock effect*). The question we are asking here is whether they are now running at different *rates* (*second clock effect*) although there is no relative velocity between them any more.

An accelerated clock is under the influence of *forces*; the resulting changes in clock rate will depend on the type of clock and the type of forces. (E.g., if the forces are of magnetic kind and the material of the clock is magnetizable, the clock might just stop ticking.) In order that the change in clock rate be negligible one has to require that the inner forces of the clock are much stronger than the exterior forces accelerating it. This requirement is not as trivial as it might appear on first sight. In the last section we mentioned the CERN experiments, where μ-mesons circulating in an accelerator were used, whose lifetime was increased by their motion. In order that the formulae of Relativity be applicable it is necessary that mesons are good clocks in the above sense, i.e., the forces responsible for the meson decay must be much stronger than the magnetic forces in the accelerator. However, since atomic, or nuclear, and, much more so, fields inside elementary particles are always much stronger than artificially generated macroscopic fields, mesons are excellent clocks in this respect: effects from acceleration are to be expected to be much smaller than the relativistic velocity effect.

This consideration shows that mesons are already good clocks that suffice for all practical purposes. But we now argue that in the framework of Special Relativity one can, in principle, construct ideal clocks without acceleration effects. One takes a good clock in the above sense and combines it with an apparatus measuring accelerations (Fig. 2.16) (as we shall show also formally, accelerations have absolute significance in Relativity!), using the result of this measurement for correcting the clock rate. An ideal clock obtained in this manner will show the proper time $\int ds$ in arbitrarily accelerated motion.

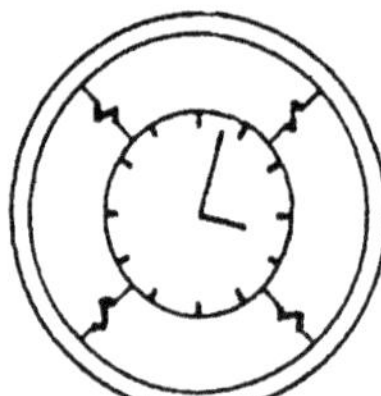

Fig. 2.16. Clock combined with an accelerometer

The considerations just made demonstrate only the consistency of the theory with the assumptions made in chap. 1, however. There we had omitted the possibility of transformations of time and length scales, granting the existence of clocks and measuring rods which are insensitive to acceleration. This then allowed us to take the scalar coefficients in eq. (1.3.3) as depending on v alone. In the resulting kinematics, governed by the Lorentz transformation, acceleration has absolute significance, and this we have just used. The empirical fact that there exist almost ideal clocks without second clock effect, thus showing proper time, is therefore taken as an explicit postulate C in addition to postulates A and B of sect. 1.1 by authors proceeding more axiomatically than was done here.

2.9   Addition of Velocities

Consider a point mass moving with velocity $\bar{\mathbf{w}}$ with respect to the system $\bar{\mathrm{I}}$. What is its velocity with respect to a system I if the former has velocity $\mathbf{v}$ against the latter?

To answer this question we take the inverse of eq. (1.4.4),

$$\mathbf{x} = \bar{\mathbf{x}} + \frac{\gamma^2}{\gamma + 1}(\bar{\mathbf{x}}\,\mathbf{v})\,\mathbf{v} + \gamma\,\mathbf{v}\,\bar{t}$$
$$t = \gamma\,\bar{t} + \gamma\,(\mathbf{v}\,\bar{\mathbf{x}}) \tag{2.9.1}$$

and insert $\bar{\mathbf{x}} = \bar{\mathbf{w}}\,\bar{t}$ to obtain for the ratio $\mathbf{u} = \mathbf{x}/t$:

$$\mathbf{u} = \frac{\frac{\bar{\mathbf{w}}}{\gamma} + \frac{\gamma}{\gamma+1}(\mathbf{v}\,\bar{\mathbf{w}})\,\mathbf{v} + \mathbf{v}}{1 + \mathbf{v}\,\bar{\mathbf{w}}} \equiv \frac{\mathbf{v} + \mathbf{w}_{\|} + (\mathbf{w}_{\perp}/\gamma)}{1 + \mathbf{v}\,\bar{\mathbf{w}}}, \tag{2.9.2}$$

where in the second version we have introduced components parallel and perpendicular to the relative velocity between the inertial systems.

In considering the possibility that the velocities are not proportional we enter the domain where all three space dimensions become important. We already handled this situation before, using ordinary Euclidean vector algebra. But now some warnings will be appropriate. If one wants to interpret the formal scalar product $\mathbf{v}\,\bar{\mathbf{w}}$ geometrically, one has to observe that the component triple $\mathbf{v}$ resp. $\bar{\mathbf{w}}$ refers to I resp. $\bar{\text{I}}$ so that due to the relativity of simultaneity it is meaningless to talk about the angle between them. However, we have $\mathbf{v}\,\bar{\mathbf{w}} = -(-\mathbf{v})\,\bar{\mathbf{w}}$, and $\bar{\mathbf{v}} = -\mathbf{v}$ are the components of the velocity of I against $\bar{\text{I}}$ (by reciprocity), so that the product may be related to the angle between $\bar{\mathbf{v}}$ and $\bar{\mathbf{w}}$ in $\bar{\text{I}}$. Despite this possibility the vector addition appearing in eq. (2.9.2) remains formal, and one has to be extremely careful in applications to make sure which are the reference frames formal vectors (i.e., component triples) are referring to. Otherwise one might run into paradoxes like the following. 'According to reciprocity, $\bar{\text{I}}$ has velocity $\bar{\bar{\mathbf{w}}} = -\bar{\mathbf{w}}$ against the rest system $\bar{\bar{\text{I}}}$ of the mass point, I has velocity $\bar{\bar{\mathbf{u}}} = -\mathbf{u}$ against $\bar{\text{I}}$, so $-\mathbf{u}$ should result when $-\bar{\mathbf{w}}$ replaces $\mathbf{v}$ and $-\mathbf{v} = \bar{\mathbf{v}}$ replaces $\bar{\mathbf{w}}$ on the right of eq. (2.9.2). Cancelling a minus sign one concludes that the right-hand side of eq. (2.9.2) should be symmetric in $\mathbf{v}, \bar{\mathbf{w}}$, which (excepting $\mathbf{v} \times \bar{\mathbf{w}} = \mathbf{0}$) is obviously not the case—contradiction!' In fact, the velocity addition formula (2.9.2) is neither commutative nor associative. The resolution of the resulting paradoxes will come from the considerations of the next section. A characteristic feature is that some higher geometry (Lobachevski space) would have to be introduced into the simple space-time diagrams for them to continue to be as helpful as before. (See the appendix to sect. 4.1 for an indication of this.)

We mention two special cases of this *general velocity addition theorem.*

1. If $\bar{\mathbf{w}}$ and $\mathbf{v}$ are proportional, we get

$$\mathbf{u} = \frac{\mathbf{v} + \bar{\mathbf{w}}}{1 + \mathbf{v}\,\bar{\mathbf{w}}}. \qquad (2.9.3)$$

This is the special case considered usually.

2. For a mass point moving orthogonally to the relative velocity $\mathbf{v}$ of the inertial systems we have $\mathbf{v}\,\bar{\mathbf{w}} = 0$, and therefore

$$\mathbf{u} = \mathbf{v} + \bar{\mathbf{w}}/\gamma = \mathbf{v} + \bar{\mathbf{w}}\sqrt{1 - v^2}. \qquad (2.9.4)$$

Motion orthogonal to the direction of relative velocity is slowed down as a pure consequence of time dilation, since spatial distances orthogonal to $\mathbf{v}$ are equal in both systems.

For the square of $\mathbf{u}$ we can verify (exercise) that

$$\mathbf{u}^2 = 1 - \frac{(1 - \bar{\mathbf{w}}^2)(1 - \mathbf{v}^2)}{(1 + \mathbf{v}\,\bar{\mathbf{w}})^2} \leq 1. \qquad (2.9.5)$$

$\mathbf{u}^2 = 1$ results only for $\bar{\mathbf{w}}^2 = 1$ or $\mathbf{v}^2 = 1$, the latter case being actually forbidden due to our interpretation of $\mathbf{v}$ as a velocity between inertial frames. Thus, eq. (2.9.5) expresses the Lorentz invariance of the speed of light and of the domains of subluminal resp. superluminal speeds. The information about the angular relations contained in eq. (2.9.2) is particularly simple when expressed in terms of the tangents of the angles θ, $\bar{\theta}$ that the particle's relative velocity vectors $\mathbf{u}$, $\bar{\mathbf{w}}$ make with the direction $\mathbf{v}$ of relative motion of the inertial systems I, $\bar{\text{I}}$: one has

$$\tan\theta = \frac{1}{\gamma}\frac{\sin\bar{\theta}}{\cos\bar{\theta} + (v/\bar{w})}, \qquad (2.9.6)$$

as may also be verified as an exercise.

Lorentz (1909; Note 86) was not able to prove the Lorentz invariance of Maxwell's equations—and thus the validity of the principle of relativity—exactly for the reason that he had obtained a wrong version of the velocity addition theorem from his considerations.

Exercises

1. Verify eqs. (2.9.5,6)!

2. Formulate a paradox corresponding to the nonassociativity of eq. (2.9.2)!

3. Show that one gets the addition theorem for the phase velocity v_{ph} of plane waves if one replaces $\mathbf{u}$, $\bar{\mathbf{w}}$, $\tilde{w}$ in this section by $\mathbf{n}/v_{ph}$, $\bar{\mathbf{n}}/\bar{v}_{ph}$, $1/\bar{v}_{ph}$, respectively, where $\mathbf{n}$, $\bar{\mathbf{n}}$ are the unit wave normals and v_{ph}, $\bar{v}_{ph}$ are the phase velocities relative to I, $\bar{\text{I}}$.

   Hint: Proceed as indicated in the last paragraph of sect. 1.4.

2.10   Thomas Precession

We now replace the mass point of last section by an inertial frame $\bar{\bar{\text{I}}}$ obtained from $\bar{\text{I}}$ by boosting with $\bar{\mathbf{w}}$. Then $\bar{\bar{\text{I}}}$ has, against I, the relative velocity $\mathbf{u}$ given by eq. (2.9.2), but surprisingly it is *not* obtained from I by boosting except in the special case 1 mentioned in sect. 2.9. Rather, we have $\bar{\bar{x}} = L_{\bar{\mathbf{w}}}\,\bar{x} = L_{\bar{\mathbf{w}}}\,L_{\mathbf{v}}\,x$, where

$$L_{\mathbf{v}} = \begin{pmatrix} \gamma_v & -\gamma_v\,\bar{\mathbf{v}}^{\mathsf{T}} \\[2mm] -\gamma_v\,\mathbf{v} & 1 + \dfrac{\gamma_v^2}{1+\gamma_v}\,\mathbf{v}\,\mathbf{v}^{\mathsf{T}} \end{pmatrix}, \qquad L_{\bar{\mathbf{w}}} = \begin{pmatrix} \gamma_{\tilde{w}} & -\gamma_{\tilde{w}}\,\bar{\mathbf{w}}^{\mathsf{T}} \\[2mm] -\gamma_{\tilde{w}}\,\bar{\mathbf{w}} & 1 + \dfrac{\gamma_{\tilde{w}}^2}{1+\gamma_{\tilde{w}}}\,\bar{\mathbf{w}}\,\bar{\mathbf{w}}^{\mathsf{T}} \end{pmatrix},$$

$$\gamma_v := \dfrac{1}{\sqrt{1-\mathbf{v}^2}}, \qquad\qquad \gamma_{\tilde{w}} := \dfrac{1}{\sqrt{1-\bar{\mathbf{w}}^2}}, \tag{2.10.1}$$

hence by matrix multiplication

$$L := L_{\bar{\mathbf{w}}}\,L_{\mathbf{v}} = \begin{pmatrix} \gamma & -\mathbf{a}^{\mathsf{T}} \\ -\mathbf{b} & M \end{pmatrix}, \tag{2.10.2}$$

where

$$\gamma = \gamma(\mathbf{v},\bar{\mathbf{w}}) := \gamma_v\,\gamma_{\tilde{w}}\,(1 + \mathbf{v}\,\bar{\mathbf{w}}) \equiv \gamma(\bar{\mathbf{w}},\mathbf{v}),$$

$$\mathbf{a} = \gamma(\mathbf{v},\bar{\mathbf{w}})\,\bar{\mathbf{w}} \circ \mathbf{v}, \qquad \mathbf{b} = \gamma(\bar{\mathbf{w}},\mathbf{v})\,\mathbf{v} \circ \bar{\mathbf{w}}, \tag{2.10.3}$$

$$M = M(\bar{\mathbf{w}},\mathbf{v}) :=$$

$$:= 1 + \dfrac{\gamma_v^2}{1+\gamma_v}\,\mathbf{v}\,\mathbf{v}^{\mathsf{T}} + \dfrac{\gamma_{\tilde{w}}^2}{1+\gamma_{\tilde{w}}}\,\bar{\mathbf{w}}\,\bar{\mathbf{w}}^{\mathsf{T}} + \gamma_v\,\gamma_{\tilde{w}}\left(1 + \dfrac{\gamma_v\,\gamma_{\tilde{w}}}{(1+\gamma_v)(1+\gamma_{\tilde{w}})}\,\mathbf{v}\,\bar{\mathbf{w}}\right)\bar{\mathbf{w}}\,\mathbf{v}^{\mathsf{T}}.$$

Here

$$\bar{\mathbf{w}} \circ \mathbf{v} := \left(\gamma_{\tilde{w}}\,\gamma_v\,\mathbf{v} + \gamma_{\tilde{w}}\,\bar{\mathbf{w}} + \gamma_{\tilde{w}}\,\dfrac{\gamma_v^2}{1+\gamma_v}\,(\bar{\mathbf{w}}\,\mathbf{v})\,\mathbf{v}\right) / \gamma(\mathbf{v},\bar{\mathbf{w}}) \tag{2.10.4}$$

is the velocity sum $\mathbf{u}$ of eq. (2.9.2). The first of eqs. (1.5.5) now verifies the claimed eq. (2.9.5), i.e.,

$$\gamma_u = \gamma(\mathbf{v}, \bar{\mathbf{w}}). \tag{2.10.5}$$

However, for $\mathbf{v} \times \bar{\mathbf{w}} \neq \mathbf{0}$ the matrix (2.10.2) is not symmetric as would be necessary for a boost. According to eq. (1.5.13) we can split L into a product $L_\mathrm{R}\,L_\mathbf{u} = L_\mathrm{Ru}\,L_\mathrm{R}$, where

$$\mathrm{R} = \mathrm{R}(\bar{\mathbf{w}}, \mathbf{v}) := \mathrm{M}(\bar{\mathbf{w}}, \mathbf{v}) - \frac{\mathbf{b}\,\mathbf{a}^\top}{1+\gamma} \tag{2.10.6}$$

is the *Thomas rotation* associated with $\mathbf{v}$, $\bar{\mathbf{w}}$. (We can see that R is proper-orthogonal either from the multiplicative property of determinants and the fact that all boosts have determinant one, or from $\det R = \pm 1$ for all orthogonal R together with the continuous dependence of R on the velocities and $\mathrm{R}(\mathbf{0}, \mathbf{0}) = \mathbf{1}$.) From the definitions of M, $\mathbf{a}$, $\mathbf{b}$ we can see that $\mathbf{v} \times \bar{\mathbf{w}}$ is an eigenvector of R for the eigenvalue 1 and thus gives the axis of rotation. The rotation angle α as calculated from $\mathrm{Tr}\,\mathrm{R} = 1 + 2\cos\alpha$ looks messy, and it is only after some tedious manipulations[1] that one arrives at the symmetric expression (McFarlane, J. Math. Phys. *3*, 1116 (1962))

$$1 + \cos\alpha = \frac{(1 + \gamma_u + \gamma_v + \gamma_{\bar{w}})^2}{(1+\gamma_u)(1+\gamma_v)(1+\gamma_{\bar{w}})} > 0. \tag{2.10.7}$$

To interpret these formulae one again has to observe that the components $\mathbf{v}$ and $\bar{\mathbf{w}}$ refer to different reference frames, so that, in analogy to what has been said in sect. 2.9 about the scalar product $\mathbf{v}\,\bar{\mathbf{w}}$, the formal vector product $\mathbf{v} \times \bar{\mathbf{w}}$ has to be suitably rewritten before geometric interpretation. Thus, to interpret it as an axis in I—corresponding to the splitting $L = L_\mathrm{Ru}\,L_\mathrm{R}$—we observe that by the definition of $\mathbf{u} = \bar{\mathbf{w}} \circ \mathbf{v}$ in eq. (2.10.4) we have

$$\mathbf{v} \times \mathbf{u} = \frac{\mathbf{v} \times \bar{\mathbf{w}}}{\gamma_v\,(1 + \mathbf{v}\,\bar{\mathbf{w}})}; \tag{2.10.8}$$

this means that the Thomas rotation of I has its axis orthogonal to the relative velocity vectors $\mathbf{v}$, $\mathbf{u}$ of $\bar{\mathrm{I}}$, $\bar{\bar{\mathrm{I}}}$ against I.

On the other hand, if it is to interpreted as an axis in $\bar{\bar{\mathrm{I}}}$, corresponding to the split $L = L_\mathrm{R}\,L_\mathbf{u}$, we observe that $L = L_\mathrm{Ru}\,L_\mathrm{R}$ says the following. $\bar{\bar{\mathrm{I}}}$ obtains from boosting a frame I' which by itself arises from rotating I by R. Therefore $\bar{\bar{\mathrm{I}}}$ has, against I or I', a velocity whose components in I' are Ru. By reciprocity, the components of the velocity of I or I' against $\bar{\bar{\mathrm{I}}}$ are given, in $\bar{\bar{\mathrm{I}}}$, by $\bar{\bar{\mathbf{u}}} = -\mathrm{Ru}$. One sees from the formula for R given above that Ru differs from $\mathbf{u}$ only by linear combinations of $\mathbf{v}$ and $\bar{\mathbf{w}}$; thus being itself of this type. $\bar{\bar{\mathbf{w}}} = -\bar{\mathbf{w}}$ is the velocity of $\bar{\mathrm{I}}$ against $\bar{\bar{\mathrm{I}}}$, so we have

$$\bar{\bar{\mathbf{u}}} \times \bar{\bar{\mathbf{w}}} = (-\mathrm{Ru}) \times (-\bar{\mathbf{w}}) \propto \mathbf{v} \times \bar{\mathbf{w}}, \tag{2.10.9}$$

i.e., the axis for the Thomas rotation of $\bar{\bar{\mathrm{I}}}$ is orthogonal to the relative velocities $\bar{\bar{\mathbf{u}}}$, $\bar{\bar{\mathbf{w}}}$ of I, $\bar{\mathrm{I}}$ against $\bar{\bar{\mathrm{I}}}$.

---

[1] A short derivation using four-vectors and Clifford algebra is given in H. K. Urbantke, Am. J. Phys. *58*, 747 (1990), *59*, 1150 (1991).

The insight that $\bar{\bar{\mathbf{u}}} = -\mathrm{R}\mathbf{u}$ and *not* $\bar{\bar{\mathbf{u}}} = -\mathbf{u}$ solves the paradox formulated in the last section (exercise), and an analogous but slightly more complicated analysis solves the associativity paradox mentioned, as was shown by A. A. Ungar (Found. Phys. *19*, 1385 (1989)—but beware of different conventions!).

To find the sense of rotation it suffices, by continuity, to restrict to the case where $\bar{\mathbf{w}}$ is small so that squares of it may be neglected. Then R becomes

$$\mathrm{R} \approx 1 + \frac{\gamma_v}{1 + \gamma_v}\left(\bar{\mathbf{w}}\,\mathbf{v}^T - \mathbf{v}\,\bar{\mathbf{w}}^T\right), \tag{2.10.10}$$

which is the matrix of a small rotation with rotation vector

$$\boldsymbol{\alpha} \approx -\frac{\gamma_v}{1 + \gamma_v}\,\mathbf{v} \times \bar{\mathbf{w}} \approx -\frac{\gamma_v^2}{1 + \gamma_v}\,\mathbf{v} \times \mathbf{u}, \tag{2.10.11}$$

as one easily sees by comparing with eq. (1.3.1,2) ($\cos\alpha \approx 1$, $\sin\alpha \approx \alpha$). The sense of rotation is therefore from the 'new' velocity $\mathbf{u}$ towards the 'old' one, $\mathbf{v}$. The angle of rotation never reaches $180°$, as eq. (2.10.7) shows.

Let us now consider the following situation. Imagine a system S in accelerated motion relative to the inertial system I, the spatial axes of S remaining parallel all the time in the sense that the instantaneous rest systems coinciding with S at times t and $t + \Delta t$ are related by a pure boost in the limit $\Delta t \to 0$. This may be achieved by orienting S with the help of rapidly spinning torque-free gyroscopes. According to the above, as judged from I, the system S seems to be rotated at each instant, and since the velocity of S varies continuously, there is a continuous rotation of S against I. This precession of the gyroscopes of S relative to I is called *Thomas precession*. We now determine its angular velocity vector.

During the small interval of time Δt (measured in I), the instantaneous velocity $\mathbf{v}(t)$ of S against I changes by $\Delta\mathbf{v}$, as measured in I; therefore eq. (2.10.11) gives the expression $\Delta\boldsymbol{\alpha} = -\gamma_v^2\,\mathbf{v} \times \Delta\mathbf{v}/(1 + \gamma_v)$ for the rotation vector during Δt, so that the angular velocity vector becomes

$$\boldsymbol{\omega}_T = -\frac{\gamma_v^2}{1 + \gamma_v}\,\mathbf{v} \times \frac{d\mathbf{v}}{dt}. \tag{2.10.12}$$

This special-relativistic precession effect had been used by Thomas to remove a discrepancy in the non-relativistic theory of the spinning electron. The gyromagnetic ratio of the electron as determined from the anomalous Zeeman effect had led to wrong theoretical values in the fine structure splittings. The Thomas precession yields a correction term to the equation of motion of the spin in an external electromagnetic field and thus a correction of the spin-orbit coupling which gives correct fine structure.[1] Within the relativistic quantum theory found later by Dirac this effect was automatic.

---

[1] L. H. Thomas, Nature *117*, 514 (1926); Philos. Mag. *3*, 1 (1927); see, in particular, W. H. Furry, Am. J. Phys. *23*, 517 (1955); for a critical discussion of the derivation see H. Bacry, Ann. Phys. (Paris) *8*, 197 (1963); N. Davidovich (Univ. Bariloche 1974, unpublished); N. Davidovich, G. Beck, Nuovo Cimento B *27*, 19 (1957); H. Mathur, Phys. Rev. Lett. *67*, 3325 (1991).

The first few sentences of Thomas' paper are of historical interest:
"It seems that Abraham [1903(!)] was the first to consider in any detail an electron with an axis. Many have since considered spinning electrons, ring electrons, and the like. Compton [1921] in particular suggested a quantized spin for the electron. It remained for Uhlenbeck and Goudsmit [1925] to show how this idea can be used to explain the anomalous Zeeman effect. The assumptions they had to make seemed to lead to optical and relativity doublet separations twice as large as those observed.

The purpose of the following paper, which contains the results mentioned in my recent letter to ‚Nature' [1926], is to investigate the kinematics of an electron with an axis on the basis of the restricted theory of relativity. The main fact used is that the combination of two ‚Lorentz transformations without rotation' in general is not of the same form."

From the historical point of view it should also be remarked that the precession effect was known by the end of 1912 to the mathematician E. Borel (C. R. Acad.Sci. *156*, 215 (1913)); it was described by him (Borel 1914) as well as by L. Silberstein (1914) in textbooks already in 1914. It seems that the effect was know to A. Sommerfeld in 1909 and before him to H. Poincaré. The importance of Thomas' work was thus not only the rediscovery of the effect but the *relevant* application to a virulent problem.

Exercises

1. For uniform circular motion, calculate the period of precession in the non-relativistic limit.

2. Will the initial orientation of a gyroscope be reached again if the accelerated motion is periodic?

3. Deduce eq. (2.10.7) from eq. (2.10.6)!

4. From $L_{\bar{\mathbf{w}}} L_{\mathbf{v}} = L_{\mathrm{R}(\bar{\mathbf{w}},\mathbf{v})} L_{\bar{\mathbf{w}} \circ \mathbf{v}} = L_{\mathrm{R}(\bar{\mathbf{w}},\mathbf{v})(\bar{\mathbf{w}} \circ \mathbf{v})} L_{\mathrm{R}(\bar{\mathbf{w}},\mathbf{v})}$, by taking transposes, deduce the relations

$$\mathrm{R}(\mathbf{v}, \bar{\mathbf{w}}) = \mathrm{R}^{-1}(\bar{\mathbf{w}}, \mathbf{v}), \tag{2.10.13}$$

$$\mathbf{v} \circ \bar{\mathbf{w}} = \mathrm{R}(\bar{\mathbf{w}}, \mathbf{v})\,(\bar{\mathbf{w}} \circ \mathbf{v}), \tag{2.10.14}$$

   whose direct verification from the definitions would be very tedious but possible (one could use symbolic computation!).

5. For any orthogonal S, show that $\gamma(\mathrm{S}\bar{\mathbf{w}}, \mathrm{S}\mathbf{v}) = \gamma(\bar{\mathbf{w}}, \mathbf{v})$ and

$$\mathrm{S}\bar{\mathbf{w}} \circ \mathrm{S}\mathbf{v} = \mathrm{S}(\bar{\mathbf{w}} \circ \mathbf{v}), \quad \mathrm{R}(\mathrm{S}\bar{\mathbf{w}}, \mathrm{S}\mathbf{v}) = \mathrm{S}\,\mathrm{R}(\bar{\mathbf{w}}, \mathbf{v})\,\mathrm{S}^{-1}. \tag{2.10.15}$$

6. Show that not every Lorentz transformation may be written as the product of two boosts!
   Hint: In eq. (1.5.13), R and $\mathbf{v}$ are independent.

2.11 On Clock Synchronization

From the point of view of space-time diagrams, the decisive difference between Einsteinian and Galilean Relativity is in the determination of the unit points and in the rotation of the x-axis. We want to analyze the latter in more detail in this section.

The equation $\bar{t} = \gamma(t - v\,x)$ tells us that for $t = 0$ clocks in the moving system $\bar{\text{I}}$ will have *pointer position* $\bar{t} = -\gamma\,v\,x$. This may be explained by the synchronization procedure used: in *each* inertial system, clocks at different locations in space are brought to the same pointer position—i.e., are *synchronized*—such that signals emitted at system time zero from two locations arrive simultaneously at an observer midway between them. (One can use light signals, sound signals, ...; see Fig. 2.17. However, if, e.g., sound is used, the gas in which it is propagating has to be at rest (on the average) in the system to be synchronized, i.e., has to be carried along with it.)

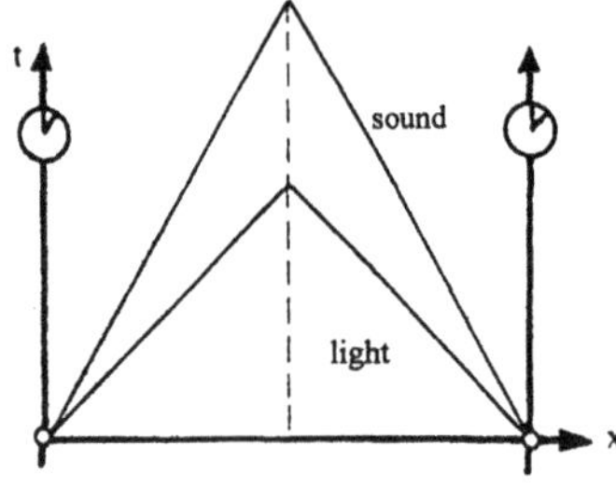

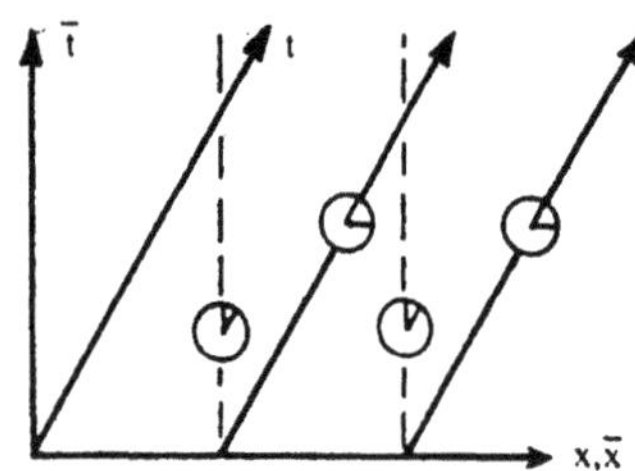

a) Einstein synchronization   b) External method

Fig. 2.17. Synchronization methods

This method of synchronization—the *Einstein synchronization*—may be characterized as being *internal* to the system, since it may be carried out within each system without reference to any other one. It therefore does not distinguish any specific system. Another method of this kind, not using signals, would just be slow transport of a standard clock from one location to the others. The essence of these internal methods lies in the fact that the symmetry between inertial systems is not destroyed if such a symmetry is enabled by the laws of nature. In our formulation of the Principle of Relativity an internal synchronization method was implicitly assumed, otherwise it would not be guaranteed that all inertial frames are on the same footing.

However, one *can* synchronize differently, other methods corresponding to a substitution of the kind $t \to t + f(x)$: this substitution just means that the pointer position zero on a clock located at x in I has been changed by the amount $f(x)$ as compared to the internal method. (The function f may vary from one system to the other.)

Writing the Lorentz transformation—we here interchange the usual I and $\bar{\text{I}}$—as

$$x = \gamma(\bar{x} - v\,\bar{t})$$
$$t = \frac{\bar{t}}{\gamma} - v\,x \tag{2.11.1}$$

we see that we can, by taking $f(x) = -v\,x$, change synchronization in I in such a way that eq. (2.11.1) becomes

$$x = \gamma\,(\bar{x} - v\,\dot{\bar{t}}) \tag{2.11.2a}$$

$$t = \bar{t}/\gamma = \bar{t}\sqrt{1 - v^2}. \tag{2.11.2b}$$

By this choice *one system*, e.g., $(\bar{t}, \bar{x})$, gets singled out by fiat, which is Einstein-synchronized; in all other systems this is not the case, and the deviation from Einstein synchronization becomes larger the faster they move relative to the $(\bar{t}, \bar{x})$-system which we may call the '*ether system*'. We thus have on purpose destroyed the physical equivalence between inertial systems by a choice of convention. The new kind of synchronization is an *external* one and may be described in physical terms as follows. We select one system $(\bar{t}, \bar{x})$ and call it 'ether'; in it, clocks are synchronized by an internal method; in every other system, clocks are synchronized by having them fly past a system of 'ether clocks', bringing them to clock position $t = 0$ right when they pass an ether clock showing time $\bar{t} = 0$ (Fig. 2.17b). This procedure obviously cannot be carried out internally but makes reference to the arbitrarily distinguished system $(\bar{t}, \bar{x})$: it is an external method.

Since $t = 0$ und $\bar{t} = 0$ agree, there is no relativity of simultaneity if this method of synchronization is used; in Fig. 2.17b, there is no relative rotation between the x- and $\bar{x}$-axes.

From this alternative choice of synchronization *convention*, a lot can be learned about the structure of physical theories in general, and about relativity theory in particular. We indicate only the simplest consequences[1]:

a. The transformation (2.11.2) only holds between the (arbitrarily chosen) ether system $(\bar{t}, \bar{x})$ and some other inertial system (t, x). The transformation connecting two non-ether systems does not have this form. (The transformations (2.11.2) *do not* form a *group*, whereas the Lorentz transformations (2.1.1) do— see chap. 3.)

b. It follows from eq. (2.11.2) that clocks are slower when moving relative to the ether, as $t < \bar{t}$, but judged from the moving system the ether clocks are faster: eq. (2.11.2b) is—contrary to what we emphasized in the standard formulation of the theory—indeed a transformation formula for time *coordinates*. The inverse of eq. (2.11.2b) is therefore $\bar{t} = \gamma t$ and *not*, e.g., $\bar{t} = \sqrt{1 - v^2}\, t$. It is easily seen, however, that all observable consequences, like the clock effect, etc., are the same in this version of the theory as they were before. To emphasize again, the only difference is in the convention on clock synchronization.

c. The transformation (2.11.2) closely corresponds to the pre-Einsteinian view. There is the rest system of the ether, $\bar{I}$, with coordinates $(\bar{t}, \bar{x})$. Measuring rods moving relative to it are shortened by the factor $\sqrt{1 - v^2}$, corresponding to the older ideas of Lorentz. Rods at rest in the ether are longer as judged from moving systems, in analogy to the phenomenon discussed around Fig. 2.10. The difference between Lorentz contraction and dilation is again due to the differences in the definition of simultaneity of spatially separated events, since they enter the method of determining lengths (*simultaneous* determination of the positions of the rod's ends).

---

[1] See R. Mansouri, R. U. Sexl, Gen. Relat. Gravit. *8*, 497, 515, 809 (1977); P. Havas, Gen. Relat. Gravit. *10*, 135 (1987); Mittelstaedt (1989); Zhang (1997) for more details.

From the kinematical point of view, we thus have an equivalence between the standard formulation of Special Relativity and the ether variant described here: it is impossible to distinguish between the two by measuring space-time intervals.

An excellent confrontation of ether theory and relativity theory is found in the 1913 inaugural lecture by P. Ehrenfest, held at Leiden where he became the successor of the retired H. A. Lorentz:

"We first discuss the point of view of Lorentz in his paper of 1904, without, however, being able to go into the step-by-step development of that point of view.

The hypothesis of the resting ether as well as the other basic hypotheses of Lorentz' older theory are retained in the 1904 paper. Therefore none of the successes of Lorentz' older theory which led to the victory over its competitors gets lost.

What is new in the 1904 paper is the systematic use of two formally very simple hypotheses. Namely [hypotheses] about the changes, as a consequence of their motion through the ether, of
1. the forces between molecules, and
2. the geometrical shape of the electrons

Curiously, these hypotheses completely remove the contradiction that had existed between the hypothesis of the ether at rest and the definitively negative result of all etherwind experiments. These contradictions vanished completely. Namely, starting from those basic assumptions, the 1904 paper arrives, in a purely deductive manner and for a wide class of experiments, at the following theorem: Assume a laboratory moves through the ether with arbitrarily large speed (but not faster than light itself). Then, if an experimenter in this laboratory carries out an experiment, he will observe exactly the same processes as he would observe if his laboratory were at rest relative to the ether.—In what follows, allow me to call this theorem the '1904-theorem', for short.

It recommends itself to think about this theorem in its application to very special cases. One then can grasp in a coherent picture why it is indeed possible, thanks to those hypotheses, to hide the etherwind from the experimenter.

Allow me to sketch, with a few flashy touches, the picture that results: the etherwind disturbs the course of the processes the experimenter is operating with; but the same etherwind spoils—if we may say so—the measuring instruments of the experimenter: it deforms the measuring rods, it changes clock rates and the forces in spring balances etc. All that is taken care of by those basic hypotheses, in particular by the hypothesis that the motion through the ether will change the attraction between molecules. And if the experimenter observes the processes disturbed by the etherwind using his instruments which are spoiled by the same etherwind, he will see exactly what the observer at rest observes in the undisturbed processes with unspoiled instruments.

It is astonishing that this result admitted a rigorous proof from so few basic assumptions, for such a comprehensive class of experiments. It is miraculous that it was possible at all to generate such a gapless chain of conclusions. It would be immodest on my part if I wished to value, by whatever epitheton, the special method by which Mr. Lorentz was able to master this task ...

We thus see that here Einstein's etherless theory requires precisely the same as does Lorentz' ether theory. This is why an observer will, according to Einstein's theory, observe precisely the same contractions and rate changes on rods and clocks running past him as he would according to Lorentz' theory. And quite generally: there is in principle no experimentum crucis between both theories."

Although the standard formulation of the theory differs from the formulation based on eq. (2.11.2) only by a change in conventions, one is led to other hypotheses concerning possible tests of the theory if the ether formulation is adopted. For instance, the Michelson-Morley experiment was repeated in 1904 by Morley and Miller[1] with an apparatus supported by pine in order to see whether this material when moving through the ether would contract in the same way as sandstone, the material used originally.

---

[1]E. Morley, D. Miller, Philos. Mag. *8*, 753 (1904); *9*, 680 (1905).

In their own words: "...If the FitzGerald-Lorentz effect exists, it may affect all materials to the same amount, independently of the nature of the material. But it is also possible that the effect is one which depends on the physical properties of the material, so that pine might be affected more than sandstone. In this case, if sandstone gives no displacement in an experiment like that of 1887, an apparatus supported by pine, which would be compressed more than sandstone, would give an effect of the sign opposite to that suggested by the original simple theory. ..."

Another experiment whose basic idea comes from the pre-relativistic conception of Lorentz contraction was carried out in 1937 (!) by Wood, Tomlinson, and Essen[1]. In it, a rod, vibrating longitudinally with is eigenfrequency, is set into rotation. Then due to length contraction a change in the eigenfrequency should result for some orientations—unless the effect is precisely compensated by a change in the elastic constants of the rod. The experiment yielded an upper bound of 4×10^{-11} for the relative frequency change.

On the basis of Einsteinian relativity, this result is evident. In the theory used by these authors, this is not so—otherwise, the experiment would not have been done. They rather assumed the ether version (2.11.2) which is kinematically equivalent to relativity. What they did not take into account is that also the proper vibrations of the rod constitute a periodic process which could be used as a clock. If the experiment had had a positive result, this would have meant that in a system moving relative to the ether there are classes of clocks that are influenced differently by the motion—the authors wanted to find a change of eigenfrequencies by comparing with clocks which were likewise in motion. In an ether theory, this is possible, but it would drastically reduce the significance of the transformation (2.11.2), because it must then be specified with which kind of clock the time is being measured. To get agreement with relativity, one has to postulate in the ether theory that *every* kind of clock is slowed down by the factor $\sqrt{1 - v^2}$ and that every kind of rod shrinks by this factor. This *kinematical* postulate must then be shown to be consistent with the *dynamics* of the inner structure of rods and clocks. For the Lorentz contraction, that proof was carried out at least partially by Lorentz himself (see sect. 5.8). In the theory of relativity, one always formulates the dynamical laws in a Lorentz covariant fashion (as we will do in the chapters to follow), guaranteeing that kinematics and dynamics never get into conflict.

The problem of clock synchronization had already been discussed extensively before Einstein, e.g., by S. Newcomb in 1880 and by A. Michelson in 1887. Other early works on this subject stem from Poincaré[2], Wien[3], and Brillouin[4]. However, it was only Einstein who saw the significance of the problem clearly. Modern Philosophy of Science also dedicates large amounts of discussion to the theme—see, e.g., Grünbaum (1973) or the "Panel Discussion of Simultaneity by Clock Transport" in *Philosophy of Science 36*, No. 1 (1969).

Einstein's clock synchronization for a long time appeared very abstract and was illustrated usually by lightening strokes in front of trains and behind, and the like. Today, this synchronization procedure has become routine, since atomic clocks have been developed to an accuracy of a few microseconds per year. Such cesium clocks are placed at several locations all over the world and are synchronized with accuracy about 5×10^{-7}, using either clock transport or Einstein synchronization by radio signals. One of the uses is the satellite system GPS (Global Positioning System),

---

[1] A. Wood, G. Tomlinson, L. Essen, Proc. R. Soc. Lond. Ser. A *158*, 606 (1937).
[2] H. Poincaré, Rev. Metaphys. Morales *6*, 1 (1888).
[3] W. Wien, Phys. Z. *5*, 603 (1904).
[4] M. Brillouin, C. R. Acad. Sci. *140*, 1674 (1905).

having clocks also mounted in a number of satellites, allowing for a determination of position from signal travel times with an accuracy of 5 m and better, being open for everybody carrying an appropriate receiver. The number of applications is growing daily.

The *historical* significance of such networks lies in the fact that they function the way they do just because the classical concept of absolute time is invalid, while the invariance of the speed of light holds: on the basis of the old concepts, the results would be off their correct values by the order of kilometers! Actually, in the GPS also the General Theory of Relativity—i.e., Einstein's relativistic theory of gravitation— has to be taken into account and is integrated into the computer programs of the system (see, e.g., N. Ashby in Dadhich and Narlikar (1998) for details). The deviations from Newton's concept of absolute time—first directly demonstrated experimentally 33 years after the creation of Special Relativity, by Ives and Stilwell—thus have now reached the realm of everyday technological routine. Only a few decades ago, nobody would have imagined such a practical application of the space-time concept of Einstein's theory. (It is interesting that Einstein used atoms as clocks in a Gedanken experiment to rule out a second clock effect like the one discussed in sect. 2.8.)

3 Lorentz Group, Poincaré Group, and Minkowski Geometry

As a consequence of the Principle of Relativity, the set $\mathcal{P}$ of transformations between inertial systems has a certain mathematical structure: composing two transformations from $\mathcal{P}$ gives a transformation from $\mathcal{P}$ again, and for each transformation from $\mathcal{P}$ there is a unique inverse in $\mathcal{P}$. The set $\mathcal{P}$ therefore forms a *group*, where the group multiplication law is given by the composition of transformations.

Generally, by a group $\mathcal{G}$ one means a set of elements, $\{g, h, \dots\}$, where to each ordered pair (g, h) of elements a 'product' gh in $\mathcal{P}$ is assigned such that the following rules (*group axioms*) hold:

1. $(g_1\, g_2)\, g_3 = g_1\, (g_2\, g_3)$ (associativity)

2. There exists an element $e \in \mathcal{G}$ such that
   $eg = ge = g$ for all $g \in \mathcal{G}$ (unit element)

3. For each $g \in \mathcal{G}$ there is an element
   $g^{-1} \in \mathcal{G}$ such that $g^{-1}g = g\, g^{-1} = e$.   (Inverse)

In our case $\mathcal{G} = \mathcal{P}$, e is the identical transformation and g^{-1} is the inverse transformation. Two things are to be observed:

- A group is given abstractly by its 'multiplication table' which registers the product gh for each pair g, h of elements. The group is called *Abelian* or *commutative* if throughout $\mathcal{G}$ one has $gh = hg$. The group $\mathcal{P}$ is not commutative, and its elements are 'numbered' or 'indexed' by 10 parameters that can vary continuously—cf. sect. 1.1.

- The group $\mathcal{P}$ is not given abstractly but as a group of transformations *acting* on the set $\mathcal{I}$ of inertial frames or on the set $\mathbf{R}^4$ of event coordinates. We shall see that the same abstract group acts (or *is realized*) in various different ways as a group of transformations on sets of elements (physical objects) of various kinds (inertial frames, event coordinates, events, four-vectors, tensors, spinors, fields, state vectors in Hilbert spaces, ...), so that it will soon become evident that the abstract point of view is very useful.

Although we shall verify the group property of $\mathcal{P}$ explicitly in the excercises to sect. 3.1, let us sketch here an argument why it must be a group on the basis of the Principle of Relativity. (A reader unable to appreciate this kind of 'abstract nonsense' argument should not be discouraged at this point!) Write again $\mathcal{I}$ for the set of all inertial frames and write $\mathcal{E}$ for the set of all space-time events. Then every $I \in \mathcal{I}$ gives, by definition of a frame of reference, a bijective map between $\mathbf{R}^4$ (the set of event coordinates) and $\mathcal{E}$ which we denote by the same letter; thus $I : \mathbf{R}^4 \to \mathcal{E}$, $\bar{I} : \mathbf{R}^4 \to \mathcal{E}$, etc. Associated to any pair I_i, I_j of frames is a *transition map* $f_{ij} = I_i^{-1} \circ I_j : \mathbf{R}^4 \to \mathbf{R}^4$. (These are the transformations written so far, beginning with eq. (1.1.1).) They obviously satisfy

$$f_{ij} \circ f_{jk} = f_{ik}, \qquad f_{ij}^{-1} = f_{ji}, \qquad f_{ii} = \mathrm{id}.$$

Let $\mathcal{P}(I)$ be the set of all transition maps $I^{-1} \circ J$ connecting I to all other frames J. Then the Principle of Relativity implies that this set is the same for all I, i.e., $\mathcal{P}(I) = \mathcal{P}(\bar{I}) = \ldots =: \mathcal{P}$. It is easy to deduce from this and the relations for the f_{ij} just written that $\mathcal{P}$ is a group (of bijections $\mathbf{R}^4 \to \mathbf{R}^4$) under composition of maps as the multiplication. Namely, to show that the composition $f_{ij} \circ f_{mn}$ also belongs to $\mathcal{P}$ although the adjacent indices do not agree as in the relation above, conclude from $\mathcal{P}(I_m) = \mathcal{P}(I_j)$ that there must exist a system I_k such that $f_{mn} = f_{jk}$, which makes the relation above applicable.

The group $\mathcal{P}$ acts on event coordinates (i.e., on $\mathbf{R}^4$) but can also be thought of as acting on inertial frames (i.e., on $\mathcal{I}$) 'from the right' as $I \mapsto I \circ f$ for $f \in \mathcal{P}$. Note that after singling out any inertial frame $I_0 \in \mathcal{I}$ we have a bijective correspondence between $\mathcal{I}$ and $\mathcal{P}$ by assigning to every I the unique transition map by which it is obtained from I_0; but only $\mathcal{P}$ is a group (one cannot meaningfully multiply inertial systems)!

We therefore have an action of the group on the product space $\mathcal{I} \times \mathbf{R}^4$, and calling the pairs $(I,(x^i))$ and $(\bar{I}, (x^{\bar{i}}))$ equivalent iff $\bar{I} = I \circ f^{-1}$, $x^{\bar{i}} = f(x^i)$ for some $f \in \mathcal{P}$ allows to identify $\mathcal{E}$ with the quotient $(\mathcal{I} \times \mathbf{R}^4)/\mathcal{P}$ by this equivalence relation. This construction will allow to transfer properties of $\mathbf{R}^4$ relative to the group $\mathcal{P}$ to the event space $\mathcal{E}$ (differentiable structure, affine structure, pseudo-metric, ...). We will then also consider *active* versions of the transformations, i.e., transformations of $\mathcal{E}$ described by I as $I \circ f \circ I^{-1}$, where $f \in \mathcal{P}$; they can also be characterized as leaving invariant the structures just mentioned.

The basic idea behind using the abstract group is that there are systematic mathematical methods for constructing and classifying other realizations once the abstract group structure has been found from one realization as a transformation group. The new objects on which the new realizations act can be used as building blocks in attempts to construct new physical theories such that the Principle of Relativity will automatically hold in them.

In this book our aim is to go on with such a program step by step, becoming acquainted with some of the pertinent methods and kinds of arguments, without however putting too much stress on rigor or completeness.

3.1   Lorentz Group and Poincaré Group

In sect. 1.5 we characterized the general Poincaré transformations as being those coordinate transformations

$$x^{\bar{i}} = f^i(x^k) \tag{3.1.1}$$

leaving invariant the line element (1.5.1),

$$ds^2 = (dx^0)^2 - (dx^1)^2 - (dx^2)^2 - (dx^3)^2 = \eta_{ik}\, dx^i dx^k. \tag{3.1.2}$$

Here we have introduced the component matrix of the so-called *metric tensor*[1],

$$\eta = (\eta_{ik}) := \mathrm{diag}\,(1, -1, -1, -1) = (\eta_{ki}), \tag{3.1.3}$$

---

[1]This name will be explained later.

which turns out indispensible in all further manipulations. With its help, the conditions of invariance of ds^2 under the transformations (3.1.1) takes the form

$$\eta_{ik}\, dx^i dx^k = \eta_{mn}\, dx^{\bar{m}} dx^{\bar{n}}, \tag{3.1.4}$$

i.e., since the dx^j are arbitrary:

$$\eta_{mn}\frac{\partial f^m}{\partial x^i}\,\frac{\partial f^n}{\partial x^k} = \eta_{ik}. \tag{3.1.5}$$

We are now in a position to supply the proof, promised in sect. 1.5, that indeed it follows from this invariance that the transformations (3.1.1) have to be invertible and linear. When we read eq. (3.1.5) as a matrix equation and take determinants we find at once $\det(\partial f^m/\partial x^i) = \pm 1 \neq 0$. Next we differentiate eq. (3.1.5) for x^j, permute the indices i, j, k cyclically and add two of the arising equations but subtract the third: because of $\eta_{mn} = \eta_{nm}$ we obtain

$$2\eta_{mn}\frac{\partial f^n}{\partial x^k}\,\frac{\partial^2 f^m}{\partial x^j \partial x^i} = 0.$$

From $\det(\partial f^n/\partial x^k) \neq 0$ it now follows that all second derivatives of f^m vanish, so that f^m is linear[1],

$$x^{\bar{\imath}} = f^i(x^k) = L^i{}_k\, x^k + a^i. \tag{3.1.6}$$

Here, according to eq. (3.1.5), the coefficients of the homogeneous transformations

$$x^{\bar{\imath}} = L^i{}_k\, x^k \tag{3.1.7}$$

are restricted by

$$\eta_{mn} L^m{}_i\, L^n{}_k = \eta_{ik} \quad \Rightarrow \quad \det L = \pm 1. \tag{3.1.8}$$

It is trivial that all invertible transformations (3.1.1) leaving ds^2 invariant form a group; but this means that all transformations (3.1.6) satisfying eq. (3.1.8) form a group, the *Poincaré group* $\mathcal{P}$. The proof that $\mathcal{P}$ coincides with the group of all transformations connecting inertial systems (*cum grano salis*—see our remarks on time reversals) is now complete.

The homogeneous transformations (3.1.7) satisfying eq. (3.1.8) form a *subgroup* of $\mathcal{P}$, called the *Lorentz group*[2] $\mathcal{L}$.

Equations (3.1.6, 7, 8) may be rewritten in matrix form as

$$\bar{x} = Lx + a \tag{3.1.6'}$$

$$\bar{x} = Lx \tag{3.1.7'}$$

$$L^\top \eta L = \eta, \tag{3.1.8'}$$

---

[1]Linear-inhomogeneous, or *affine*, according to the more modern terminology.

[2]Other nomenclature: inhomogeneous Lorentz group for $\mathcal{P}$, homogeneous Lorentz group for $\mathcal{L}$; and correspondingly for the transformations.

where $L^\top$ is the transpose of L. Eq. (3.1.8′) is completely analogous to the condition $O^\top E O = E$ for orthogonal matrices O, where E is the unit matrix diag(1,1,1,...). Equation (3.1.8′) may therefore be termed a pseudo-orthogonality relation, and the ds^2 defined on space-time correspondingly as *pseudo-Euclidean metric*. (Note that it is not a metric in the sense of topology!) We may thus describe $\mathcal{L}$ as a matrix group, i.e., as the group of all 4×4 matrices L satisfying eq. (3.1.8′). The group axioms may be verified for this form of the definition (see exercise).

Similarly, $\mathcal{P}$ may be described as the set of all pairs (a, L) formed from a column vector a and a Lorentz matrix L; the rule to form the product of two such pairs is taken from the composition of two transformations of type (3.1.6′):

$$(\bar{a}, \bar{L})(a, L) = (\bar{a} + \bar{L}a, \bar{L}L). \tag{3.1.9}$$

In later chapters we shall extensively deal with the properties and realizations of both groups. Here we just want to get acquainted with the simplest objects and concepts which are necessary to formulate relativistic mechanics.

Exercises

1. Recapitulate the basic concepts of group theory from some standard text on algebra (cf. also Appendix A). Try to complete the 'abstract nonsense' argument given in the smallprint paragraph of the introduction to this chapter for the group property of $\mathcal{P}$.

2. Verify the group axioms for the matrix group $f\mathcal{L} = \{L : L^\top \eta L = \eta\}$.

3. Verify eq. (3.1.9).

4. Verify the group axioms for $\mathcal{P} = \{(a, L) : L^\top \eta L = \eta\}$ with product given by eq. (3.1.9).

5. Recapitulate the the concept of *invariant subgroup* of a group; then show that the set $\mathcal{T}$ of all pure translations (a, E) forms an Abelian invariant subgroup in $\mathcal{P}$.

6. The *(external) direct product* of two groups $\mathcal{G}_1$, $\mathcal{G}_2$ is the set $\mathcal{G} = \mathcal{G}_1 \times \mathcal{G}_2$ of all ordered pairs (g_1, g_2), (h_1, h_2), ... where $g_i \in \mathcal{G}_i$, $h_i \in \mathcal{G}_i, \ldots$, equipped with the multiplication rule $(g_1, g_2)(h_1, h_2) = (g_1 h_1, g_2 h_2)$. Show that this makes $\mathcal{G}$ into a group. Form the direct product of the translation group $\mathcal{T}$ with $\mathcal{L}$ and compare with $\mathcal{P}$ ('*semidirect product*', see Appendix A). In which case is $\mathcal{T}$ an invariant subgroup, in which case is $\mathcal{L}$ invariant?

3.2   Minkowski Space. Four-Vectors

Already in sect. 1.1 the space-time coordinates of an event x as referred to an inertial system I were lumped together into a quadruple x^i. These quadruples taken together form the four-dimensional vector space $\mathbf{R}^4$ (column vectors in the matrix formalism

(3.1.6′, 7′)). The same holds for the quadruples $x^{\bar{i}}$ relative to $\bar{\text{I}}$. Since the transformations between the x^i and $x^{\bar{i}}$ are affine—cf. eq. (3.1.6), the set $\mathcal{E}$ of events itself receives the structure of a 4-dimensional[1] affine space $\mathbf{X}_4$, the set of connecting vectors Δx between pairs of events becoming a four-dimensional vector space $\mathbf{V}_4$ over the real numbers.

The line element introduced in sect. 1.5 assigns a 'length square'

$$(\Delta x)^2 := \eta_{ik}\,\Delta x^i \Delta x^k \tag{3.2.1}$$

to the finite connecting vectors—the right-hand side of eq. (3.2.1) being independent of the special system I in which it is evaluated. Space-time together with this affine pseudo-metric structure is called *Minkowski space*. Observe that for its construction the Einsteinian version of relativity is necessary. It precisely embodies all implications of Lorentz transformations. (Similarly, space-time with an affine structure and a system of parallel hyperplanes of absolute simultaneity with Euclidean geometry in them codifies the Galilean version of relativity.)

Under the Poincaré transformations (3.1.6) the components of connecting vectors Δx transform homogeneously, i.e., according to the Lorentz transformation

$$\Delta x^{\bar{i}} = L^{\bar{i}}{}_k\,\Delta x^k. \tag{3.2.2}$$

It turns out that there is a lot of physical objects u which are given in each inertial system I by four components u^i such that on passing to another inertial system $\bar{\text{I}}$ according to eq. (3.1.6) these components are related by

$$u^{\bar{i}} = L^{\bar{i}}{}_k\,u^k. \tag{3.2.3}$$

Such objects are termed *four-vectors*; the connecting vectors Δx are their prototypes. Four-vectors (of the same physical dimension) may be added and multiplied by numbers: let u, v be four-vectors and a, b real numbers, then $\mathrm{a}\,u + \mathrm{b}\,v = w$ is defined by $w^i = \mathrm{a}\,u^i + \mathrm{b}\,v^i$, where the w^i obviously transform in the correct manner (3.2.3). Therefore four-vectors (of a given physical dimension) form a vector space.

Note that the well-known distinction between an abstract four-dimensional real vector space and the vector space $\mathbf{R}^4$ continues to be present in the concrete physical examples in two respects. One is that $\mathbf{R}^4$ has a 'canonical' basis—the usual one consisting of columns with zeros everywhere except at one place—whereas the abstract space has no canonical basis (no preferred frame of reference, just as required by the principle of relativity). But there is another difference, hardly ever mentioned in mathematical texts: the elements of $\mathbf{R}^4$ have physical dimension zero, while physical four-vectors in general will have nonzero dimension, e.g., length, inverse length, momentum, etc. Clearly then, only four-vectors of the same physical dimension can constitute the elements of a four-vector space. Although one cannot add two four-vectors of different dimension, there is of course a concept of proportionality between them, with a dimensionful factor of proportionality. In geometrical terms, two four-vector spaces whose elements differ in physical dimension define the same projective space of directions.

---

[1] Numbers indicating dimensions will be written as subscripts except where they at the same time indicate Cartesian powers, as in $\mathbf{R}^2 = \mathbf{R} \times \mathbf{R}$, etc.

In a four-vector space, we can define a 'length square' (*four-square*) in analogy to eq. (3.2.1) by

$$u^2 := \eta_{ik}\, u^i u^k = (u^0)^2 - (\mathbf{u})^2 \tag{3.2.4}$$

and by it a *scalar product*

$$u\,w := \frac{1}{2}\big((u+w)^2 - u^2 - w^2\big) = \eta_{ik}\, u^i w^k = u^0 w^0 - \mathbf{u}\,\mathbf{w} = w\,u. \tag{3.2.5}$$

The right-hand sides of eqs. (3.2.4,5) are prototypes of Lorentz invariant expressions (*four-scalars*). A 4-dimensional vector space $\mathbf{V}_4$ equipped with a scalar product of the kind (3.2.5) will be called a *Minkowski vector space*[1]. Vectors with vanishing scalar product are called *orthogonal*. Note that it of course makes sense to form scalar products between four-vectors of different physical dimensions.

Despite the suggestive symbol u^2, the quadratic form given by expression (3.2.4) is not definite: it may take positive as well as negative values and may vanish without u vanishing itself.

A 'length square' of this kind is, of course, unsuitable for defining on $\mathbf{V}_4$ a metrical topology; the topology is rather the one inherited from $\mathbf{R}^4$, which is Lorentz invariant since Lorentz transformations are homeomorphisms of $\mathbf{R}^4$, as are all invertible linear transformations. Below we will mention a way to define the topology directly in terms of the four-square (3.2.4).

The vectors $u \neq 0$ from $\mathbf{V}_4$ therefore fall into one of the following classes:

$$
\begin{array}{lll}
u^2 > 0 & \textit{timelike} & \\
u^2 = 0 & \textit{lightlike} & \text{four-vectors.} \\
u^2 < 0 & \textit{spacelike} &
\end{array}
\tag{3.2.6}
$$

Lightlike vectors are also called *null vectors*. The nomenclature (3.2.6) becomes clear if we interpret u as a connecting vector between two events (Fig. 3.1):

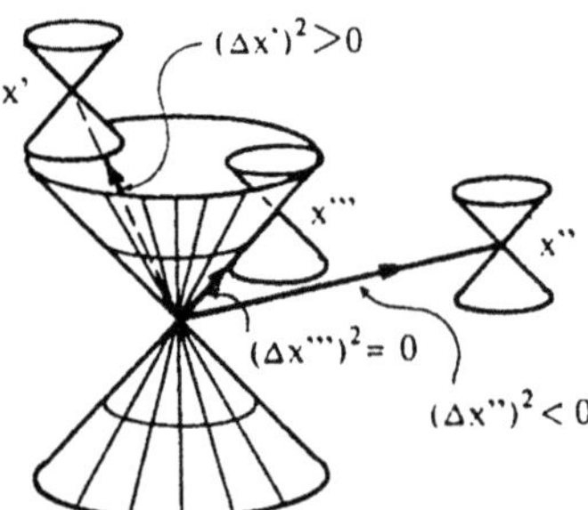

Fig. 3.1. Spacelike, timelike, and lightlike connecting vectors

If their separation is timelike, i.e., if the connecting vector is timelike (x, x' in Fig. 3.1), then x' is inside the light cone of x, thus belonging to its future or past. In

---

[1] The alternative terminology 'vector space with a Lorentzian structure' is becoming more and more established.

case of spacelike separation (x, x'' in Fig. 3.1) x'' belongs to the present of x; in case of lightlike separation (x, x''' in Fig. 3.1) x''' is on the light cone of x.

In making these distinctions, the roles of the two events may be interchanged. Since we were restricting to Lorentz transformations without time reversal, however, a Lorentz invariant *time orientation* for non-spacelike vectors becomes definable in $\mathbf{V}_4$ as follows. If a nonzero four-vector u has $u^2 \geq 0$, we have $|u^0| > 0$, and if $u^0 > 0$ in one system I, then $u^{\bar{0}} > 0$ holds in any other system $\bar{\text{I}}$, so this expresses indeed a property of the four-vector itself: it will be termed *future-oriented = future-directed*. Similarly, if $u^0 < 0$, the vector is called *past-oriented = past-directed)*. To formally prove the statements just made we may restrict to boosts (1.4.4). First from $u \neq 0$, $u^2 > 0$, $u^0 > 0$ we conclude

$$(u^0)^2 - \mathbf{u}^2 > 0 \;\Rightarrow\; u^0 > |\mathbf{u}| \geq 0,$$

and then because of $|\mathbf{v}| < 1$ for relative velocities between inertial systems I, $\bar{\text{I}}$, using Cauchy's inequality,

$$|\mathbf{u}\,\mathbf{v}| \leq |\mathbf{u}|\,|\mathbf{v}| \leq |\mathbf{u}| < u^0 \;\Rightarrow\; u^{\bar{0}} = \gamma\,(u^0 - \mathbf{v}\,\mathbf{u}) > 0.$$

There is a similar treatment of the case $u^2 = 0$.

Corresponding to the terminology employed in sect. 2.2 we call the set of future-directed lightlike, resp. timelike, four-vectors the *future light cone* of $\mathbf{V}_4$, resp. its interior. The *past light cone* of $\mathbf{V}_4$ is defined correspondingly.

The intersections of the interiors of past and future light cones may be taken as the basis of a topology ('Alexandrov topology'). This topology in fact agrees with the standard one mentioned before, but it is interesting because of its manifest invariance as well as its generalization to the curved spaces of General Relativity.

Given a timelike vector u there is always a reference frame, unique up to spatial rotations and reversals, in which its components take the normal form

$$u^{\bar{i}} = \left(\pm\sqrt{u^2}, 0\right)^{\top} \tag{3.2.7}$$

($\pm$ depending on time orientation): only its time component is different from zero, explaining our terminology. For proof we interpret u as a connecting vector of two events, one of them at the origin of some reference frame. We now change the time axis by a boost such that it passes through the other event, choosing $\mathbf{v} = \mathbf{u}/u^0$ in eq. (1.4.4). Since u^2 is invariant, the time component must be $\pm\sqrt{u^2}$ in the new frame where $\bar{\mathbf{u}}$ vanishes.

Similarly one can achieve a normal form such as

$$u^{\bar{i}} = \left(0, \sqrt{-u^2}, 0, 0\right)^{\top} \tag{3.2.8}$$

for spacelike vectors by applying a space rotation and a boost.

For lightlike vectors u we can rotate the frame such as to make its 2- and 3-component equal zero, thus $u^i = (\pm a, a, 0, 0)^{\top}$ because of its vanishing four-square. However, just because the latter is the vector's only invariant (besides $\text{sign}(u^0)$), a is

not invariant and is easily seen to get multiplied by a positive factor on applying a
boost in the 1-direction. (In physical terms, this will reappear in the Doppler effect,
sect. 4.3.) Thus we can achieve the normal form

$$u^{\vec{i}} = (\pm 1, 1, 0, 0)^{\top}, \tag{3.2.9}$$

depending on time orientation. (The degree of nonuniqueness of the frame in this
case will be considered in sect. 9.4, case b.)

As has been already mentioned, Minkowski's metric (3.2.1) is absolute in Einsteinian relativity
in the same sense as are time intervals in Galilean relativity. The formalism of four-vectors and
-tensors to be developed in the following sections will permit a very efficient use of that absolute
structure. This concerns the basic insight into the theory as well as practical manipulations—we just
mention the dangers inherent in the formal use of 3-vector algebra as shown in sects. 2.9 and 2.10. By
contrast, the formalism of four-vectors and Minkowski geometry tends to avoid such pitfalls almost
automatically, so that it pays off to develop a certain amount of ability to visualize this geometry.

A trick to visualize the orthogonality relations among subspaces of Minkowski vector space
without sacrificing dimensions is to go over to the corresponding projective space $P(\mathbf{V}_4)$, where the
light cone of $\mathbf{V}_4$ defines an oval quadric ('sphere') and where orthogonality means polarity with
respect to that surface. (The reader is advised to discuss exercise 2 in the light of this picture!)

Exercises

1. Show that the sum of non-spacelike future-directed four-vectors is non-spacelike
   and future-directed (convexity of the light cone) and that the scalar product of
   two of these is nonnegative.

2. Show that vectors orthogonal to a given lightlike vector are either spacelike or
   proportional to it. What can you say about vectors orthogonal to a given space-
   or timelike vector?

3. Consider an observer whose worldline has direction given by the timelike future-
   directed vector u. Show that two events x, y are simultaneous for this observer
   iff $u(x - y) = 0$.

4. Let the event z be lightlike with respect to two events x, y. Show that the vector
   connecting x and y is orthogonal to the vector connecting z to the midpoint of
   x and y. Interpret this result in some cases in view of the result of exercise 3 in
   the sense of Einstein synchronization.

5. Let two particles move abreast with the speed of light, i.e., let them move on
   straight parallel orbits, hitting simultaneously any hypothetical screen orthogonal to the orbits. Show that this *abreastness* property is in fact independent
   of the observer stating it and is expressed geometrically by the orthogonality
   $kv = 0$ between any four-vector along the woldline(s) and any connecting vector
   between them. Convince yourself that abreastness is not observer-independent
   if the motion is subluminal!

Hint: Since you are dealing in these exercises with frame-independent statements,
you may verify them in any frame. Suitably choosing this, you may use the normal
forms given above.

3.3 Passive and Active Tranformations. Reversals

In a Minkowski vector space $\mathbf{V}_4$ we can introduce a *basis* consisting of four linearly independent vectors e_i $(i = 0, 1, 2, 3)$ and decompose any four-vector as

$$u = \mathrm{u}^i e_i. \tag{3.3.1}$$

For clarity, *in this section* symbols for numerical components will appear *not* in italics, whereas four-vectors and active transformations will do so.

In later sections this will not be strictly obeyed to. In particular, then, u^i will either mean the components of u in some unspecified frame I, or will simply mean the four-vector u itself, where the index i only announces a vector quantity but does not take numerical values. In most cases it should emerge from the context whether u^2 means the four-square or the component u^2. (We should also mention here that some authors insist on calling coordinates of a vector what we most of the time will be calling (numerical) components, while they would call the vectors $\mathrm{u}^0 e_0$, ... the (vectorial) components of u.) In the present section such a sloppy procedure would be confusing, since we are also using indexed vectors (the e_i). A systematic notational distinction between indices of both kinds (indicator of a vector vs. number of component) is made in the abstract index formalism of R. Penrose (see Penrose and Rindler 1984).

The four-square of u is then

$$u^2 = \mathrm{u}^i \mathrm{u}^k e_i e_k, \tag{3.3.2}$$

which will agree with eq. (3.2.4) iff the basis vectors form an *orthonormal* system in the sense of Minkowski geometry:

$$\begin{aligned} e_i e_k &= \eta_{ik} \\ e_0 e_0 &= +1, \quad e_1 e_1 = e_2 e_2 = e_3 e_3 = -1. \end{aligned} \tag{3.3.3}$$

In what follows we shall use orthonormal bases only, corresponding to our choice of using, in each inertial system, Cartesian orthogonal coordinates, the Einstein synchronization and $c = 1$ (cf. exercises 3, 4 of the last section).

The transition (3.2.3) to a new frame corresponds to the transition to a new orthonormal basis $\{\bar{e}_i\}$:

$$u = \mathrm{u}^k e_k = \mathrm{u}^i \bar{e}_i = \mathrm{L}^i{}_k \mathrm{u}^k \bar{e}_i. \tag{3.3.4}$$

The transformation coefficients $\mathrm{L}^i{}_k$ here appear as the components of the e_k with respect to the new basis $\{\bar{e}_i\}$:

$$e_k = \mathrm{L}^i{}_k \bar{e}_i. \tag{3.3.5}$$

Since we have been restricting to Lorentz transformations without time reversal $(\mathrm{L}^0{}_0 > 0)$, it follows that $\bar{e}_0$ and e_0 have the same time orientation, and it is sensible to restrict to future-directed e_0, $\bar{e}_0$, ... If we also restrict to right-handed spatial bases we obtain a *total orientation* for the $\{e_i\}$, $\{\bar{e}_i\}$, ... , which is invariant under Lorentz transfomations without reversals.

For the inverse of relation (3.3.5) we write

$$\bar{e}_i = \mathrm{L}_i{}^j e_j, \tag{3.3.6}$$

where

$$\mathrm{L}^i{}_k \mathrm{L}_i{}^j = \delta_k{}^j = \mathrm{L}_k{}^i \mathrm{L}^j{}_i, \tag{3.3.7}$$

i.e., the matrices $(\mathrm{L}^i{}_k)$ und $(\mathrm{L}_i{}^j)$ are *contragredient*, one of them is the transposed inverse of the other.

The transformations performed so far are *passive*; four-vectors are not changed but only referred to a new orthonormal basis. We now also consider *active* transformations, because such were already used in sect. 1.2 in a nonformal way.

Under an active Lorentz transformation L the whole vector space $\mathbf{V}_4$ is linearly mapped onto itself, preserving all scalar products:

$$u \to \bar{u} = Lu \quad \text{with} \quad \bar{u}^2 = u^2. \tag{3.3.8}$$

We can associate to L a matrix $(\mathrm{L}_i{}^j)$ in the usual way: L maps the vectors of a basis $\{e_i\}$ to those of a basis $\{\bar{e}_i\} = \{Le_i\}$ whose vectors each may be decomposed with respect to the original one:

$$\bar{e}_i = \mathrm{L}_i{}^j e_j. \tag{3.3.9}$$

If so, the image vector $\bar{u}$ has components with respect to the original basis $\{e_i\}$ which can be read off from

$$\bar{u} = L u = \mathrm{u}^i L e_i = \mathrm{u}^i \mathrm{L}_i{}^j e_j = \bar{\mathrm{u}}^j e_j \tag{3.3.10}$$

as

$$\bar{\mathrm{u}}^j = \mathrm{L}_i{}^j \mathrm{u}^i \tag{3.3.11}$$

with the inverse relation

$$\mathrm{u}^i = \mathrm{L}^i{}_k \bar{\mathrm{u}}^k \tag{3.3.12}$$

(cf. eq. (3.3.7)). The contrast between eqs. (3.2.3) and (3.3.12) should be clear from their geometrical significance. Of course, with respect to the new basis $\{\bar{e}_i\}$ the vector $\bar{u}$ has the same components as u has with respect to the original one, $\{e_i\}$.

In complete analogy we distinguish *passive* and *active* Poincaré transformations on space-time. In place of the linear or vector bases $\{e_i\}$ of $\mathbf{V}_4$ we have here the *affine orthonormal bases* of $\mathbf{X}_4$ consisting of some point $o \in \mathbf{X}_4$ (the 'origin') together with a vector basis $\{e_i\}$ of the associated space of connecting vectors. The events x get coordinatized with respect to an affine basis by decomposing the connecting vector from o to x—the position vector of the event with respect to the chosen origin—as $\mathrm{x}^i e_i$. Therefore our mathematical model for an inertial frame I is simply an affine time- (and perhaps space-) oriented orthonormal basis $\{o, e_i\}$ for $\mathbf{X}_4$. (Cf. Appendix B.14.) Let us, with this new terminology, come back to the situation in chap. 1! In sects. 1.3 and 1.4 we determined the passive form of the transformations after pointing out in sect. 1.2 that the laws of nature are invariant under the active form of the transformations: It is an active transformation if we set up an experiment in the system $\bar{\mathrm{I}}$ in the same manner as it is set up in I; but it is a passive transformation if we refer the same event or the same process—such as propagation with speed of light, considered in sect. 1.4—to two different frames I and $\bar{\mathrm{I}}$.

It is now possible to give a short consideration to the *reversals* which were excluded so far. *Space reversals* may be performed passively without problems: this just means to go from a right-handed to a left-handed frame. The question is, however, whether these transformations are admitted in the formulation of the principle of relativity, and for this they have to be performed actively. The difficulties that arise in attempts to set up experiments in the 'same' manner with respect to reference frames which are mirror reflections of each other may be illustrated in the well-known Ørsted experiment. If the magnet is mirror-reflected in a naive geometric manner, the experiment does not appear to be reflection-invariant; however, if the magnetization is imagined as being produced by elementary circular currents and the reflection is applied to these, the experiment is reflection-invariant. This shows that it is nontrivial to perform an active reversal. Elementary particle physics has shown[1] that, in a nontrivial sense, not all processes in nature are invariant under space reversals.

Still more complicated is the situation concerning *time reversals*. It is obviously impossible to realize this transformation passively, there are no observers for which time is running backwards. It may be realized actively in the form of *reversal of motion*. The inherent difficulties may be illustrated again in Ørsted's experiment. Elementary particle physics has discovered also processes that may be interpreted as being noninvariant under time reversal.[2]

We shall take up discussing reversals again only in chap. 6.

3.4 Contravariant and Covariant Components. Fields

Alongside the vector components introduced so far, which transform according to eq. (3.2.3) and are called *contravariant components*, it is useful to introduce so-called *covariant components* by the definition

$$u_i := \eta_{ik}\, u^k = (u^0, -u^1, -u^2, -u^3) \tag{3.4.1}$$

which uses, in every orthonormal frame, the same matrix (η_{ik}) appearing in eq. (3.1.3). By means of these components the scalar product (3.2.5) appears as

$$u\,w = u_i w^i. \tag{3.4.2}$$

The contravariant components are reobtained from the covariant ones by the formula

$$u^i = \eta^{ik} u_k. \tag{3.4.3}$$

which uses the inverse matrix (η^{ik}) of (η_{ik}):

$$\eta^{ik}\eta_{kj} = \delta^i_j; \tag{3.4.4}$$

---

[1] Cf. Källén (1964); for the violation of space reversal symmetry in the organic world, where no dynamical law breaks the symmetry, see A. McDermott, Nature *323*, (Sept. 4, 1986); Janoschek (1991).

[2] See Kabir (1968); Davies (1974).

numerically we easily check that

$$\eta^{ik} := \eta_{ik}, \tag{3.4.5}$$

where—it must be stressed again—everything is referred to orthonormal bases.

The *transformation law* of covariant components results from

$$u_{\bar{\imath}} = \eta_{ik}\, u^{\bar{k}} = \eta_{ik}\, L^k{}_j\, u^j = \eta_{ik}\, L^k{}_j\, \eta^{jm} u_m \tag{3.4.6}$$

as

$$u_{\bar{\imath}} = L_{\bar{\imath}}{}^m u_m, \tag{3.4.7}$$

where

$$L_{\bar{\imath}}{}^m := \eta_{ik}\, L^k{}_j\, \eta^{jm}. \tag{3.4.8}$$

As follows easily from $u\,w = u_i w^i = u_{\bar{\imath}} w^{\bar{\imath}}$ or from eq. (3.1.8), the matrix in eq. (3.4.8) agrees with the matrix contragredient to $(L^i{}_k)$ which was introduced in eq. (3.3.6).

Up to now, the introduction of covariant components looks as a secondary, slightly superfluous step. The point here is, however, that there are objects for which they are the more natural ones in that they arise primarily, characterized by the transformation law (3.4.7), whereas contravariant ones are then defined via eq. (3.4.3) in a secondary step. One example of this is the *four-gradient* to be considered below.

An example where the transformation law (3.4.7) shows up primarily arises in the description of space- and time-periodic wave motion. Assume some observer in I describes a certain periodic plane wave by $\cos(\omega t - \mathbf{k}\,\mathbf{x})$: this wave propagates in the direction of the *wave vector* $\mathbf{k}$ with *phase velocity* $v_{Ph} = \omega/|\mathbf{k}|$ and *angular frequency* ω ($\Rightarrow$ reduced wavelength $= 1/|\mathbf{k}|$). Then this process is space-time periodic for observers in all other inertial systems $\bar{\text{I}}$ as well: if we put $\omega = k^0$ and define $k_i = \eta_{ij} k^j$, then $\omega t - \mathbf{k}\,\mathbf{x} = k_i\, x^i$, and the Lorentz transformation $x^i = L^i_j\, x^{\bar{\jmath}}$ gives $\cos k_i\, x^i = \cos k_i\, L^i_j\, x^{\bar{\jmath}} = \cos k_{\bar{\jmath}}\, x^{\bar{\jmath}}$, thus an expression of the same form, with

$$k_{\bar{\jmath}} = L_{\bar{\jmath}}{}^i\, k_i$$

as in eq. (3.4.7), where now $k^{\bar{0}} = \bar{\omega}$ and $\bar{\mathbf{k}} = (k^{\bar{1}}, k^{\bar{2}}, k^{\bar{3}})$ are the angular frequency and vectorial wave number registered in $\bar{\text{I}}$. So we see how frequency and wave number get united into the *wave number four-vector* k whose covariant components are more basic in establishing its four-vector nature. An immediate consequence of the transformation law will be the relativistic versions of the Doppler effect and aberration; however, we postpone their discussion to the next chapter.

The wave number four-vector, or wave vector for short, of a plane wave yields a linear functional on $\mathbf{V}_4$ by assigning to each space-time displacement vector Δx the corresponding change in phase $k\,\Delta x$ which is independent of the observer (just like the number of wave maxima registered along Δx). For the notion of *dual space* $\tilde{\mathbf{V}}$ for a given vector space $\mathbf{V}$ as the set of all linear functionals (=*covectors*) on $\mathbf{V}$ see Appendix B.2.)

Using the basis vector e_0 of I we have $\omega = k^0 = e_0\, k$ and $\mathbf{k}^2 = (e_0\, k)^2 - k^2$, thus

$$v_{Ph}^2 = \frac{(e_0\, k)^2}{(e_0\, k)^2 - k^2}\,.$$

This is explicitly observer-dependent except for the case $k^2 = 0$ where $v_{Ph} = 1$ equals the speed of light. For $k^2 > 0$ and $k^2 < 0$ we have $v_{Ph} > 1$ and $v_{Ph} < 1$, respectively, and these statements are also observer-independent. Also note that if one wants to associate *rays* to a wave as in geometrical optics, an observer-independent way suggests itself by taking k as their four-direction: this gives the usual thing if $v_{Ph} = 1$ but corresponds to motion with speed $|\mathbf{k}|/k^0 = 1/v_{Ph}$ otherwise!

Beside scalars and four-vectors, *scalar fields* and *vector fields* will play an important role, assigning to every space-time point x a number $\varphi(x)$ and a four-vector $u(x)$, respectively. In an inertial frame I we have coordinates x^i for the event x and components $u^i(x)$ for $u(x)$, so that these fields get specified by functions of the coordinates:

$$\Phi(x^k) = \varphi(x) = \bar{\Phi}(x^{\bar{k}})$$
$$U^i(x^k)e_i = u(x) = U^{\bar{i}}(x^{\bar{k}})\bar{e}_i.$$

$$(3.4.9)$$

Here we have written on the right-hand sides the corresponding specifications for a frame $\bar{\text{I}}$; this gives immediately the transformation laws

$$\bar{\Phi}(x^{\bar{k}}) = \Phi(x^k)$$
$$U^{\bar{i}}(x^{\bar{k}}) = L^i{}_j U^j(x^k) \qquad\qquad x^{\bar{k}} = L^k{}_m x^m + a^k \qquad (3.4.10)$$
$$U_{\bar{i}}(x^{\bar{k}}) = L_i{}^j U_j(x^k).$$

We now consider the *four-gradient* field of a scalar field, which is given by the components

$$\partial_i\varphi := \frac{\partial\Phi}{\partial x^i} =: \Phi_{,i}, \qquad \partial_{\bar{i}}\varphi = \frac{\partial\bar{\Phi}}{\partial x^{\bar{i}}} = \bar{\Phi}_{,\bar{i}}. \qquad (3.4.11)$$

By the chain rule,

$$\frac{\partial\bar{\Phi}}{\partial x^{\bar{i}}} = \frac{\partial\Phi}{\partial x^k}\frac{\partial x^k}{\partial x^{\bar{i}}}, \qquad (3.4.12)$$

and since by eq. (3.3.7) the transformation of the coordinate differentials, $dx^{\bar{i}} = L^i{}_j\,dx^j$, has the inverse

$$dx^k = L_i{}^k dx^{\bar{i}} \Rightarrow \frac{\partial x^k}{\partial x^{\bar{i}}} = L_i{}^k, \qquad (3.4.13)$$

we see that eq. (3.4.11) indeed defines covariant components of a four-vector field. As a first example we of course have the (constant) gradient k of the phase kx of a plane wave—the wave vector discussed above. We shall write eq. (3.4.12) symbolically as

$$\partial_{\bar{i}} = L_i{}^k \partial_k. \qquad (3.4.14)$$

Note the difference: If we want to specify some vector field we just write down four component functions in some frame; its component functions in any other frame may then be computed from our formulae; however, if we are given ahead four functions in every frame, we must check the validity of the transformation law if we want to claim that these data define one and the same vector field.

The four-vector $\nabla\varphi$ defined by the covariant components (3.4.12) has the contravariant components $\partial^i\varphi = \eta^{ik}\partial_k\varphi$. The (inverse) metric η is thus indispensable in assigning a space-time direction to $\nabla\varphi$. As η does not possess the usual definiteness properties enjoyed by the Euclidean metric, the direction of the four-gradient is *not* always the direction of fastest increase of φ! (See exercise; note that some concept of metric is necessary to normalize the various displacement vectors for a 'fair' comparison of the pertinent changes of φ, and here of course the Minkowski metric suggests itself for Lorentz invariance of normalization.)

The central role of four-vectors and other objects that transform in a linear-homogeneous manner under Poincaré transformations will emerge more and more in the following sections. As announced before, we shall not be very strict in distinguishing between indices referring to some frame and 'abstract' indices; also, we shall frequently use the words four-vector, or simply vector, where actually four-vector field would be in place. With scalar fields, we shall not always distinguish notationally between the function φ defined on abstract Minkowski space and the functions Φ, $\bar{\Phi}$ defined on the coordinate space $\mathbf{R}^4$; similarly for vector fields. This is in keeping with the older mathematics literature as well as with most of the physics literature and avoids lengthy expressions like 'component functions of the four-current density vector field with respect to frame I'. In most cases, clarity will come from the context. If not, the reader is advised to temporarily use the more exact notation.

It should, however, not be overlooked that there are these conceptual differences, which may result in differences in sign when active and passive transformations are in the play. A typical example of such differences and varying nomenclature is the following. The term 'scalar' or 'invariant' is used in various ways. In the context of vector space theory scalars are simply numbers (elements of the relevant ground field), with which the vectors can be multiplied or which are assigned to one or more vectors by certain operations. If the vectors are described in terms of components, the assigned scalars must not change upon changing the special basis to which components refer, and this is stressed by calling the assigned number an invariant. But one also says scalar or invariant in place of scalar field, although there are Lorentz-invariant scalar fields $\varphi(x)$, i.e., fields that take the same value at x and the actively transformed event Lx ...

A similar terminological problem exists in the use of the terms 'invariant' and 'covariant'. We do not want to suggest a solution here since the pertaining physical facts are sufficiently explained in Anderson (1967), whereas modern mathematics nowadays uses unambiguous concepts which, however, tend to sound quite differently.

Exercises

1. Recapitulate the proof of the statement that a function on $\mathbf{R}^n$ has maximum rate of change in the direction of its gradient, and try to give the necessary modifications for a correct statement in Minkowski space!
   Hint: To compare various directions, the displacement vectors must be normalized. Distinguish the cases where the four-gradient is timelike, spacelike, or lightlike.

2. Find the transformation behavior of a wave vector under a Galilean boost!

4 Relativistic Mechanics

In this chapter we will formulate the basic concepts of kine(ma)tics and the basic dynamical laws, taking care to satisfy the Einsteinian version of the principle of relativity. The formulation thus should be compatible with the postulate that inertial frames connected by Poincaré transformations be on equal footing. Mathematically this means that the laws are to be *Lorentz covariant*, i.e., we should be able to formulate them in such a way that they take the same mathematical form in all inertial frames. This postulate is certainly fulfilled if we are able to write these laws as equalities between four-vectors.

Thus, technically, we shall illustrate in this chapter the use of four-vectors and their scalar products. In most applications (but not always!) this technique offers great advantages over the Lorentz transformation method used in chap. 2.

4.1 Kinematics

Consider a point mass whose motion relative to an inertial frame I is given by $\mathbf{x} = \mathbf{x}(t)$. Its velocity is

$$\mathbf{v} = \frac{d\mathbf{x}}{dt}, \tag{4.1.1}$$

and we assume that $|\mathbf{v}| < 1$. From eq. (2.9.2) we know its rather complicated behavior under Lorentz transformations

$$x^{\bar{i}} = L^i{}_k \, x^k, \tag{4.1.2}$$

stemming from the fact that the denominator in eq. (4.1.1) has also to be transformed. We cannot expect that this velocity concept will allow the formulation of manifestly Lorentz covariant laws.

However, if we parametrize the world line of the point mass by its (Lorentz invariant) proper time s as $x^i = x^i(s)$, a suitable substitute for $\mathbf{v}$ comes to mind immediately, namely the *four-velocity* u with components

$$u^i := \frac{dx^i}{ds}. \tag{4.1.3}$$

Here the coordinates enter symmetrically as they do in eq. (4.1.2), and it is obvious that the u^i form the components of a four-vector, since the dx^i were the prototype of four-vector components. We therefore can write abstractly $u = dx/ds$. Because of eq. (2.6.2) we have

$$u^i = \left(\frac{dt}{ds}, \frac{d\mathbf{x}}{ds} \right)^\top = \frac{dt}{ds}(1, \mathbf{v})^\top = \gamma \, (1, \mathbf{v})^\top. \tag{4.1.4}$$

This shows that u does not contain more information than $\mathbf{v}$; in the (so-called 'non-relativistic', N.R.) limiting case where $|\mathbf{v}| \ll 1$ relative to the frame considered, we have $\gamma \approx 1$ and therefore $u^i \approx (1, \mathbf{v})^\top$. u is just a new packing of the ordinary velocity concept with a better Lorentz transformation behavior of its components. In terms of Minkowski geometry, u is nothing but the unit tangent vector to the world line at the point considered, since we have for its four-square

$$u^2 = u^i u_i = \frac{dx^i dx_i}{ds^2} = \frac{ds^2}{ds^2} = 1. \tag{4.1.5}$$

It is timelike and future-directed ($dx^0 > 0$, $ds > 0$). The fact that there is no absolute speed smaller than 1 here appears in the mathematical fact that the only independent Lorentz invariant quantities associated with a timelike vector u are its four-square and $\text{sign}(u^0)$—and those are the same for all four-velocities.

Our definition suggests associating with our point mass a *four-momentum*

$$p := mu, \quad p^i = (p^0, \mathbf{p})^\top, \tag{4.1.6}$$

where m is the (inertial) mass as measured in the usual ways in low velocity situations. N.R. we have $p^i \approx (m, m\mathbf{v})^\top$, so that the space components then agree with the momentum components used in Newtonian mechanics.

This definition gives the so-called kinetic momentum, to be distinguished—even in Newtonian mechanics—from the canonical momentum that arises in a Lagrangian formulation, despite the fact that the two agree in many situations. Their conceptual difference implies that the canonical momentum is a covariant vector, in that its covariant components arise primarily in its definition, just as in the case of the gradient; while the kinetic momentum is a contravariant vector 'by birth'. We shall consider only the latter here.

For the four-square of the four-momentum we have from eq. (4.1.5)

$$p^2 = (p^0)^2 - \mathbf{p}^2 = m^2, \tag{4.1.7}$$

a relation of fundamental significance for relativistic kinematics. Geometrically this relation means that four-momenta of particles of mass m are timelike and future-directed and form one sheet of a hyperboloid in 4-dimensional momentum space, called the *mass shell* for particles of mass m; its asymptotes form the light cone in momentum space. This is illustrated for two space dimensions in Fig. 4.1.

In analogy to the four-velocity we now form the *four-acceleration* a as

$$a^i := \frac{d^2 x^i}{ds^2} = \frac{du^i}{ds}. \tag{4.1.8}$$

Differentiating eq. (4.1.5) for s we get

$$0 = \frac{d}{ds}\left(\eta_{ik}\, u^i u^k\right) = \eta_{ik}\left(u^i a^k + a^i u^k\right) = 2u^i a_i. \tag{4.1.9}$$

Thus a is, in the sense of Minkowski geometry, orthogonal to u and therefore a spacelike vector. The quantity $(-a^2)^{1/2}$, geometrically speaking a Lorentz-invariant curvature of the world line, equals the absolute value of the Newtonian acceleration as measured in the instantaneous rest system (exercise). This shows the sense in which

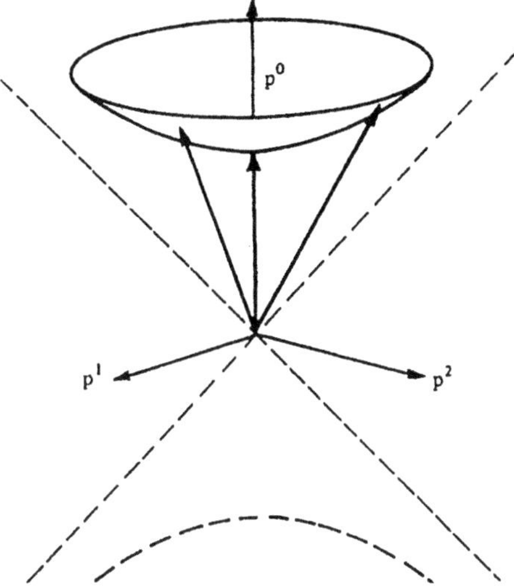

Fig. 4.1. The mass shell $(p^0)^2 - (p^1)^2 - (p^2)^2 = m^2$, $p^0 > 0$

accelerations—in contrast to velocities—do have an *absolute* character in Special Relativity.

It is clear now that a possible way to reconcile Newton's second axiom $\mathbf{F} = m\mathbf{a}$ with Einsteinian relativity is to modify it as

$$F^i = ma^i = m\frac{du^i}{ds} = \frac{dp^i}{ds}, \tag{4.1.10}$$

where F^i are the components of a four-vector F, the *four-force*. It has been introduced by eq. (4.1.10) only formally; in order that this equation acquire physical significance, F has to be taken from some theory such as electrodynamics. If so, eq. (4.1.10) may be integrated to yield the motion of the particle.

The F^i cannot, however, be prescribed arbitrarily. First, if an expression for them suggests itself in every frame, one must first check the Lorentz transformation law. When this is satisfied, we see on multiplying eq. (4.1.10) by $L^j{}_i$ that it takes the same form $F^{\bar{i}} = ma^{\bar{i}}$ in all frames $\bar{\text{I}}$. So this is an example of a Lorentz-(and Poincaré-) covariant equation, which we can also write as an equation between four-vectors:

$$F = ma = m\frac{d^2x}{ds^2} = m\frac{du}{ds} = \frac{dp}{ds}. \tag{4.1.10'}$$

If a physical law can be expressed as an equality of two four-vectors it automatically satisfies the principle of relativity. We shall later look systematically for all quantities that might play a similar role in the formulation of other laws of Nature.

But there is a second restriction on F, even if we specify its components in one frame only and obtain them in all others by the transformation law: from eq. (4.1.9) we have

$$Fu = 0, \tag{4.1.11}$$

F is a vector orthogonal to u, hence spacelike. In an instantaneous rest frame we have $u^{\bar{i}} = (1, \mathbf{0})^{\top}$, so because of eq. (4.1.11) the components of F there are $F^{\bar{i}} = (0, \mathbf{f})^{\top}$. Here $\mathbf{f}$ is the force acting on the particle in its rest frame, which may be measured by static or dynamic methods as usual. Lorentz transforming to the system I where the particle has velocity $\mathbf{v}$ we obtain

$$F^i = \left(\gamma \mathbf{v}\,\mathbf{f}, \mathbf{f} + \frac{\gamma^2}{\gamma+1}(\mathbf{v}\,\mathbf{f})\,\mathbf{v} \right)^{\top}. \tag{4.1.12}$$

Its zero component

$$F^0 = \gamma\,\mathbf{f}\,\mathbf{v} = \gamma\,\mathbf{f}\frac{d\mathbf{x}}{dt} = \mathbf{f}\frac{d\mathbf{x}}{ds} =: \frac{dA}{ds} \tag{4.1.13}$$

is the work done by $\mathbf{f}$ in unit proper time. Equation (4.1.10) for $i = 0$ is

$$\frac{dp^0}{ds} = F^0 = \frac{dA}{ds}. \tag{4.1.14}$$

Thus the work done on the particle increases the component p^0 of four-momentum, which therefore represents the *energy* of the particle—possibly up to an additive constant. For this reason, p is also called the particle's *energy-momentum vector*. (J. A. Wheeler has suggested here the new expression 'momenergy' to underline the unification effected by Relativity, in addition to writing 'spacetime' without hyphen.) In fact, we have from eqs. (4.1.4,6), expanding the γ factor,

$$p^0 = \gamma m = m + \frac{mv^2}{2} + \dots \ . \tag{4.1.15}$$

For small speeds $v \ll 1$, p^0 thus equals the kinetic energy of the particle, up to the constant m.

The considerations on energy conservation in the next section will show that p^0 has to be regarded as a total energy of the particle, consisting of the *kinetic energy* T (translational energy) of the particle, and its *rest energy* m ($= mc^2$ in conventional units). The relativistic expression of the kinetic energy thus results from

$$p^0 =: m + T \tag{4.1.16}$$

as

$$T = (\gamma - 1)\,m = \frac{mv^2}{2} + \frac{3}{8}mv^4 + \dots \ . \tag{4.1.17}$$

Appendix: Geometry of Relativistic Velocity Space

The velocity hyperboloid in four-vector space $\mathbf{V}_4$ given by $u^2 = 1$, $u^0 > 0$ is analogous to the hyperboloid of four-momenta shown in Fig. 4.1 and is a *homogeneous space* of the Lorentz group (active interpretation): every point of it may be transformed into any other, none of them is distinguished in a Lorentz invariant fashion. One may introduce four-velocities also in Galilean Relativity by $u^i := (1, \mathbf{v})$, filling the affine hyperplane $u^0 = 1$ of an analogous 4-dimensional vector space; this hyperplane is a homogeneous space of the Galileo group, and the ordinary vectors of relative velocities are its connecting vectors in the sense of affine geometry (whereas the four-velocities are 'points' of the four-velocity space, as in the Lorentzian case). However, contrary to the flat affine nature of Galilean velocity space, the relativistic velocity hyperboloid is *curved*—more exactly, the Lorentz invariant metric defined on it by $d\sigma^2 := -du^2$ makes it into a Riemannian space of constant negative curvature. (This is the *Weierstrass model* of *Lobachevski space* (cf. Fock 1959), which is used in cosmological models (cf. Sexl and Urbantke 1995).) Projecting the hyperboloid from the origin of $\mathbf{V}_4$ onto one of its tangent hyperplanes, we obtain the *Klein projective model*; projecting onto the same hyperplane but from the antipode of its point of contact we obtain the *Poincaré conformal model*, where angles are as they look for Euclidean eyes, which may be useful in semi-quantitative considerations.

The velocity hyperboloid allows to visualize general Lorentz transformations. For this purpose one identifies reference frames I,... with orthonormal bases $\{e_i\}$,... and interprets e_0,... as the four-velocities of I,... and thus as points of the hyperboloid. Now the remaining basis vectors e_α,... may be interpreted as tangent vectors of the hyperboloid at those points, forming an orthonormal tangent frame there. Every orthonormal tangent frame of the hyperboloid may be transformed into any other one, by precisely one Lorentz transformation in each case. This is expressed by saying that the Lorentz group *acts simply-transitively (or freely and transitively) on the bundle of all orthonormal frames* of the hyperboloid. Singling out one of these frames therefore yields a bijection between this bundle and the Lorentz group. It is, e.g., not hard then to see in this picture that the Thomas angle (2.10.7) is nothing but the defect (π minus sum of angles) of the triangle formed by the geodesic lines joining the points that represent the four-velocities of the three inertial systems involved.

For each world line $x(s)$ we obtain a curve on the hyperboloid traced by the four-velocities $u(s)$—the *relativistic hodograph* of the motion. The tangents to the hodograph are just the vectors $a(s)$ of four-acceleration. The velocity hyperboloid was already considered by Minkowski, and then by Varičak, Borel, and others.

Exercises

1. Show that the relative speed of two particles with four-velocities u', u'' is given by $(1 - (u'u'')^{-2})^{1/2}$.

2. Show that $(-a^2)^{1/2}$ equals the amount of acceleration as measured in the instantaneous rest frame.

4.2 Collision Laws. Relativistic Mass Increase

Collision experiments are of basic importance in mechanics since they test conservation of energy and momentum without requiring a detailed knowledge of the forces that act during the collision. We shall go into the problem of forces between particles only in chap. 5.

Figure 4.2 shows symbolically the collision between two particles. The central circle indicates the region of interaction, about which in many cases no details may be available.

Quite independently of the nature of the forces in that region we have in the nonrelativistic case equality between the sums of momenta

$$\mathbf{p}_1 + \mathbf{p}_2 = \mathbf{p}_3 + \mathbf{p}_4 \qquad \text{(N.R.)} \qquad (4.2.1)$$

and of energies ($T_A := \mathbf{p}_A^2/2m_A$ N.R.)

$$T_1 + T_2 = T_3 + T_4 \qquad \text{(N.R.)} \qquad (4.2.2)$$

before and after collision. Since the momenta $\mathbf{p}_3$, $\mathbf{p}_4$ constitute six independent quantities, while there are only four equations (4.2.1,2), the final state is not determined uniquely without knowing the interaction. However, the conservation laws decisively restrict the set of final states.

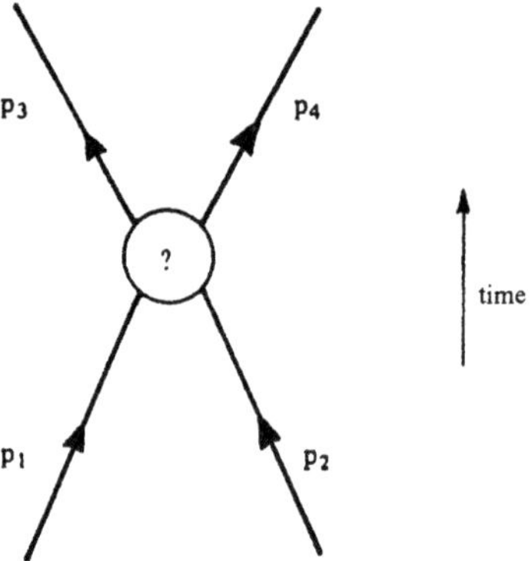

Fig. 4.2. Two particles in collision

The set of kinetically possible final states—i.e., those compatible with the conservation laws—for a given initial state is called in particle physics the *phase space* of the process, in analogy to statistical mechanics. From it the specific dynamics of the process, as given by the interaction, will select the actual final state—uniquely according to classical physics, while quantum mechanics allows only to calculate the probability with which it will fall into a given region of phase space. Usually in particle physics geometric considerations in this phase space are called kinematics instead of kinetics.

Since there are no four-vectors of the correct dimension available for the particles outside the interaction region (where they are force-free) other than their four-momenta p_A (A numbers particles), the relativistic version of the *conservation laws* (4.2.1,2) must be

$$p_1 + p_2 = p_3 + p_4. \qquad (4.2.3)$$

Equation (4.2.3) contains four laws, so one suspects that in the N.R. limiting case conservation of both, energy and momentum, will result. However, for $|\mathbf{v}| \ll 1$ we have $p^i \approx (m, m\mathbf{v})$, and eq. (4.2.3) specializes as

$$m_1 + m_2 = m_3 + m_4 \qquad \text{(N.R.)} \qquad (4.2.4)$$

$$m_1\mathbf{v}_1 + m_2\mathbf{v}_2 = m_3\mathbf{v}_3 + m_4\mathbf{v}_4. \qquad \text{(N.R.)} \qquad (4.2.5)$$

Thus instead of the expected conservation of energy we obtained in eq. (4.2.4) the law of conservation of mass, which in Newtonian mechanics is considered as self-understood and is not written down separately! Looking at eq. (4.1.16), however, we can obtain from eq. (4.2.3) a statement of energy conservation:

$$m_1 + T_1 + m_2 + T_2 = m_3 + T_3 + m_4 + T_4, \qquad (4.2.6)$$

where T_A is the relativistic kinetic energy of the A-th particle, as before.

The occurrence of the summand m in $p^0 = m + T$ thereby acquires more than formal significance: according to eq. (4.2.6) it is only the sum of kinetic and rest energy that is conserved. Conservation of kinetic energy alone such as in eq. (4.2.2) is *not* required by it. So there may exist processes where one form of energy is converted into the other one. This surprising possibility, admitted by the relativistic form of the conservation laws, indeed shows up in numerous experiments and observations, some of which will be discussed in sect. 4.5 because of their theoretical significance.

One further consequence from eq. (4.2.3) together with eqs. (4.1.6,4) is that a moving particle behaves in collisions as having—compared to Newtonian mechanics— an increased inertial mass γm, a *dynamical mass*. This fact is called relativistic mass increase. The rest mass m has therefore to be determined by experiments in the N.R. velocity domain.

Another essential feature is that the total energy $p^0 = \gamma m$ of a particle increases without limit as $v \to 1$. Therefore an infinite amount of energy is necessary to accelerate a particle to the velocity of light. This is the dynamic reason for the unattainability of the speed of light for massive particles that was announced earlier.

Finally, from the relation $p^2 = m^2$ we get the useful relation for the total energy in terms of momentum

$$p^0 = {}_+\sqrt{m^2 + \mathbf{p}^2}. \qquad (4.2.7)$$

Subtracting the rest energy m gives the kinetic energy as

$$T = \sqrt{m^2 + \mathbf{p}^2} - m, \qquad (4.2.8)$$

which reduces to the N.R. expression $\mathbf{p}^2/2m$ when $|\mathbf{p}| \ll m$; the velocity of the particle is, in terms of $\mathbf{p}$,

$$\mathbf{v} = \frac{\mathbf{p}}{p^0} = \frac{\mathbf{p}}{\sqrt{m^2 + \mathbf{p}^2}}. \qquad (4.2.9)$$

Exercises

1. Let p, P be the 4-momenta of two particles with nonzero rest masses m, M. Prove the 'reversed Cauchy-Schwarz inequality' $pP \geq mM$. When does equality hold?

2. Deduce from the previous result the 'reversed triangle inequality' $(p + P)^2 \geq (m + M)^2$ and the condition for equality therein. Sketch the domain in 4-momentum space which is available to the total 4-momentum of a system of two massive particles.

3. Taking space-time connecting vectors instead of momenta in the last inequality, explain its role in the twin paradox.

4.3   Photons: Doppler Effect and Compton Effect

The considerations made so far cannot be applied to the quanta of light—*photons*—as particles moving with the speed of light have $ds = 0$, so that $p^i = m\,dx^i/ds$ has a chance to be meaningful only if we also have $m = 0$: photons are *massless particles*. In this case one can only conclude that $p^i \propto dx^i$, where the factor of proportionality remains undetermined; but we certainly have $p^2 = (p^0)^2 - (\mathbf{p})^2 = m^2 = 0$, so that the energy-momentum vector p of a photon is a *lightlike, future-directed* vector with components

$$p^i = (|\mathbf{p}|, \mathbf{p})^{\top}. \tag{4.3.1}$$

The relation between p and the wave vector of the corresponding wave is given by quantum mechanics:

$$p = \hbar k, \qquad k^i = (\omega, \mathbf{k})^{\top}, \tag{4.3.2}$$

where $h = 2\pi\hbar$ is Planck's constant.

The original 1900 Planckian quantization $E = n\hbar\omega$ of the energy of rather formal 'field oscillators' was converted in 1905 by Einstein into the hypothesis of quanta of light which were to carry the energy $\hbar\omega$. Relativistic symmetry was one of the reasons to generalize Planck's relation to eq. (4.3.2) (Einstein, Stark, ...). It seems that A. H. Compton assumed eq. (4.3.2) independently and, in fact, very reluctantly. For de Broglie, the relativistic version (4.3.2) was the starting point for his idea of waves of matter, which was first successful, however, in its N.R. version, in the hands of Schrödinger.

We shall now illustrate the properties of k and p by some characteristic examples which at the same time will demonstrate the advantages of working with four-vectors. Note that the conclusions to be drawn from the four-vector nature alone do not really involve the relation (4.3.2) and could have been discussed already in sect. 3.4. That relation will be essential in collisions between photons and massive particles, however.

Doppler effect and *aberration* of light will follow from the transformation law for k^i. Consider a photon with wave vector given by

$$k^{\bar{i}} = \bar{\omega}\,(1, \cos\bar{\Theta}, \sin\bar{\Theta}, 0)^{\top} \tag{4.3.3}$$

relative to an inertial system $\bar{\mathrm{I}}$: so it propagates in the $(\bar{x}, \bar{y})$-plane, making an angle $\bar{\Theta}$ with the $\bar{x}$-axis. With respect to a system I which moves at speed v in the direction of the $\bar{x}$-axis, k has components

$$k^i = \omega\,(1, \cos\Theta, \sin\Theta, 0)^{\top}, \tag{4.3.4}$$

where the relation between eqs. (4.3.3) and (4.3.4) is given by

$$(k^i) = \omega \begin{pmatrix} 1 \\ \cos\Theta \\ \sin\Theta \\ 0 \end{pmatrix} = \bar{\omega} \begin{pmatrix} \gamma & -\gamma v & 0 & 0 \\ -\gamma v & \gamma & 0 & 0 \\ 0 & 0 & 1 & 0 \\ 0 & 0 & 0 & 1 \end{pmatrix} \begin{pmatrix} 1 \\ \cos\bar{\Theta} \\ \sin\bar{\Theta} \\ 0 \end{pmatrix}. \tag{4.3.5}$$

We therefore can read off the *relativistic Doppler effect*

$$\omega = \frac{\sqrt{1 - v^2}\,\bar{\omega}}{1 + v\cos\Theta} \tag{4.3.6}$$

and the relation between Θ and $\bar{\Theta}$ (*aberration*)

$$\cos\Theta = \frac{\cos\bar{\Theta} - v}{1 - v\cos\bar{\Theta}}, \qquad \sin\Theta = \frac{\sqrt{1-v^2}\,\sin\bar{\Theta}}{1 - v\cos\bar{\Theta}}. \tag{4.3.7}$$

Let us first consider the Doppler effect for $\Theta = \bar{\Theta} = 0$:

$$\omega = \sqrt{\frac{1-v}{1+v}}\,\bar{\omega}, \qquad \bar{\omega} = \sqrt{\frac{1+v}{1-v}}\,\omega. \tag{4.3.8}$$

The relation between ω and $\bar{\omega}$ results from the one between $\bar{\omega}$ und ω applying the substitution $v \to -v$, as must be the case by the principle of relativity. The nonrelativistic Doppler effect for sound is different in this respect: the square root of eq. (4.3.6) is absent here, and one has to distinguish the cases where the source or the receiver is moving relative to the gas in which sound is propagating. With light in vacuum, only the relative velocity between I (rest system of the receiver, say) and $\bar{\text{I}}$ (rest system of the source of light) is important.

The *transverse Doppler effect* $\Theta = \pi/2$ is of basic importance. In this case the receiver moves at right angle relative to the direction of the incoming light, so that classically no effect would be expected at all. The decrease in frequency predicted by relativity,

$$\omega = \bar{\omega}\,\sqrt{1-v^2}, \tag{4.3.9}$$

is a pure effect of time dilation. Its measurement in 1938 by Ives and Stilwell constitutes the first quantitative confirmation of time dilation, thus being of importance in the history of science. (Details may be found, e.g., in French (1971), p. 146.) Recent measurements of the transversal Doppler effect use the Mössbauer effect.[1] A γ-ray source is surrounded by a rotating cylinder-shaped absorber made from the same material. By the rotation the agreement between emission and absorption frequency is destroyed according to eq. (4.3.9), so that the cylinder becomes transparent to the γ-rays, which is checked by a detector outside the cylinder. This allows to test eq. (4.3.9) within a few percent of accuracy.

The importance of the aberration formula (4.3.7) for the observation of stars from the moving earth may be found in most elementary introductions to relativity (see, e.g., French 1971, Kacser 1970).

Let us discuss here the relation between aberration and the invisibility of the Lorentz contraction or the rotated appearance of moving objects in snapshots. Light emitted by the object at an angle $\bar{\Theta}$ with respect to the direction of motion as measured in its rest frame $\bar{\text{I}}$ is observed in the camera system I at the angle Θ, so that the object must appear in I as rotated through the angle $\alpha = \Theta - \bar{\Theta}$. If $\Theta = \pi/2$, corresponding to observation at a right angle with respect to the direction of motion, we get $\sin\bar{\Theta} = \sqrt{1-v^2}$, thus $\cos\alpha = \sqrt{1-v^2}$ in agreement with the result obtained in sect. 2.5. The effect is quite drastic for extremely relativistic motion, $\gamma \gg 1$. Figure 4.3 shows the relation between Θ, $\bar{\Theta}$ and α when $\gamma = 2$, while Fig. 4.4 illustrates the resulting apparent rotation of a cube-shaped object passing past a camera at large distance (snapshots for a range of observation angles).

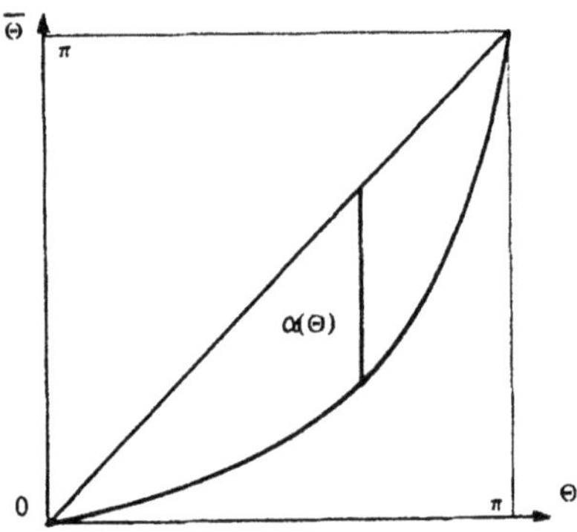

Fig. 4.3. Aberration for $\gamma = 2$

Fig. 4.4. Apparently rotated positions of a cube flying past a camera

We still have to supply the general proof of the invisibility of Lorentz contraction by showing that there is indeed *only* a rotation of the object, which is assumed to be far away from the camera or the observer so that the photographic mapping is by parallel rays. Then all photons involved have the same wave vector k. Consider any two of them: their world lines are given by

$$x_A = k\,\lambda_A + d_A, \qquad x_B = k\,\lambda_B + d_B, \tag{4.3.10}$$

where λ_A and λ_B are parameters varying along the world lines. (Since for photons we have $ds = 0$, we cannot parametrize these world lines by proper time as we did for massive particles; in contradistinction to more complicated parametrizations λ is called an *affine parameter*.) They will arrive simultaneously at a photographic plate orthogonal to their direction of propagation if $k\,(d_A - d_B) = 0$, as is best seen in the rest system of the plate (cf. exercise 5 of sect. 3.2). As we also have $k^2 = 0$, the spatial distance between the two rays is given by the Lorentz invariant expression $(x_A - x_B)^2 = (d_A - d_B)^2$; therefore this distance, which is relevant for the snapshot, is the same in the rest system of the object and the rest system of the camera. This proves our assertion.

As a last example we investigate the kinematics of *Compton scattering*, i.e., the scattering of light off electrons (Fig. 4.5).

---

[1]H. Hay, J. Schiffer, T. Cranshaw, P. Engelstaff, Phys. Rev. Lett. **4**, 165 (1960).

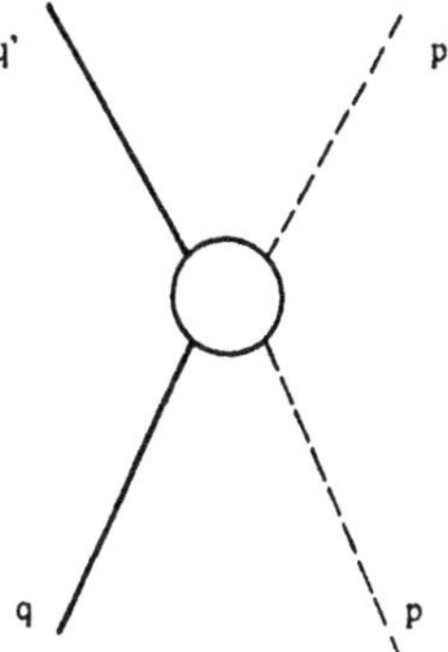

Fig. 4.5. Compton scattering

Let the 4-momenta of the photon before and after scattering be p and p', and those of the electron q and q'; then energy-momentum conservation requires

$$p + q = p' + q'. \tag{4.3.11}$$

To calculate the energy change of the photon during the process (which is what is usually measured), we eliminate q' from eqs. (4.3.11). Its is useful here to work with invariants first instead of specializing to some definite inertial system. So we bring p' to the left side and then form the four-square:

$$m^2 = q'^2 = (p - p' + q)^2 = \underbrace{p^2}_{=0} + \underbrace{p'^2}_{=0} - 2pp' + 2q(p - p') + \underbrace{q^2}_{=m^2}. \tag{4.3.12}$$

In the resulting formula

$$q\,(p - p') = p\,p' \tag{4.3.13}$$

the momentum of the electron after scattering has been eliminated. We now use the relation $p = \hbar k$ between momentum and wave vector of the photon and specialize to the rest system of the incoming electron, where we have $q^i = (m, \mathbf{0})^\top$, $k^i = (\omega, \mathbf{k})^\top$, $k'^i = (\omega', \mathbf{k}')^\top$. Equation (4.3.13) then gives

$$\hbar m(\omega - \omega') = \hbar^2 \,\omega\,\omega'\,(1 - \cos\Theta), \tag{4.3.14}$$

where Θ is the angle between the directions of the incoming and the scattered photon (the scattering angle). Writing $2\pi/\omega = \lambda$ we obtain from eq. (4.3.14)

$$\Delta\lambda := \lambda' - \lambda = \frac{h}{m}(1 - \cos\Theta). \tag{4.3.15}$$

This is the well-known Compton relation. For $\Theta = \pi/2$ the change in wavelength is given by the *Compton wavelength* $h/mc = 2.426 \times 10^{-10}$ cm of the electron.

From the point of view of the particle theory of light, the decrease in energy (frequency) is not surprising since energy is transferred to the electron (recoil). From the point of view of the classical electromagnetic wave theory, however, this effect is ununderstandable, since there the scattering process is interpreted as follows. The incoming electromagnetic wave causes the electron to oscillate and thus to emit electromagnetic waves on its part which have the same frequency as the incoming wave (but different direction).

The historical significance of Compton's experiment lies in the quantitative confirmation of the relation $p = \hbar k$ by measuring eq. (4.3.15). Since 1912 a reduced ability for penetration had been observed in scattered X-rays, which had been interpreted as a reduction in frequency; and for this, several classical explanations had been looked for. In 1922 Compton deduced eq. (4.3.15) and confirmed it experimentally as well. The recoil electrons were made visible one year later by Wilson, using his cloud chamber.

The intensity and angular distribution of the scattered light cannot be calculated from eqs. (4.3.11), since those represent, together with $q'^2 = m^2$, $p'^2 = 0$, only 6 equations for the 8 unknowns p', q'. In the limit of long wavelengths—so that $\Delta\lambda$ becomes negligible—the intensity of the scattered light is given by the *Thomson cross section σ_T*. The equation of motion of the electron in the incoming electromagnetic wave is $m\ddot{\mathbf{x}} = e\mathbf{E}$, and the energy radiated by the accelerated electron per unit time is

$$\frac{dE}{dt} = \frac{2}{3}\frac{e^2}{c^3}\ddot{\mathbf{x}}^2 = \frac{2}{3}\frac{e^2}{c^3}\frac{e^2}{m^2}\mathbf{E}^2. \tag{4.3.16}$$

The flux of energy of the incoming wave is $c\overline{\mathbf{E}^2}/4\pi$ (where the bar indicates an average over a period), so dividing eq. (4.3.16) by it gives us the scattering cross section

$$\sigma_T = \frac{8\pi}{3}\frac{e^4}{m^2c^4} = \frac{8\pi}{3}r_e^2 = 6.65 \times 10^{-25}\,\text{cm}^2. \tag{4.3.17}$$

Here $r_e = e^2/mc^2 = 2.818 \times 10^{-13}$ cm is the *classical electron radius* (cf. chap. 5). At photon energies comparable to the electron rest energy the cross section is given by the Klein-Nishina formula (see Bjørken and Drell 1966).

In astrophysics, the *inverse Compton effect* is of importance, in which a high-energy electron from cosmic rays scatters off a low energy photon from starlight or cosmic background radiation. If we restrict, for simplicity, to a head-on collision in x-direction, we have $q^i = (\gamma m, \gamma m v, 0, 0)^\top$, $p^i = \hbar(\omega, -\omega, 0, 0)^\top$, $p'^i = \hbar(\omega', \omega', 0, 0)^\top$, and eq. (4.3.13) together with the approximations $1 + v \approx 2$, $1 - v \approx 1/2\gamma^2$ leads to

$$\omega' = \frac{4\omega\gamma^2}{1 + 4\hbar\omega\gamma/m}. \tag{4.3.18}$$

Inverse Compton scattering is an important source of X-rays (see, e.g., D. W. Sciama in Sachs (1971)).

Exercises

1. Consider a particle emitting light isotropically in all directions in its rest system $\bar{\text{I}}$, i.e., the angular distribution is $L(\bar{\Theta}) = L = const$. What is the distribution $L(\Theta)$ of this radiation as observed in a system I in which the particle is moving extremely relativistically ($\gamma \gg 1$)? Discuss the maximum of $L(\Theta)$ in forward direction in connection with the radiation of extremely relativistic particles (see, e.g., Jackson 1999, sect. 14). Show that the Doppler effect makes an additional contribution to increase the maximum.

   Hint: $L(\Theta)\sin\Theta\,d\Theta = L(\bar{\Theta})\sin\bar{\Theta}\,d\bar{\Theta}$.

2. From eq. (4.3.7) a simple relation between $\operatorname{tg}\Theta/2$, $\operatorname{tg}\bar{\Theta}/2$ may be derived.

For its application to contours of moving spheres see R. Penrose, Proc. Cambridge Philos. Soc. *55*, 137 (1959).

3. For a periodic wave with phase velocity $v_P \neq 1$, find the formulae for the Doppler effect, aberration and the transformation law of phase velocities. Compare the latter to eq. (2.9.5)!

   Hint: Use the invariance of k^2!

4. Repeat the last exercise using the Galilean boost (1.3.12) instead of the Lorentzian one! (Cf. exercise 2 of sect. 3.4.)

5. Why can a single free photon never create an electron-positron pair?

6. Use the result of exercise 2 of the last section to find the maximum energy of a photon created in a bremsstrahlung process (collision of two massive charged particles) (i) in the CM frame, where, by definition, the total momentum of the incoming particles vanishes, (ii) in the rest frame of one of the incoming particles (the lab frame) as dependent on the emission angle, (iii) the maximum over all angles in the latter case.

4.4 Conversion of Mass into Energy. Mass Defect

The relativistic version (4.2.3) of the conservation laws has shown that only the sum of kinetic energy and rest energy is required to be conserved. If there are no further conservation laws implying further restrictions, then the conversion of rest mass to energy (or the other way round) will have to be expected in collisions. Figure 4.6 symbolically represents some of the wealth of examples furnished by elementary particle physics.

a. A *creation process* is observed, e.g., in proton-proton scattering, where frequently one or more π-mesons are produced: $p + p \rightarrow p + p + \pi^0$, or also $p + p \rightarrow p + n + \pi^+$. The kinetic energy of the incoming proton supplies the required rest energy of the pion.

b. The historically most important example of this kind is the *pair annihilation* $e^+ + e^- \rightarrow 2\gamma$, where rest mass is converted completely into energy. It allowed, in 1932, precision measurements to test the validity of eq. (4.2.3).

c. The *decay* $\pi^0 \rightarrow 2\gamma$ also allows a detailed check of the conversion of mass into energy. One also can test velocity addition here by determining the velocity of the γ-quanta stemming from pions decaying in flight ($v = 0.98c$).

These examples should suffice to demonstrate that conversion between mass and energy may be observed and tested in many kinds of experiments in the domain of elementary particles. In everyday life, however, relativistic mass-energy conservation practically separates into two separate conservation laws: mass and energy are

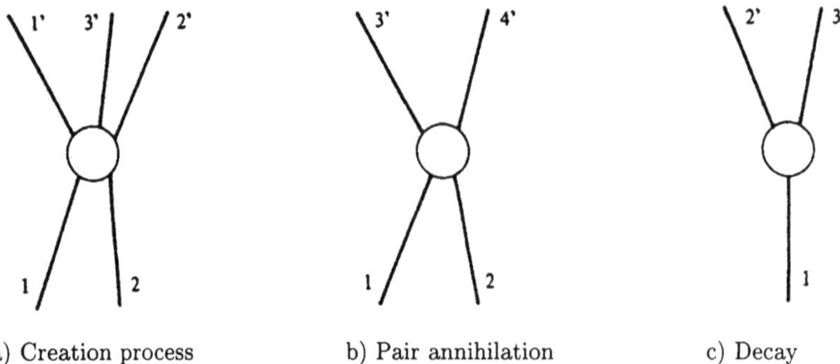

a) Creation process   b) Pair annihilation   c) Decay

Fig. 4.6. Interaction of elementary particles

separately conserved to a high degree of accuracy. The main reason for this is the existence of the further conservation laws for charge, lepton number and baryon number.[1] Thus electrons cannot decay, being the lightest charged particles; protons cannot decay, being the lightest baryons. For neutrons, the situation is more complicated: free neutrons undergo β-decay

$$n \to p + e + \bar{\nu}_e \tag{4.4.1}$$

with a lifetime of approx. 1000 sec. On the other hand, in stable nuclei the neutrons contained cannot decay because of the Pauli exclusion principle, the energy levels left over by it for the decay protons lying unfavorably, making the process (4.4.1) energetically impossible. The conservation of rest mass in the absence of antimatter is thus a consequence of the laws of quantum mechanics, of the form of the mass spectrum, and of the mentioned nongeometrical conservation laws (the latter term will be explained in chap. 10).

However, this conservation is only approximate. Let us analyze some chemical reaction in more detail, e.g., the formation of hydrogen from proton and electron in the reaction

$$p + e \to H + 13.55 \text{ eV}. \tag{4.4.2}$$

The binding energy set free in this reaction is $E_B = 13.55$ eV $= \frac{1}{2}m\alpha^2$, where m is the electron mass and $\alpha = 1/137$ is the fine structure constant. Assume for simplicity that E_B shows up as two photons emitted in opposite direction as shown in Fig. 4.7.

Taking electron and proton (mass M) as approximately at rest, their 4-momenta are $p_1^i = (M, \mathbf{0})^{\mathsf{T}}$, $p_2^i = (m, \mathbf{0})^{\mathsf{T}}$, whereas we have for the photons emitted in the process $p_4^i = (\omega, \mathbf{p})^{\mathsf{T}}$, $p_5^i = (\omega, -\mathbf{p})^{\mathsf{T}}$, where $2\omega = 2|\mathbf{p}|$. The energy-momentum balance

$$p_1 + p_2 = p_3 + p_4 + p_5 \tag{4.4.3}$$

yields

$$p_3^i = (m + M - 2\omega, \mathbf{0})^{\mathsf{T}}. \tag{4.4.4}$$

---

[1] See textbooks on particle physics for appropriate definitions of these concepts.

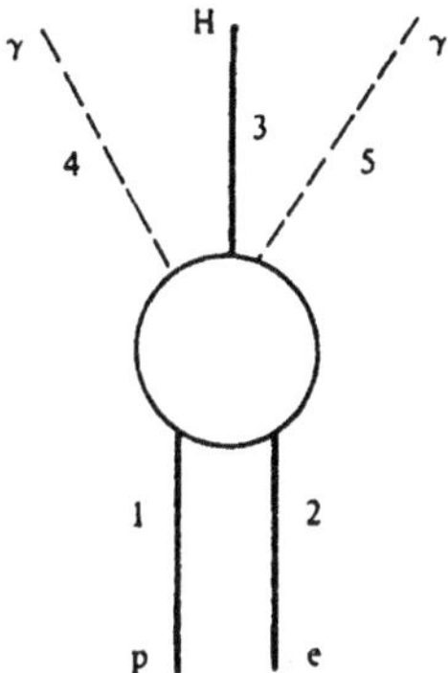

Fig. 4.7. Formation of hydrogen

The hydrogen atom results at rest, but its mass μ is not given by $m + M$ but is smaller: $\mu = m + M - 2\omega$, where the *mass defect* $\Delta\mu = (m + M) - \mu = 2\omega = E_B$ is due to the binding energy. The relative mass defect

$$\frac{\Delta\mu}{\mu} = \frac{E_B}{m + M - E_B} \approx \frac{E_B}{M} = \frac{1}{2}\alpha^2 \frac{m}{M} \approx 10^{-8} \tag{4.4.5}$$

is small on account of the smallness of the fine structure constant and the electron-to-proton mass ratio, and is still smaller in other chemical reactions, since in most cases heavier molecules with lower binding energies are formed.

From the point of view of relativity, chemical reactions may be taken as examples for the conversion of rest mass into energy, where the relative mass defect remains unmeasurably small, however. This seems to be at variance with the usual interpretation, according to which the reaction causes a change in binding energy which may be set free. This is also correct in the sense of Relativity; but this theory in addition predicts that to this energy loss $\Delta E = E_B$ there corresponds a mass loss $\Delta\mu = E_B/c^2$. Newton's theory, on the contrary, makes no statement as to change or conservation of mass in collisions or chemical reactions. Such statements would have to be postulated there in addition to the conservation of energy and momentum and have no logical connection to the structure of the theory.

Large relative mass defects $(\Delta\mu/\mu \lesssim 1\%)$ are well known to occur in atomic nuclei; they allow to test the relation between energy and mass defect to an accuracy of 10^{-3} (see, e.g., Kacser 1970).

The *largest* binding energies, and therefore the most essential mass defects, occur in astrophysics. The gravitational binding energy of a homogeneous ball of mass M and radius R is well known to be

$$E_B = \frac{3}{5}\frac{GM^2}{R}, \tag{4.4.6}$$

where G is Newton's gravitational constant. Now when a star is formed from a cloud of gas, this amount of energy is radiated away, so that the star remains with mass

$M_1 = M - E_B$. According to Newton's theory M_1 could be even negative if R is taken small enough. The General Theory of Relativity, i.e., Einstein's relativistic theory of gravitation, shows that this is not the case, since in it eq. (4.4.6.) holds only for small values of E_B/M. However, even according to the latter theory gravitational binding energies can go up to 40% of the preassembly rest mass. Thus gravitative phenomena involve the largest amounts of mass converted into energy, of course apart from matter-antimatter annihilation where 100% of the rest mass gets converted.

Exercises

1. What is the significance of the invariant $s = (q_1 + q_2)^2$ in the scattering process $q_1 + q_2 \to p_1 + p_2$? (We have written 4-momenta instead of particles.) Discuss s in the lab system ($q_2^i = (m, \mathbf{0})^\top$) and in the CM frame ($\mathbf{q}_1 + \mathbf{q}_2 = \mathbf{0}$). What is the significance of $t = (q_1 - p_1)^2$?

2. What is the minimum energy required for a proton in the lab frame to generate a pion upon interacting with a proton at rest (e.g., in a hydrogen bubble chamber)?

4.5 Relativistic Phase Space

The calculation of transition probabilities between quantum states allows to find lifetimes of excited states (e.g., of atoms) and scattering cross sections. Quantum mechanical perturbation theory yields, as a first approximation, Fermi's *Golden Rule*, according to which the transition probability per unit time from state A to state B is given by

$$w_{BA} = \frac{2\pi}{\hbar}\, \rho(E)\, |H_{BA}|^2. \tag{4.5.1}$$

Here $|H_{BA}|^2$ is the matrix element of the interaction Hamiltonian and $\rho(E)$ is the density of final states (see, e.g., Schiff 1968, p. 285).

The splitting of the transition probability appearing in eq. (4.5.1) into the factors ρ and $|H|^2$ is also of importance in relativistic quantum (field) theory. For instance, the decay of the neutron, $n \to p + e + \bar{\nu}_e$, as well as that of the muon, $\mu \to e + \bar{\nu}_e + \nu_\mu$, is caused by weak interaction. However, the lifetimes of these particles ($\tau_n \approx 1000$ sec, $\tau_\mu \approx 2 \times 10^{-6}$ sec) differ by nine orders of magnitude, since in μ-decay more rest mass is converted into energy and there is a larger number of final states available for the outgoing particles than is the case in neutron decay. This fact is described by the *phase space factor*, the relativistic generalization of $\rho(E)$.

To introduce this factor (which actually should be called momentum space factor), consider a creation process

$$q_1 + q_2 \to p_1 + p_2 + p_3, \tag{4.5.2}$$

where as before we wrote 4-momenta instead of particles. The transition probability for this process has to be of the following form, analogous to eq. (4.5.1):

$$w \propto \int d^4p_1\, d^4p_2\, d^4p_3\, \delta^4(p_1 + p_2 + p_3 - q_1 - q_2)\, \delta(p_1^2 - m_1^2)\, \delta(p_2^2 - m_2^2)\cdot$$
$$\cdot \delta(p_3^2 - m_3^2)\, h^2(p_1, p_2, p_3, q_1, q_2). \tag{4.5.3}$$

Here the first δ-function secures energy-momentum conservation in the process—i.e., if it is violated the probability is zero—and the other δ-functions put all 4-momenta onto their respective mass shell; the integration is only over future-directed vectors. The factor h^2 corresponds to $|H_{AB}|^2$ in eq. (4.5.1) and is an invariant function of the 4-momenta of the particles involved, to be obtained from the rules of quantum field theory. (The missing details to put the Lorentz invariance of expression (4.5.3) into evidence will be supplied in the appendix to this section.)

Sometimes little is known about h^2, as in the case of strong interactions. When as a first ansatz $h^2 = const.$ is *tried*, the distribution of particles in the final state is, in this approximation, determined by the *phase space factor*

$$R_3(q) := \int d^4p_1\, d^4p_2\, d^4p_3\, \delta^4(p_1 + p_2 + p_3 - q)\, \delta(p_1^2 - m_1^2)\, \delta(p_2^2 - m_2^2)\, \delta(p_3^2 - m_3^2), \tag{4.5.4}$$

an *invariant function* of the total 4-momentum $q := q_1 + q_2$. (The relativistic invariance of $R_3(q)$—and of an analogous $R_n(q)$ for n particles in the final state—follows from the invariance of d^4q and $\delta^4(q)$ under Lorentz transformations, as explained in the appendix to this section.) This ansatz is analogous to the basic assumptions of statistical mechanics, and the theory based on it is called *statistical theory*. It was introduced in 1950 by Fermi in order to explain observations at high energies (cosmic rays), since the large number of particles involved seemed to justify statistical arguments. But even with only a few particles in the final state, phase space considerations are an important aid, since one can draw conlusions either from statistical or from nonstatistical behavior. We shall illustrate this in an example.

As a first step in the calculation of $R_3(q)$ we use the δ-functions $\delta(p^2 - m^2)$ to carry out the p^0-integration: we have

$$\delta(p^2 - m^2) = \delta(p_0^2 - E^2(\mathbf{p})) = \frac{1}{2E(\mathbf{p})}[\delta(p_0 - E(\mathbf{p})) + \delta(p_0 + E(\mathbf{p}))]$$
$$E(\mathbf{p}) := \sqrt{\mathbf{p}^2 + m^2}, \tag{4.5.5}$$

and because of $p^0 > 0$ the argument of the second δ-function is always positive, so that this term does not contribute to the integral. We thus obtain

$$\int d^4p\, \delta(p^2 - m^2)\, f(p_0, \mathbf{p}) = \int \frac{d^3p}{2E(\mathbf{p})}\, f(E(\mathbf{p}), \mathbf{p}) \tag{4.5.6}$$

for arbitrary functions f.

This relation gives us the transition from the manifestly covariant 4-dimensional momentum space integral to an integration over the 3-momenta as it is known from nonrelativistic theory; the factor $1/2E(\mathbf{p})$ distinguishes between the 'noninvariant' momentum space volume element $\int d^3p$ and the 'invariant' one which is nothing but the volume element of the mass shell considered as a curved Riemannian space analogous to the velocity hyperboloid (cf. appendix to sect. 4.1).

In the manner described we can now evaluate all integrals over the 0-components of the momenta of the outgoing particles in eq. (4.5.4):

$$R_3(q) = \int \frac{d^3p_1}{2E_1} \int \frac{d^3p_2}{2E_2} \int \frac{d^3p_3}{2E_3} \, \delta^4(p_1 + p_2 + p_3 - q), \qquad (4.5.7)$$

where the p_i^0 now stand for the corresponding E_i.

At this place we can make the assumptions of the statistical theory more precise. It is assumed that not only the total transition probability is proportional to $R_3(q)$, but that also the distribution of the particles is given by the integrand of eq. (4.5.7). For instance, the probability to find particle 1 in the volume d^3p_1 is assumed to be proportional to

$$w(p_1)\, d^3p_1 \propto \frac{d^3p_1}{2E_1} \int \frac{d^3p_2}{2E_2} \int \frac{d^3p_3}{2E_3} \, \delta^4(p_1 + p_2 + p_3 - q). \qquad (4.5.8)$$

Other probabilities are calculated similarly.

From the numerous applications statistical theory has found, in particular in the realm of strong interactions (see Hagedorn 1963), we pick out the discovery of the ρ^0-meson by Erwin, March, Walker, and West in 1961[1]. In the scattering of negative π-mesons off protons, one observes, amongs others, the reaction

$$\pi^- + p \to \pi^+ + \pi^- + n. \qquad (4.5.9)$$

It had been conjectured on grounds which cannot be explained here that the reaction proceeds at least in part as shown in Fig. 4.8.

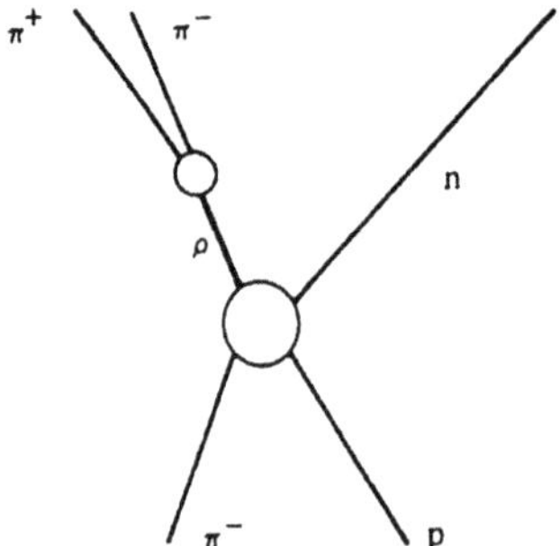

Fig. 4.8. Discovery of the ρ-meson

In the scattering process a ρ-meson is formed first, which then decays into π^+ and π^- but is much too short-lived—its lifetime is estimated to be about 10^{-23} sec—to

<hr>

[1] A. R. Erwin, R. March, W. D. Walker, E. West, Phys. Rev. Lett. 6, 628 (1961).

be observable (e.g., by leaving a trace in a bubble chamber). Now if the hypothesis of the existence of the ρ-meson is true the momenta p_1 and p_2 of the decay pions must satisfy

$$(p_1 + p_2)^2 = M^2, \qquad (4.5.10)$$

where M is the mass of the ρ-meson. This mass will only be defined within an uncertainty ΔM related to the lifetime $\tau = \Delta t$ of the ρ-meson by the uncertainty relation $\Delta M = \Delta E \approx \hbar/\Delta t$. To test this hypothesis we must ascertain that more pion pairs do satisfy the condition (4.5.10) within the mass uncertainty ΔM than would be expected statistically.

For this we use the relation

$$\int dM^2\, \delta((p_1 + p_2)^2 - M^2) = 1 \qquad (4.5.11)$$

(here M^2 is used as an integration variable) to rewrite eq. (4.5.7):

$$R_3(q) = \int dM^2 \int \frac{d^3p_1}{2E_1} \int \frac{d^3p_2}{2E_2} \int \frac{d^3p_3}{2E_3}\, \delta((p_1 + p_2)^2 - M^2)\, \delta^4(p_1 + p_2 + p_3 - q) =$$

$$=: \int dM^2\, w(M^2, q), \qquad (4.5.12)$$

where $w(M^2, q)dM^2$ is the statistical probability for the mass square (4.5.10) of the π^+-, π^--pair to lie in the interval dM^2 around M^2.

The calculation of $w(M^2, q)$ now offers the opportunity to introduce some of the standard methods of evaluating momentum space integrals. We first use

$$\int d^4k\, \delta^4(p_1 + p_2 - k) = 1 \qquad (4.5.13)$$

to rewrite $w(M^2, q)$:

$$w(M^2, q) = \int \frac{d^3p_1}{2E_1} \int \frac{d^3p_2}{2E_2} \int \frac{d^3p_3}{2E_3} \int d^4k\, \delta^4(p_1 + p_2 - k)\, \delta(k^2 - M^2)\, \delta(k + p_3 - q)$$

$$(4.5.14)$$

(because of the factor $\delta^4(p_1 + p_2 - k)$ we were able to write $\delta(k^2 - M^2)$). Changing the order of the integrations yields

$$w(M^2, q) = \int \frac{d^3p_3}{2E_3} \int d^4k\, \delta(k^2 - M^2)\, \delta^4(k + p_3 - q)\, R_2(k), \qquad (4.5.15)$$

where

$$R_2(k) = \int \frac{d^3p_1}{2E_1} \int \frac{d^3p_2}{2E_2}\, \delta^4(p_1 + p_2 - k) \qquad (4.5.16)$$

is just the invariant phase space factor for two particles (π^+, π^-). We evaluate eq. (4.5.16) for two particles of differing masses m_1, m_2, as we shall need that result immediately. $R_2(k)$ is a scalar that depends only on k; thus it is a function of k^2 alone.

Since $R_2(k)$ vanishes unless k is timelike and future-directed, it may be evaluated by going to the rest system of k, where $k^i = (\sqrt{k^2}, \mathbf{0})^\top$:

$$R_2(k) = \int \frac{d^3p_1}{2E_1} \int \frac{d^3p_2}{2E_2}\, \delta\left(E_1 + E_2 - \sqrt{k^2}\right) \delta^3(\mathbf{p}_1 + \mathbf{p}_2) =$$

$$= \frac{1}{4} \int \frac{d^3p}{\sqrt{\mathbf{p}^2 + m_1^2}\,\sqrt{\mathbf{p}^2 + m_2^2}}\, \delta\left(\sqrt{\mathbf{p}^2 + m_1^2} + \sqrt{\mathbf{p}^2 + m_2^2} - \sqrt{k^2}\right) = \frac{\pi \mathrm{p}}{\sqrt{k^2}},$$

where p is defined as the solution of

$$\sqrt{\mathbf{p}^2 + m_1^2} + \sqrt{\mathbf{p}^2 + m_2^2} = \sqrt{k^2}. \tag{4.5.17}$$

The left-hand side of eq. (4.5.17) is greater than or equal to $m_1 + m_2$, so the integral vanishes below the threshold value $k^2 = (m_1 + m_2)^2$:

$$R_2(k) = \begin{matrix} \dfrac{\pi \mathrm{p}}{\sqrt{k^2}} \\[2mm] 0 \end{matrix} \quad \text{for} \quad \begin{matrix} k^2 > (m_1 + m_2)^2 \\[2mm] k^2 < (m_1 + m_2)^2. \end{matrix} \tag{4.5.18}$$

This result is now inserted into eq. (4.5.15) with $m_1 = m_2 = m$ (pion mass):

$$w(M^2, q) = \frac{\pi}{2} \int \frac{d^3p_3}{2E_3} \int d^4k\, \delta(k^2 - M^2)\, \delta^4(k + p_3 - q)\sqrt{1 - 4m^2/k^2} =$$

$$= \frac{\pi}{2} \sqrt{1 - 4m^2/M^2} \int \frac{d^3p_3}{2E_3} \int \frac{d^3k}{2E(\mathbf{k})}\, \delta^4(k + p_3 - q). \tag{4.5.19}$$

The remaining integral is again of the form (4.5.16), and with eq. (4.5.18) and $m_3 = \mu$ (neutron mass) we obtain

$$w(M^2, q) = \begin{matrix} \dfrac{\pi^2}{2} \sqrt{1 - 4m^2/M^2}\, \dfrac{\mathrm{k}}{\sqrt{q^2}} & \text{for } 2m < M < \sqrt{q^2} - \mu \\[3mm] 0 & \text{otherwise,} \end{matrix} \tag{4.5.20}$$

where k is to be taken as the solution of

$$\sqrt{\mathrm{k}^2 + \mu^2} + \sqrt{\mathrm{k}^2 + M^2} = \sqrt{q^2}. \tag{4.5.21}$$

Here $q = q_1 + q_2$ is the sum of the momenta of the proton and of the pion. In the lab system, the proton is at rest, so that upon neglecting the proton-neutron mass difference we have $q_1^i = (\mu, \mathbf{0})^\top$. For q^2 it follows

$$q^2 = q_1^2 + 2q_1 q_2 + q_2^2 = \mu^2 + 2\mu E + m^2 \approx \mu(\mu + 2E), \tag{4.5.22}$$

where E is the energy of the incoming pion. Now all quantities in eq. (4.5.20) are known and $w(M^2, q)$ may be calculated.

Figure 4.9 compares $W(M)$—which is related to our $w(M^2, q)$ by $W(M)\,dM = w(M^2, q)\,dM^2$—with the result of the experiment of Erwin et al. It shows quite clearly

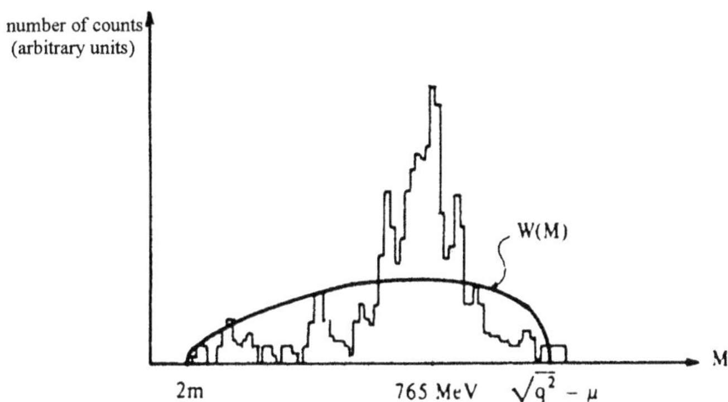

Fig. 4.9. $W(M)$—Comparison between statistical theory and experiment

that the distribution of the momenta of the pions is not statistical but corresponds to the existence of a ρ-meson with mass 765 MeV. From the figure one can also read off ΔM and calculate the ρ-meson's lifetime (exercise!).

With this application we must conclude the discussion of relativistic phase space. Numerous other applications—of particular elegance is the *Dalitz plot*—are found in Hagedorn (1963), Byckling and Kajantie (1973), Pietschmann (1974) and in all recent texts on elementary particle physics.

Appendix: Invariance of $R_n(q)$

If in momentum space we use coordinates p^i and $p^{\bar{i}}$ related by a Lorentz transformation

$$p^{\bar{i}} = L^i{}_k\, p^k, \tag{4.5.23}$$

we may form the coordinate volume elements $dp^0 dp^1 dp^2 dp^3$ and $dp^{\bar{0}} dp^{\bar{1}} dp^{\bar{2}} dp^{\bar{3}}$. Their ratio is well-known to be given by the Jacobian of the tranformation (4.5.23), for which from eq. (3.1.8) by forming determinants we obtain $|\det L| = 1$. Therefore

$$d^4 p := dp^0 dp^1 dp^2 dp^3 = dp^{\bar{0}} dp^{\bar{1}} dp^{\bar{2}} dp^{\bar{3}} \tag{4.5.24}$$

defines a volume element on momentum space itself.

The 4-dimensional delta function $\delta(p)$ is defined by

$$\int d^4 p\, \delta^4(p)\, f(p) = f(0), \tag{4.5.25}$$

thus independently of the reference frame. Because of eq. (4.5.24), it may be expressed in coordinates as

$$\delta^4(p) = \delta(p^0)\, \delta(p^1) \ldots = \delta(p^{\bar{0}})\, \delta(p^{\bar{1}}) \ldots . \tag{4.5.26}$$

The quantities $d^3 x$, $d^3 p$ are not invariant because of Lorentz contraction. Since $d^4 x = d^3 x\, dx^0$, they are just the 0-components of a covariant vector, $d^3 x = d\sigma_0$. (See sects. 5.6 and 5.7 for a more comprehensive introduction of scalar, vectorial and tensorial volume elements for lower-dimensional submanifolds.)

Exercises

1. Calculate the angular distribution of the γ-quanta for the process $e^+ + e^- \to 2\gamma$ according to the statistical theory and compare with the result of exercise 1 of sect. 4.3.

2. Determine the lifetime of the ρ-meson.

5 Relativistic Electrodynamics

The origin of Relativity Theory is strongly tied to electrodynamics, and also the wealth of applications makes relativistic electrodynamics an important part of Einstein's theory. Quantum electrodynamics, which unites Relativity, electrodynamics and quantum theory, is perhaps the most precise physical theory we have, and its successes dominated our thinking about elementary particles during the period 1945–1960. Its predictions about the magnetic moments of the electron and the muon, accurate to eight decimal places, and the calculations of the spectral lines of hydrogen with a similar precision are at the same time our best confirmations of Relativity and of electrodynamics. They also show that the relativistic space-time concept is valid down to distances of about 10^{-15} cm.

We shall here only touch upon some of the most important aspects of relativistic electrodynamics, leaving aside numerous applications of the theory, for which we refer the reader to, e.g., Jackson (1999) or Landau and Lifshitz (1961).

The formal development of the theory will be supplemented in this chapter by the introduction of the tensor concept.

5.1 Forces

In the last chapter we wrote down the relativistic version $F = ma$ of Newton's basic law of dynamics. However, for this equation to have physical content it is necessary to specify the four-force F occurring therein. What can be inserted for it?

On the phenomenological level of *macrophysics*, F could be a pressure or frictional force as in relativistic hydrodynamics, which will be sketched in chap. 10; for relativistic continuum mechanics see, e.g., Schwartz (1968). The domain of applicability of such theories is, however, quite narrow (except in astrophysical or cosmological situations, where general-relativistic versions are needed, however), since fluid flow and other macroscopic processes hardly ever reach ('relativistic') velocities close to $c(=1)$.

If we now turn to *microphysics*, we there encounter four kinds of interactions:

electrodynamics   strong interactions
gravitation   weak interactions.

The interactions on the left are characterized by *infinite range* and may be described classically by fields of (velocity-dependent) forces. The interactions on the right become pronounced only when particles approach each other closer than about 10^{-13} cm. At these short distances, however, the classical orbit concept becomes meaningless, so that a particle's acceleration cannot be defined. Consequently, in the processes illustrated in Fig. 4.6 it is not possible to use cassical concepts like force and acceleration, and one can measure and calculate only interaction cross sections, i.e., probabilities for particle scattering, production, decay, etc.

Among the two classical forces, gravitation turns out to require a special treatment also, since gravitational fields change the space-time structure: this is the subject of General Relativity.[1] Thus the electrodynamical forces remain as the only ones to be inserted into $F = ma$. Nonrelativistically, they are given by the Lorentz force $\mathbf{F} = e\,(\mathbf{E} + \mathbf{v} \times \mathbf{B})$, and the question therefore is how to convert this into a 4-vector, knowing that instead of $\mathbf{v}$ we can use the 4-velocity u, to which F then should be orthogonal. But what about $\mathbf{E}$ and $\mathbf{B}$? Is $\mathbf{E}$ to be converted into a 4-vector similarly to $\mathbf{v}$? The answer is no, and the correct way to handle this question comes from looking at Maxwell's equations.

In enumerating possible candidates for F we left out an apparently obvious possibility: relativistic theories of *action at a distance*, where the interparticle force is, e.g., proportional to $1/r^2$, r being a retarded distance in order to account for the finite speed at which the interaction is to propagate. One expects a picture like the one sketched in Fig. 5.1, in which the lines with arrows indicate the transfer of force between particles A and B. (For more details see, e.g., Anderson 1967).

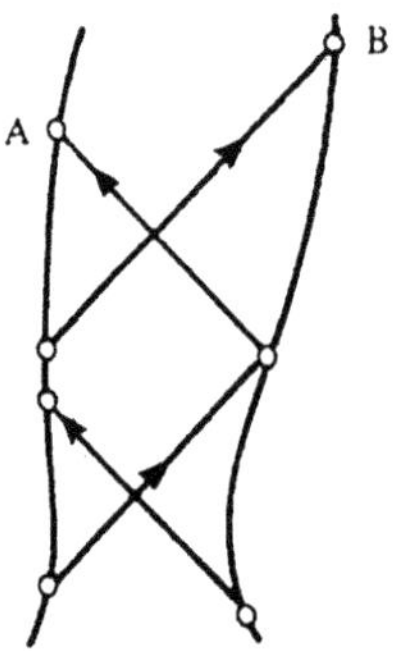

Fig. 5.1. Relativistic action at a distance

An obstacle to the construction of such theories are so-called 'no-interaction' theorems such as the following one proved by H. Leutwyler (Nuovo Cimento *37*, 556 (1965)): "A (nondegenerate) Hamiltonian theory for a finite number of interacting classical particles cannot describe interactions (i.e., the particles will move freely) if the theory is relativistically invariant and the particle coordinates transform correctly under the Poicaré group". The discussion of the consequences of this theorem and of the possibilities to loosen the assumptions made for it is not closed; see the reprint collection Kerner (1972) and, e.g., H. P. Künzle, J. Math. Phys. *15*, 1033 (1974); A. Kracklauer, J. Math. Phys. *17*, 693 (1976); Trump and Schieve (1999).

5.2 Covariant Maxwell Equations

We had to change Newton's Second Law and the definitions of energy and momentum in the conservation laws in order to adapt them to the requirements of Einsteinian Relativity. What about Maxwell's electrodynamics? Do we have to modify it also? There is one consequence of Maxwell theory which we used as a postulate in the

---

[1]See, e.g., Sexl and Urbantke (1995).

derivation of the Lorentz transformation: the way electromagnetic radiation is predicted to propagate. We actually used this only in the geometric optics limit where it says that light propagates along rectilinear rays at speed $c = 1$ in every inertial system, irrespective of its source. So not only is this consequence of Maxwell theory not modified, but its unmodified validity was at the basis of our theory. Then what about the full Maxwell eqations? One good guess—indeed the correct one—is that they also are not modified at all; this certainly guarantees compatibility with the geometric optics limit. The latter not only allows high precision tests for its validity in all frames, but gave us also a formally easy handle in the final derivation of the Lorentz transformation. With the full Maxwell equations, there is the added complication of the transformation behavior of the field strengths, which is stripped off by the geometric optics limit. It is clear that the electric and magnetic field components have to get mixed up upon change of the reference frame: consider a charge at rest in I with the corresponding electrostatic field bot no magnetic one: when viewed from $\bar{\text{I}}$ there will be an electric current and a corresponding magnetic field as well. Thus the issue here is: can we write down a transformation law for $\mathbf{E}, \mathbf{B}$ which when combined with the Lorentz transformation $\bar{x} = Lx$ carries the unbarred Maxwell equations into the barred Maxwell equations, and is this transformation law obeyed in experiment?

We shall restrict to charges and fields in vacuum, since a dielectric/magnetic medium ($\varepsilon, \mu \neq 0$) would distinguish a rest system (see, e.g., Schwartz 1968). Then the equations to be investigated are[1]

$$\text{div}\,\mathbf{B} = 0, \qquad\qquad \text{rot}\,\mathbf{E} = -\frac{\partial \mathbf{B}}{\partial t}, \qquad\qquad (5.2.1a,b)$$

$$\text{div}\,\mathbf{E} = 4\pi\rho, \qquad\qquad \text{rot}\,\mathbf{B} = \frac{\partial \mathbf{E}}{\partial t} + 4\pi\mathbf{j}. \qquad\qquad (5.2.2a,b)$$

A necessary condition for their consistency (integrability) is the *equation of continuity*

$$\text{div}\,\mathbf{j} + \frac{\partial \rho}{\partial t} = 0, \qquad\qquad (5.2.3)$$

which follows from eq. (5.2.2) using the identity $\text{div}\,\text{rot}\,\mathbf{B} \equiv 0$.

Equations (5.2.1–3) together with the Lorentz force law (sect. 5.3) in principle contain all of classical electrodynamics and are, without modification, the starting point of our considerations. Any possible distribution of charge, current and field strengths, described in I by $\rho(\mathbf{x}, t)$, $\mathbf{j}(\mathbf{x}, t)$, $\mathbf{E}(\mathbf{x}, t)$, $\mathbf{B}(\mathbf{x}, t)$, has to satisfy these equations; similarly, any distribution of sources and fields, described in $\bar{\text{I}}$ by $\bar{\rho}(\bar{\mathbf{x}}, \bar{t})$, $\bar{\mathbf{j}}(\bar{\mathbf{x}}, \bar{t})$, $\bar{\mathbf{E}}(\bar{\mathbf{x}}, \bar{t})$, $\bar{\mathbf{B}}(\bar{\mathbf{x}}, \bar{t})$, has to satisfy the analogous system of equations with $\mathbf{x}$, t replaced by $\bar{\mathbf{x}}$, $\bar{t}$. In particular, the configuration described in I by $\rho(\mathbf{x}, t), \ldots$ will have in $\bar{\text{I}}$ a description $\bar{\rho}(\bar{\mathbf{x}}, \bar{t}), \ldots$ which must be calculable in terms of $\rho(\mathbf{x}, t), \ldots$ and $\bar{x} = Lx$ in such a way that eqs. (5.2.1,2,3) imply the analogous barred versions.

Our guessing of the relations between $\bar{\rho}, \ldots$ and $\rho, \ldots$ becomes easier if the homogeneous equations (5.2.1) are satisfied identically by the well-known ansatz

$$\mathbf{E} = -\text{grad}\,V - \frac{\partial \mathbf{A}}{\partial t}, \qquad\qquad \mathbf{B} = \text{rot}\,\mathbf{A} \qquad\qquad (5.2.4a,b)$$

---

[1]We are using Gaussian units with $c = 1$.

Here the *potentials* V, $\mathbf{A}$ are determined only up to a *gauge transformation*[1]

$$V \to V - \frac{\partial \Lambda}{\partial t}, \qquad\qquad \mathbf{A} \to \mathbf{A} + \operatorname{grad} \Lambda, \qquad\qquad (5.2.5)$$

and Λ may be chosen such as to satisfy the *Lorenz condition*[2]

$$\operatorname{div} \mathbf{A} + \frac{\partial V}{\partial t} = 0; \qquad\qquad (5.2.6)$$

even then, Λ is determined only up to the addition of a solution of the equation $\frac{\partial^2}{\partial t^2} \Lambda - \triangle \Lambda = 0$. Accepting eq. (5.2.6), insertion of the ansatz (5.2.4) into eqs. (5.2.2) gives the inhomogeneous equations in the simple form

$$\Box V = 4\pi\rho, \qquad\qquad \Box \mathbf{A} = 4\pi\mathbf{j}, \qquad\qquad (5.2.7a,b)$$

where the *d'Alembertian operator* $\Box$ is defined as

$$\Box := \frac{\partial^2}{\partial t^2} - \triangle \equiv \eta^{ik} \partial_i \, \partial_k \,. \qquad\qquad (5.2.8)$$

We have indicated here already that $\Box$ is an invariant operator, being the 4-square of the 4-gradient operator; so it has the same form in all inertial frames.

If in I we now define a *four-potential* A by

$$A^i = (V, A^1, A^2, A^3)^\top, \qquad\qquad (5.2.9)$$

the Lorenz gauge condition (5.2.6) becomes

$$\partial_0 V + \partial_1 A^1 + \partial_2 A^2 + \partial_3 A^3 = \partial_i A^i = 0, \qquad\qquad (5.2.10)$$

which would be a covariant equation if the $A^i(x)$ were related to the analogous quantities in $\bar{\mathrm{I}}$ as are components of a 4-vector field. Similarly, we collect the densities of charge and current together as the components

$$j^i = (\rho, j^1, j^2, j^3)^\top \qquad\qquad (5.2.11)$$

of a *four-current (density)* j, so that the continuity equation (5.2.3) and the inhomogeneous field equations (5.2.7) become

$$\partial_i \, j^i = 0 \qquad\qquad (5.2.12)$$

$$\Box A^i = 4\pi j^i. \qquad\qquad (5.2.13)$$

Because of the invariance of $\Box$ it follows from the last equation that *assuming* that the $A^i(x)$ and their barred counterparts transform into each other like components

---

[1] This terminology stems from H. Weyl's (unsuccessful) first attempt at a 'geometrization' of the electromagnetic field where it literally referred to the gauging of rods and clocks; see, e.g., Sexl and Urbantke (1995).

[2] Not Lorentz!

of a 4-vector is compatible with the *fact*—to be shown immediately—that the same is true about the j^i.

We now show the latter, using the model of a point charge e with world line $\mathbf{z}(t)$ or $z^i(s)$, for which

$$\rho(\mathbf{x}, t) = e\,\delta^3(\mathbf{x} - \mathbf{z}(t)), \qquad \mathbf{j}(\mathbf{x}, t) = e\,\frac{d\mathbf{z}}{dt}\,\delta^3(\mathbf{x} - \mathbf{z}(t)). \tag{5.2.14}$$

We can bring this into a manifestly covariant form by using proper time as a parameter on the world line of the charge and by artificially inserting a δ-function $\delta(x^0 - z^0(s))$:

$$\rho(x) = e\int dz^0\,\delta(x^0 - z^0)\,\delta^3(\mathbf{x} - \mathbf{z}) = e\int ds\,\frac{dz^0}{ds}\,\delta^4(x - z(s))$$

$$\mathbf{j}(x) = e\int dz^0\,\frac{d\mathbf{z}}{dz^0}\,\delta(x^0 - z^0)\,\delta^3(\mathbf{x} - \mathbf{z}) = e\int ds\,\frac{d\mathbf{z}}{ds}\,\delta^4(x - z(s)).$$

Since $(dz^0/ds, d\mathbf{z}/ds) = u^i(s)$ are just the components of 4-velocity, while ds and $\delta^4(x - z)$ are invariants,

$$j^i(x) = e\int ds\,u^i(s)\,\delta^4(x - z(s)) \tag{5.2.15}$$

indeed are 4-vector components. Other charge-current distributions may be thought of as being composed of point charges. So the covariance of eqs. (5.2.10,13) is now clear if we postulate that the A^i are 4-vector components.

The 4-vector character of j may also derived from other assumptions: if in every inertial frame four functions $j^i(x)$ are given such that the equations $j^i = 0$, $\partial_i j^i = 0$ in one system imply the corresponding equations in the other systems, and if the principle of relativity holds, then the j^i define a 4-vector field. The first assumption means that the (classical) vacuum is Lorentz invariant, while the second assumption means the validity of local charge conservation in all systems. (See, e.g., Robertson and Noonan 1968, p. 84.)

The gauge transformations (5.2.5) then also appear in covariant form if we pass from A^i to covariant components $A_i = (A_0, A_1, A_2, A_3) = (A^0, -A^1, -A^2, -A^3)$:

$$A_i \to A_i - \partial_i \Lambda. \tag{5.2.16}$$

Using them, the relation (5.2.4) between the potentials and the field strengths takes the transparent form

$$\begin{aligned}
E_1 &= -\partial_1 A_0 + \partial_0 A_1 & B_1 &= -\partial_2 A_3 + \partial_3 A_2 \\
E_2 &= -\partial_2 A_0 + \partial_0 A_2 & B_2 &= -\partial_3 A_1 + \partial_1 A_3 \\
E_3 &= -\partial_3 A_0 + \partial_0 A_3 & B_3 &= -\partial_1 A_2 + \partial_2 A_1.
\end{aligned}$$

This suggests defining an *electromagnetic field tensor*[1] F with components

$$F_{ik} := \partial_i A_k - \partial_k A_i = -F_{ki}, \tag{5.2.17}$$

---

[1] The notion of *tensor* will be explained in sect. 5.4.

which we also list in matrix form as

$$(F_{ik}) = \begin{pmatrix} 0 & E_1 & E_2 & E_3 \\ -E_1 & 0 & -B_3 & B_2 \\ -E_2 & B_3 & 0 & -B_1 \\ -E_3 & -B_2 & B_1 & 0 \end{pmatrix}, \tag{5.2.18}$$

so that, e.g., $F_{01} = E_1$, $F_{12} = -B_3$.

The structure of Maxwell's equations thus suggests to unite $\mathbf{E}$ and $\mathbf{B}$ into a matrix and not to try and complete them to give two separate 4-vectors. The consequences of the transformation behavior of the field strengths resulting from eq. (5.2.17) and the assumed 4-vector nature of A will be investigated in sect. 5.8.

The contravariant components of the field tensor are defined as

$$F^{ik} := \partial^i A^k - \partial^k A^i = \eta^{il}\eta^{km} F_{lm} \tag{5.2.19}$$

and are explicitly obtained from eq. (5.2.18) to be

$$(F^{ik}) = \begin{pmatrix} 0 & -E_1 & -E_2 & -E_3 \\ E_1 & 0 & -B_3 & B_2 \\ E_2 & B_3 & 0 & -B_1 \\ E_3 & -B_2 & B_1 & 0 \end{pmatrix}. \tag{5.2.20}$$

Taking the divergence of eq. (5.2.19), using eq. (5.2.10), we get

$$\partial_k F^{ik} = \partial_k\,\partial^i A^k - \partial_k\,\partial^k A^i = -\Box A^i, \tag{5.2.21}$$

and thus the 4-dimensional form of the inhomogeneous Maxwell equations is

$$\partial_k F^{ik} = -4\pi j^i. \tag{5.2.22}$$

When a further divergence is formed, the local conservation law

$$0 = \partial_i\,\partial_k F^{ik} = -4\pi\partial_i\,j^i$$

results as an integrability condition.

The homogeneous Maxwell equations (5.2.1) may be expressed by the field tensor as well, namely by writing

$$F_{ik,j} + F_{ji,k} + F_{kj,i} = 0, \tag{5.2.23}$$

as should be verified as an exercise. We shall encounter an elegant alternative version in sect. 5.7.

Exercises

1. The covariant form of Maxwell's equations hides the fact that the equations $\operatorname{div}\mathbf{E} = 4\pi\rho$, $\operatorname{div}\mathbf{B} = 0$ contain no time derivatives and therefore are just conditions on the initial values of the fields. Show that these conditions are propagated by the remaining time development equations; i.e., they will hold at all times if they do at one time.

2. Verify that eq. (5.2.23) produces all of the homogeneous Maxwell equations on specializing the indices!

5.3 Lorentz Force

It remains to find the relativistic version of the *Lorentz force*[1]

$$\mathbf{F} = e\,(\mathbf{E} + \mathbf{v} \times \mathbf{B}) \tag{5.3.1}$$

upon a charged particle. The right-hand side is linear in the field strengths and the second term is also linear in the velocity. This suggests the ansatz

$$F^i = e\,F^{ik}u_k, \tag{5.3.2}$$

which indeed has all desired properties. With eq. (5.2.20) and $u_k = \gamma\,(1, -\mathbf{v})$ we get

$$F^i = \gamma\,e\,(\mathbf{E}\,\mathbf{v}, \mathbf{E} + \mathbf{v} \times \mathbf{B})^\top. \tag{5.3.3}$$

In the N.R. approximation the space parts of eqs. (5.3.3) and (5.3.1) thus agree. Because of the antisymmetry of F^{ik} also the condition $F^i u_i = 0$ is satisfied. Finally, from writing

$$F^i/e = (\partial^i A^k - \partial^k A^i)\,u_k \tag{5.3.4}$$

we see that the F^i constitute the components of a 4-vector F, since A^i, ∂^i are 4-vector components and $A^k u_k$, $u_k \partial^k$ are 4-scalar products.

With eq. (5.3.3), the equation of motion $F = ma = dp/ds = \gamma\,dp/dt$ gives

$$\frac{dp^0}{dt} = e\,\mathbf{E}\,\mathbf{v}, \qquad\qquad \frac{d\mathbf{p}}{dt} = e\,(\mathbf{E} + \mathbf{v} \times \mathbf{B}) : \tag{5.3.5}$$

the work done per unit time (not unit proper time) is $e\,\mathbf{E}\,\mathbf{v}$, and the change in momentum is exactly the Lorentz force (5.3.1).

Equation (5.3.2) gives the 4-force on a point particle. In case of a *continuous distribution of current* $j(x)$, $e\,u_k$ has to be replaced by the 4-current density $j_k(x)$, and we obtain the *4-force density* (force per unit volume)

$$f^i(x) = F^{ik}(x)\,j_k(x). \tag{5.3.6}$$

The transition between point particle and continuous distribution requires some care: if eq. (5.3.6) is integrated over a volume containing a point charge one does *not* obtain eq. (5.3.3), but

$$\tilde{F}^0 = \int d^3x\,\mathbf{E}\,\mathbf{j} = e\,\mathbf{E}\,\mathbf{v} = K^0/\gamma \tag{5.3.7a}$$

$$\tilde{\mathbf{F}} = \int d^3x\,(\rho\mathbf{E} + \mathbf{j} \times \mathbf{B}) = e\,(\mathbf{E} + \mathbf{v} \times \mathbf{B}) = \mathbf{F}/\gamma, \tag{5.3.7b}$$

differing by a factor γ from eq. (5.3.3). Indeed $\tilde{F}^i = (\tilde{F}^0, \tilde{\mathbf{F}})$ *does not* define a 4-vector, since the volume d^3x, by Lorentz contraction, is not invariant. Nevertheless, the $\tilde{F}^i$

---

[1] It is slightly unfortunate that the usual symbol F for force coincides with the symbol F for the Maxwell(-Faraday) field tensor, so that it is only the number of indices that will distinguish them in writing; however, we will not introduce an index-free way of writing equations like (5.3.2).

do have physical significance, giving the change in energy and momentum per unit time for the current distribution contained in some volume (where unit time refers to the inertial system in which the integration (5.3.7) is carried out). For more point particles we have

$$\tilde{F}^i = \sum_A F^i_A/\gamma = \sum_A dp^i_A/dt, \tag{5.3.8}$$

where the sum is over the particles contained in the volume.

5.4   Tensor Algebra

The definition of the quantities F_{ik} in eq. (5.2.17) together with the postulated 4-vector component transformation law for the 4-potential leads to a definite transformation law for the F_{ik} and thus to a new kind of quantities. We shall now write down this transformation law explicitly, generalize it and by abstraction introduce a new class of quantities—tensors—that can be used in writing down Lorentz-covariant laws.

From eqs. (5.2.17), (3.4.7) and (3.4.14) we obtain for the components $F_{\bar{i}\bar{k}}$ in some inertial frame $\bar{\text{I}}$:

$$F_{\bar{i}\bar{k}} = \partial_{\bar{i}} A_{\bar{k}} - \partial_{\bar{k}} A_{\bar{i}} = L_i{}^m L_k{}^n \left(\partial_m A_n - \partial_n A_m\right) = L_i{}^m L_k{}^n F_{mn}. \tag{5.4.1}$$

The F_{ik} transform like the products $b_i\, c_k$ of covariant components of two arbitrary 4-vectors b, c:

$$b_{\bar{i}} = L_i{}^m b_m \qquad c_{\bar{k}} = L_k{}^n c_n \qquad b_{\bar{i}}\, c_{\bar{k}} = L_i{}^m L_k{}^n b_m c_n. \tag{5.4.2}$$

An object F which is specified with respect to every frame I by components F_{ik} transforming like products of covariant components of two 4-vectors is called a *tensor*[1] of degree 2 with covariant components F_{ik}.

From eqs. (5.4.1) and (5.2.19) follows the transformation law of the contravariant components:

$$F^{\bar{i}\bar{k}} = L^i{}_m L^k{}_n F^{mn}. \tag{5.4.3}$$

They transform like the products of contravariant vector components.

The field tensor is just a special case of the general tensor concept which we are going to formulate now. This could be done in at least two different abstract algebraic ways,[2] but we prefer to introduce them as objects associated with a vector space $\mathbf{V}$ (such as Minkowski vector space) being given with respect to every reference frame (basis) for that space by a number of components which transform in a well-defined way, to be specified below, when we pass from one frame $\{e_i\}=\text{I}$ to another one, $\{\bar{e}_i\} = \bar{\text{I}}$, according to eqs. (3.3.5,6). We remember that vector components with respect to these bases transform as

$$
\begin{aligned}
x^{\bar{i}} &= L^{\bar{i}}{}_k\, x^k & L^{\bar{i}}{}_k L_i{}^{\bar{j}} &= \delta_k^j \\
x^k &= L_i{}^k x^{\bar{i}} & L_i{}^k L^j{}_k &= \delta_{\bar{i}}^{\bar{j}}
\end{aligned}
\, , \tag{5.4.4}
$$

---

[1]This term is taken from elasticity: tensio = stress; its use for the general type of quantities considered below is due to A. Einstein and M. Grossmann: see Reich (1994).

[2]See Appendix B.8 for one of them.

and we assume for the moment that the matrices $(L^i{}_k)$ and $(L_i{}^k)$ are just contragredient to each other, i.e., one is the transposed inverse of the other, without necessarily satisfying any further conditions such as our pseudo-orthogonality relations (3.1.8); for the moment the dimension of the vector space is not restricted to be 4 (but is to be finite).

Then an object T is called a *tensor of type (or bidegree)* (a, b) if in every frame I it is specified by a system of numerical components

$$T^{i_1 \dots i_a}{}_{k_1 \dots k_b}$$

such that the components in I and the ones in $\bar{\text{I}}$ are related by the linear transformation law

$$T^{\bar{i}_1 \dots \bar{i}_a}{}_{\bar{k}_1 \dots \bar{k}_b} = L^{i_1}{}_{m_1} \dots L^{i_a}{}_{m_a} L_{k_1}{}^{n_1} \dots L_{k_b}{}^{n_b} T^{m_1 \dots m_a}{}_{n_1 \dots n_b}. \tag{5.4.5}$$

Tensors of type $(a, 0)$ are called *contravariant* of degree a, those of type $(0, b)$ are called *covariant* of degree b, the others are called *mixed*. Scalars are included as tensors of type $(0, 0)$, the original vectors (*contravariant vectors*) as type $(1, 0)$ and covectors (linear functionals on the original vector space = *covariant vectors*) as type $(0, 1)$ tensors. Since eq. (5.4.5) is linear, the vanishing of all components in one basis implies the vanishing in all other bases—one then says that the tensor T itself vanishes.

We now come to the algebraic manipulations of tensors. Given two tensors A, B of the *same* type we can define linear combinations $C = \alpha A + \beta B$ with numerical coefficients (from the same field of numbers over which the original vector space $\mathbf{V}$ is defined) as tensors C with components

$$C^{i \dots}{}_{kj \dots} := \alpha\, A^{i \dots}{}_{kj \dots} + \beta\, B^{i \dots}{}_{kj \dots} \tag{5.4.6}$$

relative to I, since it follows immediately from eq. (5.4.5) as written down for A, B that the components of C thus defined transform in the required way. Therefore, *tensors of a fixed type* (a, b) (and a fixed physical dimension) over the same vector space (of dimension n, say) *form a vector space* (of dimension n^{a+b}).

Symmetry resp. *antisymmetry* in any pair of index positions of the same kind (upper or lower), e.g.,

$$A^{i \dots}{}_{k \dots j \dots} = A^{i \dots}{}_{j \dots k \dots} \quad \text{resp.} \quad B^{i \dots}{}_{k \dots j \dots} = -B^{i \dots}{}_{j \dots k \dots}, \tag{5.4.7}$$

is a property of the tensor itself, as is easily checked from eq. (5.4.5). All tensors of a fixed type which are symmetric or antisymmetric in a fixed pair of index positions each form a subspace of the tensor space under consideration.[1]

Besides addition of tensors of the same type one can define a multiplication of tensors A, B of arbitrary types (a, a'), (b, b'): the result is the *tensor product* $D = A \otimes B$ of type $(a + b, a' + b')$ with components

$$D^{i \dots mn \dots}{}_{kj \dots l \dots} := A^{i \dots}{}_{kj \dots}\, B^{mn \dots}{}_{l \dots}. \tag{5.4.8}$$

---

[1] A systematic treatment of more complicated symmetry types needs methods from combinatorics; for its relation to the representation theory of the linear group see Boerner (1955) or Fulton and Harris (1991).

The correct transformation law is readily seen.

For a single mixed tensor T of type (a, b) one can define an operation of *contraction* with respect to a pair of an upper and a lower index position. The result is a tensor E of type $(a - 1, b - 1)$ with components

$$D^{k\cdots}{}_{jm\ldots} := T^{ik\cdots}{}_{jim\ldots} \tag{5.4.9}$$

if the contraction is between the first upper and the second lower position; the correct transformation law follows using the relations (5.4.4). It is important that always one upper index is taken equal to one lower index and is summed over its range. A special case is the *trace* T^i_i of a tensor T of type $(1,1)$, a scalar quantity. One also can combine the formation of tensor products and contraction on a pair of indices belonging to different factors (*transvection*); a special example here is the scalar product of a covariant and a contravariant vector, i.e., of tensors of types $(0,1)$ and $(1,0)$:

$$c := a_i b^i \tag{5.4.10}$$

is a tensor of type $(0,0)$, a scalar.

A basic theorem about contractions, useful in many considerations, is the *quotient theorem*: If D is an object that with respect to all frames I is specified by components $D^{\cdots}{}_{\ldots}$ such that the quantities

$$B^{ik\cdots}{}_{jm\ldots} := D^{ik\ldots rs}{}_{jm\ldots n\ldots}\, A^{n\cdots}{}_{rs\ldots} \tag{5.4.11}$$

formed in all frames I with the help of arbitrary tensors A of type (a, a') turn out to behave as components of a tensor B of type (b, b'), then the object D is a tensor of type $(b + a', b' + a)$.

This theorem is frequently invoked to demonstrate the tensorial nature of some object. It also allows to regard tensors as linear maps between tensor spaces in many ways (D maps the type (a, a') tensor space into the type (b, b') tensor space by contraction as in eq. (5.4.11)). Reading eq. (5.4.10) in this manner, the covariant vector appears as a map of the space of contravariant vectors into the space of scalars: this is the way covariant vectors = covectors (making up the dual[1] vector space— see Appendix B.2) are introduced in abstract linear algebra. Other examples are found in elasticity theory, where, e.g., the stress tensor $P_{\mu\nu}$ assigns to a given vectorial surface element $d\mathbf{O}$ the force $\mathbf{F}$ acting on it according to $F_\mu = P_{\mu\nu}\, dO^\nu$: hence the name tensor. Similarly, according to eq. (5.3.2) the Lorentz 4-force is linearly assigned to the particle's 4-velocity by the electromagnetic field tensor.

Exercises

1. Find the dimension of the type (a, b) tensor space.

2. Prove the quotient theorem, starting from the simplest cases.

---

[1] This name stems from projective geometry, where contravariant x^i appear as homogeneous point coordinates and covariant a_i appear as homogeneous hyperplane coordinates, the equation of a hyperplane being given by $a_i x^i = 0$, and the 'principle of duality' of this geometry appearing as an interchange between vectors and covectors.

5.5 Invariant Tensors, Metric Tensor

The identity map $x \mapsto x$ of a vector space $\mathbf{V}$ onto itself is written in components as

$$x^i \mapsto x'^i = x^i \equiv \delta^i_{\ k} x^k. \tag{5.5.1}$$

This shows that the $\delta^i_{\ k}$ form the components of a tensor of type $(1,1)$—the *unit tensor*—whose components take the same values, given by the Kronecker symbol, in all frames. This is an example of numerically *invariant tensors*, whose components behave like scalars. We may ask ourselves whether there are other such tensors.

It is trivial that linear combinations of products

$$a\,\delta_i^{\ k}\delta_j^{\ m}\ldots + b\,\delta_i^{\ m}\delta_j^{\ k}\ldots + \ldots \tag{5.5.2}$$

give further invariant tensors of all types (p,p). One can show that this exhausts all invariant tensors if—as has been assumed in the last section and up to this point—the transformations in eqs. (5.4.4) are completely general (invertible) linear transformations. We also remark that if these linear transformations and the corresponding transformation laws (5.4.5) are interpreted actively as in sect. 3.3, then the numerical invariance of a tensor means that all the linear transformations defined by it—see end of last section—*commute* with the transformations (5.4.4,5).

Particularly important among the tensors (5.5.2) are those mediating projections of the type $(p,0)$ or $(0,p)$ tensor spaces onto subspaces of a certain symmetry type. We shall consider here only the subspaces of tensors totally symmetric or totally antisymmetric with respect to all pairs of index positions. (The latter are sometimes called p-vectors when contravariant and (exterior) p-forms when covariant.)

The projection onto the latter is given by the *antisymmetrizer*

$$T_{ijk\ldots} \mapsto T_{[ijk\ldots]} := \frac{1}{p!}\,\delta^{lmn\ldots}_{ijk\ldots}\,T_{lmn\ldots}, \tag{5.5.3}$$

where

$$\delta^{lmn\ldots}_{ijk\ldots} := \begin{vmatrix} \delta_i^l & \delta_j^l & \delta_k^l & \ldots \\ \delta_i^m & \delta_j^m & \delta_k^m & \ldots \\ \delta_i^n & \delta_j^n & \delta_k^n & \ldots \\ \ldots\ldots\ldots\ldots\ldots \end{vmatrix} = \delta_i^l\,\delta_j^m\delta_k^n\ldots - \delta_i^m\,\delta_j^l\,\delta_k^n\ldots + \ldots \tag{5.5.4}$$

is the *generalized Kronecker symbol*. The factor $1/p!$ in eq. (5.5.3) makes the map into a projection, i.e., it acts like the identity on the subspace of totally antisymmetric tensors.

Similarly one defines the *symmetrizer*

$$T_{ijk\ldots} \mapsto T_{(ijk\ldots)}, \tag{5.5.5}$$

where instead of the $\delta^{\cdots}_{\cdots}$ a tensor has to be used that arises from the expanded version of eq. (5.5.4) replacing all minus signs by plus signs. Analogous operations can be performed on type $(p,0)$ tensors.

Further invariant tensors arise when the group of transformations L in eq. (5.4.4) is restricted to some subgroup—we finally want to get back to our Lorentz group! In the active interpretation we are looking for tensors whose associated linear maps between tensor spaces commute with the action of the transformations of the subgroup in those spaces. Passively, we consider a subclass of bases in our original vector space whose elements are related by the transformations of the group—so the whole subclass is distinguished, but no basis in the subclass is distinguished among the other ones there. (For the Lorentz group, this will be a class of bases called orthonormal, as earlier.) What we are looking for here are tensors whose components are the same with respect to all bases of the subclass; such tensors will be called (numerically) invariant with respect to the subgroup.

We shall describe our first step in shrinking our group for dimension 4 only, mainly for typographical reasons, most generalizations to other dimensions being evident, except that there is a basic difference between even and odd dimensions. The space of all totally antisymmetric tensors of type $(4,0)$ or $(0,4)$ is 1-dimensional, as these tensors have, in a given basis, only one essential component: the components have to be given as a numerical multiple of the permutation symbol

$$\epsilon(ijkm) := \begin{array}{l} 0, \text{ if two indices are equal} \\ +1, \text{ when } ijkm = \text{even permutation} \\ -1, \text{ when } ijkm = \text{odd permutation} \end{array} \text{ of 0123.} \qquad (5.5.6)$$

This numerical factor changes upon change of basis by the factor $\det L$ or $(\det L)^{-1}$, due to the definition of the determinants of the matrices $L^i{}_k$ or $L_k{}^i$, respectively:

$$\begin{aligned} \epsilon(abcd)\, L^i{}_a\, L^j{}_b\, L^k{}_c\, L^m{}_d &= \epsilon(ijkm)\det(L) \\ \epsilon(abcd)\, L_i{}^a\, L_j{}^b\, L_k{}^c\, L_m{}^d &= \epsilon(ijkm)\det(L^{-1}) \end{aligned} \qquad (5.5.7)$$

This shows that such tensors are invariant under the subgroup of *unimodular transformations* L (i.e., $\det L = 1$). If nonzero, they are called *determinant tensors* or *ϵ-tensors*. If we distinguish one of them, $\epsilon_{abcd} \neq 0$, we can consider the subclass of bases, called *unimodular* with respect to ϵ and related to each other by unimodular transformations, in which $\epsilon_{abcd} = \epsilon(abcd)$ holds. In addition, we now also choose a contravariant determinant tensor ϵ^{abcd} by postulating

$$\epsilon^{abcd} = -\epsilon_{abcd} = -\epsilon(abcd) \quad \text{for unimodular bases.} \qquad (5.5.8)$$

(The minus sign is, for the moment, a convention only.)

Contracting four contravariant vectors with $\epsilon_{....}$ gives, in a real vector space, the *oriented volume* of the parallelopiped spanned by them relative to a unit parallelopiped as defined by one of the unimodular bases. $\epsilon_{....}$ and the class of unimodular bases determine each other according to eq. (5.5.8). We will come back later to the choice which is relevant to *physics*.

The tensor product $\epsilon_{abcd}\epsilon^{ijkm}$ and its contractions are invariant even under the full linear group because of $\det L \cdot \det L^{-1} = 1$; thus they must be expressible in terms of products $\delta_a{}^i \delta_b{}^k \ldots$. The following formulae, plausible from the antisymmetry of the $\epsilon \ldots$, are useful:

$$\epsilon_{ikjm}\epsilon^{ikjm} = -4!$$ (5.5.9a)

$$\epsilon_{ikjm}\epsilon^{ikjn} = -3!\,\delta^n_m$$ (5.5.9b)

$$\epsilon_{ikjm}\epsilon^{ikrn} = -2!\,\delta^{rn}_{jm}$$ (5.5.9c)

$$\epsilon_{ikjm}\epsilon^{irsn} = -1!\,\delta^{rsn}_{kjm}$$ (5.5.9d)

$$\epsilon_{ikjm}\epsilon^{abcd} = -0!\,\delta^{abcd}_{ikjm}.$$ (5.5.9e)

Using these ϵ-tensors one may associate to a given antisymmetric tensor of type $(p,0)$ its so-called *dual tensor*[1] of type $(0, n - p)$. Let T^{ikjm}, T^{ikj}, T^{ik} be antisymmetric tensors, T^i a vector, T a scalar; then we form (* *operation*)

$$*T = \frac{1}{4!}\,\epsilon_{ikjm}\,T^{ikjm}$$ (5.5.10a)

$$*T_m = \frac{1}{3!}\,\epsilon_{ikjm}\,T^{ikj}$$ (5.5.10b)

$$*T_{jm} = \frac{1}{2!}\,\epsilon_{ikjm}\,T^{ik}$$ (5.5.10c)

$$*T_{kjm} = \frac{1}{1!}\,\epsilon_{ikjm}\,T^{i}$$ (5.5.10d)

$$*T_{ikjm} = \frac{1}{0!}\,\epsilon_{ikjm}\,T.$$ (5.5.10e)

Analogously, using ϵ^{abcd} we define a * *operation* mapping $(0, p)$ tensors linearly to $(n - p, 0)$ tensors. It emerges from eqs. (5.5.9) that both operations are essentially inverses of each other; we have, e.g.,

$$*_*T^{ab} = \frac{1}{2!}\epsilon^{jmab}\,*T_{jm} = -T^{ab}.$$ (5.5.11)

It must be pointed out here that the definitions (5.5.8,10) contain conventions that vary from one author to the other. Also observe that on using x^4 in place of x^0 the natural ordering of indices—for which the permutation symbol was defined to equal $+1$—is 1234, so that $\epsilon(4123) = -1$, while we here chose $\epsilon(0123) = +1$.

From the group of unimodular transformations let us now come back to the *Lorentz group*. In this case, the matrices $(L^i{}_k)$ and $(L_i{}^k)$ were not only contragredient to each other but satisfied eq. (3.1.8). Multiplying that equation by $L_a{}^i L_b{}^k$ gives

$$\eta_{ab} = L_a{}^i L_b{}^k \eta_{ik},$$ (5.5.12)

which means that upon restriction of the basis transformations to the Lorentz group the η_{ik} form the components of a numerically invariant symmetric tensor

---

[1]This name again comes from projective geometry; in particular, the product tensors $x^{[i}y^{k]}$ resp. $a_{[i}b_{k]}$ are there interpreted as the Plücker-Grassmann coordinates of a straight line that connects two points x, y resp. is, for $n = 4$, the intersection of two planes a, b in projective 3-space, and $*$, $\,^*$ carry one description of the line into the other.

of type (0,2)—the *metric tensor*. More precisely: if *one* basis is declared to be orthonormal, and a metric tensor η of type (0,2) is defined by having components $\eta_{ik} = \mathrm{diag}\,(1,-1,-1,-1)$ with respect to that basis, then η will have the same components in all Lorentz-transformed bases (which together form the *class of orthonormal bases*; cf. Appendix B.14). If interpreted as a linear map, the metric tensor maps the space of contravariant vectors bijectively onto the space of covariant vectors according to $x^i \mapsto a_i = \eta_{ik}\,x^k$, since $\det \eta_{ik} = -1 \neq 0$. The inverse matrix η^{ik} introduced in sect. 3.4 yields the inverse map and forms, according to the quotient theorem, the components of a numerically invariant symmetric type (2,0) tensor.

In the theory of relativity, the metric tensor η and the map associated with it plays a fundamental role. One therefore identifies covariant vectors and contravariant vectors related by this map, calling them just (four-)vectors; keeping only the distinction between covariant and contravariant components, related by eqs. (3.4.1,5). The map, the identification and the resulting index transport are also extended to tensors. For example,

$$F^{ik} = \eta^{im}\,\eta^{kn}\,F_{mn}, \qquad F^i{}_j = \eta^{im}\,F_{mj}, \qquad F_{ik} = \eta_{im}\,\eta_{kn}\,F^{mn}$$

are considered as the contravariant, mixed, and covariant components of one and the same 4-tensor, so that instead of its type, or bidegree, (a,b) we only have to consider its (total) *degree* $p = a + b$ (=2 in our example). Note, however, that if indices are written in a mixed position they have to be *staggered*, so that, e.g., the antisymmetry of F can be appropriately expressed: $F^i{}_j = -F_j{}^i$. (This would be unnecessary for mixed tensors under the general linear group where upper and lower indices are totally unrelated.)

In writing η^{ik}, η_{ik} we are indicating what can be checked explicitly (exercise): these may be viewed as the contravariant and covariant components of one and the same tensor, due to the symmetry of η.

So far we have restricted our transformations by eqs. (3.1.8) or (5.5.12) only; but we already pointed out that from these conditions we may obtain

$$\det\left(L^i{}_k\right) = \pm 1. \tag{5.5.13}$$

Now the transformations L with $\det L = +1$ obviously form a subgroup (by the multiplication law for determinants); they are called *proper Lorentz transformations* (cf. chap. 6). For them, besides η also the ϵ-tensors are at our disposal, and the conventions chosen in eq. (5.5.8) are just to indicate that the two ϵ's are assumed to be related by η, which fixes one of them in terms of the other, and the covariant one is fixed as follows. We choose some orthonormal basis with future-directed e^0 and with $\{e^1, e^2, e^3\}$ forming a right-handed system, and just decree it to be unimodular. Then according to the rules of index transport

$$\epsilon^{ijkm} = \eta^{ia}\,\eta^{jb}\,\eta^{kc}\,\eta^{md}\,\epsilon_{abcd} = \epsilon(ijkm)\det(\eta^{pq}) = -\epsilon(ijkm),$$

which agrees with eq. (5.5.8).

Since we do not distinguish any more between $\epsilon^{\cdots}$ and $\epsilon_{\cdots}$ and between vectors and covectors, we have instead of the maps *, $_*$ just one map, called the (*Hodge-*)

$*$-operation, mapping totally antisymmetric tensors of degree p to those of degree $4 - p$—the dimension of both tensor spaces is $\binom{4}{p} = \binom{4}{4-p}$. A repeated application gives

$$**T = (-1)^{p-1}\,T, \tag{5.5.14}$$

as may be shown using eqs. (5.5.9); we thus have an explicit inverse. The case $p = 2$ will be of special importance due to its application to the electromagnetic field.

Under 'improper' Lorentz transformations ($\det L = -1$) the components of ϵ only change sign; such quantities are called *pseudoscalars* under the full Lorentz group. Analogously one can define *pseudotensors*, for which the transformation law (5.4.5) contains the extra factor sign $\det L$ (see sect. 8.5). It should be mentioned that, in the older physics literature in particular, it is sometimes usual to define $\epsilon_{....}$ as a pseudotensor, such that it is invariant under the full Lorentz group including improper transformations; we will express this choice by using the symbol $\varepsilon_{....}$. If this is used, dualization is written $\star T$; it yields a pseudotensor when T is a tensor; in cases where the difference is relevant, these duals cannot be added to tensors. It may be debated whether this disadvantage is compensated by the full invariance of $\varepsilon_{....}$. In the mathematical literature and in modern physics texts ϵ is preferred, since then $*$ does not lead out of the proper tensors, thus enabling to define the important concept of *selfduality* (cf. sect. 6.6), which, however, is not reflection invariant. The use of $\varepsilon_{....}$, $\star$ corresponds to the traditional way of proceeding in three dimensions: under a reflection one does not change the right-hand rule into a left-hand rule.

The algebraic manipulations on tensors developed so far allow to form, for any given set of tensors $(A_i,\ B_{jk},\ \dots)$, an infinity of further tensors of various degrees, such as

$$A_i A^i, \quad B^k{}_k, \quad B_{jk} B^{jk}, \quad A_i A_k B^{jk}, \quad A_i B^i{}_k, \quad \alpha\,A_i A_k + \beta\,B_{ik},$$
$$\delta_i{}^j A_k B_{jl}\,\epsilon^{kjlm}, \quad \dots\ ,$$

so-called *concomitants* of the tensor system. Generally, a tensorial concomitant of a given tensor system is a tensor whose components are functions of the components of the tensors of the given system, the same functions occurring whatever frame the components are referring to. Of particular importance are the scalar concomitants, the *invariants* of the tensor system. There are infinitely many of them, but only finitely many of them are independent, one may always choose a 'fundamental system' of invariants and express any further invariant as a function of the members of the fundamental system. However, apart from the simplest cases it is a rather complicated problem to do this explicitly.

For the concomitants obtained by the manipulations of tensor algebra as indicated by the selection written above, the functions just mentioned are homogeneous *polynomials* of some given degree. This, together with the restriction to the linear group, is the realm of the classical *theory of invariants*. Basic here is a famous theorem of Hilbert (see Weitzenböck 1923, Weyl 1946, Dieudonné and Carrell 1971) saying that there is a finite system of polynomial invariants such that all polynomial invariants are polynomials of the members of that finite system. The latter are in general functionally—indeed polynomially—dependent, the polynomial relations among them being called *syzygies*; it is not possible in general, however, to work with a functionally independent set if all polynomial invariants are to be expressed polynomially. There is a corresponding statement for other concomitants.

For the purposes of physics, the restriction to polynomial invariants and concomitants would be too narrow, of course. For instance, the phase space factor $R_2(k)$ considered in eq. (4.5.16) is a scalar concomitant of k but is not polynomial, and is nonpolynomially expressible in terms of the only fundamental polynomial invariant—the 4-square—as eqs. (4.5.17,18) show. A polynomial fundamental system is also not sufficient if discontinuous functions of the components are admitted: $\mathrm{sign}\,k^0$ is invariant under Lorentz transformations without time reversal but cannot be expressed in any way, polynomially or not, in terms of the 4-square.

There are analogous statements about polynomial concomitants. In physics this offers the possibility of making phenomenological ansätze; as an example we mention the introduction of form factors in the calculation of matrix elements of 4-current operators in particle physics (see Källén 1964).

Consider, for example, the tensor system consisting just of one single antisymmetric tensor F_{ik} of degree two; then a fundamental system of invariants under proper Lorentz transformations is

$$I_1 := \frac{1}{4} F_{ij}\, F^{ji}, \qquad\qquad I_2 := \frac{1}{4}\, {*F}_{ij}\, F^{ji}. \qquad (5.5.15)$$

All further polynomial invariants such as

$$F_i{}^k\, F_k{}^j\, F_j{}^i\; (\equiv 0), \qquad\qquad F_i{}^k\, F_k{}^j\, F_j{}^m\, F_m{}^i,\; \dots\,, \qquad (5.5.16)$$

may be expressed polynomially in terms of them (cf. exercise 6). Their physical significance will be illustrated later.

Exercises

1. Show that the map (5.5.3) is indeed a projection, i.e., that

$$\frac{1}{p!}\, \delta^{abc\dots}_{ijk\dots}\; \frac{1}{p!}\, \delta^{lmn\dots}_{abc\dots} \;=\; \frac{1}{p!}\, \delta^{lmn\dots}_{ijk\dots}\,. \qquad (5.5.17)$$

2. Prove eqs. (5.5.9,11) and find the remaining inverses for eqs. (5.5.10)! What would be the generalization to n dimensions?

3. Show that writing η^{ik} for the inverse matrix of the metric tensor components is compatible with the rule of index transport.

4. Let F_{ik} be an antisymmetric tensor. Show that the following equations are equivalent:

$$v_{[j}\, F_{ik]} = 0, \quad v_j\, F_{ik} + v_i\, F_{kj} + v_k\, F_{ji} = 0, \quad v_k\, {*F}^{ik} = 0. \qquad (5.5.18)$$

5. Let F_{ik}, G_{ik} be antisymmetric tensors. Prove the relation

$$G_{ij}\, F^{jk} - {*F}_{ij}\, {*G}^{jk} = \frac{1}{2}\, \delta_i{}^k\, G_{mj}\, F^{jm}. \qquad (5.5.19)$$

6. Let F_{ik} be an antisymmetric tensor. Denote the component matrices $(F^i{}_k)$, $({*F}^i{}_k)$ by F, $*F$.

    a. Using eqs. (5.5.14,19), prove the relations

$$F^2 - (*F)^2 = 2I_1\,E, \qquad F\,{*F} = I_2\,E,$$
$$F^4 - 2I_1\,F^2 - I_2^2\,E = 0, \tag{5.5.20}$$

in which E denotes the unit matrix and I_1, I_2 are as in eqs. (5.5.15). Now express the invariants (5.5.16) by I_1, I_2; in particular, show that

$$\det F = \det *F = -I_2^2.$$

    b. Show that if $I_2 = 0$ but $F \not\equiv 0$, then the rank of the matrix F equals 2 and that in this case there are two linearly independent 4-vectors p, q so that $F_{ik} = p_i\,q_k - q_i\,p_k$. Verify that I_1 is now proportional to the *Gram determinant* $p^2 q^2 - (pq)^2$ of these vectors, whose sign accounts for the number of lightlike directions $k = \lambda p + \mu q$ $(k^2 = 0)$ contained in the 2-dimensional space spanned by p, q.

    Hint: Exclude first the case of rank 1 from antisymmetry and then rank 3 using the second of eqs. (5.5.20), according to which the columns of $*F$ are now in the 1-dimensional kernel of F, which would imply rank 1 for $*F$, again impossible from antisymmetry.

7. Let $A_j\,(x) = \mathrm{Re}\{a_j \exp(-ik_l x^l)\}$ be the 4-potential of a plane electromagnetic wave in vacuum, with complex amplitude a and wave vector k.

    a. Find the conditions on a, k implied by the field equations and the Lorenz gauge condition.

    b. The field tensor takes the form $F_{mn} = \mathrm{Re}\{f_{mn} \exp(-ikx)\}$. Find the complex amplitude f_{mn} and show that

$$f^{mn}\,k_n = 0 = *f^{mn}\,k_n, \qquad f^{mn}\,f_{nm} = 0 = f^{mn}\,{*f_{nm}},$$
$$F^{mn}\,k_n = 0 = *F^{mn}\,k_n, \qquad F^{mn}\,F_{nm} = 0 = F^{mn}\,{*F_{nm}}. \tag{5.5.21}$$

(Here $*$ means dualization, not complex conjugation!) Interpret these equations, splitting k as $(k^0, \mathbf{k})^\top$ and F into $\mathbf{E}, \mathbf{B}$. What are the consequences of the fact that we have here the vanishing of 4-vectors and 4-scalars?

    c. Show that the wave is *circularly polarized* if and only if the complex field amplitude is *selfdual* or *antiselfdual* in the sense that

$$*f_{mn} = \pm i f_{mn}. \tag{5.5.22}$$

Which sign corresponds to right circular polarization, according to all conventions made so far? What are the consequences of the fact that circular polarization has been characterized by a tensorial relation under proper Lorentz transfomations?

5.6   Tensor Fields and Tensor Analysis

The electromagnetic 4-potential and field tensor actually provide examples for 4-vector *fields* and 4-*tensor fields*. Generally, a tensor field of degree p on Minkowski space $\mathbf{X}_4$ associates to every point x of that space a tensor of degree p. It is specified with respect to each reference frame I by component functions which upon change of that frame according to Poincaré transformations

$$x^{\bar{i}} = L^i{}_k\, x^k + a^i \tag{5.6.1}$$

transform in a way completely analogous to eq. (3.4.10):

$$T^{\bar{i}\cdots}{}_{\bar{k}\dots}(x^{\bar{l}}) = L^i{}_m \dots L_k{}^n \dots T^{m\cdots}{}_{n\dots}(x^l). \tag{5.6.2}$$

It should be stressed that the component functions on the left and on the right have to be thought of as depending on different arguments, related by eq. (5.6.1). The consequences of this will be discussed in detail for the example of the electromagnetic field tensor.

The *differentiation of tensor fields* is simple: since the differential operators $\partial_i = \partial/\partial x^i$ with respect to our Cartesian inertial coordinates behave, according to sect. 3.4, like 4-vector components, their application to the component functions of a tensor field T of degree p will lead to components of a tensor field D of degree p+1:

$$\partial_j\, T^{i\cdots}{}_{k\dots}(x) = D^{i\cdots}{}_{k\dots j}(x). \tag{5.6.3}$$

By repeated application of ∂ and the tensor operations discussed above one obtains expressions which, when equated to zero, yield Poincaré-covariant field equations—i.e., field equations that take the same form in all Cartesian inertial systems. When a law of nature is formulated in this manner, as are the laws (5.2.12,13,22,23), it automatically satisfies the principle of relativity.

This already concludes the essentials of the process of differentiating tensor fields. It should be noted, however, that if tensorial field equations are written down at will, their consistency has to be checked. Tensorial field equations are systems of partial differential equations, among which in general there will exist integrability conditions, to be obtained by applying further operations ∂_i and taking into account their commuting property $\partial_i\,\partial_j = \partial_j\,\partial_i$. A simple example of this procedure is furnished by our deduction of 4-current conservation from eq. (5.2.22); an example of a slightly more refined procedure will follow eq. (5.9.29); the general theory is very complicated but is fortunately not needed in ordinary applications.

We come to the *integration of tensor fields* over regions of Minkowski space or over lower-dimensional submanifolds therein, e.g., hyperplanes such as $t = const.$, light cones $(x - x_0)^2 = 0$, ... For this we first need suitable volume elements. Our starting point is the formula for the invariant volume of a parallelopiped spanned by four 4-vectors A, B, C, D:

$$\mathcal{V}(A, B, C, D) = \epsilon_{ijkm}\, A^i\, B^j\, C^k\, D^m. \tag{5.6.4}$$

This (pseudo)scalar enjoys all properties one expects from a 4-volume in the sense of Minkowski geometry: If one edge, A, say, is replaced by λA, $\mathcal{V}$ gets replaced by $\lambda \mathcal{V}$; it vanishes if any two spanning edges A, B, C, D are parallel; it is invariant under (active) proper Lorentz transformations of these edges; and it is normalized in the sense that for all orthonormal bases we have $\mathcal{V}(e_0, e_1, e_2, e_3) = \pm 1$, the $+$ sign referring to positively oriented bases as discussed earlier.

The volume element spanned by the infinitesimal vectors $e_0 dx^0$, $e_1 dx^1$, $e_2 dx^2$, $e_3 dx^3$ is therefore

$$dV = dx^0 \, dx^1 \, dx^2 \, dx^3 = d^4 x. \tag{5.6.5}$$

This allows arbitrary tensor fields $T(x)$ to be integrated over 4-dimensional domains Γ of Minkowski space, the result being a tensor t given by

$$t^{i\cdots}{}_{k\ldots} = \int_G d^4 x \, T^{i\cdots}{}_{k\ldots}(x^l), \qquad t^{\bar{i}\cdots}{}_{\bar{k}\ldots} = \int_{\bar{G}} d^4 x \, T^{\bar{i}\cdots}{}_{\bar{k}\ldots}(x^{\bar{l}}). \tag{5.6.6}$$

$(G, \bar{G}$ are the coordinate domains describing Γ.) We must stress here that t depends on Γ: if in the second integral one integrated over a domain of the $x^{\bar{i}}$ which is given in numerically the same way as G is given in terms of the x^i, the result would be components of *another* tensor, corresponding to the integral of $T(x)$ over a domain obtained from Γ by active Poincaré transformation. An exception occurs when Γ is Poincaré invariant, i.e., coincides with all of $\mathbf{X}_4$.

Regarding integrals over 3-dimensional submanifolds, we shall primarily need the generalization of the flux integrals $\int \mathbf{v} \, dO$ known from $\mathbf{R}^3$. The domains to be integrated over are hypersurfaces σ, given in parametric form as $x = x(u, v, w)$. The analog to the infinitesimal flux $\mathbf{v} dO$ for a 4-vector field $A(x)$ is the 4-volume of the parallelopiped spanned by A and the tangent vectors $B = (\partial x/\partial u) du$, $C = (\partial x/\partial v) dv$, $D = (\partial x/\partial w) dw$ to the hypersurface:

$$\epsilon_{ijkm} A^i \frac{\partial x^j}{\partial u} \frac{\partial x^k}{\partial v} \frac{\partial x^m}{\partial w} \, du \, dv \, dw = A^i \, d\sigma_i, \tag{5.6.7}$$

where we have introduced the *vectorial hypersurface element*

$$d\sigma_i := \epsilon_{ijkm} \frac{\partial x^j}{\partial u} \frac{\partial x^k}{\partial v} \frac{\partial x^m}{\partial w} \, du \, dv \, dw = \epsilon_{ijkm} \, dx^j \, dx^k \, dx^m \tag{5.6.8}$$

(the second way of writing is to indicate its independence of the special parametrization used). $d\sigma_i$ is orthogonal to the hypersurface, since for the tangent vectors B, C, D we have

$$d\sigma_i \, B^i = d\sigma_i \, C^i = d\sigma_i \, D^i = 0.$$

This enables us to form integrals of the kind

$$t^{k\cdots} = \int_\sigma d\sigma_i \, T^{ik\cdots}(x), \tag{5.6.9}$$

giving tensors from tensor fields, again depending on σ in general: If σ is given in I parametrically by $x^i = \varphi^i(u, v, w)$, then the parameter representation of it in $\bar{\text{I}}$ will employ other functions of u, v, w, while $x^{\bar{i}} = \varphi^i(u, v, w)$ describes an actively Poincaré transformed hypersurface. An important exceptional case is described below.

Hypersurfaces are called spacelike, timelike, lightlike, respectively, if their normals, and thus $d\sigma_i$, are timelike, spacelike, lightlike, respectively:

$$
\begin{aligned}
d\sigma_i\, d\sigma^i &> 0 \qquad & \sigma \text{ spacelike} \\
d\sigma_i\, d\sigma^i &< 0 \qquad & \sigma \text{ timelike} \\
d\sigma_i\, d\sigma^i &= 0 \qquad & \sigma \text{ lightlike.}
\end{aligned}
\tag{5.6.10}
$$

Of fundamental importance will be *Gauss' theorem* which allows to convert integrals of tensor fields over closed hypersurfaces into integrals over the 4-dimensional domain bounded by the hypersurface as

$$
\int_{\partial\Gamma} d\sigma_i\, T^{ik\cdots} = \int_\Gamma d^4x\, \partial_i\, T^{ik\cdots},
\tag{5.6.11}
$$

where Γ is the domain with boundary hypersurface $\partial\Gamma$.

We now come to a case where hypersurface integrals of a tensor field take the same value for different hypersurfaces. Assume that on a 4-dimensional region Γ the 4-divergence of a tensor field T vanishes:

$$
\partial_i\, T^{ik\cdots} = 0;
\tag{5.6.12}
$$

then

$$
\int_\sigma d\sigma_i\, T^{ik\cdots} = \int_{\sigma'} d\sigma_i'\, T^{ik\cdots}
\tag{5.6.13}
$$

will hold for any two hypersurfaces σ, σ' that coincide outside Γ (see Fig. 5.2); altenatively, σ may be deformed arbitrarily inside Γ without changing the value of the integral.

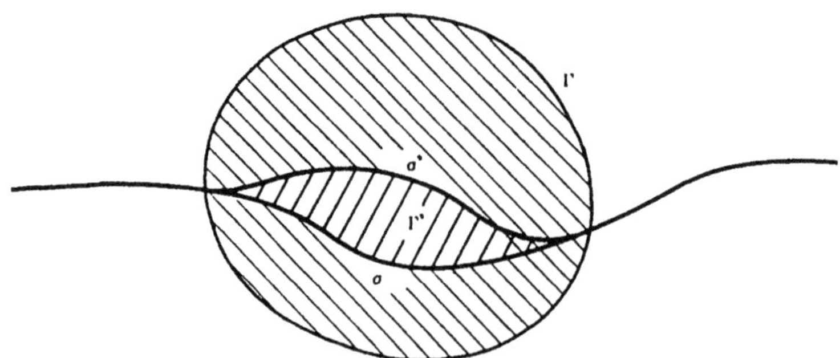

Fig. 5.2. Deforming a hypersurface

The proof obtains by changing the orientation of σ, such that $\Gamma\cap\sigma$ and $\Gamma\cap\sigma'$ taken together give a closed uniformly oriented hypersurface bounding a domain $\Gamma' \subset \Gamma$.

Transforming the integral over this hypersurface into a volume integral by eq. (5.6.11) gives zero by assumption (5.6.12):

$$\left(\int_{\sigma'} - \int_{\sigma} \right) d\sigma_i \, T^{ik\cdots} = \int_{\Gamma'} d^4x \, \partial_i \, T^{ik\cdots} = 0.$$

For an exact and extended presentation of integration in higher dimensions and the integral theorems we must refer, e.g., to Spivak (1965) or other modern texts on differential and integral calculus; in particular, all matters concerning orientation are to be found there. Note in particular that, because of the peculiarities of Lorentzian orthogonality, orientation is better formulated in terms of differential forms rather than directions of vectorial normals; see exercise 2f below for illustration.

Exercises

1. From the vectorial volume element $d\sigma_i$ one may also form a *scalar volume element* $d\sigma := |d\sigma_i \, d\sigma^i|^{1/2}$ for hypersurfaces. Using eqs. (5.6.8), (5.5.9) show that $d\sigma = \sqrt{|\Delta|} \, du \, dv \, dw$, where Δ is the Gram determinant

$$\Delta := \begin{vmatrix} x_u^2 & x_u \, x_v & x_u \, x_w \\ x_u \, x_v & x_v^2 & x_v \, x_w \\ x_u \, x_w & x_v \, x_w & x_w^2 \end{vmatrix} \tag{5.6.14}$$

of the vectors $x_u := \partial x/\partial u$, x_v, x_w. (The matrix appearing in eq. (5.6.14) is the component matrix of the induced metric tensor in the hypersurface with respect to the tangential basis x_u, x_v, x_w.) Calculate $d\sigma$ for a mass shell $p^2 = m^2$ in momentum space ($x \to p$, $u = p^1$, $v = p^2$, $w = p^3$) and compare with eq. (4.5.6). What happens when $m = 0$?

2. If a hypersurface is given implicitly by an equation $F(x^i) = 0$, then its normal has the direction of the 4-gradient $\partial^i F$. Decide in this way whether and where the following hypersurfaces are spacelike, timelike, or lightlike.
(a) $x^0 = const.$, (b) $x^1 = const.$, (c) light cone: $(x - x_0)^2 = 0$, (d) unit hyperboloids $(x - x_0)^2 = \pm 1$, (e) hypersurfaces of constant phase for a plane electromagnetic wave as in exercise 7a of sect. 5.5, (f) $t^2 + \mathbf{x}^2 = 1$. (This hypersurface bounds a compact region: is the contravariant gradient everywhere directed towards its interior?)

5.7 The Full System of Maxwell Equations. Charge Conservation

In sect. 5.2 we wrote down Maxwell's equations in covariant form, using the field tensor F. The homogeneous equations (5.2.23) may be rewritten in a different manner if we make use of the dual $*F$ with components

$$*F^{ik} = \frac{1}{2!} \, \epsilon^{ikmn} \, F_{mn}, \qquad (*F^{ik}) = \begin{pmatrix} 0 & B_1 & B_2 & B_3 \\ -B_1 & 0 & -E_3 & E_2 \\ -B_2 & E_3 & 0 & -E_1 \\ -B_3 & -E_2 & E_1 & 0 \end{pmatrix}. \tag{5.7.1}$$

We see that $(*F^{ik})$ arises from (F_{ik}) by interchanging $\mathbf{E}$ and $\mathbf{B}$ and from (F^{ik}) by $\mathbf{E} \to -\mathbf{B}, \mathbf{B} \to \mathbf{E}$. Therefore the inhomogeneous equations may be written in analogy to eq. (5.2.22) as (cf. exercise 4 of sect. 5.5)

$$\partial_i *F^{ik} = 0. \tag{5.7.2}$$

The dual field tensor is therefore source-free. In principle one could think here of a magnetic 4-current to achieve perfect symmetry between electricity and magnetism. At present, there is no experimental evidence whatsoever for magnetic charges (monopoles). There is, however, continued interest in searching for them, the main reason being that it was pointed out by Dirac (P. A. M. Dirac, Proc. R. Soc. Lond. Ser. A *133*, 60 (1931)) that their existence automatically leads, in the framework of quantum mechanics, to quantization of the product of electric and magnetic charge. See also J. Schwinger, Science *165*, 757 (1969); P. Price et al., Phys. Rev. Lett. *35*, 487 (1975).

We can now write down the basic equations of electromagnetism in the following covariant form:

$$\begin{aligned}
\partial_k F^{ik} &= -4\pi\, j^i & \partial_k *F^{ik} &= 0 \\
k^i &= F^{ik} j_k & \partial_i j^i &= 0.
\end{aligned} \tag{5.7.3}$$

This elegant formulation was given for the first time in 1908 by H. Minkowski. We still show that the ansatz

$$F_{ik} = \partial_i A_k - \partial_k A_i \tag{5.7.4}$$

satisfies eqs. (5.7.2) identically:

$$*F^{\,ik} = \frac{1}{2}\, \epsilon^{mnik}\,(\partial_m A_n - \partial_n A_m) = \epsilon^{mnik}\,\partial_m A_n$$

$$\partial_k *F^{\,ik} = \epsilon^{mnik}\,\partial_k\,\partial_m A_n \equiv 0$$

because of the commutativity of partial derivatives, $\partial_k\,\partial_m = \partial_m\,\partial_k$, and the antisymmetry of $\epsilon^{\cdots}$. Equation (5.7.2) is an integrability condition for eq. (5.7.4).

The equation of continuity for the 4-current $\partial_i j^i = 0$ is valid in all of space-time; this leads to the Poincaré *invariance* of the *total charge*

$$Q_\sigma = \int_\sigma d\sigma_i\, j^i(x) \tag{5.7.5}$$

associated with a charge distribution which is spatially finite or decreases sufficiently rapidly towards spatial infinity. To see that Q_σ is indeed the total charge measured by observers in the inertial system I we take the hypersurface σ to be the spacelike hyperplane $x^0 = t = const.$, on which we use x^1, x^2, x^3 as parameters. Equation (5.6.8) then gives

$$d\sigma_i = \epsilon_{ijkm}\,\delta_1^j\,\delta_2^k\,\delta_3^m\, dx^1\, dx^2\, dx^3 = \epsilon_{i123}\, d^3x = (d^3x, \mathbf{0}) \tag{5.7.6}$$

(cf. the remark made on d^3x in the appendix to sect. 4.5!) and thus

$$Q_\sigma = \int_\sigma d^3x\, j^0(\mathbf{x}, x^0) = \int d^3x\, \rho(\mathbf{x}, t). \tag{5.7.7}$$

The total charge measured by these observers at some other time t', and the total charge measured by observers in a boosted system $\bar{\mathrm{I}}$, are

$$\int_{\sigma'} d^3x\, \rho(\mathbf{x}, t') = \int_{\sigma'} d\sigma_i\, j^i(x) \text{ and } \int d^3\bar{x}\, \bar{\rho}(\bar{\mathbf{x}}, \bar{t}) = \int_{\bar\sigma} d\sigma_{\bar\imath}\, j^{\bar\imath} = \int_{\bar\sigma} d\sigma_i\, j^i,$$

thus equal to $Q_{\sigma'}$ and $Q_{\bar\sigma}$, respectively. Outside the world tube of the charge distribution we may deform σ', $\bar\sigma$ without changing the value of the integral (Fig. 5.3). In this manner, σ', $\bar\sigma$ effectively become deformations of σ and it follows that $Q_\sigma = Q_{\sigma'} = Q_{\bar\sigma}$, since $\partial_i\, j^i = 0$ in all of spacetime.

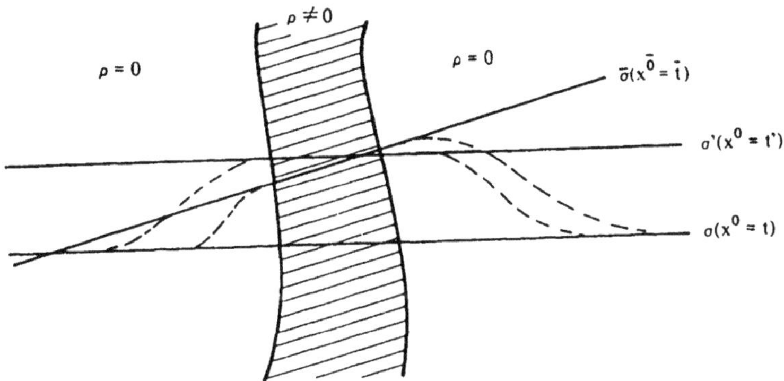

Fig. 5.3. Invariance of total charge

We have therefore shown the invariance of total charge under all active Poincaré transformations, in particular its time independence (conservation of charge) and observer independence.

We must stress that this is true only for the *total* charge; the charge contained in smaller volumes will in general be neither time-independent nor observer-independent. The usual local form of the law of charge conservation as a balance equation results from the equation of continuity $\partial_i\, j^i = 0$ by integrating over a spatial volume, using Gauss' theorem:

$$\frac{d}{dt} \int d^3x\, \rho = -\oint \mathbf{j}\, d\mathbf{O}. \tag{5.7.8}$$

This also has a 4-dimensional generalization, in which the *piece* of hypersurface σ to be integrated over in eq. (5.7.5) is displaced infinitesimally along the flow lines of a *deformation vector* field $\delta^i(x)$. The points of σ' are then given by $x^i + \delta^i(x)$, where x runs over σ. From the 4-dimensional version of Gauss' theorem and $j^i{}_{,i} = 0$ it then follows

$$\delta \int_\sigma d\sigma_i\, j^i = \left(\int_{\sigma'} - \int_\sigma\right) d\sigma_i\, j^i = \int_{\text{mantle}} d\sigma_i\, j^i,$$

where the mantle is formed by pieces of field lines of δ^i which emanate from the boundary of σ. On it, we have from eq. (5.6.8), replacing $dx^m \to \delta^m$,

$$d\sigma_i = \epsilon_{ijkm}\, dx^j\, dx^k\, \delta^m = \delta^m\, d\upsilon_{im}$$

with the *tensorial 2-surface element* of the boundary surface $\partial\sigma$

$$d\sigma_{il} := \epsilon_{iljk}\, dx^j\, dx^k. \tag{5.7.9}$$

The generalization of eq. (5.7.8) therefore is

$$\delta \int_\sigma d\sigma_i\, j^i = \oint_{\partial\sigma} d\sigma_{il}\, j^i\, \delta^l. \tag{5.7.10}$$

Specializing σ as $t = const.$, $\delta^l = (\delta t, \mathbf{0})$ leads back to eq. (5.7.8).

5.8 Discussion of the Transformation Properties

Our formal considerations led us to the surprising result that the field strengths $\mathbf{E}$ and $\mathbf{B}$ become united into the field tensor F as an observer-independent concept. (We could introduce 4-vectors for $\mathbf{E}$ and $\mathbf{B}$ if we employed the 4-velocity u of an observer: $E^i = F^{ik}u_k, B^i = -*F^{ik}\,u_k$, as one checks using the rest system.) In the present section we shall study the consequences of the close relations between $\mathbf{E}$ and $\mathbf{B}$ enforced by the structure of Maxwell's equations.

In dealing with 4-vectors we found that the existence of the invariant 4-square allowed important conclusions on the transformation behavior to be drawn without using the explicit matrices of Lorentz transformations (which in the general case are complicated). We have encountered similar invariants for the antisymmetric tensor F, namely

$$I_1 = \frac{1}{4}F_{ik}\, F^{ki} = \frac{1}{2}\left(\mathbf{E}^2 - \mathbf{B}^2\right) \tag{5.8.1a}$$

$$I_2 = \frac{1}{4}\, *F_{ik}\, F^{ki} = -\mathbf{E}\,\mathbf{B}. \tag{5.8.1b}$$

One consequence of the invariance of these expressions is that the characterization of plane electromagnetic waves by $|\mathbf{E}| = |\mathbf{B}|$, $\mathbf{E}\,\mathbf{B} = 0$ is Lorentz invariant, as this may be written $I_1 = 0$, $I_2 = 0$ (cf. exercise 7 of sect. 5.5). Another consequence is that the conditions $\mathbf{E}^2 \gtrless \mathbf{B}^2$ $(I_1 \gtrless 0)$, $\cos(\mathbf{E},\mathbf{B}) \gtrless 0$ $(I_2 \gtrless 0)$ are Lorentz invariant, so that a field which is purely electric in one reference system cannot appear as purely magnetic in some other one and vice versa, and an acute angle between $\mathbf{E}, \mathbf{B}$ in one system cannot become obtuse in another one.

These were general statements, valid for all Lorentz transformations. We now study the behavior of the components of F^{ik}, generally given by the transformation law of tensor field components

$$F^{\bar{i}\bar{k}}(\bar{x}) = L^i{}_m L^k{}_n F^{mn}(x), \tag{5.8.2}$$

specializing to a boost in the 1-direction. (Applying a pure space rotation would just give us back the 3-vector character of $\mathbf{E}, \mathbf{B}$ inherent already in the 3-dimensional form of Maxwell's equations.) The boost matrix is (cf. eq. (2.1.1))

$$L = (L^i{}_m) = \begin{pmatrix} \gamma & -\gamma v & 0 & 0 \\ -\gamma v & \gamma & 0 & 0 \\ 0 & 0 & 1 & 0 \\ 0 & 0 & 0 & 1 \end{pmatrix}, \tag{5.8.3}$$

and we can evaluate eq. (5.8.2) by matrix multiplication, since upon introducing the matrices $F = (F^{mn})$, $\bar{F} = (F^{\bar{i}\bar{k}})$, that equation simply is $\bar{F} = L\,F\,L^T$. Multiplying blockwise we quickly get

$$
\begin{aligned}
\bar{E}_1 &= E_1 & \bar{B}_1 &= B_1 \\
\bar{E}_2 &= \gamma\,(E_2 - v\,B_3) & \bar{B}_2 &= \gamma\,(B_2 + v\,E_3) \\
\bar{E}_3 &= \gamma\,(E_3 + v\,B_2) & \bar{B}_3 &= \gamma\,(B_3 - v\,E_2),
\end{aligned}
\tag{5.8.4}
$$

or, in vectorial form,

$$
\overline{\mathbf{E}} = \gamma\,\mathbf{E} - \frac{\gamma - 1}{v^2}\,(\mathbf{E}\,\mathbf{v})\,\mathbf{v} + \gamma\,\mathbf{v} \times \mathbf{B}
\tag{5.8.5a}
$$

$$
\overline{\mathbf{B}} = \gamma\,\mathbf{B} - \frac{\gamma - 1}{v^2}\,(\mathbf{B}\,\mathbf{v})\,\mathbf{v} - \gamma\,\mathbf{v} \times \mathbf{E}.
\tag{5.8.5b}
$$

Here one has to add arguments for the functions occurring, i.e., $E_1(x)$, $\bar{E}_2(\bar{x})$, etc., together with $\bar{x} = Lx$.

To illustrate these formal consideration in a concrete example, consider a charged particle at rest in some inertial system I. When the electromagnetic field of this particle is measured in I, the usual Coulomb field

$$
\mathbf{B} = 0, \qquad\qquad \mathbf{E} = \frac{e\,\mathbf{x}}{r^3}
\tag{5.8.6}
$$

is found (where $r = |\mathbf{x}|$) if we assume that the particle has no magnetic moment. In the system $\bar{\text{I}}$ the situation is different: measurement of the field of the *same* particle yields not only an electric but also a magnetic field. The classical explanation is that the particle now appears moving, thus representing an electric current that generates a magnetic field. Here we obtained this result simply from the transformation law of the field tensor.

Also the electric field is affected by the transformation. Let us first investigate the electric field component in the direction of relative motion. We have

$$
\bar{E}_1(\bar{x}) = E_1(x) = \frac{e\,x}{r^3};
\tag{5.8.7}
$$

but remember that we have to express this explicitly in terms of the barred coordinates in order to get the full description of the field as registered in $\bar{\text{I}}$. We introduce the squared distance $b^2 = y^2 + z^2 = \bar{y}^2 + \bar{z}^2$ of the field point from the x-axis $= \bar{x}$-axis to write

$$
\bar{E}_1 = \frac{e\,x}{r^3} = \frac{e\,\gamma\,(\bar{x} + v\,\bar{t})}{[\gamma^2(\bar{x} + v\,\bar{t})^2 + b^2]^{3/2}}
\tag{5.8.8}
$$

and similarly, from eq. (5.8.4),

$$
\bar{E}_2 = \gamma\frac{e\,y}{r^3} = \frac{e\,\gamma\,\bar{y}}{[\gamma^2(\bar{x} + v\,\bar{t})^2 + b^2]^{3/2}}.
\tag{5.8.9}
$$

Observe in particular the occurrence of the factor γ in both formulae, unexpected from a superficial look at eq. (5.8.4); so the *field* character is quite essential for the transformation behavior. To get an idea of the field distribution (5.8.8,9), we consider the instantaneous field lines at $\bar{t} = 0$. Now eqs. (5.8.8,9) are vectorially

$$\overline{\mathbf{E}}(\bar{x}) = \frac{e\,(1 - v^2)\,\bar{\mathbf{x}}}{[\bar{r}^2 - v^2 b^2]^{3/2}}\,,\qquad\qquad (5.8.10)$$

where $\bar{r}^2 = \bar{\mathbf{x}}^2$. This shows that the field lines are straight lines as in the case of a charge at rest in $\bar{\mathrm{I}}$; i.e., they all pass through the *instantaneous*(!) position of the charge (this obviously also holds for other times as well). The absolute value

$$|\overline{\mathbf{E}}| = \frac{e\,(1 - v^2)}{\bar{r}^2(1 - v^2 \sin^2 \bar{\Theta})^{3/2}}\qquad\qquad (5.8.11)$$

$(\sin \bar{\Theta} := b/\bar{r})$ is, for a given $\bar{r}$, maximum in the plane orthogonal to the direction of motion:

$$|\overline{\mathbf{E}}| = \frac{e}{\bar{r}^2\,\sqrt{1 - v^2}}\qquad \text{for } \sin \bar{\Theta} = 1,\qquad\qquad (5.8.12)$$

and minimum on the orbit of the particle (x-axis):

$$|\overline{\mathbf{E}}| = \frac{e\,(1 - v^2)}{\bar{r}^2}\qquad \text{for } \sin \bar{\Theta} = 0.\qquad\qquad (5.8.13)$$

Thus in a sense the Coulomb field is dilated in directions orthogonal to the direction of motion and contracted along the line of motion. We can illustrate this by drawing a *pattern of field lines* (Fig. 5.4)—a procedure which makes sense whenever one has a vector field which is, in some region, divergence-free like the velocity field of an incompressible fluid. (Note that this is the case here, by the covariance of Maxwell's equations or by direct verification!)

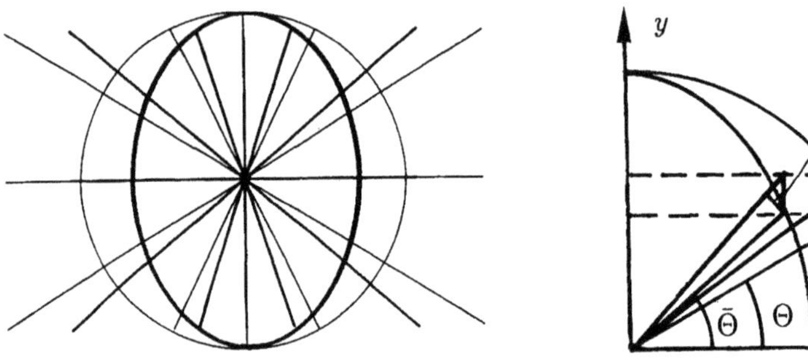

a) field pattern (thin for charge at rest)   b) calculating the contraction

Fig. 5.4. Electric field of a uniformly moving charge

As is well known, in such patterns one draws a certain number of field lines such that the number of lines drawn through a unit surface element orthogonal to them equals the absolute value $|\overline{\mathbf{E}}|$ (up to a convenient scale). It is remarkable now that the present pattern may be obtained geometrically from the pattern corresponding to a charge at rest by affinely compressing the latter by a factor $\sqrt{1-v^2}$ in the direction of motion, just as if the sheet of paper on which the pattern is drawn underwent Lorentz contraction (Fig. 5.4; we are here following Rindler (1969)). For proof, consider, at some point (x, y, z), a surface element dA at right angle to the x-axis together with those field lines of a charge e sitting at the origin which pass through it. The projection of dA orthogonal to the field lines is $dA \cos \Theta$ (see Fig. 5.4b), so their number is $dA \cos \Theta . e/r^2$. Upon affine compression of the pattern, the images of the field lines just considered will pass through a surface element of size $d\bar{A} = dA$ located at the image point $(\bar{x} = x\sqrt{1-v^2}, \bar{y} = y, \bar{z} = z)$; its projection perpendicular to the image field lines is $d\bar{A} \cos \bar{\Theta}$. From the equality of the number of field lines through dA and $d\bar{A}$ we get for the field strength represented by the latter the expression

$$\frac{e}{r^2} \frac{dA \cos \Theta}{d\bar{A} \cos \bar{\Theta}} = \frac{e}{r^2} \frac{x}{r} \frac{\bar{r}}{\bar{x}} . \tag{5.8.14}$$

With $x = \gamma \bar{x}, r^2 = \gamma^2 \bar{r}^2 (1 - v^2 \sin^2 \bar{\Theta})$ this becomes

$$|\overline{\mathbf{E}}| = \frac{e(1 - v^2)}{\bar{r}^2 (1 - v^2 \sin^2 \bar{\Theta})^{3/2}} , \tag{5.8.15}$$

in agreement with eq. (5.8.11).

This simple geometric construction of the field pattern of a moving charge was already known to Heaviside in 1889. Its existence may be concluded upon in part from the invariance of (total) electric charge, since from the moving charge the same number of field lines must emerge as do from the one at rest, so that the field lines just become redistributed. Lorentz looked at this construction as confirming—and indeed explaining—his (and Fitzgerald's) hypothesis about the contraction of bodies moving relative to the ether. Thus in 1909 he writes: "Let us come back to the hypothesis with the help of which we have tried to explain the negative result of the Michelson experiment. We can understand the possibility of the assumed differences in length if we take into account that the form of a rigid body depends on the forces between its molecules, and that these forces are transferred through the ether in between in a manner which is more or less equal to the manner in which electromagnetic interactions are transferred. From this point of view it is natural to assume that molecular attractions and repulsions get modified by translation of the body just as are electromagnetic forces; and this might well lead to a change in the dimension of the body. It is remarkable that the change in lengths postulated earlier results if we carry over to molecular interactions the findings which we obtained for the electromagnetic field."

This *dilation of the Coulomb field* obtained from the transformation behavior is also observed experimentally: particles passing through a bubble chamber leave a ionization track; as shown in Fig. 5.5, the thickness of this track, i.e., the number of ionized particles produced per unit length on the track, first *decreases* when the speed of the passing particle is increased. Roughly speaking this is because less time is left for the particle to ionize atoms. However, if the speed is increased close to the speed of light, ionization starts to increase after passing a minimum. This may in part be explained by the relativistic effect derived above: the dilated Coulomb field ionizes more atoms per unit track length.

A clear description of the connection between the dilation of the Coulomb field and the increase in ionization density at high speeds is found in Jackson (1999); see also B. Price, Rep. Prog. Phys. 18, 52 (1955) or H. A. Bethe, J. Ashkin in Segrè (1953).

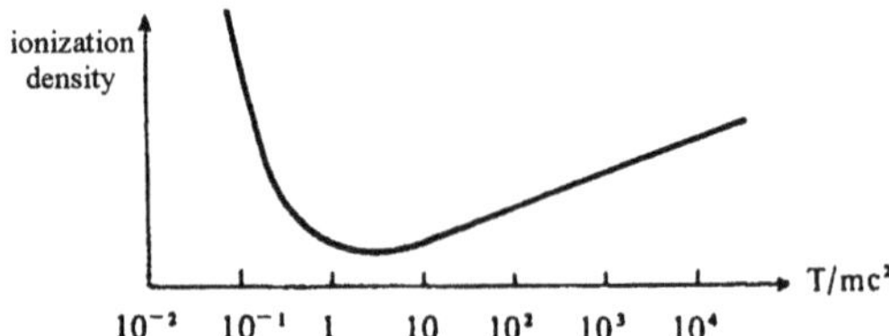

Fig. 5.5. Density of ionization as a function of speed

In recent years the minimum of the ionization density has played a role in the context of the search for *quarks*. These hypothetical particles carry only 2/3 of the elementary charge quantum and therefore should, in a suitable energy regime, leave ionization tracks with a density below the one left by particles that carry a full elementary charge.

The dilation of the Coulomb field is also of importance in connection with the Weizsäcker-Williams method (1934) for calculating the emission of *bremsstrahlung*. One uses the fact that for $\gamma \gg 1$ the dilated Coulomb field comes closer and closer to the field of a plane electromagnetic wave pulse (cf. Jackson 1999).

When a fast particle is decelerated or a particle at rest is *accelerated*, the dilated form of the Coulomb field must go over into the usual one. Since the information about the beginning deceleration or acceleration propagates with speed of light, both kinds of fields will be present with a transition zone between them which propagates as a *shock wave* through the Coulomb field and corresponds to the *radiation field* emitted in the process (Fig. 5.6).

We can see that the field lines in the shock wave are (at finite distances approxi-

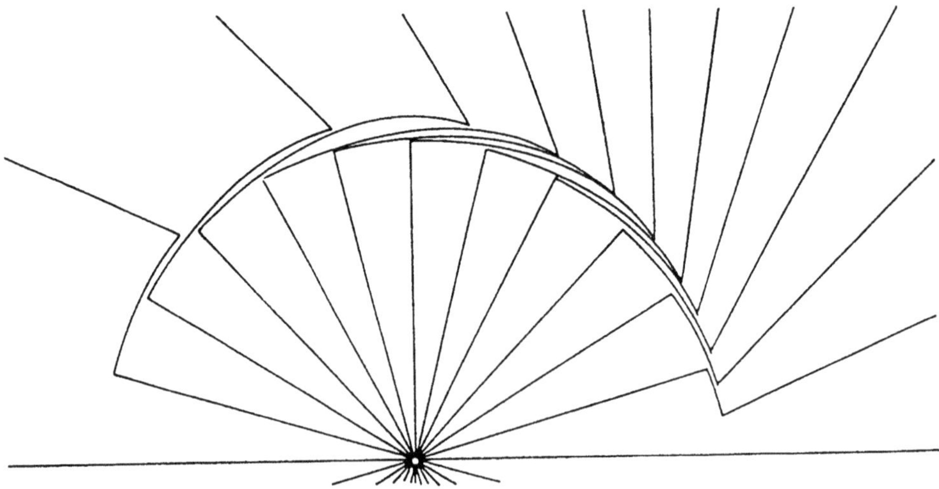

Fig. 5.6. Field pattern for a decelerated charge

mately and at infinity exactly) perpendicular to the radius vector; the shock propagates away from the particle at the speed of light. If the stoppage is from relativistic speeds the characteristic forward maximum of *bremsstrahlung* will result.

For speeds $v \ll 1$ our picture even admits a simple heuristic calculation of the radiation from an accelerated charge, which may serve as a preparation for the considerations in sect. 5.10. Figure 5.7 shows a line of the field at time t of a charged particle which was stopped down to rest from uniform rectilinear motion at speed v during the time interval from 0 to $\tau \ll t$. Up to a radius $r = t - \tau$ the Coulomb field line OP already corresponds to the stopped particle, while from $r = t$ outwards the distribution of field lines still corresponds to what would have resulted from fictitious continued uniform motion; O' is the fictitious position at time t, so that $\overline{OO'} = vt$. The position Q where the field line OP continues after passing through the transition zone of width τ to form the outer part of the straight line $O'Q$ is fixed by the equality of the electric flux through the spherical caps with axis $\mathbf{v}$ and centers O resp. O' whose boundary circles contain P resp. Q. From eq. (5.8.11), assuming $v \ll 1$, we find that OP and $O'Q$ are parallel, as indicated in Fig. 5.7. With the added assumption $t \gg \tau$ we have $\overline{OP} \gg \overline{OO'}$; under these circumstances one also has $\overline{OP} = \overline{O'Q}$, so that PQ and OO' are parallel as shown in Fig. 5.7.

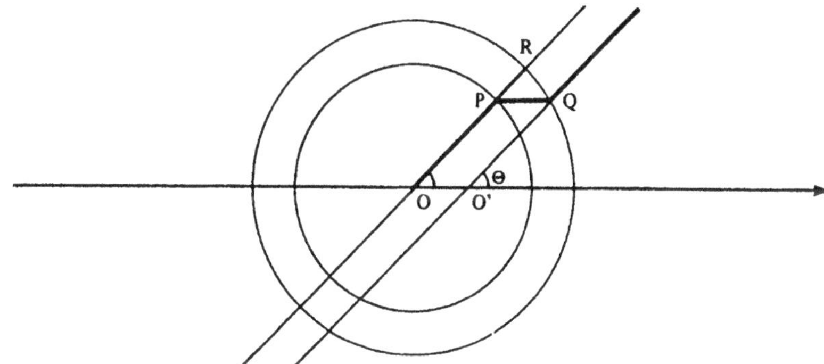

Fig. 5.7. Field line of a stopped particle

From electric flux conservation we get the radial electric field component in the transition zone as $E_r = e/r^2$. Finally from the geometric relation

$$\frac{E_\varphi}{E_r} = \frac{\overline{RQ}}{\overline{PR}} = \frac{v\,t\sin\Theta}{\tau}$$

we can read off the transversal electric field component in the transition zone to be

$$E_\varphi = \frac{v t \sin\Theta}{\tau}\,\frac{e}{r^2} = \frac{e\,v\sin\Theta}{\tau\,r} = e\,a\,\frac{\sin\Theta}{r}, \tag{5.8.16}$$

where $a = v/\tau$ is the particle's acceleration. The field E_φ propagates outwards with the speed of light, dominating E_r, accompanied by a magnetic field of equal strength orthogonal to it and also transversal. The Poynting vector of energy flow becomes

$$\mathbf{S} = \frac{e^2 b^2}{4\pi}\,\frac{\sin^2\Theta}{r^2}\,\frac{\mathbf{r}}{r}. \qquad (5.8.17)$$

This gives the angular distribution of the radiation whose total intensity (radiated energy per unit time) is

$$\frac{dE}{dt} = \int \mathbf{S}\,d\mathbf{O} = \frac{2}{3}e^2 b^2. \qquad (5.8.18)$$

In this heuristic calculation of the radiation from an accelerated particle we were following J. J. Thomson (1904). Modern texts—we know of just two notable exceptions—only give the exact but unvisualizable analytic derivation using retarded potentials (cf., e.g., Jackson 1999). We also recommend the movie series described in J. C. Hamilton, J. L. Schwartz, Am. J. Phys. *39*, 1540 (1971)), illustrating the generation of radiation in the way we have done here.

It is interesting that Thomson in the textbook mentioned above (a printed version of his Silliman Lectures, delivered in May 1903 at Yale University) gives the calculation under the heading "Effects due to acceleration of the Faraday tubes" (a concept related to—but not identical to—the field line patterns) and then adds, among other things: "This view of light as due to the tremors in tightly stretched Faraday tubes [anticipated by Faraday himself] raises a question which I have not seen noticed. The Faraday tubes stretching out through the ether cannot be regarded as entirely filling it. They are rather to be looked upon as discrete threads embedded in a continuous ether, giving to the latter a fibrous structure; but if this is the case, then on the view we have taken of a wave of light the wave itself must have a structure, and the front of the wave, instead of being, as it were, uniformly illuminated, will be represented by a series of bright specks on a dark ground, the bright specks corresponding to the places where the Faraday tubes cut the wave front."

As we see, in this paragraph Thomson comes remarkably close to the discovery of light quanta. He then continues by a tentative explanation of the experimental observation that X-rays (still being called Röntgen rays by him, while X-rays is the original name given by Röntgen!) upon penetrating matter ionize only a small fraction of all atoms, which does not seem compatible with the idea of a continuous wave. (Remember that the wave character of X-rays was also not yet established experimentally at that time.)

Exercises

1. Consider the electromagnetic field $\mathbf{E}(x)$, $\mathbf{B}(x)$ at a fixed space-time point. Show that

   a. if $I_2 = 0$, $I_1 \neq 0$, then there is a reference frame where either $\mathbf{E} = \mathbf{0}$ or $\mathbf{B} = \mathbf{0}$, depending on the sign of I_1;

   b. if $I_2 \neq 0$ one can achieve $\mathbf{E} \propto \mathbf{B}$ in a suitable frame.

   The 'normal forms' for F_{ik} thus arising correspond to the normal forms (3.2.7,8,9) of 4-vectors, and may be used similarly in simplifying calculations.

2. Figure 5.4 also shows a sphere and its affine image, an ellipsoid. Interpret the former as a level surface of the Coulomb potential $V = A^0 = e/r$. Supplement this by $\mathbf{A} = \mathbf{0}$ to obtain a 4-potential satisfying the Lorenz condition.

   a. Boost this to obtain a 4-potential for the field of a moving charge.

    b. Show that the level surfaces of $A^{\bar{0}}(\bar{x})$ are ellipsoids of the above kind.

    c. Why is there no conflict in the fact, obvious from Fig. 5.4, that the electric field lines of the transformed field are still orthogonal to the sphere but not to the ellipsoid?

    Remark: The condition on a vector field $\mathbf{E}$ to admit for a family of surfaces orthogonal to it is well known to be $\mathbf{E}\,\mathrm{rot}\,\mathbf{E}{=}0$.

3. Using conservation of electric flux, find the position of Q in Fig. 5.7—i.e., the angle at O'—without assuming $v \ll 1$. Show that it is determined by requiring $O'Q$ to be parallel to the affine transform of OP as described in Fig. 5.4!

5.9 Conservation Laws. Stress-Energy-Momentum Tensor

When we were setting up relativistic mechanics we started from asking for the most natural form of the conservation laws for energy and momentum; from this we drew conclusions like the relativistic increase of mass, etc. In electrodynamics, on the other hand, we already have in hands the covariant form of the dynamics, so that the formulation of conservation laws will be a matter of mathematical deduction, and the covariant form of the conservation laws will just round off the formal structure, giving us the opportunity to introduce the *energy-momentum-stress tensor*, a concept that will prove, in chap. 10, to be of fundamental importance in relativistic field theories. Let us start with a review of the noncovariant formulation of the conservation laws of electrodynamics.

To derive the law of *conservation of energy*, we use the identity

$$\mathrm{div}\,(\mathbf{E} \times \mathbf{B}) \equiv \mathbf{B}\,\mathrm{rot}\,\mathbf{E} - \mathbf{E}\,\mathrm{rot}\,\mathbf{B},$$

valid for two arbitrary vector fields, together with eqs. (5.2.1,2), to arrive, with the definitions

$$\mathcal{E} := \frac{1}{8\pi}(\mathbf{E}^2 + \mathbf{B}^2) \tag{5.9.1}$$

$$\mathbf{S} := \frac{1}{4\pi}\,\mathbf{E} \times \mathbf{B}, \tag{5.9.2}$$

at the almost-continuity equation

$$\frac{\partial \mathcal{E}}{\partial t} + \mathrm{div}\,\mathbf{S} = -\mathbf{j}\,\mathbf{E}. \tag{5.9.3}$$

When this is integrated over a domain in space, using eqs. (5.3.7a,8) and Gauss' theorem, we obtain the balance equation

$$\frac{d}{dt}\left(\sum_A p_A^0 + \int d^3x\,\mathcal{E}\right) = -\oint \mathbf{S}\,d\mathbf{O}. \tag{5.9.4}$$

Since the first term in the bracket on the left is the sum of the energies of the charged particles constituting the current distribution, it is natural to identify the second term with the energy E_F of the electromagnetic field and $\mathcal{E}$ as its *energy density* (which thus is *positive-definite*). The *Poynting vector* $\mathbf{S}$ must therefore be interpreted as the *energy current* density of the field.

To derive the law of *conservation of momentum*, we introduce an auxiliary *constant* vector field $\mathbf{a}$ and use the identities

$$\mathbf{v} \times \mathrm{rot}\, \mathbf{v} \equiv \mathrm{grad}\, \frac{\mathbf{v}^2}{2} - (\mathbf{v}\, \nabla)\, \mathbf{v}$$

$$\mathbf{a}\, \mathrm{grad}\, \frac{\mathbf{v}^2}{2} \equiv \mathrm{div}\, \left(\frac{\mathbf{v}^2}{2}\, \mathbf{a} \right), \qquad (\mathbf{v}\, \nabla)\, (\mathbf{a}\, \mathbf{v}) \equiv \mathrm{div}((\mathbf{a}\, \mathbf{v})\mathbf{v}) - (\mathbf{a}\, \mathbf{v})\, \mathrm{div}\, \mathbf{v},$$

valid for arbitrary vector fields $\mathbf{v}$, together with eqs. (5.2.1,2) to arrive at the almost-continuity equation

$$\frac{\partial}{\partial t}\, (\mathbf{a}\, \mathbf{S}) + \frac{1}{4\pi}\, \mathrm{div}\, \left[\frac{1}{2}(\mathbf{E}^2 + \mathbf{B}^2)\mathbf{a} - (\mathbf{a}\, \mathbf{E})\mathbf{E} - (\mathbf{a}\, \mathbf{B})\mathbf{B} \right] = -\mathbf{a}(\rho\mathbf{E} + \mathbf{j} \times \mathbf{B}). \quad (5.9.5)$$

When this is integrated over a domain in space, using eqs. (5.3.7b,8) and Gauss' theorem, we obtain

$$\mathbf{a}\, \frac{d}{dt}\, \left(\sum_A \mathbf{p}_A + \int d^3x\, \mathbf{S} \right) = -\frac{1}{4\pi}\, \oint \left[\frac{1}{2}(\mathbf{E}^2 + \mathbf{B}^2)\mathbf{a} - (\mathbf{a}\, \mathbf{E})\mathbf{E} - (\mathbf{a}\, \mathbf{B})\mathbf{B} \right] dO =: \mathbf{a}\, \mathbf{G}$$

$$(5.9.6)$$

or—since $\mathbf{a}$ was arbitrary—the balance equation

$$\frac{d}{dt}\, \left(\sum_A \mathbf{p}_A + \int d^3x\, \mathbf{S} \right) = \mathbf{G}. \qquad (5.9.7)$$

The *momentum of the electromagnetic field* $\mathbf{p}_F$ is therefore identified as

$$\mathbf{p}_F = \int d^3x\, \mathbf{S}. \qquad (5.9.8)$$

The surface integral $\mathbf{G}$ gives the net momentum flowing out of the domain per unit time, i.e., gives the force acting on that domain. The components G_α of $\mathbf{G}$ are, from eqs. (5.9.5,6),

$$4\pi\, G_\alpha = \int d^3x\, \partial_\beta \left[\mathbf{E}_\alpha\, \mathbf{E}_\beta + \mathbf{B}_\alpha\, \mathbf{B}_\beta - \frac{1}{2}\delta_{\alpha\beta}\, (\mathbf{E}^2 + \mathbf{B}^2) \right] =$$

$$= -4\pi \int d^3x\, \partial_\beta\, T_{\alpha\beta} = -4\pi \int dO_\beta\, T_{\alpha\beta},$$

$$(5.9.9)$$

where

$$-4\pi\, T_{\alpha\beta} := E_\alpha\, E_\beta + B_\alpha\, B_\beta - \frac{1}{2}\, \delta_{\alpha\beta}\, (\mathbf{E}^2 + \mathbf{B}^2) =: P_{\alpha\beta} \qquad (5.9.10)$$

are the components of the *Maxwell stress tensor* $P_{\alpha\beta}$. Its interpretation is analogous to the one of the stress tensor in elasticity theory: $dG_\alpha = -dO_\beta\, T_{\alpha\beta}$ is the element of force acting on a surface element of the domain. (One has to be cautious with this interpretation, however, as we are trying here to draw a conclusion from the integral about the integrand, which is not an admissible procedure in general. This remark also applies to the interpretation of $\mathbf{S}$, which is apparently wrong, e.g., in a crossed electrostatic and magnetostatic field. However, if one restricts attention to the total force $\mathbf{G}$ on a domain, the nonuniqueness of $dG_\alpha = -dO_\beta\, T_{\alpha\beta}$ becomes inessential; and on the other hand, as Maxwell demonstrated in 1873, the use of dG_α enables to visualize the situation in many cases to the extent that forces between charges, dipoles, etc., may be read off, given the pattern of field lines. (Cf. also the discussion of localization at the end of sect. 10.2!) Consider a surface element of size dO orthogonal to the x-axis, thus $dO_\alpha = (1, 0, 0)\, dO$, then the force on it is

$$dG_1 = -T_{11}\, dO = \frac{1}{8\pi}\left(E_1^2 + B_1^2 - E_2^2 - E_3^2 - B_2^2 - B_3^2\right) dO$$

$$dG_2 = -T_{12}\, dO = \frac{1}{8\pi}\left(E_1\, E_2 + B_1\, B_2\right) dO \qquad\qquad (5.9.11)$$

$$dG_3 = -T_{13}\, dO = \frac{1}{8\pi}\left(E_1\, E_3 + B_1\, B_3\right) dO.$$

This force is interpreted in Fig. 5.8.

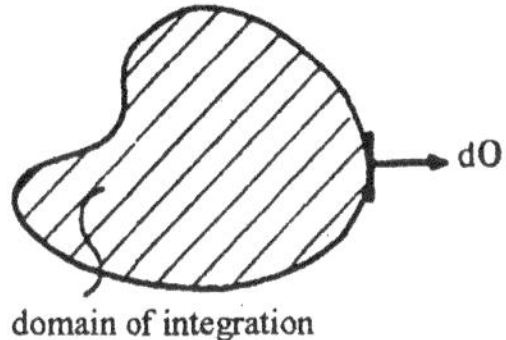

a) The volume and surface element considered

b) Tension along the field lines

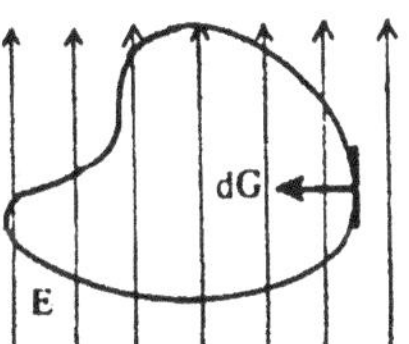

c) Pressure transversal to field lines

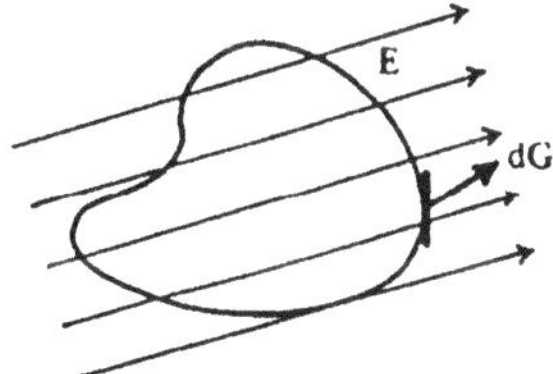

d) Shear forces at oblique angles

Fig. 5.8. Interpreting Maxwell's stress tensor

Observe that $T_{\alpha\beta}$ is quadratic in $\mathbf{E}$ and $\mathbf{B}$, so that a reversal of the field lines does not change $d\mathbf{G}$. The tension along the field lines and the pressure transversal to them allow to read off forces from field line patterns as the ones shown in Fig. 5.9 (the domains of integration are indicated by dashed lines). Only the integral over the plane of symmetry has to be performed, the hemispheres at infinity do not contribute because of $T_{\alpha\beta} \propto 1/r^4$.

The *symmetry* $T_{\alpha\beta} = T_{\beta\alpha}$ seen from eq. (5.9.10) corresponds to the symmetry of the stress tensor in elasticity theory; it is shown there that in a static situation this symmetry entails the vanishing of torque on the domain under consideration.

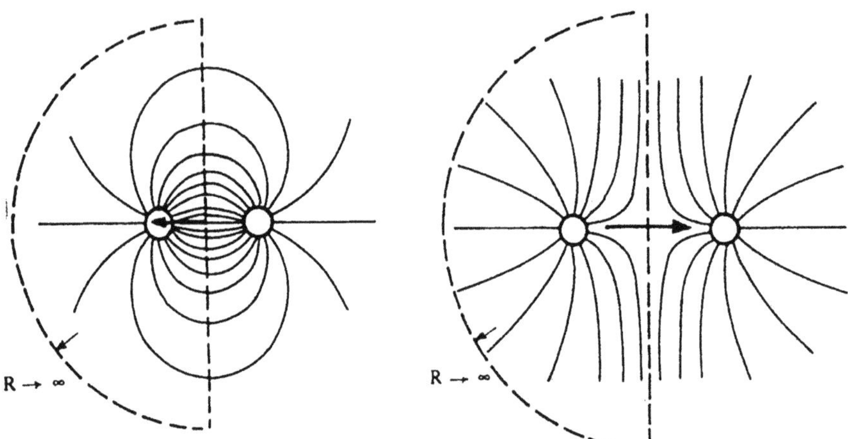

Fig. 5.9. Forces between charges of equal or opposite signs

We now pass on to the 4-dimensional *covariant formulation of the conservation laws*. We first express $T_{\alpha\beta}$ in terms of the field tensor F and then formally complement it as a tensor T_{ik} $(i, k = 0, 1, 2, 3)$. Since $T_{\alpha\beta}$ is quadratic in F_{ik} and symmetric in α and β, there are only a few possibilities for its construction from F, and the correct one turns out to be

$$4\pi\, T_{ij} := F_{ik}\, F^k{}_j - \frac{1}{4}\, \eta_{ij}\, F_{lk}\, F^{kl}. \tag{5.9.12}$$

We now have to find out about the physical significance of the additional components T_{00}, $T_{0\alpha}$ in eq. (5.9.12). Inserting from eqs. (5.2.18,20) we get from a short calculation

$$T_{00} = \frac{1}{8\pi}\,(\mathbf{E}^2 + \mathbf{B}^2) = \mathcal{E}, \tag{5.9.13}$$

$$T_{0\alpha} = -\frac{1}{8\pi}\,(\mathbf{E} \times \mathbf{B})_\alpha = -S_\alpha. \tag{5.9.14}$$

Thus the *stress-energy-momentum tensor* field $T(x)$ of the electromagnetic field comprises energy density, energy current density, momentum density and Maxwell's stresses according to the matrix of its contravariant components

$$T_F{}^{ik} = \left(\begin{array}{c|c} \mathcal{E} & S^\alpha \\ \hline S^\alpha & T^{\alpha\beta} \end{array} \right), \tag{5.9.15}$$

where the index F is to indicate that we are dealing with the stress-energy-momentum tensor of the field. (We shall see how to associate a corresponding object to particles and other fields as well.)

It may be confusing that the usual Cartesian components of a spatial vector sometimes occur with a change of sign as space components of a 4-vector and sometimes without; and similarly for tensors (e.g. $\Delta\mathbf{x} \to \Delta x^\alpha$, $\nabla \to \partial_\alpha = -\partial^\alpha$, $\mathbf{S} \to T^{0\alpha}$, $P_{\alpha\beta}$ (Maxwell stress tensor)$\to -4\pi T_{\alpha\beta}$). This shows the disadvantage of the signature choice (1.5.1) for the space-time metric. For the opposite signature $\eta_{ik} = \mathrm{diag}\,(-1,1,1,1)$ there is, on the other hand, the disadvantage that some important quantities which are 'physically' positive, like p^0, $T_0{}^0$, are positive only if a definite position of the index 0 is chosen. Although one can work out a concept of 'natural index position' to bring some order into this problem, the expenditure of doing so does not pay off for us (cf. Post 1962). Our choice makes formulae of 2-component spinor algebra somewhat simpler (see sect. 8.4).

The energy-momentum tensor (we shall prefer this over the more complete version stress-energy-momentum tensor or the alternative versions stress-energy tensor, stress tensor, energy tensor, mass tensor, matter tensor, ... found in the literature, for no particular good reason) was written down first by Minkowski in 1908. E. T. Whittaker (1960) regards this unification of the energy density (Lord Kelvin, 1853), of the Poynting vector (Poynting, Heaviside, 1884), and of the stress tensor (Maxwell, 1873) as Minkowski's greatest discovery. It perhaps shows to the maximum extent the intrinsic beauty of the 4-dimensional formalism, the discovery of which led Minkowski to his famous words in the opening sentences of the talk given on 21 September 1908: "Gentlemen! The views upon space and time which I am going to develop for you grew on experimental ground. Therein lies their strength. Their tendency is a radical one. From now on space by itself and time by itself shall totally sink down as shadows, and only a kind of union of both shall keep its independence...".(See, e.g., Lorentz, Einstein, Minkowski 1958.)

Using the energy-momentum tensor (5.9.12) we can now write the conservation laws for energy and momentum together in (partially) covariant fashion:

$$\int d^3x \left(T_F{}^{i0}{}_{,0} + T_F{}^{i\alpha}{}_{,\alpha} \right) = \int d^3x\, T_F{}^{ik}{}_{,k} = -\sum_A \frac{dp_A^i}{dt}. \tag{5.9.16}$$

For $i = 0$ this agrees with eq. (5.9.3), for $i = \beta$ it agrees with eq. (5.9.7), the terms $T^{i\alpha}{}_{,\alpha}$ giving the corresponding surface integrals by Gauss' theorem. By eqs. (5.3.6,7,8) the sum on the right may be replaced by $-\int d^3x\, F^{ik} j_k$, and since the domain of integration is arbitrary the integrands must be equal:

$$T_F{}^{ik}{}_{,k} = -F^{ik} j_k. \tag{5.9.17}$$

(This equation follows also on a purely differential level from the definition (5.9.12), using Maxwell's equations (5.2.22,23), as may be shown as an exercise.)

The asymmetry between the description of the field by its energy-momentum tensor field and of the particles by their 4-momentum vector in eq. (5.9.16) may be removed by introducing an energy-momentum tensor field

$$T_P{}^{ik}(x) := \sum_A m_A \int ds_A\, \delta^4(x - z_A(s_A))\, u_A^i\, u_A^k \tag{5.9.18}$$

for the particles as well. Here the $z_A(s_A), A = 1, 2, ...$ are the world lines of the particles, each parametrized by its proper time, and the u_A are their 4-velocities.

Restrict, for simplicity, to one particle only; the components

$$T_P{}^{0i}(x) = m \int ds\, \delta^4(x - z(s))\, \frac{dz^0}{ds}\, u^i = m\, u^i\, \delta^3(\mathbf{x} - \mathbf{z}(s)) \tag{5.9.19}$$

then give, in analogy with the electromagnetic field, the density of energy-momentum of the particle, whose integral

$$\int T_P{}^{0i}\, d^3x = m\, u^i = p^i \tag{5.9.20}$$

is indeed the energy-momentum vector (4-momentum) of the particle. We further have

$$T_P{}^{ik}{}_{,k} = m \int ds \, \frac{\partial}{\partial x^k} \delta^4(x - z(s)) \, u^i u^k = -m \int ds \, u^i \frac{dz^k}{ds} \frac{\partial}{\partial z^k} \delta^4(x - z(s)) =$$

$$= -m \int ds \, u^i \frac{d}{ds} \delta^4(x - z(s)) = m \int ds \, \frac{du^i}{ds} \delta^4(x - z(s)).$$

$$(5.9.21)$$

Integration over a spatial domain containing the particle yields

$$\int d^3x \, T_P{}^{ik}{}_{,k} = \int ds \, \frac{dp^i}{ds} \delta(x^0 - z^0(s)) = \int dt \, \frac{dp^i}{dt} \delta(t - z^0(s)) = \frac{dp^i}{dt}$$

and for several particles

$$\int d^3x \, T_P{}^{ik}{}_{,k} = \sum_A \frac{dp^i_A}{dt}. \qquad (5.9.22)$$

With this, eq. (5.9.16) becomes

$$\int d^3x \left(T_F{}^{ik} + T_P{}^{ik} \right)_{,k} = 0. \qquad (5.9.23)$$

Since the domain of integration is arbitrary we can again conclude that the integrand must vanish, i.e., that the *total stress-energy-momentum density*

$$T^{ik} := T_P{}^{ik} + T_F{}^{ik} = T^{ki} \qquad (5.9.24)$$

satisfies the equation

$$T^{ik}{}_{,k} = \left(T_P{}^{ik} + T_F{}^{ik} \right)_{,k} = 0. \qquad (5.9.25)$$

(Again, this may be obtained on a purely differential level, substituting for $m \frac{du^i}{ds}$ in the last term of eq. (5.9.21) from the equation of motion (4.1.10), (5.3.2) and using eq. (5.2.15) to obtain

$$T_P{}^{ik}{}_{,k} = +F^{ik} j_k.) \qquad (5.9.26)$$

Equation (5.9.25) is the *differential* version of the conservation laws in covariant form.

The relation between the differential and the *fully* covariant integral form of the conservation laws is now quite analogous to the situation we had when discussing charge conservation; more precisely, we can formally reduce it to that earlier case by what on a first sight just appears as a mathematical trick whose deeper significance will emerge below. Namely, we introduce a *constant* auxiliary vector field a^k (whose space components actually appeared already in eqs. (5.9.5,6)!) and consider the 4-vector field $a^k T_k{}^i$: this obviously has vanishing 4-divergence and may thus formally replace j^i in our considerations on the Poincaré invariance of electrical charge. The result is the Poicaré invariance of $a^k p_k$, where $p^k = p_F^k + \sum_A p_A^k$ and

$$p_F^k[\sigma] = \int_\sigma T_F^{ki} \, d\sigma_i = (E_F, \mathbf{p}_F)[\sigma] \qquad (5.9.27)$$

is the 4-momentum of the field associated with the hypersurface σ (e.g., a constant time surface), while each p_A^k is to be evaluated at the proper time that corresponds to the intersection of the world line $z_A(s_A)$ with σ. Since a^k was an arbitrary 4-vector, we obtain the σ-independence (hence time independence for all observers) and the 4-vector nature of the *total energy-momentum vector* p^k of the combined system (particles + field). Note the difference to the situation considered in sect. 4.2 where we had assumed no field except in the interaction region. Also note the difference in the ways $p^k, p_F^k[\sigma]$ and $p_A^k[\sigma]$ are to be looked at as 4-vector components: while the latter two behave in the correct manner only if we Lorentz-transform the basis of 4-vectors but *do not* Lorentz-transform the hypersurface involved in their definitions, this restriction may be ignored for the total 4-momentum.

The trick we applied to reduce the situation of energy-momentum conservation formally to the one of charge conservation depended on two things: the differential conservation law (5.9.25) and the assumption $a^k = const$. If the latter is not made, i.e., if we replace a by a not necessarily constant vector field ξ but still require $(\xi^k T_k{}^i)_{,i} = 0$, the restriction on ξ is only, from eqs. (5.9.25,24),

$$0 = \xi^k{}_{,i} T_k{}^i = \xi_{k,i} T^{ki} = \xi_{(k,i)} T^{ki}. \tag{5.9.28}$$

If this is to hold for all field configurations, it is sufficient to require the *Killing equation*

$$\xi_{i,k} + \xi_{k,i} = 0. \tag{5.9.29}$$

While obviously satisfied by $\xi_i = a_i = const.$, this equation has the further solutions $\xi_i(x) = \epsilon_{ik} x^k$ with arbitrary constant $\epsilon_{ik} = -\epsilon_{ki}$. The conservation laws deriving from the latter solutions of the Killing equation are of the form $\frac{1}{2}\epsilon_{ab} J^{ab}$ and comprise angular momentum conservation and motion of the center of mass; they will be discussed further in sect. 10.2.

To see that we now have exhausted all solutions of the Killing equation, differentiate eq. (5.9.29) for x^j, cyclically permute the three indices, add two of the equations thus obtained and subtract the third: the result is $\xi_{i,kj} = 0$, therefore $\xi_{i,k} = \epsilon_{ik}$ as above. Remarkably, the general solution $\xi^i = \epsilon^i{}_k x^k + a^i$ with infinitesimal $\epsilon^i{}_k$, a^i constitutes the displacement vector field δx^i under a general infinitesimal Poincaré transformation $x^i \mapsto x^i + \delta x^i = L^i{}_k x^k + a^i$ with $L^i{}_k = \delta^i{}_k + \epsilon^i{}_k$ (where $\epsilon_{ik} = -\epsilon_{ki}$ takes care of the orthogonality condition (3.1.8) up to $O(\epsilon^2)$). This is the relativistic version of the *connection between symmetries and conservation laws*, which will be discussed in greater detail in chap. 10.

If $T_P{}^{ik} = 0$, then because of $T_F{}^i{}_i \equiv 0$ (see exercise) eq. (5.9.28) will be satisfield under the weaker assumption (*conformal Killing equation*)

$$\xi_{i,k} + \xi_{k,i} - \frac{1}{4}\xi^j{}_{,j}\,\eta_{ik} = 0. \tag{5.9.30}$$

This equation has a wider class of solutions, involving 15 independent constants rather than the $4 + 6 = 10$ constants appearing above; however, the corresponding conservation laws have been of minor importance so far.

Exercises

1. Show that

$$T_F{}^i{}_i = 0, \tag{5.9.31}$$

$$4\pi T_F{}_i{}^j = \frac{1}{2}\left(F_{ik}F^{kj} + {}^*F_{ik}\,{}^*F^{kj}\right).$$
(5.9.32)

2. Show eqs. (5.9.17,26) directly.

3. a. Show that

$$\mathcal{E}^2 - \mathbf{S}^2 = \frac{1}{4\pi}\left[\frac{1}{2}(\mathbf{E}^2 - \mathbf{B}^2)^2 + (\mathbf{E}\,\mathbf{B})^2\right] \geq 0.$$
(5.9.33)

   b. Interpret this physically in terms of an energy flow velocity!

   c. Although the quantities $\mathcal{E} = T^0{}_0$, $S^\alpha = T^\alpha{}_0$ taken together obviously do not form the components of a 4-vector, the right-hand side of eq. (5.9.33) is an invariant. How can that be? (Cf. eq. (8.4.29)!)

4. Show that the 4-velocity of the frame determined in exercise 1 of sect. 5.4 is an eigenvector of $T_F^j{}_i$. What is the eigenvalue? What happens, on the other hand, when $T_F^j{}_i$ is boosted with the velocity of the energy flow? Explain! (The algebraic structure of the eigenvalue-eigenspace situation can be analyzed using eq. (8.4.29) which gives directly the eigenprojections.)

5. Show that eq. (5.9.29) arises from eq. (3.1.5) upon substituting $f^m(x) = x^m + \xi^m(x)$ and neglecting quadratic terms in ξ.

5.10   Charged Particles

We shall now apply the insights gained from the discussion of conservation laws to an apparently simple situation, namely the electromagnetic field of a slowly moving point charge, which, according to eq. (5.8.4), assuming $v \ll 1$, is

$$\mathbf{E} = \frac{e\,\mathbf{x}}{r^3} \qquad\qquad \mathbf{B} = \frac{e\,\mathbf{v}\times\mathbf{x}}{r^3}.$$
(5.10.1)

Then, because of $\mathbf{B}^2 \propto \mathbf{v}^2 \approx 0$, only the electric field will contribute to the energy E_F of the field:

$$E_F = \int \frac{d^3x}{8\pi}\,\mathbf{E}^2 = \frac{4\pi\,e^2}{8\pi}\int \frac{dr\,r^2}{r^4} = \frac{e^2}{2R}.$$
(5.10.2)

The radial integration was taken here from some finite radius R—and not from zero—out to infinity, since otherwise we would have obtained an infinite *self-energy* E_F. Cutting off the integral at R corresponds to assuming a charge distribution concentrated on a spherical shell of radius R: then the interior is field-free and does not contribute to the integral. (Other distributions of the charge would only change the numerical factor $1/2$.)

The energy E_F also contributes to the mass of the charged particle. If the mass of the particle without its electromagnetic field (i.e., the mass of the uncharged particle) is m_0, the total mass will be

$$P^0 = m = m_0 + E_F.$$
(5.10.3)

The electromagnetic field does not cause a mass defect but an increase in mass as compared to the uncharged case.

The momentum of the field surrounding the particle is, from eq. (5.9.8),

$$\mathbf{p}_F = \int \frac{d^3x}{4\pi}\, \mathbf{E} \times \mathbf{B} = e^2 \int \frac{d^3x}{4\pi} \left(\frac{\mathbf{v}}{r^4} - \frac{(\mathbf{v}\,\mathbf{x})\,\mathbf{x}}{r^6} \right). \tag{5.10.4}$$

An elementary calculation yields

$$\mathbf{p}_F = \frac{2}{3}\frac{e^2}{R}\,\mathbf{v} = \left(\frac{4}{3} E_F \right) \mathbf{v}. \tag{5.10.5}$$

The total momentum of the charged particle is therefore

$$\mathbf{P} = m_0\,\mathbf{v} + \mathbf{p}_F = \left(m_0 + \frac{4}{3} E_F \right) \mathbf{v} \neq m\,\mathbf{v}. \tag{5.10.6}$$

The manifest discrepancy between eqs. (5.10.3) and (5.10.6) was the subject of numerous publications for decades. Before going into the history of this puzzle and its consequences, we shall give its resolution as it follows from taking into account all conservation laws. The energy-momentum vector of the particle is calculated from the total energy-momentum tensor as

$$P^i = \int d\sigma_k\, T^{ki}. \tag{5.10.7}$$

According to this, P^i is a 4-vector. If we put $d\sigma_k = (d^3x, \mathbf{0})$ we get

$$P^i = \int d^3x\, T^{0i}. \tag{5.10.8}$$

Specializing further to the rest system of the particle, in which $P^i = (m, \mathbf{0})$, we have there

$$m = \int d^3x\, T^{00} = m_0 + E_F \tag{5.10.9}$$

and

$$\int d^3x\, T^{0\alpha} = 0, \qquad \alpha = 1, 2, 3. \tag{5.10.10}$$

From the transformation behavior (4.1.6) of the 4-momentum it then follows that with respect to a system in slow relative motion P has the components

$$P^{\bar{i}} = (m, m\mathbf{v}), \tag{5.10.11}$$

which contradicts eq. (5.10.6). We thus also formally obtained a contradiction to the basic transformation properties, although the calculation of P in cqs. (5.10.3,6) exactly conforms to the prescription (5.10.7).

Having formulated the contradiction in a clear-cut way we now have in hands the fundament for its resolution. Explicitly, we have

$$P^{\bar{i}} = \int_{\bar{t}=0} d^3x\, T^{0\bar{i}}(x^{\bar{m}}) = L^i{}_k\, L^0{}_l \int d^3\bar{x}\, T^{kl}(x^m), \tag{5.10.12}$$

where $T^{kl}(x^m)$ refers to the rest system and where the $L^i{}_k$ for a boost in the 1-direction are given by eq. (2.1.1), i.e., the coordinates x^m in the rest system are related to the $x^{\bar{m}}$ by $t = \gamma\,(\bar{t} + v\,\bar{x}) = \gamma\,v\,\bar{x}$, $x = \gamma\,(\bar{x} + v\,\bar{t}) = \gamma\,\bar{x}$, $y = \bar{y}$, $z = \bar{z}$, since the integration is to be executed at the time $\bar{t} = 0$. Because of the time independence of the energy-momentum tensor in the rest system we further have

$$T^{kl}(x^m) = T^{kl}(\mathbf{x}) = T^{kl}(\gamma\bar{x}, \bar{y}, \bar{z}) \tag{5.10.13}$$

and

$$\int d^3\bar{x}\, T^{kl}(\gamma\bar{x}, \bar{y}, \bar{z}) = \frac{1}{\gamma} \int d^3x\, T^{kl}(x, y, z), \tag{5.10.14}$$

which obviously takes into account the Lorentz contraction of the volume element.

We now substitute this and eq. (2.1.1) into eq. (5.10.12), taking into account eq. (5.10.10). The result is

$$\begin{aligned}
P^{\bar{0}} &= \frac{1}{\gamma}\,(L^0{}_0)^2 \int d^3x\, T^{00} + \frac{1}{\gamma}\,(L^0{}_1)^2 \int d^3x\, T^{11} = \\
&= \gamma\,m + \gamma\,v^2 \underline{\int d^3x\, T^{11}}
\end{aligned} \tag{5.10.15}$$

and analogously

$$\begin{aligned}
P^{\bar{1}} &= \gamma\,v\,m + \gamma\,v \underline{\int d^3x\, T^{11}} \\
P^{\bar{2}} &= P^{\bar{3}} = 0.
\end{aligned} \tag{5.10.16}$$

This is the correct transformation law up to the underlined terms proportional to $\int d^3x\, T^{11}$. To show that these must vanish in a consistent theory we form

$$\left(T^{ik}\,x^l\right)_{,k} = T^{ik}{}_{,k}\,x^l + T^{ik}\,\delta^l_k = T^{il} \tag{5.10.17}$$

and integrate this equation for $l = i = \alpha$ over all space:

$$\int d^3x\,\left(T^{\alpha k}\,x^\alpha\right)_{,k} = \int d^3x\, T^{\alpha\alpha}, \qquad \alpha = 1, 2, 3 \tag{5.10.18}$$

(no sum over α!). Because of the time independence of T^{lk} in the rest system we have

$$\int d^3x\,\left(T^{\alpha k}\,x^\alpha\right)_{,k} = \int d^3x\,\left(T^{\alpha\beta}\,x^\alpha\right)_{,\beta} = \int dO_\beta\, T^{\alpha\beta}\,x^\alpha, \qquad \alpha = 1, 2, 3. \tag{5.10.19}$$

This surface integral vanishes for a localized particle if the domain of integration extends over all space, so that eq. (5.10.18) becomes

$$\int d^3x\, T^{\alpha\alpha} = 0, \qquad \alpha = 1, 2, 3 \tag{5.10.20}$$

(no sum). The terms underlined in eqs. (5.10.15,16) thus indeed must vanish as a consequence of the conservation law (5.9.25).

For point particles in the rest system we have $T_P{}^{\alpha\alpha}(x) = 0$, so that eq. (5.10.20) is satisfied for uncharged particles. However, for the electromagnetic field it follows from eq. (5.9.12) that

$$0 \equiv T_F{}^i{}_i = T_F{}^{00} - \sum_\alpha T_F{}^{\alpha\alpha} = 0, \tag{5.10.21}$$

i.e., the trace of the field energy-momentum tensor vanishes. For a sphere-shaped particle no direction is distinguished, so that

$$T_F{}^{11} = T_F{}^{22} = T_F{}^{33} = \frac{1}{3} \sum_\alpha T_F{}^{\alpha\alpha} = \frac{1}{3} T_F{}^{00} \tag{5.10.22}$$

and consequently

$$\int d^3x\, T_F{}^{11} = \frac{1}{3}\, E_F. \tag{5.10.23}$$

Neglecting all terms proportional to v^2 as in eqs. (5.10.1–6) we get from eqs. (5.10.15,16)

$$P^{\bar 0} = m = m_0 + E_F \tag{5.10.24}$$

$$P^{\bar 1} = (m_0 + E_F)\, v + \frac{1}{3}\, v\, E_F. \tag{5.10.25}$$

The factor 4/3 therefore results from the fact that eq. (5.10.23) does not satisfy the restrictions (5.10.20) that follow from the differential conservation laws. The reason for this is easy to see: in calculating the self energy integral E_F, eq. (5.10.2), we were forced to make a cutoff at some radius $R \neq 0$ to obtain a finite value. This corresponds—as stated before—to a charge distribution concentrated on a spherical shell. Such a distribution cannot, however, remain stable (static), as was assumed above, but would explode without the action of further cohesive forces, since the charges distributed on the shell would repel each other. This can also be read off formally from eq. (5.10.18): writing this formula as

$$\frac{d}{dt} \int d^3x\, \mathbf{S}\, \mathbf{x} = \sum_\alpha \int d^3x\, T^{\alpha\alpha} + (\text{vanishing surface integral}) \tag{5.10.26}$$

we see that for $\int d^3x\, T^{\alpha\alpha} > 0$ a stable energy distribution is impossible and a radial flux $\mathbf{S}$ of energy has to be present. Stable charged particles are only possible if the energy-momentum tensor field of the particle matter allows to satisfy eqs. (5.10.20). One can try to arrange for this in two different ways.

If the *model of an extended particle* is retained, its energy-momentum tensor has to be supplemented by a phenomenological cohesion tensor T_C that avoids the explosion of the particle. This achieves $\int d^3x\, T^{\alpha\alpha} = 0$, removes the factor 4/3, and resolves all problems as far as *uniform rectilinear particle motion* is concerned. The problem of accelerated motion will turn out to be very difficult for this model, however.

On the other hand, passing to the limit $R \to 0$ of a *point particle* makes $E_F = \infty$, i.e., the self energy of the particle diverges. The total mass $m = m_0 + E_F$ of the particle will remain finite only if we assume $m_0 \to -\infty$ at the same time. Since a point particle has no parts that could repel each other, the problem of the instability of a (classical) charged particle seems to have been resolved without phenomenological cohesive forces. However, the formal manipulation of the expressions for energy and momentum is then very delicate, always involving terms like $E_F + m_0 = \infty - \infty$. Such terms are made unique by *postulating* that energy and momentum are components of a 4-vector P. This makes the correct transformation behavior—and thus the relativistic invariance of the theory—into a basic principle for obtaining finite and physically meaningful quantities from formally meaningless expressions like $\infty - \infty$. This way of proceeding became a very successful one after 1945 in the renormalization procedures of quantum field theory.

Before entering into the dynamics of charged particles we want to sketch here the historical development of the ideas about charged particles and their relation to $E = mc^2$. The first calculation of the energy $E_F(v)$ of the electromagnetic field of a moving charged hollow sphere was by J. J. Thomson 1881 (Philos. Mag. *11*, 227 (1881)); he substituted eq. (5.10.1) into eq. (5.9.1) to obtain

$$E_F(v) = \frac{e^2}{2R} + \left(\frac{4}{3}\frac{e^2}{2R}\right)\frac{v^2}{2}. \tag{5.10.27}$$

The occurrence of the second term in eq. (5.10.27) was interpreted by him as a mass increase for the particle, $m = m_0 + \frac{4}{3}(e^2/2R)$. Note that the factor 4/3 arises here in the calculation of the energy, not of momentum. From hindsight, his calculation is incorrect in two ways: there are no cohesive forces taken into account, and the result is given to second order in v although eq. (5.10.1) is correct only to first order.

This calculation was improved later to all orders in v by an exact calculation of the field of a moving charge according to Maxwell's equations. The result was

$$E_F(v) = E_F + m'(v)\,\frac{v^2}{2}, \tag{5.10.28}$$

where the *longitudinal mass* $m'(v)$ is given by

$$m'(v) = \frac{4}{3}\frac{e^2}{2R}\left[\frac{1}{v}\ln\frac{1+v}{1-v} - 1\right] = \frac{4}{3}\left(\frac{e^2}{2R}\right)\left(1 + \frac{6}{5}v^2 + \dots\right) \tag{5.10.29}$$

and is relevant for the inertia of the particle against acceleration in the direction of motion.

The momentum $\mathbf{p}_F$ of the particle's field was calculated by M. Abraham to be $\mathbf{p}_F = m''(v)\,\mathbf{v}$, wherein the *transversal mass*

$$m''(v) = \left(\frac{e^2}{2R}\right)\left[\frac{1+v^2}{v^3}\ln\frac{1+v}{1-v} - \frac{2}{v^2}\right] = \frac{4}{3}\left(\frac{e^2}{2R}\right)\left[1 + \frac{2}{5}v^2 + \dots\right] \tag{5.10.30}$$

is relevant for the inertia against acceleration orthogonal to the direction of motion. As we see, expressions (5.10.29,30) agree for small speeds; the (incorrect) factor 4/3 now occurring in the energy as well as in the mass.

The first measurements of a possible speed dependence of the mass by Kaufmann (Gött. Nachr. (1901), p. 143, (1902), p. 92; deflection of electrons in electric and magnetic fields) were just made for the purpose of finding out which part of the mass m of an electron would be the 'electromagnetic mass' $m''(v)$, i.e., its was attempted to separate m_0 and $m''(v)$ from $m = m_0 + m''(v)$. The measurements—in which changes of m by a factor of 2 were observed—seemed to agree with the hypothesis that $m_0 = 0$, so that the structure of the electron would be entirely of electromagnetic nature.

It is interesting to compare these calculations (which still neglected the problem of cohesive forces) with the analogous results of relativistic theory: eqs. (5.10.15,16) yield

$$P^{\bar{0}} = \gamma\left(m_0 + \frac{e^2}{2R} + v^2\,\frac{e^2}{6R}\right) \tag{5.10.31}$$

$$P^{\bar{1}} = \gamma v \left(m_0 + \frac{4}{3}\frac{e^2}{2R} \right) \approx v\,(m_0 + m''). \tag{5.10.32}$$

Expanding this for $v \ll 1$ gives

$$P^{\bar{0}} = m_0 + \frac{e^2}{2R} + \frac{v^2}{2} \left(m_0 + \frac{5}{3}\frac{e^2}{2R} \right). \tag{5.10.33}$$

In the energy, the term proportional to $\frac{v^2}{2}$ contains a mass $m_0 + \frac{5}{3}\frac{e^2}{2R}$, in contradistinction to eq. (5.10.28); the constant term $m_0 + e^2/2R$ in eq. (5.10.33) was, in prerelativistic calculations, interpreted as a change in origin on the energy scale and was not taken into account any further.

The reason for this discrepancy is that in prerelativistic times the moving charge ('electron') was considered as a rigid sphere with no Lorentz contraction. The field energy was integrated only over the exterior space of the electron shown in Fig. 5.10a. In 1905 Lorentz applied the hypothesis of length contraction (Lorentz contraction) to the electron itself. He integrated over the exterior domain of the contracted electron shown in Fig. 5.10b. This additional energy gives just the factor 5/3 in eq. (5.10.33) instead of 4/3, which he tried to explain by an ad hoc argument not important enough to mention. What is important is that he wrote $m_0 = 0$ in eq. (5.10.33) and

$$P^{\bar{1}} = \gamma v \,\frac{4}{3}\frac{e^2}{2R} =: m_L''(v)\, v \tag{5.10.34}$$

for the momentum of the electron. The transversal mass calculated by Lorentz shows the correct dependence on the speed, and the wrong factor 4/3 was not disturbing because of the immeasurability of the electron radius.

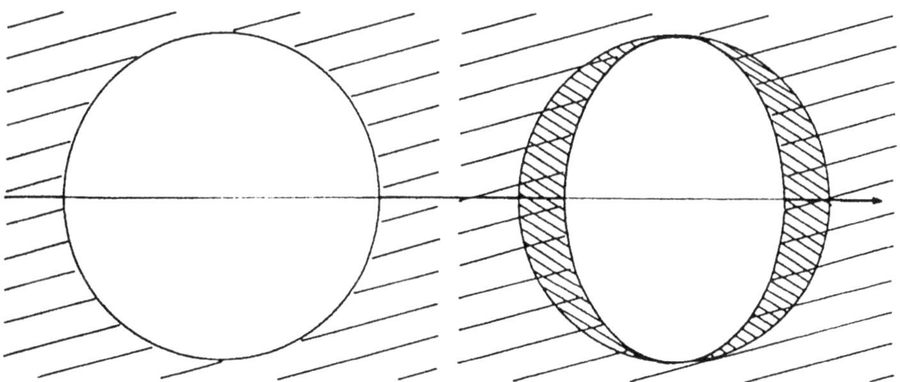

a) Domain of integration for the *rigid* electron   b) Domain of integration for the *deformable* electron

Fig. 5.10. Calculating the energy of the moving electron

In 1906 Kaufmann (Ann. Phys. (Leipzig) *19*, 487 (1906)) repeated his experiments in order to distinguish between Lorentz' hypothesis of the deformable electron m_L'' and Abraham's theory of the rigid electron $m''(v)$. The experiments—whose accuracy he overestimated—seemed to prefer Abraham, and it was only Bucherer (Phys. Z. *9*, 755 (1908)) who was able, with more precise methods of measurement, to decide on the mass formula of Lorentz.

With this, the difficulties stemming from the factor 4/3 (or 5/3) were not removed, however. Also Hasenöhrl obtained an additional mass $\Delta m = 4/3\,E_F$ when he calculated the inertia of a moving cavity filled with electromagnetic radiation of field energy E_F. (Without cohesive forces, the cavity would have to explode under the pressure of the radiation.) Only in 1922 some papers of Fermi resolved the puzzle of the factor 4/3. The further development is closely related to the one of the radiation of accelerating charges and the quantum mechanical description of electrons.

We now turn to the problems that occur when charges are accelerated. In such a process radiation is emitted as described by eq. (5.8.18), causing a *radiation reaction* upon the particle that corresponds to the energy loss; there will be a Lorentz force $\mathbf{F}_{rad}$ from the field of the particle to the particle itself, changing the equation of motion. For an extended distribution $\rho(\mathbf{x}, t)$ of charge this will be in N.R. approximation, according to eq. (5.3.1,6),

$$\mathbf{F}_{rad} = \int d^3x \, \rho \, (\mathbf{E} + \mathbf{v} \times \mathbf{B}). \tag{5.10.35}$$

If one calculates $\mathbf{E}$ and $\mathbf{B}$ from ρ, using eqs. (5.2.13,17), one finds after some long calculation (first done by Lorentz in 1909 and to be found in Jackson (1999))

$$\mathbf{F}_{rad} = \frac{1}{3} E_F \, \mathbf{a} + \frac{2}{3} e^2 \, \dot{\mathbf{a}} - \frac{2}{3} \sum_{n=2}^{\infty} \frac{(-)^n}{n!} \frac{d^n \mathbf{a}}{dt^n} O(R^{n-1}), \tag{5.10.36}$$

where $\mathbf{a}$ is the particle's acceleration and $\dot{\mathbf{a}}$ its time derivative. The terms $O(R^{n-1})$ are of the order of the corresponding power of the particle radius R and thus vanish in the limit of a point particle. The field energy E_F is given by

$$E_F = \frac{1}{2} \int \frac{\rho(\mathbf{x}) \, \rho(\mathbf{x}')}{|\mathbf{x} - \mathbf{x}'|} = \int \frac{d^3x}{8\pi} \mathbf{E}^2. \tag{5.10.37}$$

The equation of motion of the particle is

$$m_0 \mathbf{a} = \mathbf{F} + \mathbf{F}_{rad}, \tag{5.10.38}$$

where m_0 is again the 'mechanical' mass, i.e., the mass of the uncharged particle, and $\mathbf{F} = -\mathrm{grad}\, V(x)$ is an external force that causes the acceleration $\mathbf{a}$. Inserting from eq. (5.10.36) we get

$$\left(m_0 + \frac{4}{3} E_F \right) \mathbf{a} = \frac{2}{3} e^2 \, \dot{\mathbf{a}} + \mathbf{F} - \frac{2}{3} \sum_{n=2}^{\infty} \frac{(-)^n}{n!} \frac{d^n \mathbf{a}}{dt^n} O(R^{n-1}). \tag{5.10.39}$$

This equation of motion contains the acceleration $\mathbf{a}$ and all higher derivatives of it, so that the motion cannot be calculated from a knowledge of the usual initial conditions $\mathbf{x}(0)$, $\mathbf{v}(0)$ alone.

At first sight, this seems peculiar indeed, since the basic equations of the theory were of second differential order throughout, whereas now we have an equation containing derivatives of arbitrary order! We must not forget, however, that the system (particle + field) contains infinitely many degrees of freedom, of which in eq. (5.10.39) all except the one of the particle appear eliminated to obtain its equation of motion. The degrees eliminated reemerge in the form of higher derivatives, making

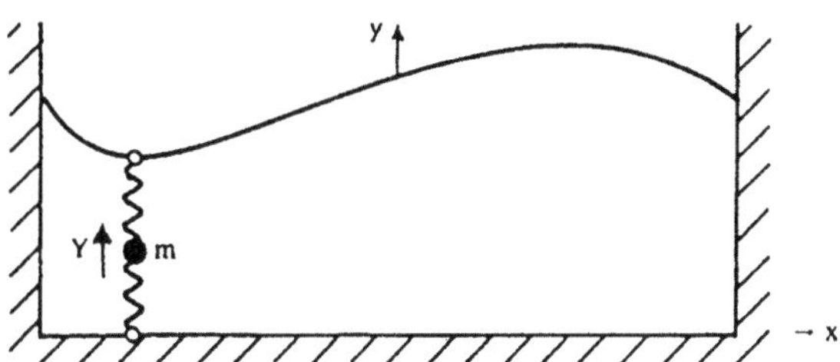

Fig. 5.11. A point mass coupled to a string

eq. (5.10.39) completely useless for all practical purposes, except in cases where one can approximately break off the infinite sum after a few terms.

There is a simple mechanical analog to this situation (Fig. 4.11), consisting of a mass point m coupled to an elastic string by a spiral spring (and similarly coupled to a rigid wall).

This mechanical system may be described by a system of differential equations of second order in time for the displacement $y(x,t)$ of the string and the amplitude $Y(t)$ of the mass point; given the initial conditions $y(x,0)$, $\dot{y}(x,0)$, $Y(0)$, $\dot{Y}(0)$, the time evolution is determined. One can eliminate the degrees of freedom of the string and find an equation of motion for the mass point *alone*; but this equation will then contain all higher time derivatives. (If the string is replaced by a chain of N mass points with elastic nearest neighbor coupling, the elimination process is known to lead to an equation of derivative order $2N + 2$.)

If we again write

$$m = m_0 + \frac{4}{3}\, E_F = m_0 + \frac{4}{3}\left(\frac{e^2}{2R}\right) \tag{5.10.40}$$

(the factor $4/3$ is to make up for omitted cohesive forces, as above), we can consider the limit $R \to 0$, $m_0 \to -\infty$, since only the observable total mass m of the particle is relevant. In this limiting case of a *point particle* a relatively simple equation of motion results:

$$m\,\mathbf{a} = \frac{2}{3}\, e^2\,\dot{\mathbf{a}} + \mathbf{F}, \tag{5.10.41}$$

since the infinite sum in eq. (5.10.39) does not contribute.

It was necessary in our procedure to consider an extended particle first and take the limit of zero radius only after the *mass renormalization* (5.10.40) to come to grips with the infinite self energy $E_F = \lim e^2/2R = \infty$.

The first thing to be remarked about eq. (5.10.41) is that for $\dot{\mathbf{a}} = \mathbf{0}$—i.e., for uniform acceleration—we get *no* radiation reaction, although the radiation (5.8.18) does not vanish at all. To investigate this more closely, let us derive an energy theorem for eq. (5.10.41) in the usual way by scalarly multiplying with $\mathbf{v}$:

$$m\,\mathbf{a}\,\mathbf{v} = \frac{2}{3}\, e^2\,\dot{\mathbf{a}}\,\mathbf{v} + \mathbf{F}\,\mathbf{v} = \frac{2}{3}\, e^2\,\dot{\mathbf{a}}\,\mathbf{v} - \operatorname{grad} V \cdot \mathbf{v}$$

or

$$\frac{d}{dt}\left(m\,\frac{\mathbf{v}^2}{2} + V(\mathbf{x})\right) = \frac{2}{3}\, e^2\,\dot{\mathbf{a}}\,\mathbf{v}. \tag{5.10.42}$$

This does not seem to indicate that the energy of the particle decreases according to eq. (5.8.18). However, when this is rewritten as

$$\frac{d}{dt}\left(m\,\frac{\mathbf{v}^2}{2} + V(\mathbf{x}) - \frac{2}{3}\,e^2\,\mathbf{a}\,\mathbf{v}\right) = -\frac{2}{3}\,e^2\,\mathbf{a}^2 \le 0, \tag{5.10.43}$$

then the right-hand side exactly corresponds to the radiated energy (5.8.18). Therefore the energy of the accelerated particle should be identified with

$$E = m\,\frac{\mathbf{v}^2}{2} + V(\mathbf{x}) - \frac{2}{3}\,e^2\,\mathbf{a}\,\mathbf{v}. \tag{5.10.44}$$

The *Schott term*, proportional to $\mathbf{a}\,\mathbf{v}$, is present only during acceleratory periods of the motion, to be understood as a reversible deformation of the electromagnetic field accompanying the particle, under the influence of the acceleration. (A detailed visualizable interpretation of the expression (5.10.44) is, however, not known to us.) During periods of uniform acceleration, $\dot{\mathbf{a}} = 0$, the Schott term exactly compensates for the radiative energy loss, so that there is no radiation reaction.[1] Since for point particles the field energy E_F is infinite, arbitrary amounts of energy for the radiation field may be borrowed from the field of the near zone during arbitrarily long periods of constant acceleration (which must flow back at the end of such periods).

We now come to the relativistic versions of the radiative power (5.8.18) and the equation of motion (5.10.41). Since the radiated energy E_{rad} must be the zero component of a 4-vector p_{rad} (energy-momentum of radiation), the relativistic version of eq. (5.8.18) must be

$$\frac{dp^i_{rad}}{ds} = -\frac{2}{3}\,e^2\,a^2\,u^i. \tag{5.10.45}$$

Here u is the 4-velocity of the radiating particle and a is its 4-acceleration, satisfying $a^2 = -\mathbf{a}^2$ in the instantaneous rest system (cf. exercise 2 of sect. 4.1); dp_{rad}/ds is the 4-momentum radiated per unit proper time.

If one now tried to write the equation of motion in the plausible form

$$\frac{d\,(p^i + p^i_{rad})}{ds} = m\,a^i - \frac{2}{3}\,e^2\,a^2\,u^i = F^i, \tag{5.10.46}$$

one encountered at once a contradiction, since the 4-force $\frac{2}{3}e^2\,a^2\,u$ of radiation reaction would not be orthogonal to u unless $a^2 = -\mathbf{a}^2 = 0$. Thus eq. (5.10.46) has to be modified to become

$$m\,a^i = \frac{2}{3}\,e^2\left(\frac{da^i}{ds} + a^2\,u^i\right) + F^i. \tag{5.10.47}$$

Indeed there is now no contradiction upon multiplication by u, since by differentiating $b\,u = 0$ we have $u\,db/ds + b^2 = 0$. The 4-vector of radiation reaction

$$F_{rad} = \frac{2}{3}\,e^2\left(\frac{da}{ds} + a^2\,u\right) \tag{5.10.48}$$

---

[1] Detailed arguments may be found in Rohrlich (1965).

is often called *Abraham 4-vector*, because it was M. Abraham who in 1905 derived eq. (5.10.47), albeit at that time from other considerations. The zero component of eq. (5.10.47) gives the relativistic version of the energy theorem (5.10.43), since the zero component a^0 of a is given by $a^0 = \gamma^4\, \mathbf{a}\, \mathbf{v}$, as is easily checked ($\mathbf{a} = d\mathbf{v}/dt$). The zero component of the Abraham 4-vector therefore is just the relativistic version of the Schott term.

Setting up the equation of motion (5.10.47) does, however, not solve all problems of charged particles: this equation is of higher differential order—what initial conditions are to be chosen? Furthermore, even if external forces are absent, $F^i = 0$, there are solutions of eq. (5.10.47) having $b \neq 0$ ('runaway solutions'), such as

$$u^i(s) = (\cosh(A\,e^{s/\tau} + B), \sinh(A\,e^{s/\tau} + B), 0, 0)^\top, \tag{5.10.49}$$

where $\tau = 2e^2/3m$ is a characteristic time (of the order 10^{-23} sec for electrons). The particle seems to take the energy necessary for the acceleration from the infinite reservoir of field energy E_F. Unphysical solutions such as the ones given by eq. (5.10.49) are the price to be paid for allowing quantities like $m = m_0 + E_F = -\infty + \infty$.

One can avoid runaway solutions by adding the boundary condition $a \to 0$ as $s \to \infty$ to the equation of motion,[1] which then may be converted into the integro-differential equation (see Rohrlich 1965)

$$m\,a^i(s) = \int_0^\infty d\alpha\, F^i(s + \alpha\,\tau)\, e^{-\alpha}, \tag{5.10.50}$$

where

$$F^i(s) = F^i - \frac{2}{3}\,e^2\,a^2\,u^i. \tag{5.10.51}$$

On differentiating eq. (5.10.50) for s we return to eq. (5.10.47).

Equation (5.10.50) shows another disturbing phenomenon, however: the acceleration at proper time $s = 0$ is determined by forces (5.10.51) at later times; in particular, we have $a(0) \neq 0$ even if the force begins to act only later ('*pre-acceleration*'). An electron will, because of the exponential function in eq. (5.10.50), begin to accelerate about 10^{-23} sec before the force starts acting. This effect cannot, of course, be measured by the methods of classical physics and is therefore of no practical relevance; but it shows the difficulty of a consistent formulation of the equation of motion of a charged particle.

The consideration of *extended charge distributions*, i.e., of particles with structure, does not simplify the situation. To the contrary, some new complications appear, as sketched in Fig. 5.12.

The charge distribution cannot be assumed rigid and thus may not be described by a given function $\rho(\mathbf{x})$. Rather, a dynamical description by equations of motion for each volume element of the particle is necessary, in which—again because of the problem of cohesion—electromagnetic interactions alone do not suffice. It has then to be specified upon which volume element the additional force will act—which requires ad hoc assumptions—and the 'particles within the particle' in general will start oscillating

---

[1] This procedure eliminates sometimes—e.g., in the Coulomb problem—physical solutions also.

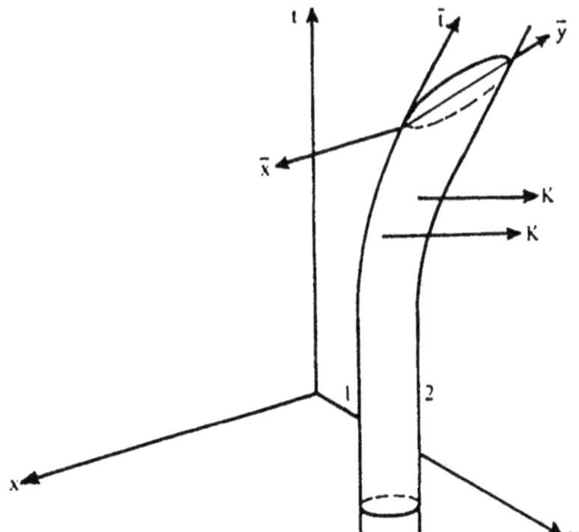

Fig. 5.12. An accelerated charge distribution

against each other, as indicated in Fig. 5.12, causing additional radiation. The world lines labeled 1 and 2 in Fig. 5.12 will have different lengths, so that the particle has no uniquely defined proper time associated with it. These remarks are just to illustrate some of the technical problems to be overcome in setting up the dynamics of extended charges.[1]

Relativistic speeds are observed practically only for *elementary particles*, for which the concepts used so far cannot be applied meaningfully. They are to be described by *quantum field theory*, which, e.g., yields a prediction for the self energy of a particle which is basically different from the classical one:

$$E_F \approx \frac{e^2}{2R} \qquad \text{(classical physics)}$$

$$E_F \approx \frac{e^2}{\hbar}\, m \ln \frac{\hbar}{m\,R} \qquad \text{(perturbative quantum field theory)}. \tag{5.10.52}$$

In the limiting case $R \to 0$ the result from quantum field theory is much less divergent than the classical one, as first shown by V. Weisskopf in 1939. The divergence difficulties have not yet been removed even here entirely, and a situation prevails that reminds of the Zeno paradox: the particle is neither allowed to be pointlike nor is it allowed to be not pointlike. A highly developed scheme of renormalization—one of the greatest achievements in physics after 1945—admits passing over these difficulties without really solving the basic paradoxes of the 'infinitely small'.

The derivation and interpretation of the equations of motion of acclerated charges as indicated in eqs. (5.10.35–41) and (5.10.45–47) is more of a phenomenological nature. An approach which

[1]See, e.g., H. Hönl, Ergeb. Exakten Naturwiss. *26*, 291 (1952); J. S. Nodvik, Ann. Phys. (N.Y.) *29*, 225 (1964).

is consistently founded on the basic equations of relativistic electrodynamics has been attempted as late as 1938 by P. A. M. Dirac, Proc. R. Soc. Lond. Ser. A *167*, 148 (1938), stimulated by the divergence problems of quantum field theory, then in its initial stages. Since then, the discussion about the 'exactness' of the Lorentz-Dirac equation (5.10.47), and about its unphysical solutions, has not found its end.

For a closer study of this and related problems, see the extensive work of T. Erber, Fortschr. Phys. *9*, 343 (1961) and the references given there, as well as the more recent investigations by C. Teitelboim, Phys. Rev. D *1*, 1572 (1970); *3*, 297 (1971); *4*, 345 (1971).

With this sketch of the problems of motion of charged particles—an excellent and detailed presentation is given in Rohrlich (1965)—we arrived at some borderline of the non-quantum mechanical application of Special Relativity. Further studies require the construction of a consistent *relativistic quantum field theory* of interacting fields—one of the most important, most challenging, and most difficult tasks of modern physics. To show up the potentialities and problems of the field theoretical description of elementary particles, it is necessary to systematically find out all kinds of fields (scalar fields, vector fields, spinor fields, ...) which can serve as the basic ingredients for such a description. This is the task of the theory of *representations* and *realizations* of the Poincaré group, contained in chap. 9. The basic philosophy of group symmetry will be characterized there in terms of abstract nonsense (sect. 9.2), continuing the one given in the smallprint paragraph of the introduction to chap. 3. In the following chaps. 6, 7, 8, some group theoretical tools will be prepared concerning the simpler theories of the rotation and Lorentz groups. The reader is now advised to gradually familiarize himself or herself with the basic definitions and concepts from group theory and abstract linear algebra, a condensed account of which is given in Appendices A and B. It is not necessary to do this all at once, one can proceed stepwise as the main text proceeds.

Exercises

1. Show, in analogy to eqs. (5.10.17–19), that the *moments* of the energy-momentum tensor

$$\int d^3x \, T^{00} \, x^{\alpha_1} \ldots x^{\alpha_\ell} =: E^{\alpha_1 \ldots \alpha_\ell}$$

$$\int d^3x \, T^{0\alpha} \, x^{\alpha_1} \ldots x^{\alpha_\ell} =: P^{\alpha\alpha_1 \ldots \alpha_\ell} \tag{5.10.53}$$

$$\int d^3x \, T^{\beta\alpha} \, x^{\alpha_1} \ldots x^{\alpha_\ell} =: \Pi^{\beta\alpha\alpha_1 \ldots \alpha_\ell}$$

satisfy the *Laue identities*

$$\frac{d}{dt} E^{\alpha_1 \ldots \alpha_\ell} = \ell \, P^{(\alpha_1 \ldots \alpha_\ell)}, \qquad \frac{d}{dt} P^{\alpha\alpha_1 \ldots \alpha_\ell} = \ell \, \Pi^{\alpha(\alpha_1 \ldots \alpha_\ell)}. \tag{5.10.54}$$

   Which (combinations) of these are conservation laws?

2. In what time would an electron on the first Bohr radius spiral into the hydrogen nucleus on account of radiation reaction if considered as a classical particle?

6 The Lorentz Group and Some of Its Representations

As we have seen in the last chapter, all laws of nature that can be written as the vanishing of some 4-tensor (field) manifestly satisfy the principle of (Einsteinian) Relativity. The essential point here is that the 4-tensor spaces are linear spaces on which the Lorentz group acts as a group of linear transformations. This will be characterized formally in sect. 6.4 where we introduce the concept of *representation* of a group. From a more systematic point of view we may then ask whether tensors are the only type of quantities that allow such a linear action. In chap. 8, we will answer this question in the positive—but in this investigation a new type of quantities will emerge that on the one hand turns out, in chap. 9, to be essential if the question is asked from a quantum mechanical point of view, and on the other hand also proves very helpful even in the classical, tensorial regime. These are the *spinors* and *spinor fields*.

For a systematic investigation of all possible representations it is, however, necessary to study the Lorentz group itself more closely, since the structure of the group has implications on the structure of the set of all its representations. The study of the group itself as well as the introduction of the basic notions of representation theory is the purpose of the present chapter. At this point it becomes advisable to gradually refresh the basic concepts of abstract group theory (a short account of which is found in Appendix A).

6.1 The Lorentz Group as a Lie Group

In chap. 3 we found all transformations leaving invariant the 4-dimensional line element ds^2. Apart from space-time translations (which will be taken up again only in chap. 9) these are homogeneous transformations

$$x^{\bar{i}} = L^i_{\ k}\, x^k \qquad \text{or} \qquad \bar{x} = L\, x \tag{6.1.1}$$

satisfying the (pseudo)orthogonality relations

$$L^i_{\ m}\, L^k_{\ n}\, \eta_{ik} = \eta_{mn} \qquad \text{or} \qquad L^T \eta\, L = \eta. \tag{6.1.2}$$

Because of $\eta_{mn} = \eta_{nm}$ these are 10 relations restricting the 16 matrix elements of L; and these relations are independent from each other, so that only 6 matrix elements can be chosen independently. This follows, e.g., from the fact that we were able, in sect. 1.5, to associate to any L satisfying eq. (6.1.2) the 6 components $\mathbf{v}$, $\boldsymbol{\alpha}$ which uniquely characterize L and are allowed to vary arbitrarily over the admissible domain $|\mathbf{v}| < 1$, $|\boldsymbol{\alpha}| \le \pi$. (Note that the latter restrictions are *in*equalities, whereas the former restrictions by orthogonality are equalities!)

A slightly more direct argument which at the same time is characteristic of Lie groups (to be defined below) would be as follows. Let L be a solution of eq. (6.1.2); then for every infinitesimal change $L \to L + \delta L$ it follows from eq. (6.1.2)

$$(\delta L)^{\mathsf{T}} \eta L + L^{\mathsf{T}} \eta \delta L = 0, \tag{6.1.3}$$

or, because of $\eta^{\mathsf{T}} = \eta$:

$$L^{\mathsf{T}} \eta \delta L \equiv \eta L^{-1} \delta L \text{ is an antisymmetric matrix.}$$

As such, it has 6 independent elements; and conversely, *every* infinitesimal antisymmetric matrix δA defines, via

$$\delta L = \left(L^{\mathsf{T}} \eta\right)^{-1} \delta A = L \eta^{-1} \delta A, \tag{6.1.4}$$

increments for L that will satisfy eq. (6.1.3).

The elements L of the Lorentz group thus depend on 6 continuously varying parameters. We will write these as $L(\mathbf{v}, \boldsymbol{\alpha})$ in the case of transformations not containing a time or space reversal, to which we were able in eqs. (1.5.13,10) to associate a vector $\mathbf{v}$ of relative velocity and a rotation vector $\boldsymbol{\alpha}$. If we now compose two transformations $L(\mathbf{v}_1, \boldsymbol{\alpha}_1)$ and $L(\mathbf{v}_2, \boldsymbol{\alpha}_2)$, we obtain again a transformation without reversals to which parameters $\mathbf{v}_3$, $\boldsymbol{\alpha}_3$ are associated according to eqs. (1.5.13,10):

$$L(\mathbf{v}_1, \boldsymbol{\alpha}_1) \, L(\mathbf{v}_2, \boldsymbol{\alpha}_2) = L(\mathbf{v}_3, \boldsymbol{\alpha}_3). \tag{6.1.5}$$

This gives, in principle, the 'continuous multiplication table' for the Lorentz group without reversals, i.e., the *composition functions*

$$\mathbf{v}_3 = \mathbf{v}_3(\mathbf{v}_1, \mathbf{v}_2; \boldsymbol{\alpha}_1, \boldsymbol{\alpha}_2), \qquad \boldsymbol{\alpha}_3 = \boldsymbol{\alpha}_3(\mathbf{v}_1, \mathbf{v}_2; \boldsymbol{\alpha}_1, \boldsymbol{\alpha}_2) \tag{6.1.6}$$

for the parameters of the product element, and thus the *abstract structure* of the Lorentz group. In particular, the formulae $(2.9.2) = (2.10.4)$ and $(2.10.6,7)$ now turn out to be those parts of this multiplication table in which $\boldsymbol{\alpha}_1 = \boldsymbol{\alpha}_2 = \mathbf{0}$. In fact, these formulae together with $L_{\mathrm{R}\mathbf{v}} \, L_{\mathrm{R}} \equiv L_{\mathrm{R}} \, L_{\mathbf{v}}$ (cf. exercise 2 of sect. 1.3) enable us to make the whole multiplication table somewhat more explicit (exercise):

$$\mathbf{v}_3 = (\mathrm{R}^{-1}(\boldsymbol{\alpha}_2) \, \mathbf{v}_1) \circ \mathbf{v}_2, \qquad \mathrm{R}(\boldsymbol{\alpha}_3) = \mathrm{R}(\boldsymbol{\alpha}_1) \, \mathrm{R}(\boldsymbol{\alpha}_2) \, \mathrm{R}(\mathrm{R}^{-1}(\boldsymbol{\alpha}_2) \, \mathbf{v}_1, \mathbf{v}_2). \tag{6.1.6'}$$

Here $\circ$ and $\mathrm{R}(\,\cdot\,, \cdot\,)$ indicate relativistic velocity addition and Thomas rotation, respectively. (An even more perspicuous version will result from the spinor representation; see sects. 7.6 and 8.2.)

The parameter values $\mathbf{v} = \mathbf{0}$, $\boldsymbol{\alpha} = \mathbf{0}$ yield the identical transformation $L = E$; for $\mathbf{v} = \mathbf{0}$ we get pure rotations $L(\mathbf{0}, \boldsymbol{\alpha})$, and for $\boldsymbol{\alpha} = \mathbf{0}$ we get pure boosts $L(\mathbf{v}, \mathbf{0})$. The decomposition (1.5.13) then writes

$$L(\mathbf{v}, \boldsymbol{\alpha}) = L(\mathbf{0}, \boldsymbol{\alpha}) \, L(\mathbf{v}, \mathbf{0}) = L(\mathrm{R}(\boldsymbol{\alpha}) \, \mathbf{v}, \mathbf{0}) \, L(\mathbf{0}, \boldsymbol{\alpha}). \tag{6.1.7}$$

The inverses for $L(\mathbf{v}, 0)$ and $L(0, \boldsymbol{\alpha})$ simply are $L(-\mathbf{v}, 0)$ and $L(0, -\boldsymbol{\alpha})$, respectively; therefore we have

$$L^{-1}(\mathbf{v}, \boldsymbol{\alpha}) = L(-\mathrm{R}(\boldsymbol{\alpha})\mathbf{v}, -\boldsymbol{\alpha}). \tag{6.1.8}$$

We can now characterize the situation encountered here in general terms as follows. Each group element is given by a finite number n of parameters—here we have $n = 6$—playing the role of 'coordinates' for the group elements and varying over a certain domain of $\mathbf{R}^n$. (In our case this is the domain $0 \leq |\boldsymbol{\alpha}| \leq \pi, 0 \leq |\mathbf{v}| < 1$ of $\mathbf{R}^6$, where $(\mathbf{v}, \boldsymbol{\alpha})$ and $(\mathbf{v}, -\boldsymbol{\alpha})$ for $|\boldsymbol{\alpha}| = \pi$ correspond to the same group element.) The whole group may be decomposed into subsets for which a bijective correspondence between group elements and points in some parameter domain is possible. (In our case this is effected in part by distinguishing the cases $\mathrm{sign}(L^0{}_0) = \pm 1$, $\det L = \pm 1$—see sect. 6.3.) The parametrization is not unique (we could have used polar coordinates instead of Cartesian components v^1, v^2, v^3, α^1, α^2, α^3 or Euler angles instead of $\boldsymbol{\alpha}$) but can always be chosen so that the composition functions and the parameters of inverse elements become analytic functions (i.e., functions that can be expanded into convergent power series), as exemplified here by eqs. (6.1.6',8).

The abstract group forms an n-dimensional manifold such that group multiplication and formation of inverses have analyticity properties. Such a group is called a finite-dimensional (n-dimensional or n parameter) *Lie group*. We are not going to fashion these mathematically still imprecise statements into a precise definition of a Lie group—for this the reader is advised to consult suitable mathematical textbooks, e.g., Chevalley (1946), Pontrjagin (1966), Dieudonné (1972), Warner (1983), Kirillov (1976); and we also do not discuss the weakest assumptions under which a group can be shown to be a Lie group.[1] What is essential for us is that the concept and the mathematical theory of Lie groups furnish a framework where many of the groups that occur in physics, like the Poincaré, Lorentz and rotation groups, fit in and can be treated systematically.

Let us denote the n-tuple of parameters associated with an element of some arbitrary Lie group by β—in our example of the Lorentz group this was the 6-tuple of components $\boldsymbol{\alpha}$, $\mathbf{v}$; denote the parameters of the unit element by 0 and the parameters of the inverse element by β^{-1} and let us imagine the parameters of the product of two elements being given as a function $f(\beta_1; \beta_2)$ of the parameters of the two factors: then the composition functions f have to satisfy certain functional equations that follow from the group axioms:

$$f(f(\beta_1; \beta_2); \beta_3) = f(\beta_1; f(\beta_2; \beta_3)) \qquad \text{(associativity)}$$

$$f(\beta; 0) = f(0; \beta) = \beta \qquad \text{(unit element)} \qquad (6.1.9)$$

$$f(\beta^{-1}; \beta) = f(\beta, \beta^{-1}) = 0. \qquad \text{(inverse)}$$

The basic idea of the theory of Lie groups is to first restrict attention to an *infinitesimal neighborhood of the unit element*, i.e., to expand f and other functions depending on group parameters into Taylor series near $\beta = 0$. It turns out that the

---

[1] An overview about this '5th Hilbert problem' is given, e.g., by Skljarenko in Alexandrow et al. (1971) or by Yang in Browder (1976).

relations (6.1.9) are so restrictive that it suffices to know these expansions only up to second order to fix them completely. Considerations of this kind will be sufficient for most of the problems treated in this book, so in particular for the classification of all types of quantities on which the Lorentz group can act linearly.

It is nevertheless of some merit to imagine the whole group as a manifold: group manifolds are used in cosmological models of General Relativity (cf. Ryan and Shepley 1975) or in statistical mechanics for ergodicity or mixing properties. For the Lorentz group we may interpret the $L^i{}_k$ as Cartesian coordinates in a 16-dimensional Euclidean space and the 10 orthogonality conditions as the equations of 10 algebraic hypersurfaces of second degree therein. The intersection of these hypersurfaces is the 6-dimensional group manifold of the Lorentz group, which in this sense is an 'algebraic group', of which $L^i{}_k = L^i{}_k(\mathbf{v}, \boldsymbol{\alpha})$ is a parametric description. The increments (6.1.4), interpreted as vectors in Euclidean space, are then *tangent* to the group manifold at the points L.

With this picture in mind we can visualize two important concepts. One of them is connectedness, which we can understand as follows. From elementary geometry we know that the intersection of two surfaces in real Euclidean space may consist of two or more separate pieces (Fig. 6.1).

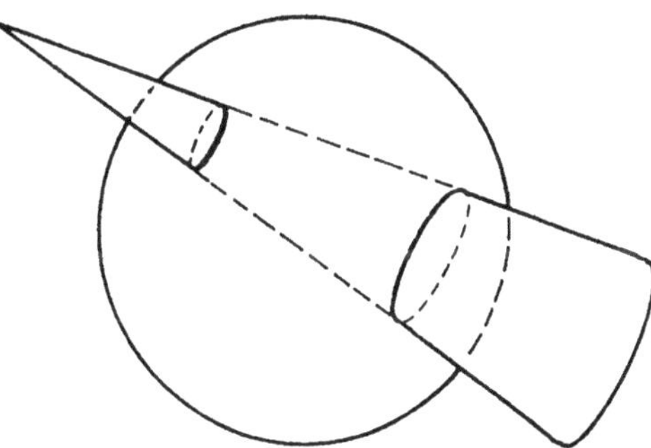

Fig. 6.1. Separate pieces of the intersection of a sphere and a cone

In the same sense, the Lorentz group (including all reversals) consists of 4 *separate* connected pieces (*connected components*, to be discussed in sect. 6.3). This is one of the reasons why we can require a bijection between group elements and points in parameter domains only for pieces of the group—but these pieces taken together are to cover the whole group.

The second important concept is that of *compactness* (resp. noncompactness) of the group manifold: the intersection of surfaces may be 'closed up onto itself' (compact without boundary) or noncompact, as indicated in Fig. 6.2. In this sense each of the 4 connected components of the Lorentz group is *noncompact*, because the parameter domain for $\mathbf{v}$ is open, $0 \le |\mathbf{v}| < 1$, on account of the restriction on the relative speed between inertial frames.

We emphasize that these concepts may be introduced and made precise without

Fig. 6.2. Compact and noncompact intersections of surfaces

using an embedding of the group manifold into a Euclidean space; but we must refer
the reader to one of the pertaining mathematical texts for doing this in all generality.

Exercises

1. Derive eq. (6.1.6$'$).

2. The general Lorentz transformation, not restricted by the absence of reversals,
   according to sect. 1.5 may be written as

   $$L = P^{\epsilon_P}\, L(\mathbf{0}, \boldsymbol{\alpha})\, T^{\epsilon_T}\, L(\mathbf{v}, \mathbf{0}) = L(\mathbf{v}, \boldsymbol{\alpha}, \epsilon_P, \epsilon_T),$$

   where the exponents ϵ_P, ϵ_T independently take the values 0 or 1. Prove, interpret, and use the relations

   $$T^{-1} L(\mathbf{v}, \boldsymbol{\alpha})\, T = L(-\mathbf{v}, \boldsymbol{\alpha}) = P^{-1} L(\mathbf{v}, \boldsymbol{\alpha})\, P \qquad (6.1.10)$$

   to obtain a multiplication table for the $L(\mathbf{v}, \boldsymbol{\alpha}, \epsilon_P, \epsilon_T)$ that generalizes eq.
   (6.1.6$'$).

3. Prove eq. (6.1.8).

4. As a first example of applying the infinitesimal method in Lie groups, consider
   the group GL(n) of general linear nonsingular transformations $\mathbf{V}_n \rightarrow \mathbf{V}_n$, resp.
   of nonsingular $n \times n$ matrices L. Write

   $$L^i{}_k = \delta^i{}_k + \epsilon\, \ell^i{}_k \qquad \text{or} \qquad L = E + \epsilon\, \ell$$

   (where ℓ is an arbitrary $n \times n$ matrix and $\epsilon^2 \approx 0$) for transformations deviating
   only infinitesimally from the identity $L = E$ ('infinitesimal transformations').

   a. Show that up to the same accuracy in ϵ

   $$\det L \approx 1 + \epsilon\, \mathrm{Tr}\, \ell \qquad (6.1.11a)$$

$$L_i{}^k = \delta^k{}_i - \epsilon\, \ell^k{}_i \qquad \text{or} \qquad L^{-1} = E - \epsilon\, \ell. \qquad (6.1.11b)$$

Now look for invariant tensors under $GL(n)$ by restricting attention to infinitesimal transformations, taking advantage of the simple form $(6.1.11b)$ of the inverse of L:

b. Show that there are no purely covariant or contravariant tensors invariant under $GL(n)$.

Hint: From $T_{\bar{i}\bar{k}...} = L_i{}^m L_k{}^n \ldots T_{mn...} = T_{ik...}$ it follows, using eq. $(6.1.11)$:

$$-\epsilon\left(T_{mk...}\ell^m{}_i + T_{im...}\ell^m{}_k + \ldots\right) = 0$$
$$\left(T_{mk...}\delta^j{}_i + T_{im...}\delta^j{}_k + \ldots\right)\ell^m{}_j = 0. \qquad (6.1.12)$$

Since the $\ell^m{}_j$ are arbitrary, the bracket must vanish. Contracting $m = j$ then yields $T_{ik...} = 0$.—Proceed analogously in the contravariant case.

c. Show that mixed tensors invariant under $GL(n)$ must be of type (p,p). In simple cases show that these tensors are in fact linear combinations of p-fold tensor products $\delta^i{}_j\, \delta^k{}_m \ldots$ (The general proof is more difficult.)

5. For the unimodular group $(\det L = 1)$ and the Lorentz group $(L^\mathsf{T}\eta L = \eta)$ one cannot argue as in the last step of exercise 4, since the $\ell^i{}_k$ are not all independent. What are the restrictions upon them?

6. By the result just obtained for infinitesimal Lorentz transformations, make it clear that the conclusion from eq. $(6.1.12)$ is now that the bracket antisymmetrized in m, j must vanish. Evaluate this condition further by a suitable contraction, assuming degree $p = 2$ and either symmetry or antisymmetry of the tensor in question. Note the exceptional case of dimension 2!

6.2 The Lorentz Group as a Quasidirect Product[1]

The basic kinematic differences between Galilean and Einsteinian relativity can be seen also on the more abstract level of the Galileo and Poincaré groups, the essential distinction becoming apparent already on the level of the homogeneous groups without reversals.

The homogeneous Galileo group is generated by space rotations $G_\mathrm{R} \equiv L_\mathrm{R}$, eq. $(1.5.8)$, and Galilean boosts $(1.3.12)$ in arbitrary number and order. Writing eq. $(1.3.12)$ in matrix notation as $x' = G_\mathbf{v}\, x$, where

$$G_\mathbf{v} := \begin{pmatrix} 1 & \mathbf{0}^T \\ -\mathbf{v} & \mathbf{1} \end{pmatrix}, \qquad (6.2.1)$$

---

[1]This section, less important for later sections, is based on papers by A. A. Ungar; see, e.g., Found. Phys. *27*, 881 (1997). To compare his formulae with ours it would be necessary, however, to observe that he is using different notation which makes his formulae somewhat simpler but does not fit in with the conventions of this book. For the group theoretical terminology used in this section see Appendix A.

we have the relations

$$G_R\, G_v = G_{Rv}\, G_R, \qquad G_{R_1}\, G_{R_2} = G_{R_1 R_2}, \qquad (6.2.2a)$$

$$G_{v_1}\, G_{v_2} = G_{v_1 + v_2} = G_{v_2}\, G_{v_1} \qquad (6.2.2.b)$$

with $\mathbf{v}$, $\mathbf{v}_1$, $\mathbf{v}_2 \in \mathbf{R}^3$ for their domain. On account of these relations, each group element may be brought to the uniquely determined form $G_R\, G_v$.

For the Lorentz group, eqs. (6.2.2a) persist, while eq. (6.2.2b) is changed in an essential way (cf. sects. 2.9, 2.10):

$$L_R\, L_v = L_{Rv}\, L_R, \qquad L_{R_1}\, L_{R_2} = L_{R_1 R_2}, \qquad (6.2.3a)$$

$$L_{v_1}\, L_{v_2} = L_{R(v_1, v_2)}\, L_{v_1 \circ v_2}, \qquad (6.2.3b)$$

where $R(\mathbf{v}_1, \mathbf{v}_2)$ is the Thomas rotation and $\circ$ indicates relativistic velocity addition; and the domain for $\mathbf{v}$, $\mathbf{v}_1$, $\mathbf{v}_2$ is given by $|\mathbf{v}| < 1$. Note that the operations $\mathbf{v} \to R\,\mathbf{v}$ and $\mathbf{v} \to \mathbf{v} + \mathbf{v}_1$, resp. $\mathbf{v} \to \mathbf{v} \circ \mathbf{v}_1$, do not lead out of the respective domains.

Both groups contain as a subgroup the group of all spatial rotations $G_R \equiv L_R$. The *set* of boosts G_v, resp. L_v, is invariant under 'conjugation' by $G_R = L_R$ in the sense that $G_R\, G_v\, G_R^{-1} = G_{Rv}$, resp. $L_R\, L_v\, L_R^{-1} = L_{Rv}$, are boosts again.

In the Galileo group, the set of boosts forms an (Abelian) subgroup (see eq. (6.2.2b)) and thus an *invariant subgroup*; if we form the factor group with respect to it we get a group which is isomorphic to the subgroup of space rotations. Equations (6.2.2) yield the multiplication law

$$\underbrace{G_{R_1}\, G_{v_1}}\ \underbrace{G_{R_2}\, G_{v_2}} = G_{R_1 R_2}\, G_{R_2^{-1} v_1 + v_2} = G_{R_3}\, G_{v_3},$$
$$R_3 = R_1 R_2, \qquad v_3 = R_2^{-1} v_1 + v_2. \qquad (6.2.4)$$

This shows that the homogeneous Galileo group (without reversals) is a *semidirect product* of the rotation group with the Abelian group $\mathbf{R}^3$ of all $\mathbf{v}$.

Within the Lorentz group, however, the set of boosts does *not* form a subgroup. One can define a multiplication $\circ$ in this set by

$$L_{v_1} \circ L_{v_2} := L_{v_1 \circ v_2} \qquad (6.2.5)$$

on account of the uniqueness of the decomposition (6.2.3b), but this does *not* define a group structure on this set; it is just called a *groupoid*. Similarly, the domain $|\mathbf{v}| < 1$ becomes a groupoid under the multiplication $\mathbf{v}_1 \circ \mathbf{v}_2$. In particular, associativity does not hold for this multiplication; however, a weaker form of it is true, involving the Thomas rotation (exercise 1 below):

$$(\mathbf{v}_1 \circ \mathbf{v}_2) \circ \mathbf{v}_3 = (R^{-1}(\mathbf{v}_2, \mathbf{v}_3)\, \mathbf{v}_1) \circ (\mathbf{v}_2 \circ \mathbf{v}_3). \qquad (6.2.6)$$

Similarly, the multiplication is not commutative but satisfies a weaker form of commutativity given by eq. (2.10.14). While in general groupoids one has to distinguish between right and left unit elements (if any), we have here $\mathbf{v} = \mathbf{0}$ as a two-sided unit element for the velocity groupoid. In the same vein, while in general groupoids right and left inverses may differ (if any), $-\mathbf{v}$ is the unique two-sided inverse of $\mathbf{v}$ relative to the $\circ$ multiplication. It is nevertheless nontrivial (because of nonassociativity), albeit true, that the equation

$$\mathbf{v}_1 \circ \mathbf{v}_2 = \mathbf{v}_3, \qquad (6.2.7)$$

given $\mathbf{v}_1$, $\mathbf{v}_3$, may be solved uniquely for $\mathbf{v}_2$ and, given $\mathbf{v}_2$, $\mathbf{v}_3$, may be solved uniquely for $\mathbf{v}_1$. A groupoid in which eq. (6.2.7) can be solved in the manner just described is called a *quasi-group*; a quasi-group with a two-sided unit element is called a *loop*.[1] The claimed solvability of eq. (6.2.7) for $\mathbf{v}_1$ follows from eq. (6.2.6), but the solvability for $\mathbf{v}_2$ (exercise 2 below) needs a further property of $\circ$, which for this reason is called the *loop property*:

$$R(\mathbf{v}_1, \mathbf{v}_2) = R(\mathbf{v}_1, \mathbf{v}_1 \circ \mathbf{v}_2). \qquad (6.2.8)$$

(Just as for eqs. (6.2.6) and (2.10.14), an indirect argument recommends itself to derive this equation (exercise 3 below); alternatively, it may be verified from the definitions using symbolic computing.)

Orthogonal S act as automorphisms of the velocity groupoid:

$$(S\,\mathbf{v}_1) \circ (S\,\mathbf{v}_2) = S(\mathbf{v}_1 \circ \mathbf{v}_2)$$

(cf. eq. (2.10.15)). In the present groupoid therefore the Thomas rotation gives us a map $(\mathbf{v}_1, \mathbf{v}_2) \mapsto R(\mathbf{v}_1, \mathbf{v}_2)$ into the automorphism group which satisfies

$$R(\mathbf{0}, \mathbf{v}) = \mathrm{id} = R(-\mathbf{v}, \mathbf{v}) \qquad (6.2.9)$$

and eq. (6.2.6). A groupoid with a left-sided unit and left-sided inverses with these properties has been called *weakly associative*; if also eq. (2.10.14) holds it is *weakly associative-commutative*; if eq. (6.2.8) holds in addition it has been called *complete*.

The multiplication rule (6.1.6′) of the Lorentz group without reversals makes it, in the sense of all these definitions, into the *quasidirect* product of the rotation group (as a subgroup of the automorphism group of the velocity groupoid) with the weakly associative groupoid of velocities. The Lorentz group is, in contrast to the Galileo group, in no way a semidirect product, being a *simple group*, i.e., noncommutative and having no nontrivial invariant subgroup whatsoever, as we shall prove as an appendix to the next section.

The algebraic structure just presented—a groupoid whose composition $\circ$ satisfies eqs. (6.2.6–9) and (2.10.14), thus being a loop—has an interesting history[2] of multiple discovery, with ensuing multiple terminology.

In 1988, A. A. Ungar abstracted the properties of the above composition $\circ$ and automorphisms $R(.,.)$ from the example of the Lorentz group, thus discovering the abstract structure together with an example. In fact, well hidden behind the Iron Courtain, A. Nesterov and coworkers in the Soviet Union had studied, since 1986, the same quasigroup (report no. 400, Kircnsky Institute of Physics of the Soviet Academy of Sciences, Krasnoyarsk).

---

[1]Supplementary terminology: an associative groupoid is called a *semigroup*; a semigroup with a two-sided unit element is called a *monoid*.

[2]We are indepted to P. Kuusk (Univ. Tartu), H. Pflugfelder (Univ. Berkeley), A. A. Ungar (Univ. of North Dakota), and H. Wcfclacheid (Univ. Duisburg) for help in tracing this history.

On the other hand, 20 years before Ungar, H. Karzel had postulated a version of the same abstract structure as integrated into a richer one with *two* compositions, called 'near-domain', where the automorphisms R(.,.) were to be realized by the (distributive) left multiplication (in the sense of the second composition) with suitable elements of the near-domain (Abh. Math. Sem. Univ. Hamburg 1968). Despite the endeavours of some researchers— among them H. Wefelscheid—no concrete example of a near-domain was found to demonstrate the consistency of the axioms postulated, and none exists today. This led them to giving up the second composition; but even then no example was forthcoming until Ungar's discovery. At the suggestion of Wefelscheid, Ungar also introduced the term *K-loop* as an alternative to his 'complete weakly associative-commutative group[oid]' structure in his first publications on the subject. Later he developed a more systematic, descriptive terminology, according to which the structure is called '*gyrocommutative gyrogroup*', alluding to the Thomas gyration (Ungar, loc. cit.). On the other hand, Wefelscheid and coworkers used some intermediate nomenclature while checking the independence of axioms, to come up with their final definition of K-loop around 1993.

But there are at least two more sources! In Japan, M. Kikkawa had studied certain loops with a compatible differentiable structure which he called 'homogeneous symmetric Lie loops' (Hiroshima Math. J. *5*, 141 (1975)). Although he did not discuss any concrete example, it is quite obvious that the loop encountered above is among Kikkawa's objects. In view of this, the designation K-loop may be interpreted as honoring Karzel as well as Kikkawa. However, again similar ideas had been expressed by L. Sabinin and coworkers since 1972 (Sov. Math. Dokl. *13*, 970 (1972)). (The relation to 'symmetric homogeneous spaces of noncompact type' has been discussed recently by W. Krammer and H. K. Urbantke, Res. Math. *33*, 310 (1998).)

Finally, while the approaches mentioned so far were motivated from geometry and physics, there is also an approach from the purely algebraic loop theory side. Here, a certain type of loops (Bruck loops), introduced by D. A. Robinson in 1966, was shown in 1995 by A. Kreuzer (Math. Proc. Camb. Philos. Soc. *123*, 53 (1998)) to be identical to K-loops, thus again effectively increasing the multiplicity of nomenclature.

We close this rather formal section with the following comment. In the appendix to sect. 4.1, we mentioned the geometrical distinction between the flat affine velocity space of Galilean Relativity and the curved Lobachevski (=hyperbolic) velocity space of Einsteinian Relativity. (This geometrical distinction was probably known earlier to Poincaré, but was certainly pointed out as early as 1908 by V. Varičak: Jahrb. Dt. Math.-Ver. *17*, 70.) Now the formalism of Euclidean 3-vectors has immediate geometrical and physical significance for the former, but not for the latter, as we already pointed out in sects. 2.9, 2.10. When these vectors are used nevertheless, there arise the somewhat unusual structures discussed in this section, which may be of interest by themselves. Conversely, the velocity groupoid furnishes an explicit example for them. Also, as shown in sect. 2.10, the formalism is useful in the discussion of some paradoxes and might be helpful in comparing Einsteinian Relativity with rivalling theories.

In the framework of Einsteinian Relativity, the formalism of 4-vectors appears to be better adapted generally, enabling formulations and considerations without using a frame to which the Lorentz matrices are referred. (In this context, we may mention the so-called *intrinsic* decomposition of active Lorentz transformations and their eigenvalue structure, which will be treated in sect. 8.4.)

Exercises

1. Starting from the associativity of matrix multiplication, $(L_{\mathbf{v}_1} L_{\mathbf{v}_2}) L_{\mathbf{v}_3} = = L_{\mathbf{v}_1} (L_{\mathbf{v}_2} L_{\mathbf{v}_3}))$, prove eq. (6.2.6) by first decomposing the bracketed products into boost and Thomas rotation, then treating the remaining products of boosts in the same way until both sides are of the form $L_{\mathrm{R}} L_{\mathbf{v}}$, and finally comparing both sides on the basis of the uniqueness of the decomposition. In this way one also gets, in addition, the identity

$$\mathrm{R}(\mathbf{v}_1, \mathbf{v}_2)\, \mathrm{R}(\mathbf{v}_1 \circ \mathbf{v}_2, \mathbf{v}_3) = \mathrm{R}(\mathbf{v}_2, \mathbf{v}_3)\, \mathrm{R}(\mathrm{R}^{-1}(\mathbf{v}_2, \mathbf{v}_3)\mathbf{v}_1, \mathbf{v}_2 \circ \mathbf{v}_3). \qquad (6.2.10)$$

2. Show that eqs. (6.2.6,8) despite nonassociativity permit solving eq. (6.2.7) uniquely for $\mathbf{v}_1$ or $\mathbf{v}_2$.
   Hint: Observe the difference between the uniqueness and the existence part! For the uniqueness of the solution for $\mathbf{v}_2$ it suffices to use eq. (6.2.8); for its existence one needs the specialization $\mathbf{v}_2 = -\mathbf{v}_1$ of eq. (6.2.10).

3. Derive eq. (6.2.8), by uniqueness of the rotation-boost decomposition, from the fact that the product $L_{\mathbf{v}_1} L_{\mathbf{v}_2} L_{\mathbf{v}_1}$ must be a pure boost on account of the criterion found at the end of sect. 1.5. (A 180°-rotation, admitted by that criterion, is excluded because of the continuous dependence on $\mathbf{v}_1$, $\mathbf{v}_2$.)

6.3   Some Subgroups of the Lorentz Group

The transformations L that are restricted only by the orthogonality relations (6.1.2) form the *full Lorentz group* $\mathcal{L}$ containing also space and time reversals, whose peculiar role in the context of the principle of relativity we were addressing several times. Transformations not changing the sense of time (*orthochronous* transformations) have $L^0{}_0 \geq 1$, since from eq. (6.1.2) it follows that $(L^0{}_0)^2 = 1 + L^\alpha{}_0\, L^\alpha{}_0 \geq 1$. Performed actively, they map the future light cone into itself. As this is true also for the product of two orthochronous transformations, these form a subgroup of the full Lorentz group which is called the *orthochronous Lorentz group* $\mathcal{L}^\uparrow$. When orthochronous transformations are composed with the time reversal operation (1.5.7), general transformations with time reversal (*antichronous transformations*) are obtained ($L^0{}_0 \leq -1$), and conversely. We therefore have the decomposition

$$\mathcal{L} = \mathcal{L}^\uparrow \cup T\mathcal{L}^\uparrow, \qquad\qquad \mathcal{L}^\uparrow \cap T\mathcal{L}^\uparrow = \emptyset \qquad\qquad (6.3.1)$$

of $\mathcal{L}$, where $\emptyset$ is the empty set and $T\mathcal{L}^\uparrow$ is the only *coset* of $\mathcal{L}^\uparrow$ in $\mathcal{L}$.

We can also divide $\mathcal{L}$ into two disjoint sets according to the value of the determinant $\det L = \pm 1$ (cf. sect. 5.5): the *proper Lorentz group* $\mathcal{L}_+$ comprises the transformations having $\det L = 1$. The intersection $\mathcal{L}_+^\uparrow = \mathcal{L}_+ \cap \mathcal{L}^\uparrow$ is called the *proper orthochronous* or *restricted Lorentz group*; it does not contain reversals.

Composing $\mathcal{L}_+^\uparrow$ with the space reversal P, eq. (1.5.9), we obtain the orthochronous group as the disjoint union

$$\mathcal{L}^\uparrow - \mathcal{L}_+^\uparrow \cup P\mathcal{L}_+^\uparrow. \qquad\qquad (6.3.2)$$

From this we get the decomposition

$$\mathcal{L} = \mathcal{L}_+^\uparrow \cup P\mathcal{L}_+^\uparrow \cup T\mathcal{L}_+^\uparrow \cup PT\mathcal{L}_+^\uparrow \tag{6.3.3}$$

of the full group into cosets of the restricted group $\mathcal{L}_+^\uparrow$. We also have

$$\mathcal{L}_+ = \mathcal{L}_+^\uparrow \cup PT\mathcal{L}_+^\uparrow, \tag{6.3.4}$$

since the *space-time reversal*

$$PT = \begin{pmatrix} -1 & \mathbf{0}^T \\ \mathbf{0} & -\mathbf{1} \end{pmatrix} = -E \tag{6.3.5}$$

has positive determinant. Finally, the union $\mathcal{L}_0 := \mathcal{L}_+^\uparrow \cup T\mathcal{L}_+^\uparrow$ forms the *orthochorous Lorentz group*.

The subgroup $\mathcal{L}_+^\uparrow$ is connected: any element can be written in the form $L(\mathbf{v}, \boldsymbol{\alpha})$, and continuously varying the parameters we can reach all elements of $\mathcal{L}_+^\uparrow$ starting at any one of it, say at the unit element $\mathbf{v} = \mathbf{0}$, $\boldsymbol{\alpha} = \mathbf{0}$; therefore $\mathcal{L}_+^\uparrow$ is called the *component of unity* of $\mathcal{L}$. In the same sense, each of the cosets $P\mathcal{L}_+^\uparrow$, $T\mathcal{L}_+^\uparrow$, $PT\mathcal{L}_+^\uparrow$ is connected, while it is impossible to get from one of these sets to another by continuous variation of parameters. Equation (6.3.3) thus is a decomposition of the group into four components, each connected.

Generally, any Lie group $\mathcal{G}$ may be decomposed into *connected components*, the component of unity $\mathcal{G}_e$ being an invariant subgroup (see exercise 1). Restricting attention to $\mathcal{G}_e$ has the advantage that one can consider first only elements close to the unit element and then reach all of $\mathcal{G}_e$ by composition; as we saw already in sect. 6.1 (cf. eq. (6.1.3) and exercise 4), the use of such elements leads to remarkable simplifications, to be exploited systematically later on.

Besides $\mathcal{L}_+^\uparrow$, also $\mathcal{L}_+$, $\mathcal{L}^\uparrow$, $\mathcal{L}_0$ are invariant subgroups in $\mathcal{L}$, the factor groups being isomorphic to the discrete subgroups $\{E, P, T, PT\}$ (the 'four-group') and $\{E, P\}$, $\{E, T\}$, $\{E, PT\}$.[1] The 'discrete' transformations P, T, PT reach their full importance only in quantum theory and elementary particle physics where they lead to important conservation laws. (The designation 'discrete' for these elements themselves is usual but incorrect—what it means is that within $\mathcal{L}$ they are not connected with the identity; they are not lying discretely within $\mathcal{L}$, of course.)

In concluding our considerations about the four components of the full Lorentz group we still remark that in some investigations also Lorentz transformations with complex coefficients $L^i{}_k$ are used. In this bigger group, the distinction $L^0{}_0 > 0$, < 0 makes no sense, so that the complexification of $\mathcal{L}_+$ is a *connected* set. The distinction according to $\det L = \pm 1$ is maintained, however, since the determinant is a well-defined continuous function also in the complex domain: the complex(ified) Lorentz group consists of two connected components.

We now turn to some subgroups of $\mathcal{L}_+^\uparrow$.

The most important subgroup is the group of all spatial rotations L_{R}, where R is a proper orthogonal matrix. This group is written $SO(3,\mathbf{R})$, in which O means

---

[1] We thus have semidirect products here; cf. sect. 8.5 and Appendix A.

orthogonal, S means special, i.e., $\det R = +1$; 3 is the dimension of space and $\mathbf{R}$ indicates the real number field.

The rotation group is a 3-parameter connected compact Lie group: $0 \leq |\alpha| \leq \pi$ gives all rotations and is, as a solid ball, a compact domain; the parameter assignment is not bijective, however, since in the cases where $|\alpha| = \pi$ the same rotation is given by α and $-\alpha$. Antipodal points of the surface of the ball are thus to be identified so that there is no boundary surface left. The occurrence of a phenomenon of this kind is the second reason why one requires a bijective parametrization only for (open) subsets of the group manifold which together cover the latter. This is related to the in general complicated topology of the group manifold which we cannot discuss here in any generality; for the rotation and Lorentz group this will be illuminated in sects. 7.6 and 8.2.

The boosts, on the other hand, do not form a subgroup, as we saw in sect. 6.2, except if we restrict to relative velocities of a fixed direction: to every direction in space there is a 1-parameter group of boosts, and, similarly, a 1-parameter group of rotations around this direction as an axis. Both subgroups are commutative and also commute with each other, thus generating a 2-parameter Abelian subgroup for each given direction.

The enumeration of subgroups given here was limited to the most accessible ones as visualization is concerned; it is by no means complete. There are subgroups of rotations in lightlike planes, subgroups leaving spacelike directions fixed, etc. We do not go into a systematic treatment here (cf. J. Patera et al., J. Math. Phys. *16*, 1597 (1975)) and describe the relevant subgroups when and where they are needed.

Appendix 1: Active Lorentz Transformations

So far we have described the full Lorentz group and its subgroups as matrix groups corresponding to a passive interpretation. In particular, we were talking about 'the' space reversal and 'the' time reversal. When $\mathcal{L}$ is defined *actively* as the set of linear transformations of a Minkowski vector space $\mathbf{V}_4$ leaving invariant the metric η, then $\mathcal{L}^{\uparrow}$ is that subgroup of it which leaves invariant each connected component of the light cone (with its vertex removed) separately. The coset $\mathcal{L}^{\downarrow}$, previously called $T\mathcal{L}^{\uparrow}$, consists of those transformations that interchange these components. Similarly, $\mathcal{L}_+$ is the subgroup that leaves invariant some chosen orientation in $\mathbf{V}_4$; $\mathcal{L}_-$ (previously $P\mathcal{L}_+ = T\mathcal{L}_+$) changes it into the opposite orientation. $\mathcal{L}_{\pm}^{\uparrow} = \mathcal{L}^{\uparrow} \cap \mathcal{L}_{\pm}$, $\mathcal{L}_{\pm}^{\downarrow} = \mathcal{L}^{\downarrow} \cap \mathcal{L}_{\pm}$ then are the four connected components of $\mathcal{L}$ as before. Our new symbols are to express the observer independence of these subsets, while the active operations P, T must now be referred to some observer with 4-velocity u: the operations P_u resp. $T_u = -P_u$ are given by $v = v_{\|} + v_{\perp} \mapsto v_{\|} - v_{\perp}$ resp. $\mapsto -v_{\|} + v_{\perp}$ (reflection in the line $\prec u \succ$ resp. in the hyperplane $\prec u \succ^{\perp}$), where $v_{\|} = (vu/u^2)u$ is the projection of v onto u. The product $P_u T_u = -\mathrm{id}_{\mathbf{V}_4}$ is, however, independent of u, as are the sets $P_u \mathcal{L}_+^{\uparrow} = \mathcal{L}_-^{\uparrow}$, $P_u \mathcal{L}_+ = \mathcal{L}_-, \ldots$

$L \in \mathcal{L}_+^{\uparrow}$ is a *boost with respect to u* if L leaves the (timelike) 2-plane $\prec u, Lu \succ$ spanned by u and Lu invariant as a whole while all vectors of the spacelike orthogonal plane $\prec u, Lu \succ^{\perp}$ remain fixed. L is a *pure rotation with respect to u* if $Lu = u$; then also the Euclidean subspace $\prec u \succ^{\perp}$ remains invariant; the proper orthogonal transformation induced there leaves invariant an axis vectorwise and the orthogonal 2-plane as a whole. One therefore calls the latter Lorentz transformations *spacelike rotations*, the former ones *timelike rotations*. The space rotations relative to any given 4-velocity u form a subgroup isomorphic to $SO(3, \mathbf{R})$ (see exercise 5).

Given two 4-velocities u, u', there is exactly one timelike rotation $\Lambda_{u,u'}$ carrying u into u' and leaving $\prec u, u' \succ$ invariant (in exercise 4 it is to be constructed as a product of two reflections).

Defining $K(L, u) := \Lambda_{u,Lu}^{-1} L$, this transformation leaves the vector u invariant and thus is a pure rotation relative to u; therefore $L = \Lambda_{u,Lu} K(L, u)$ corresponds to one of the decompositions of a Lorentz transformation as given in sect. 1.5. In sect. 9.4 we shall need a generalization of this where one further 4-velocity $\bar{u}$ is singled out and one puts $K(L, u; \bar{u}) := \Lambda_{\bar{u},Lu}^{-1} L \, \Lambda_{\bar{u},u}$ —the *Wigner rotation* belonging to L, u with respect to $\bar{u}$. For $L = \Lambda_{u,Lu}$ it goes over into the Thomas rotation.

The fact that boosts $\Lambda_{u,u'}$ can be generated by reflections, as shown in exercise 4, together with the well-known fact that spatial rotations can be generated by reflections, now shows that all $L \in \mathcal{L}$ are products of hyperplane reflections. It is possible here (exercise 6) to reduce the number of reflections needed to four or less; the four components $\mathcal{L}_{+}^{\uparrow}$, $\mathcal{L}_{+}^{\downarrow}$, $\mathcal{L}_{-}^{\uparrow}$, $\mathcal{L}_{-}^{\downarrow}$ are then distinguished by the parity (even/odd) of the number of reflections in spacelike and in timelike hyperplanes that are needed. (This is shown for pseudo-orthogonal groups in an arbitrary number n of dimensions and for an arbitrary signature of the 'metric' in Cartan (1966); the total number of needed reflections can be reduced here to be $\leq n$.)

We finally remark that the observer-dependent 'Cartan decomposition' of transformations discussed above has to be distinguished from the *intrinsic classification* and *decomposition* to be derived in sect. 8.4. There one classifies the $L \in \mathcal{L}_{+}^{\uparrow}$ *without* reference to any u into general Lorentz transformations and null rotations (lightlike rotations); the general L may be uniquely written as a product of a timelike rotation and a spacelike rotation, where these rotations are in orthogonal 2-planes and thus *commute* with each other. Special cases are purely timelike rotations (*hyperbolic* transformations), purely spacelike rotations (*elliptic* transformations) and the identity. (The general case is also called *loxodromic*.) The null rotations (*parabolic* transformations) are more complicated to describe but will concern us in sect. 9.4: they leave fixed the vectors of a lightlike 2-plane spanned by a lightlike vector and a spacelike vector orthogonal to it (cf. exercise 5 of sect. 3.2 for such a configuration); the 2-plane orthogonal to it and all 3-planes passing through the latter each remain invariant as a whole. This is easiest to visualize in the projective picture mentioned at the end of sect. 3.2.

Appendix 2: Simplicity of the Lorentz group $\mathcal{L}_{+}^{\uparrow}$

We shall prove here, after E. Wigner, that the proper orthochronous Lorentz group $\mathcal{L}_{+}^{\uparrow}$ is *simple*, i.e., that it is noncommutative (which we know already) and contains no nontrivial invariant subgroup $\mathcal{N}$ (to be shown here). We shall need two facts which we note here before entering the proof. One is the remark that every boost is the square of another boost in the same direction: one can explicitly solve the equation $\mathbf{u} = \mathbf{v} \circ \mathbf{v}$ for $\mathbf{v}$. The other is the theorem that the rotation group $SO(3, \mathbf{R})$ is simple, as will be shown in sect. 7.6. (Of course, the whole development in chap. 7 on the rotation group will be logically independent of the Lorentz group.) Our proof here proceeds in two steps: first one shows that $\mathcal{N}$ contains one nontrivial rotation and thus, by the simplicity of the rotation group, contains all rotations; and second, one shows that $\mathcal{N}$ contains all boosts and therefore all of $\mathcal{L}_{+}^{\uparrow}$.

Step one: Let $N \in \mathcal{N}, N \neq E$ and decompose it into a boost B and a rotation R relative to an observer whose 4-velocity u is *not* orthogonal to a spacelike 2-plane left invariant by N (if any): $N = RB$. Let S be a rotation through the angle π about an axis orthogonal to the boost velocity vector as seen by the observer. Then we have $S^{-1} = S$ and $SBS^{-1} = B^{-1}$ according to eq. (6.2.3a), since the rotation S reverses the direction of the relative velocity. When we conjugate N by $B^{-1}S$ and multiply the result from the left by N, we stay inside $\mathcal{N}$; using the relations just written down we find that the rotation $(RS)^2$ belongs to $\mathcal{N}$. This already concludes the first step if this rotation differs from the identity. Assuming the contrary, we would have either $RS = E$, which is impossible for all the choices for S that we still have, or else RS is a rotation through an angle of π, just as S itself. So $R = RSS$ is a product of two π-rotations. Thinking about this situation, perhaps with the help of the more detailed description of the composition of rotations given in sect. 7.6, one finds that R must have its axis orthogonal to the axis of S for all possible S, thus must have it in the direction of the relative velocity of the boost as seen by the observer. It follows that $N = RB$ must have an invariant 2-plane orthogonal to u, which, however, was excluded by assumption.

Step two: Let B be an arbitrary boost and let S be any π-rotation with axis orthogonal to the relative velocity as above. Since the latter is in $\mathcal{N}$, we can conjugate with B and multiply by S^{-1} from the right without leaving $\mathcal{N}$. From the above relations we then find that B^2 is in $\mathcal{N}$; and from our initial remarks we know that B^2 will run over all boosts if B does.

Exercises

1. A subset of a manifold is called connected if any two points of it may be connected by a continuous curve belonging to the subset. The component of unity $\mathcal{G}_e$ in a Lie group $\mathcal{G}$ is then defined to be the largest connected subset containing the unit element. Show that $\mathcal{G}_e$ is an invariant subgroup.

2. Which of the discrete subgroups $\{E, P\}$, $\{E, T\}$, $\{E, PT\}$ is an invariant subgroup of $\mathcal{L}$ or of $\mathcal{L}^\uparrow$, $\mathcal{L}_0$, $\mathcal{L}_+$ respectively?

3. Can you find isomorphisms between $\mathcal{L}$ and some direct product (cf. exercise 6 of sect. 3.1) of one of the subgroups $\mathcal{L}_+$, $\mathcal{L}^\uparrow$, $\mathcal{L}_0$ with one of the discrete groups just mentioned? Is there such a product decomposition for $\mathcal{L}_+$, $\mathcal{L}^\uparrow$, $\mathcal{L}_0$ with $\mathcal{L}_+^\uparrow$ as a factor?

4. Let T_u denote the reflection in the hyperplane orthogonal to the 4-vector u, where $u^2 \neq 0$, as given in the text. Show that, given the 4-velocities u, u', the transformation

$$\Lambda_{u,u'} := T_{u+u'}\, T_u \qquad (6.3.6)$$

is (i) proper, (ii) orthochronous, and (iii) sends u into u' while leaving invariant (iv) $\prec u, u' \succ^\perp$ vectorwise and (v) $\prec u, u' \succ$ as a whole. Find, in a similar manner, a spacelike rotation sending a given spacelike vector m into another given spacelike vector n of the same length. For which observer 4-velocities is this a pure space rotation?

5. Show that any two subgroups of $\mathcal{L}_+^\uparrow$, each leaving fixed some timelike vector, are conjugate in $\mathcal{L}_+^\uparrow$.

   Hint: Use the transformation $\Lambda_{u,u'}$!

6. Show for the cosets of $\mathcal{L}_+^\uparrow$ that the number of reflections necessary to generate any element may be made equal to four or less.

7. Show that the *centralizer* of the rotation subgroup in $\mathcal{L}_+^\uparrow$ is trivial, i.e., that any element $L \in \mathcal{L}_+^\uparrow$ that commutes with all rotations L_S, $S \in \mathrm{SO}(3,\mathbf{R})$, must equal the identity $L = E$.

   Hint: Use the unique decomposition (1.5.13) or (6.1.7) of L and an appropriate insertion of $L_S^{-1} L_S$ together with eq. (6.2.3a) and the analogous eq. (7.3.4).

8. Show that the only nontrivial invariant subgroup of $\mathcal{L}^\uparrow$ and of $\mathcal{L}_o$ is $\mathcal{L}_+^\uparrow$, while for $\mathcal{L}_+$ the only one to be added is $\{E, PT = -E\}$. Now write down the list of all invariant subgroups of $\mathcal{L}$.

Hints: Let $\mathcal{N}$ be an invariant subgroup of the group $\mathcal{G}$ under discussion. By the simplicity of $\mathcal{L}_+^\uparrow$, the invariant subgroup $\mathcal{N} \cap \mathcal{L}_+^\uparrow$ equals either $\mathcal{L}_+^\uparrow$ or $\{E\}$. It should be clear how to carry on with the first of these possibilities; for the second, study the conditions imposed on any element of $\mathcal{N}$ belonging to a coset $D\mathcal{L}_+^\uparrow$ of $\mathcal{L}_+^\uparrow$ in $\mathcal{G}$, where $D \in \{P, T, PT\}$. Write this element as DL with $L \in \mathcal{L}_+^\uparrow$. Its square must belong to $\mathcal{N} \cap \mathcal{L}_+^\uparrow$ and hence $(DL)^2 = E$. Also, for all $S \in \mathcal{L}_+^\uparrow$ we have $SDLS^{-1} \in \mathcal{N}$, and so its product with DL is again in $\mathcal{N} \cap \mathcal{L}_+^\uparrow$, thus equalling E: hence $SDLS^{-1} = DL$. Now take for S first arbitrary rotations, use eq. (6.1.10) and the result of the previous exercise; then take an arbitrary boost and use eq. (6.1.10) again for your final conclusion.

6.4   Some Representations of the Lorentz Group

It will be our aim to develop methods for systematically finding all kinds of objects that behave linearly under Lorentz transformations, just like tensors. This makes it necessary to make the property of behaving linearly more precise.

In sect. 6.1 we were describing the abstract Lorentz group as a Lie group, i.e., as a 6-dimensional manifold $\mathcal{G} = \mathcal{L}_+^\uparrow$ whose points $g_1, g_2, \ldots$ can be multiplied with each other, where the product is again a point in the manifold. The group parameters act as coordinates on this manifold, and multiplication is given in terms of the parameters of the product element as functions of the parameters of the factors. The appearance of this 'continuous multiplication table' naturally depends on the particular system of coordinates (parameters) used.

Originally we had introduced the Lorentz group as a group of transformations. Thus, to each group element there is a Lorentz transformation $L(g)$ which upon passive interpretation carries the quadruple $(u^i) = (u^0, u^1, u^2, u^3)^\top$ of components of a 4-vector relative to the frame I by matrix multiplication into the component quadruple $(u^{\bar{i}}) = (u^{\bar{0}}, u^{\bar{1}}, u^{\bar{2}}, u^{\bar{3}})^\top$ of u relative to the frame $\bar{\text{I}}$ and upon active interpretation maps the space $\mathbf{V}_4$ of 4-vectors linearly into itself:

$$u^{\bar{i}} = L^i{}_k(g)\, u^k, \qquad\qquad \bar{u} = L(g)\, u, \qquad\qquad (6.4.1)$$

where scalar products remain invariant:

$$u \mapsto L(g)\, u, \qquad v \mapsto L(g)\, v, \qquad u\,v = (L(g)\,u)(L(g)\,v). \qquad (6.4.2)$$

By definition of the product $g_1 g_2$ of two group elements, this abstract element is assigned to the product transformation $L(g_1)\, L(g_2)$:

$$g_1 \leftrightarrow L(g_1), \qquad g_2 \leftrightarrow L(g_2), \qquad g_1 g_2 \leftrightarrow L(g_1)\, L(g_2). \qquad (6.4.3)$$

The unit element e of the abstract group belongs to the identical transformation $E = L(e)$, the inverse g^{-1} of g belongs to the inverse transformation $L(g^{-1}) = L^{-1}(g)$.

If we now think of the abstract group as the primary object, then the assignment $g \mapsto L(g)$ is a *realization* of the abstract group as a *group of transformations* that *act* on a space (from the left); the assignment has the properties

$$g_1 \mapsto L(g_1), \qquad g_2 \mapsto L(g_2), \qquad g_1 g_2 \mapsto L(g_1)\, L(g_2), \qquad e \mapsto \mathrm{id}, \qquad (6.4.4)$$

characterizing, by definition, a general group action, where id is the identical transformation of the space. In our case, the space is a vector space, and the realization is by linear transformations of that vector space: one calls such a linear realization a *representation* of the group. Equation (6.4.4) is then called the *representation property* of the assignment and the vector space is called the *representation space*.[1]

If the realization is by passive transformations, the vector space in question in the case above is just the space $\mathbf{R}^4$ of coordinate quadruples $u = (u^i)$, i.e., of column vectors, and the transformations are given by matrices that multiply them. This is a *matrix representation*.

Consider now the realization by active transformations in the case above. Then the representation space is the space $\mathbf{V}_4$ of 4-vectors, and assigned to the group elements are linear transformations $L(g)$ of that space. If we choose a *fixed* basis $\{e_i\}$ we can obtain a matrix representation here as well (cf. eq. (3.3.8,9)):

$$g \mapsto (L_i{}^k(g)), \qquad\qquad (L\,u)^k = L_i{}^k\,u^i. \qquad\qquad (6.4.5)$$

If this basis $\{e_i\}$ is chosen to agree with the one with respect to which the components u^i were formed in the passive interpretation before, then the matrices $(L_i{}^k(g))$ and $(L^i{}_k(g))$ are contragredient to each other (cf. eq. (3.3.7)):

$$(L^i{}_k(g)) = L(g) \Rightarrow (L_i{}^k(g)) = (L^{-1})^{\top}(g) =: \tilde{L}(g). \qquad\qquad (6.4.6)$$

We can also verify directly: if the assignment $g \mapsto L(g)$ is a matrix representation, so is the assignment $g \mapsto \tilde{L}(g)$ (*contragredient representation*). Namely, from $L(g_1)\,L(g_2) = L(g_1\,g_2)$ we conclude

$$L^{-1}(g_1\,g_2) = L^{-1}(g_2)\,L^{-1}(g_1) \Rightarrow \tilde{L}(g_1\,g_2) = \tilde{L}(g_1)\,\tilde{L}(g_2),$$

which demonstrates the representation property.

When the active transformation $u \mapsto L\,u$ is referred to another frame $\{\bar{e}_i = S_i{}^k e_k\}$ in place of $\{e_i\}$, then the same transformation L is described by the matrix $S_i{}^j\,L_j{}^m\,S^k{}_m$, i.e., by $(S := (S_i{}^k))$

$$\tilde{L} \mapsto S\,\tilde{L}\,S^{-1} \qquad\qquad (6.4.7)$$

instead of $(L_i{}^k)$, as is easily checked. In this way we associate with the representation $g \mapsto L(g)$ infinitely many matrix representations, corresponding to the possible choices of new bases or nonsingular matrices S. Two matrix representations

$$g \mapsto L(g), \qquad g \mapsto L'(g) = S\,L(g)\,S^{-1} \qquad\qquad (6.4.8)$$

are called *equivalent representations*. Again the representation property for $L'(g)$ may be verified directly:

$$L'(g_1)\,L'(g_2) = S\,L(g_1)\,S^{-1}\,S\,L(g_2)\,S^{-1} = S\,L(g_1)\,L(g_2)\,S^{-1} = S\,L(g_1\,g_2)\,S^{-1} =$$
$$= L'(g_1\,g_2).$$

---

[1] There is a tendency in modern mathematics to use the language of $\mathcal{G}$-modules here, but we shall remain oldfashioned.

For the Lorentz transformations, the matrix representations $g \mapsto L(g)$ and $g \mapsto \tilde{L}(g)$ are equivalent since from eq. (6.1.2) it follows that

$$\tilde{L} = \left(L^{\mathsf{T}}\right)^{-1} = \left(L^{-1}\right)^{\mathsf{T}} = \eta\, L\, \eta = \eta\, L\, \eta^{-1}. \qquad (6.4.9)$$

To see that our change of point of view which takes the abstract group as the primary object and the original transformation group as a representation of it is nontrivial and useful we further consider the transformation law of tensor components under the Lorentz transformation belonging to g_1:

$$T^{\bar{i}\cdots}{}_{\bar{j}\ldots} = L^i{}_m(g_1)\, L_j{}^n(g_1) \ldots T^{m\cdots}{}_{n\ldots}\,. \qquad (6.4.10)$$

With a further transformation, belonging to g_2, we have

$$\begin{aligned}
T^{\bar{\bar{a}}\cdots}{}_{\bar{\bar{b}}\ldots} &= L^a{}_i(g_2)\, L_b{}^j(g_2) \ldots T^{\bar{i}\cdots}{}_{\bar{j}\ldots} = \\
&= L^a{}_i(g_2)\, L^i{}_m(g_1)\, L_b{}^j(g_2)\, L_j{}^n(g_1) \ldots T^{m\cdots}{}_{n\ldots} = \qquad (6.4.11)\\
&= L^a{}_m(g_2\, g_1) \ldots L_b{}^n(g_2\, g_1) \ldots T^{m\cdots}{}_{n\ldots}\,.
\end{aligned}$$

We can think of the tensor components $T^{m\cdots}{}_{n\ldots}$ as being arranged in a certain way as a column vector and similarly for the $T^{\bar{i}\cdots}{}_{\bar{j}\ldots}$; then the linear transformation (6.4.10) may be written as multiplication with a big matrix—called the *Kronecker product* of the matrices $L(g_1), \ldots, \tilde{L}(g_1), \ldots$ and symbolized as

$$L(g_1) \otimes \tilde{L}(g_1) \otimes \ldots. \qquad (6.4.12)$$

Equation (6.4.11) then shows that the product of two such matrices is given by

$$\left(L(g_2) \otimes \tilde{L}(g_2) \otimes \ldots\right) \left(L(g_1) \otimes \tilde{L}(g_1) \otimes \ldots\right) = L(g_2)\, L(g_1) \otimes \tilde{L}(g_2)\, \tilde{L}(g_1) \otimes \ldots$$
$$(6.4.13)$$

and also shows that this is equal to $L(g_2\, g_1) \otimes \tilde{L}(g_2\, g_1) \otimes \ldots$. The assignment

$$g \mapsto L(g) \otimes \tilde{L}(g) \otimes \ldots \qquad (6.4.14)$$

is therefore a representation of the abstract group which is different from the original '*defining*' representation $g \mapsto L(g)$. It is called *Kronecker product*[1] of the representations $g \mapsto L(g)$, $g \mapsto \tilde{L}(g)$ (where $L(g)$, $\tilde{L}(g)$ occur as often as the bidegree of the tensors tells us). The explicit form of these matrices depends on the basis chosen in the vector space but also on the order we choose in arranging the tensor components into a column; because of the high dimensions that may occur it would in general not be advisable to write out these matrices and multiply them in the ordinary way. Rather, one uses the multiplication rule (6.4.13). (With 4-tensors of total degree p we would have to deal with $4^p \times 4^p$ matrices!)

It is a tautology for the 'defining' representation $g \mapsto L(g)$ and it is true in $\mathcal{L}^{\uparrow}_{+}$, $\mathcal{L}^{\uparrow}$, $\mathcal{L}_0$ for the Kronecker products considered above that the assignment of the representing matrix to the group element g is bijective, so that we have an isomorphism

---

[1]Sometimes also called *direct product*, which, however, is used for other constructions as well; in active interpretation, *tensor product* of representations is most common.

between the abstract group and the assigned transformation group or matrix group. In this situation one speaks of *faithful representations*. The tensor representations of even degree are, however, not faithful for the groups $\mathcal{L}$ and $\mathcal{L}_+$ (see exercise 3).

The concept of representation is, accordingly, taken somewhat wider in the sense that the assignment of transformations to the abstract group elements is required only to be a *homomorphism* of the abstract group $\mathcal{G}$ into the group of linear (nonsingular) transformations of a linear space $\mathbf{V}$:

$$g \mapsto T_g \quad \text{with} \quad g_1\,g_2 \mapsto T_{g_1g_2} = T_{g_1}T_{g_2}. \tag{6.4.15}$$

Here T is the *representation homomorphism* and T_g is the transformation assigned to g by T. The space $\mathbf{V}$ is called the representation space, its dimension is called the dimension of the representation. As with every homomorphism, eq. (6.4.15) implies

$$e \mapsto \mathrm{id}_{\mathbf{V}}, \tag{6.4.15'}$$

the identity on $\mathbf{V}$, and

$$g^{-1} \mapsto T_g^{-1}. \tag{6.4.15''}$$

Some authors do not include nonsingularity into the concept of a linear transformation of a space, just taking it to be synonymous with a linear map of the space into itself (also called *endomorphism* or *linear operator*); the linear transformations then do not form a group (but only a monoid—cf. sect. 6.2), and the definition of representation as an assignment (6.4.15) has to be supplemented by eq. (6.4.15') as a postulate rather than consequence. Together with eq. (6.4.15) it implies nonsingularity.

The representation then is the pair $(\mathbf{V}, T)$, but colloquially one just speaks of the representation T. General T are allowed to assign the same transformation to different group elements. The subset of group elements to which the identity transformation is assigned, the *kernel* of the representation, forms an invariant subgroup, and the representation yields a faithful representation of the corresponding factor group. For faithful representations the kernel consists of the unit element alone.

As examples, let us first consider 1-dimensional representations of $\mathcal{L}$. Scalars form a 1-dimensional space and are unchanged by Lorentz transformations—so we get a 1-dimensional representation which assigns to each group element the identical transformation (the 1×1 unit matrix), which here consists in the multiplication by 1:

$$g \mapsto 1, \qquad g_1\,g_2 \mapsto 1 = 1 \cdot 1. \tag{6.4.16}$$

This representation is called the *trivial representation*; it is possible for any group $\mathcal{G}$, the kernel being the whole group. The multiples of an invariant tensor like $\delta^i{}_k$ form likewise a 1-dimensional space on which $\mathcal{L}$ acts as the identity. We say that an invariant tensor transforms according to the trivial representation, or, in quantum mechanical parlance, that it is a singlet.

A nontrivial 1-dimensional representation of $\mathcal{L}$ is obtained in the space of pseudoscalars (determinant tensors): the transformation laws

$$\begin{aligned}
\epsilon_{\bar{\imath}\bar{\jmath}\bar{m}\bar{n}} &= (\det L)^{-1}\,\epsilon_{ijmn} \\
\epsilon^{\bar{\imath}\bar{\jmath}\bar{m}\bar{n}} &= \det L\,\epsilon^{ijmn}
\end{aligned} \tag{6.4.17}$$

show that the assignment $g \mapsto \det L(g)$ is a 1-dimensional representation; the representation property is here nothing but the multiplication law of determinants. This representation becomes trivial if it is restricted to the subgroup $\mathcal{L}_+$ which is thus the kernel of this representation. The factor group $\mathcal{L}/\mathcal{L}_+$, to which the subgroups $\{E, P\}$ or $\{E, T\}$ are isomorphic, is represented faithfully.

Another nontrivial 1-dimensional representation of $\mathcal{L}$ is given by the assignment $g \mapsto \operatorname{sign} L^0{}_0(g)$; it becomes trivial upon restriction to $\mathcal{L}^\uparrow$. The factor group $\mathcal{L}/\mathcal{L}^\uparrow$, to which the subgoups $\{E, T\}$ or $\{E, PT\}$ are isomorphic, is represented faithfully.

A third nontrivial 1-dimensional representation is obtained by taking the Kronecker product of the two foregoing ones, i.e., $g \mapsto \operatorname{sign} L^0{}_0 \det L(g)$; it is trivial upon restriction to $\mathcal{L}_0$.

As shown in the last section, the restricted Lorentz group $\mathcal{L}^\uparrow_+$ is simple; as a consequence, except for the trivial representation there are only faithful representations. This and the noncommutativity of $\mathcal{L}^\uparrow_+$ imply that there are no nontrivial 1-dimensional representations for $\mathcal{L}^\uparrow_+$. In the exercises below, we shall argue that all 1-dimensional representations of $\mathcal{L}$ are exhausted by the ones encountered so far.

The examples given should suffice to show that the concept of representation offers the appropriate mathematical framework for 'quantities that transform in a linear homogeneous manner': such quantities are elements of some representation space of the group under consideration. From a systematical point of view one then will be interested in finding *all* representations of a group. For the purposes of quantum mechanics it will be necessary to envisage infinite-dimensional complex representation spaces (Hilbert spaces) as well, and, as will be indicated in sect. 9.2, even two additional generalizations of the concept of representation will be required.

So far we only considered the algebraic aspect of representations. However, as we are dealing with Lie groups, thus having a nondenumerable infinity of elements, considerations from analysis must come in, a natural requirement being that the dependence of the transformation T_g on g be continuous, from which it can be shown that there is no loss of generality if one restricts to an analytic dependence in some analytic parametrization. We will come back to this aspect only later; in the following two sections, we will consider only the algebraic aspects of representations, except at one place.

Exercises

1. From the simplicity of $\mathcal{L}^\uparrow_+$ conclude that its 1-dimensional representations are all trivial, by using the following general argument. In an arbitrary group $\mathcal{G}$ the set of all products of elements of the form $aba^{-1}b^{-1}$ generates (Appendix A) a subgroup which in 1-dimensional representations obviously gets represented trivially. Now show that this *commutator subgroup* is invariant!

   Hint: It suffices to show that $gaba^{-1}b^{-1}g^{-1}$ can be written in the above form; for this, insert $g^{-1}g$ in three suitable places.

   Remarks: Under the assumption of continuity we shall arrive at the same result for $\mathcal{L}^\uparrow_+$ as a byproduct of later developments. On the other hand, there are non-simple groups which nevertheless coincide with their commutator subgroup.

2. Show that the 1-dimensional representations of $\mathcal{L}$ given above exhaust all such representations.

   Hint: Granting the result of the previous exercise, it suffices to find the 1-dimensional representations of the subgroup $\mathcal{V}_4 = \{E, P, T, PT\}$ (the *four-group*); now use the relations $P^2 = T^2 = E$.

3. As claimed in the text, the (positive degree) tensor representations are faithful for $\mathcal{L}_+^\uparrow$, $\mathcal{L}^\uparrow$, $\mathcal{L}_0$. Prove this and investigate the case of $\mathcal{L}$, $\mathcal{L}_+$ in that respect: prove here that in even degree there is a nontrivial kernel, given by $\{E, PT = -E\}$ — which is at the same time the *center* $\mathcal{Z}(\mathcal{G})$ for both of these groups, i.e., the set of those group elements that commute with *all* elements of the group $\mathcal{G}$.

   Hint: The condition that eq. (6.4.10) should yield the identity transformation in the tensor space is

$$L^i{}_m L_j{}^n \ldots = \delta^i_m \, \delta^n_j \ldots , \tag{6.4.18}$$

implying $L = \lambda E$ for some λ. Now insert this back into eq. (6.4.18) as well as into eq. (6.1.2).

6.5   Direct Sums and Irreducible Representations

From given representations of a group one can form others by a handful of general procedures, of which we already know the formation of equivalent ones, of contragredient ones, and of Kronecker products. Before we come to discuss the kind of representations which are characteristic of a specific given group and are not obtainable by completely general procedures, we must still introduce one more such general procedure: the formation of *direct sums*.

Let us illustrate the basic idea by a simple example. From a pair of 4-vectors u, v we take components and arrange them into an 8-component column vector $(u^i, v^i)^\top$. It is then obvious that these objects transform as

$$\begin{pmatrix} u^{\bar\imath} \\ v^{\bar\imath} \end{pmatrix} = \begin{pmatrix} L^i{}_k & 0 \\ 0 & L^i{}_k \end{pmatrix} \begin{pmatrix} u^k \\ v^k \end{pmatrix}. \tag{6.5.1}$$

The resulting 8-dimensional representation

$$g \mapsto \begin{pmatrix} L(g) & 0 \\ 0 & L(g) \end{pmatrix} \tag{6.5.2}$$

is called the direct sum of the two 4-vector representations. When we pass from the basis used so far in the column vector space to another one, then the representaion in general will lose its block form (6.5.2) under the resulting equivalence transformation (eq. (6.4.7)). It is then not immediate on first sight that the representation decomposes as the direct sum of two others. If, e.g., the basis is chosen such that the new components are $(u^0 + v^1, u^1, u^2, u^3, v^0, v^1, v^2, v^3)^\top$, then the representing matrices do not any more appear in the block form (6.5.2).

Let us now define the *Kronecker* or *tensor product* and the *direct sum* of two arbitrary representations $(\mathbf{V}', T')$ and $(\mathbf{V}'', T'')$. For the Kronecker product we start from the observation that tensor components T^{ik} transform like products of vector components $u^i v^k$—tensor products of vectors are special tensors. We therefore choose in $\mathbf{V}'$ a basis $\{e'_i\}$, in $\mathbf{V}''$ a basis $\{e''_\alpha\}$, and form components v'^i, v''^α for the vectors $v' \in \mathbf{V}'$, $v'' \in \mathbf{V}''$; these components transform as

$$v'^{\bar{i}} = T'^{\ i}_{g\ k}\, v'^k, \qquad\qquad v''^{\bar{\alpha}} = T''^{\ \alpha}_{g\ \beta}\, v''^\beta. \qquad (6.5.3)$$

Their products therefore transform as

$$v'^{\bar{i}} v''^{\bar{\alpha}} = T'^{\ i}_{g\ k}\, T''^{\ \alpha}_{g\ \beta}\, v'^k v''^\beta =: (T'_g \otimes T''_g)^{i\alpha}_{\ \ k\beta}\, v'^k v''^\beta. \qquad (6.5.4)$$

(The usual matrix form of this transformation is again obtained by replacing the double indices $i\alpha$, $k\beta$ by single ones which take $\dim(\mathbf{V}') \cdot \dim(\mathbf{V}'')$ values.) One can see immediately that, in analogy to eq. (6.4.13), the following multiplication rule holds:

$$\left(T'_{g_1} \otimes T''_{g_1}\right) \left(T'_{g_2} \otimes T''_{g_2}\right) = T'_{g_1} T'_{g_2} \otimes T''_{g_1} T''_{g_2}, \qquad (6.5.5)$$

allowing to verify the representation property of the assignment $g \mapsto T'_g \otimes T''_g$.

To form the direct sum, arrange the components v'^i, v''^α as columns $(v'^i, v''^\alpha)^T$; they transform as

$$\begin{pmatrix} v'^{\bar{i}} \\ v''^{\bar{\alpha}} \end{pmatrix} = \begin{pmatrix} T'^{\ i}_{g\ k} & 0 \\ 0 & T''^{\ \alpha}_{g\ \beta} \end{pmatrix} \begin{pmatrix} v'^k \\ v''^\beta \end{pmatrix}. \qquad (6.5.6)$$

For the block matrices here one also writes the symbol $T'_g \oplus T''_g$; we obviously have the multiplication rule

$$\left(T'_{g_1} \oplus T''_{g_1}\right) \left(T'_{g_2} \oplus T''_{g_2}\right) = T'_{g_1} T'_{g_2} \oplus T''_{g_1} T''_{g_2}, \qquad (6.5.7)$$

which immediately implies the representation property of the assignment $g \mapsto T'_g \oplus T''_g$. It is also easy to verify the distributive law

$$T \otimes (T' \oplus T'') = (T \otimes T') \oplus (T \otimes T'') \qquad (6.5.8)$$

and to extend both operations, Kronecker multiplication and direct sums, to the case of more representations to be multiplied or added, the usual associative laws being valid.

In this way one obtains the *representation ring*—actually a semiring to begin with, because there is no inverse to (direct) addition; but the semiring may be enlarged abstractly to a ring by adding so-called 'virtual representations', very analogously to how one constructs the integer numbers from the natural numbers. However, we will not make use of this.

We now introduce the central notion of representation theory: the notion of *irreducible representation*. To decide whether a given representation $(\mathbf{V}, T)$ can be looked at as a direct sum $T' \oplus T''$, we observe that in eq. (6.5.1) vectors of the form

$(v'^k, 0)^\top$ are transformed to vectors of the same form. Such vectors form a subspace $\mathbf{V}'$ which is invariant under (all transformations of) the representation $T_g' \oplus T_g''$. If by an equivalence transformation the block form (6.5.2) is lost, there nevertheless exists an invariant subspace, although not given any more by vectors of the form $(v'^k, 0)^\top$. Thus necessary for the equivalence of a representation with a direct sum of representations is the existence of an *invariant subspace*. A representation is called *reducible* if there is a nontrivial (i.e., different from the whole space and different from the zero vector) invariant subspace. If such a subspace does not exist, the representation is called *irreducible*. One of the fundamental tasks of the representation theory of a group is to find all its (equivalence classes of) irreducible representations.

In eq. (6.5.1) also the vectors $(0, v''^\beta)^\top$ form an invariant subspace, and every vector is in a unique manner a sum of two vectors $(v'^k, 0)^\top$ and $(0, v''^\beta)^\top$ belonging to the subspaces. Reducible representations of this kind are called *decomposable*. Not every reducible representation of a group will be decomposable as the direct sum of two subrepresentations, as this requires two complementary invariant subspaces.

Here appears the second important task of the representation theory of a group: develop methods to decide whether a given representation is reducible, and if so, to possibly effect a decomposition into some direct summands. A representation is called *completely reducible* or *fully reducible* if in (at least) one process of continued decomposition one ends up with a direct sum of irreducible representations—or, equivalently, in finite dimensions, if every invariant subspace has an invariant complementary subspace.[1] After a suitable equivalence transformation all matrices T_g of the repesentation then simultaneously take block form:

$$T_g = \begin{pmatrix} T_g' & A_g \\ 0 & T_g'' \end{pmatrix} \qquad\qquad T_g = \begin{pmatrix} T_g' & 0 \\ 0 & T_g'' \end{pmatrix} = T_g' \oplus T_g''$$

$$(6.5.9)$$

<table>
<tr><td>Reducible:
subspace of vectors $(v', 0)^\top$
invariant</td><td>Fully reducible:
subspace of vectors $(v', 0)^\top$
subspace of vectors $(0, v'')^\top$ both invariant</td></tr>
</table>

In chap. 9 we shall encounter an example where—for the Poincaré group—reducibility does *not* imply full reducibility. For the finite-dimensional representations of the homogeneous Lorentz group, however, we will have a theorem of complete reducibility for all of them.—We cannot go into the refinements from functional analysis for the concepts just introduced in the case of infinite-dimensional representations, although we will be interested in such representations, as mentioned before—see, e.g., Naimark (1960) for them.

A frequently occurring problem of the type just described is the decomposition of Kronecker products of irreducible representations. The direct sum that arises is called a *Clebsch-Gordan decomposition*.

As a first application of the general concepts, let us consider the representation of the restricted Lorentz group $\mathcal{L}_+^\uparrow$ which follows from the transformation law of the electromagnetic field tensor F. Arranging the components of this antisymmetric tensor as a 6-vector (*sixtor*) $(\mathbf{E}, \mathbf{B})^\top$, we can easily write down the action of the Lorentz

---

[1] See Chevalley (1956), p. 61 for a proof of this.

group (we ignore the space-time point dependence of the field here, considering the frame dependence of the components only). Under spatial rotations, $\mathbf{E}$ and $\mathbf{B}$ transform separately in the well-known manner, giving a representation of the rotation subgroup $L_{\mathrm{R}} \mapsto \mathrm{R} \oplus \mathrm{R}$ in the space of sixtors as a direct sum. Now, while rotations leave separately invariant the subspaces formed by the sixtors $(\mathbf{E}, \mathbf{0})^{\top}$ and by the sixtors $(\mathbf{0}, \mathbf{B})^{\top}$, these two subspaces get, according to eq. (5.8.5), mixed up under boosts, so that the representation seems in fact irreducible under the whole group.

It is interesting and significant, however, that this irreducibility does not hold if we allow for *complex numbers* as coefficients—so far the reals were tacitly assumed to be the ground field for the representation spaces $\mathbf{V}$! Thus, allowing for *complex* vector components, it is quite easy to see that the combinations

$$\mathbf{F}_{\pm} = \mathbf{E} \pm i\mathbf{B}, \tag{6.5.10}$$

transform, according to eq. (5.8.5), as

$$\bar{\mathbf{F}}_{\pm} = \gamma\mathbf{F}_{\pm} - \frac{\gamma-1}{v^2}(\mathbf{F}_{\pm}\mathbf{v})\mathbf{v} \mp i\gamma\mathbf{v} \times \mathbf{F}_{\pm}, \tag{6.5.11$\pm$}$$

i.e., transform totally separately from each other—and this clearly also holds for rotations and therefore for the whole restricted Lorentz group. What we found is that upon use of complex coefficients, i.e., upon use of the *complexification* of our original representation space, the sixtor representation decomposes into two *complex conjugate* 3-dimensional representations. These are irreducible, already so for rotations alone.

It is also remarkable that the transformations (6.5.11) are *complex-orthogonal*— putting $\mathbf{v}/v = \mathbf{n}$ and $\gamma = \cos\alpha$, $i\gamma v = \sin\alpha$ (where α is imaginary), eq. (6.5.11+) goes over into eq. (1.3.1) with $\mathbf{F}_+$ instead of $\mathbf{x}$ and $\mathbf{n}$ instead of $\boldsymbol{\alpha}/\alpha$. Therefore also the representation of the general transformations of $\mathcal{L}_+^{\uparrow}$ are complex-orthogonal. It follows that the expressions

$$\mathbf{F}_{\pm}^2 = (\mathbf{E} \pm i\mathbf{B})^2 = \mathbf{E}^2 - \mathbf{B}^2 \pm i \cdot 2\mathbf{E}\mathbf{B} \tag{6.5.12$\pm$}$$

are invariant. Real and imaginary part are the invariants (5.8.1) of the field tensor encountered before.

The matrices of the two subrepresentations found here each belong to the complex-orthogonal group SO(3,$\mathbf{C}$). Since the latter is a 3 complex (= 6 real) parameter Lie group into which $\mathcal{L}_+^{\uparrow}$ is mapped homomorphically and real-analytically, and since $\mathcal{L}_+^{\uparrow}$ is simple, we obtain here an isomorphism between the two groups: $\mathcal{L}_+^{\uparrow} \cong \mathrm{SO}(3, \mathbf{C})$. (This isomorphism will be made even more explicit in sect. 8.2.) We can use the appearance of the complex Lie group SO(3,$\mathbf{C}$) to see that the two subrepresentations must be inequivalent. Namely, taking one of them as the defining representation of that group, it is complex-analytic (holomorphic) in its complex parameters, and so is any equivalent representation; however, the other subrepresentation is complex-conjugate and thus antiholomorphic in the complex parameters.

Generally we note that, having in hands a complex representation (i.e., a representation in a complex representation space or by complex matrices: $g \mapsto T_g$), we get another one, $g \mapsto T_g^*$, which may be equivalent to T_g or not (cf. exercise 5 of sect.

6.6); but it obviously shares with T_g the property of being reducible, decomposable or irreducible.

In terms of abstract linear algebra: to any representation in a complex vector space $\mathbf{V}$ there is the complex-conjugate representation in the complex-conjugate space $\mathbf{V}^*$ (see Appendix B.3).

Mathematically, one can consider representations in vector spaces over various number fields. When the number field is extended, irreducibility may change into reducibility, as we saw in the example above where the real sixtor representation of the restricted Lorentz group is irreducible but becomes reducible when complexified, whereas its defining representation by real 4-vectors remains irreducible upon complexification. (Cf. exercise 11 of sect. 6.6 for generalities on extending from the real to the complex number field.)

In physics, the concept of reducibility is usually referred to the field $\mathbf{C}$ of complex numbers as the ground field of representation spaces. There are two reasons for this, apparently independent of one another.

The first reason is just mathematical convenience. The theory of representations over $\mathbf{C}$ is—essentially because of $\mathbf{C}$ being algebraically closed—simpler than the one over the real number field $\mathbf{R}$. The theory of real representations is best gotten by sorting out their complexifications from complex representations (cf. exercises 5 and 11 of sect. 6.6).

The second reason is the mathematical structure of *Quantum Mechanics*, which—as is well-known—works with *complex* Hilbert spaces. Of course it is always possible by going to real and imaginary parts to arrive at a formulation that works with $\mathbf{R}$ alone, which, however, is 'crying for complex numbers' in much the same way as the manipulation of trigonometric functions is simplified by using exp with imaginary exponents: whatever the numbers used, there is a *complex structure* (cf. Appendix B.6) involved here.

It is perhaps a historical curiosity that Quantum Mechanics was discovered already in its complex version—it could have happened differently. (See J. H. D. Jensen, D. Hepp, Sitzungsber. Heidelb. Akad. Wiss., Math. Naturwiss. Kl. 1971/4, pp. 89–122, as well as R. G. Gehrenbeck, Phys. Today, *31*, No.1, 34 (1978) for the history of matter wave interference experiments.)

In order to illuminate the relevance of the combination $\mathbf{F}_+ = \mathbf{E} + i\mathbf{B}$ we point out that Maxwell's equations in vacuum, eqs. (5.2.1,2) with $\rho = 0$, $\mathbf{j} = 0$, may be combined as

$$\operatorname{div} \mathbf{F}_+ = 0, \qquad i\frac{\partial}{\partial t}\mathbf{F}_+ = \operatorname{rot} \mathbf{F}_+ \qquad\qquad (6.5.13+)$$

or as

$$\operatorname{div} \mathbf{F}_- = 0, \qquad -i\frac{\partial}{\partial t}\mathbf{F}_- = \operatorname{rot} \mathbf{F}_-, \qquad\qquad (6.5.13-)$$

which exactly corresponds to the Lorentz covariance and to the reducibility found above. It is essential here that one can get along with only one of the combinations, $\mathbf{E} + i\mathbf{B} =: \mathbf{F}_+$ (*or* $\mathbf{E} - i\mathbf{B} =: \mathbf{F}_-$) *alone*. If in the law of induction there were a plus sign instead of a minus (which would have terrible physical consequences!), both combinations would have to be used simultaneously, and there would be no simplification in using them (there would be no complex structure hidden in Maxwell's equations); at the same time, Lorentz covariance would be lost. To the direct discovery of Schrödinger's equation in complex form there would correspond a direct discovery of Maxwell's equations in vacuum in the form $\operatorname{rot} \mathbf{F}_+ = i\frac{\partial \mathbf{F}_+}{\partial t}$, $\operatorname{div} \mathbf{F}_+ = 0$, from which one could afterwards go via $\operatorname{Re} \mathbf{F}_+ = \mathbf{E}$, $\operatorname{Im} \mathbf{F}_+ = \mathbf{B}$ to their usual real form. Therefore the vacuum equations satisfy—just as Schrödinger's equation—a *complex superposition principle*: if $\mathbf{F}_+$, $\mathbf{F}'_+$ are solutions, so is $c\mathbf{F}_+ + c'\mathbf{F}'_+$, where c, $c' \in \mathbf{C}$. The only new thing we get here are the *duality rotations* $\mathbf{F}_+ \to e^{i\alpha}\mathbf{F}_+$ (α real). However, the nature of the *sources* of the Maxwell field—more precisely, the experimentally confirmed absence of magnetic charges and currents—destroys the invariance under duality rotations for the inhomogeneous equations.

Exercises

1. The differential ds of proper time is Lorentz invariant; more precisely, under $\mathcal{L}^\uparrow$ it transforms according to the trivial representation. Taking into account that $ds = dt$ in the rest system, according to which representation of $\mathcal{L}$ does it transform?

2. Under $\mathcal{L}^\uparrow$, 4-velocity and 4-current transform according to the vector representation (6.4.1). For $\mathcal{L}$ it must be observed, however, that time reversal has to be interpreted as reversal of motion. From this, or from the result of the preceding exercise, show that both transform under $\mathcal{L}$ according to the Kronecker product of the 4-vector representation and the 1-dimensional representation $g \mapsto \operatorname{sign} L^0{}_0(g)$.

3. The transformation laws of the 4-potential A and of the field tensor F were written down up to now as $A^i \mapsto L^i{}_k A^k$, $F^{ik} \mapsto L^i{}_m L^k{}_n F^{mn}$. This would automatically imply a certain behavior under space and time reversals when $g \mapsto L^i{}_m(g)$ is allowed to vary over all of $\mathcal{L}$. But if the field is coupled to its sources according to eq. (5.2.13), the behavior of the field must be adapted to the behavior of its sources under reversals (cf. exercise 2). Write down the correct behavior under all of $\mathcal{L}$. Discuss the result by way of the Ørsted experiment, effecting active reversals of space and/or motion. What is the representation for the dual field strength tensor?

4. Show that the $\mathcal{L}^\uparrow_+$-invariant subspaces $\{\mathbf{F}_\pm\}$ are also $\mathcal{L}_+$-invariant but get transformed into each other upon space or time reversal.

5. Show that the defining (real) representations of $\mathrm{SO}(3,\mathbf{R})$ and $\mathcal{L}^\uparrow_+$ are irreducible in the real sense and remain so even after complexification. Also show the real irreducibility of the real sixtor representation under $\mathcal{L}^\uparrow_+$.

6. Let $(\mathbf{V}, T)$ be a representation of the group $\mathcal{G}$. A vector $v \in \mathbf{V}$ is called a *cyclic vector* for the representation if the set of vectors $T_g\, v$, where g varies over all of $\mathcal{G}$, spans all of $\mathbf{V}$. Show that a representation is irreducible if and only if every nonzero vector in $\mathbf{V}$ is cyclic.

7. Let $\mathbf{V}' \subset \mathbf{V}$ be an invariant subspace of some representation T. Show that in addition to the subrepresentation T' defined in $\mathbf{V}'$ one can define a representation T'' in the quotient space $\mathbf{V}/\mathbf{V}'$ in a natural manner. How are both representations related to the matrices T'_g, T''_g entering eq. (6.5.9)? Also show by direct matrix multiplication that the T''_g furnish a representation.

8. Let $(\mathbf{V}, T)$ be a *real* irreducible representation. Consider its complexification, i.e., extend the operators T_g to complex-linear operators T^c_g on the complexified space $\mathbf{V}^c$. Show that the resulting complex representation is either irreducible or decomposes into two irreducible complex-conjugate subrepresentations.

Hint: Let $\mathbf{W} \subset \mathbf{V}^c$ be a subspace invariant under the T_g^c, and let $*$ be the complex conjugation *in* $\mathbf{V}^c$. Then $\mathbf{W}^*$ and therefore $\mathbf{W} \cap \mathbf{W}^*$ as well as the linear span $\prec \mathbf{W} \cup \mathbf{W}^* \succ$ are invariant. The latter two subspaces are, however, invariant under $*$ and therefore are complexifications of subspaces of $\mathbf{V}$. Conclude from this that $\mathbf{W} \cap \mathbf{W}^* = \{0\}$, $\mathbf{W} \oplus \mathbf{W}^* = \mathbf{V}^c$.

9. The last three exercises were of a more abstract nature. Discuss in which of them the ground field is arbitrary and in which of them the group structure of $\mathcal{G}$ is unimportant, so that $\mathcal{G}$ only plays the role of an indexing set.

6.6   Schur's Lemma

In this section we present Schur's lemma and some of its far-reaching consequences. It turns out here that for most of the general considerations the index notation, but even more so the explicit writing of representation matrices, becomes cumbersome, unnecessary, or even impossible (because of high or infinite dimensionality of representations). It is then preferable to use the formulations of *abstract linear algebra*[1], as we already have been doing occasionally; this will put the active interpretation of linear transformations into the foreground. For reasons explained in the last section, we will work in vector spaces over the complex numbers; but the reader should try to remain aware as to where this is essential and where it is not.

So we take up the abstract version of our definition of a representation $(\mathbf{V}, T)$ given at eq. (6.4.15); if required in concrete examples, one can, in the finite-dimensional case, always return from the abstract (transformations, operators) T_g to the matrices $\left(T_g{}^i{}_k\right)$ used earlier, by choosing a basis $\{e_i\}$ in $\mathbf{V}$ (cf. sect. 3.3). As before, a linear subspace $\mathbf{V}_1 \subset \mathbf{V}$ is called invariant under the representation if for all $g \in \mathcal{G}$ we have $T_g \mathbf{V}_1 \subset \mathbf{V}_1$, i.e., $T_g v \in \mathbf{V}_1$ whenever $v \in \mathbf{V}_1$. We always have the *trivial* invariant subspaces $\mathbf{V}$ and $\{0\}$. A representation is called reducible if there exists a nontrivial invariant subspace; decomposable if there are two nontrivial complementary invariant subspaces; irreducible if the only invariant subspaces are the trivial ones; completely (or fully) reducible if every invariant subspace has an invariant complement (or, equivalently, if decomposable as a direct sum of irreducible representations). We now give the abstract, or active, version of equivalence: two representations $(\mathbf{V}, T)$, $(\mathbf{V}', T')$ are called *equivalent*, $T \cong T'$, if there exists a bijective linear map $S \colon \mathbf{V} \to \mathbf{V}'$ such that

$$T_g = S^{-1} T_g' S \tag{6.6.1}$$

for all $g \in \mathcal{G}$—cf. the passive version (6.4.8). Writing this condition as

$$S T_g = T_g' S, \tag{6.6.2}$$

we can illustrate it by means of the *commuting diagram* shown below:

---

[1] See any modern text on linear algebra, such as Halmos (1974), Chevalley (1956), Greub (1975, 1978), Lang (1966), and also Appendix B.

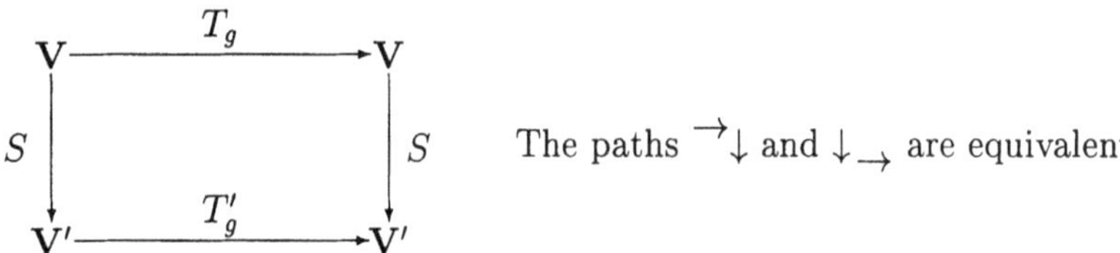

The paths $\overset{\rightarrow}{\downarrow}$ and $\downarrow_{\rightarrow}$ are equivalent

Now let $S\colon \mathbf{V} \to \mathbf{V}'$ be a linear map which is not necessarily bijective but satisfies eq. (6.6.2); one says it *intertwines* the two representations and calls S an *intertwiner*. Then the image $S\,\mathbf{V} \subset \mathbf{V}'$ is a linear subspace of $\mathbf{V}'$, which because of $T'_g S\,\mathbf{V} = S\,T_g\,\mathbf{V} = S\,\mathbf{V}$ is invariant under T'. Also, the kernel $\ker S \subset \mathbf{V}$, i.e., the set of vectors mapped onto the zero vector in $\mathbf{V}'$ by S, is an invariant subspace under T, since from $\{o'\} = T'_g S \ker S = S\,T_g \ker S$ it follows that $T_g \ker S \subset \ker S$.

From this we obtain the following *theorem (Schur's Lemma, part I)*:

Let $(\mathbf{V},T)$ and $(\mathbf{V}',T')$ be two irreducible representations and let S be an intertwiner between them, i.e., a linear map from $\mathbf{V}$ to $\mathbf{V}'$ satisfying eq. (6.6.2), then S either vanishes, or it is a bijection, in which case the representations are equivalent.

This is because, by the irreducibility of both representations, $S\mathbf{V}$ has to coincide with $\{0'\}$ or $\mathbf{V}'$ and $\ker S$ with $\{0\}$ or $\mathbf{V}$. Two possibilities remain: either $\ker S = \mathbf{V}$ and $S\mathbf{V} = \{0'\}$, implying $S \equiv 0$, or $\ker S = \{0\}$ and $S\mathbf{V} = \mathbf{V}'$, implying that S is bijective.—Note that the nature of the ground field plays no role here and that the map S could also be taken semilinear (cf. Appendix B.1); also note that the group $\mathcal{G}$ plays only the role of an indexing set for the operators T_g, its group structure is not needed.

Even more important will be the following *theorem (Schur's Lemma, part II)*:

Let $(\mathbf{V},T)$ be a representation and S a linear operator in $\mathbf{V}$ that commutes with all T_g. If S possesses an eigenvalue s, then S must be the multiple $s\cdot\mathrm{id}_\mathbf{V}$ of the identity, or else the representation is reducible.

It is understood here that the eigenvalue belongs to the ground field used. Clearly, in a finite-dimensional complex representation the existence of an eigenvalue is guaranteed by the fundamental theorem of algebra: s is a solution of the characteristic equation $\det(S^i{}_k - s\,\delta^i{}_k) = 0$. For us, this will be the most important case, but we already see here the parallel between the appearance of eigenvalues after complexification and the possible reducibility of complexified real-irreducible representations encountered in sect. 6.5; this will become more explicit below.

For real representations, see exercise 11 or Kirillov (1976), p. 119. The set of self-intertwiners of a representation is called its *commutant*; it forms an algebra in the sense that linear combinations and products of self-intertwiners are again self-intertwiners. Schur I says, in other words, that for an irreducible representation all nonzero elements of this algebra have an inverse, while Schur II implies that for finite-dimensional complex irreducible representations the commutant is isomorphic to the field of complex numbers by assigning $s \mapsto s \cdot \mathrm{id}_\mathbf{V}$ for all $s \in \mathbf{C}$.

Although we shall not use it, we give here a dictionary between the module terminology used in books on abstract algebra and the terminology used here and in many texts on representation theory:

representation space	$\mathcal{G}$-module
invariant subspace	submodule
irreducible representation	simple $\mathcal{G}$-module
completely reducible representation. . .	semisimple $\mathcal{G}$-module
intertwiner	$\mathcal{G}$-module homomorphism
self-intertwiner	$\mathcal{G}$-module endomorphism

etc.

For the proof of Schur II, consider the eigenspace $\mathbf{V}_s \subset \mathbf{V}$ consisting of (0 and) all eigenvectors v of S for the eigenvalue s, i.e., vectors v satisfying $Sv = sv$ but $v \neq 0$. Then from $ST_g = T_g S$ we have $ST_g v = T_g S v = s T_g v \in \mathbf{V}_s$; thus $\mathbf{V}_s$ is invariant under T. Assuming the representation to be irreducible we conclude that the eigenspace must coincide with $\mathbf{V}$; but $Sv = sv$ for all $v \in \mathbf{V}$ just means that $S = s \cdot \mathrm{id}_{\mathbf{V}}$.

Whether in case $S \not\propto \mathrm{id}_{\mathbf{V}}$ for some intertwiner S we have reducibility or decomposability will depend on the detailed structure of the elementary divisors of S; in particular on the question whether its eigenspaces together span $\mathbf{V}$ or not; the lemma just guarantees reducibility.

The *(outer) direct sum* $\mathbf{V} = \mathbf{V}' \oplus \mathbf{V}''$ of two vector spaces is the set of pairs $(v', v'') =: v' \oplus v''$, equipped with the vector space structure

$$\alpha(v', v'') + \beta(w', w'') = (\alpha v' + \beta w', \alpha v'' + \beta w''). \qquad (6.6.3)$$

The vectors $v' \oplus o''$ and $o' \oplus v''$ form two subspaces $\mathbf{V}_1$ and $\mathbf{V}_2$ of $\mathbf{V}$. Every vector from $\mathbf{V}' \oplus \mathbf{V}''$ is uniquely the sum of a vector v_1 from $\mathbf{V}_1$ and a vector v_2 from $\mathbf{V}_2$:

$$v' \oplus v'' = v_1 + v_2 = (v' \oplus o'') + (o' \oplus v''). \qquad (6.6.4)$$

The *projection operators* $P_1 \colon \mathbf{V} \to \mathbf{V}_1$, $P_2 \colon \mathbf{V} \to \mathbf{V}_2$ are defined by

$$P_1 v = v_1 \in \mathbf{V}_1, \qquad\qquad P_2 v = v_2 \in \mathbf{V}_2. \qquad (6.6.5)$$

They are linear operators which satisfy

$$
\begin{aligned}
P_1^2 &= P_1, & P_2^2 &= P_2, \\
P_1 P_2 &= 0, & P_2 P_1 &= 0, \\
P_1 + P_2 &= \mathrm{id}_{\mathbf{V}}. & &\text{\textit{(complementary projections)}}
\end{aligned}
\qquad (6.6.6)
$$

These relations hold in every vector space $\mathbf{V}$ in which are given two subspaces $\mathbf{V}_1$ and $\mathbf{V}_2$ such that every vector is in a *unique* manner a sum $v = v_1 + v_2$, where $v_1 \in \mathbf{V}_1$, $v_2 \in \mathbf{V}_2$. $\mathbf{V}$ is then isomorphic to the direct sum $\mathbf{V}_1 \oplus \mathbf{V}_2$ and is called their *(inner) direct sum*. We shall use the same notation $\oplus$ in both cases, although there is a logical distinction. There is an obvious construction and terminology if there are more summands. (For an infinite number of them, a conceptual refinement is needed, but we do not go into that, although we will formally deal even with 'direct integrals'.)

We stress here that the introduction of the projection operators P_i above was possible only because we started from a decomposition of $\mathbf{V}$ as a direct sum. If only a *single* subspace were distinguished, this would *not* suffice to define a projection onto it (except in the case where a scalar product and the corresponding concept of orthogonality is available—cf. sect. **7.5**).

On the other hand, given a single *idempotent operator* $P_1 \colon \mathbf{V} \to \mathbf{V}$, $P_1^2 = P_1$, we can define a subspace $P_1 \mathbf{V} = \mathbf{V}_1$, and similarly, $P_2 := \mathrm{id}_{\mathbf{V}} - P_1$ defines a subspace $\mathbf{V}_2$; P_2 is idempotent as well:

$$P_2^2 = (\mathrm{id}_{\mathbf{V}} - P_1)\,(\mathrm{id}_{\mathbf{V}} - P_1) = \mathrm{id}_{\mathbf{V}} - P_1 - P_1 + P_1^2 = P_2$$

and satisfies

$$P_1 P_2 = P_1 - P_1^2 = P_1 - P_1 = 0, \qquad \text{similarly} \qquad P_2 P_1 = 0,$$
$$P_1 + P_2 = \mathrm{id}_{\mathbf{V}}.$$

It is easy to see that $\mathbf{V} = \mathbf{V}_1 \oplus \mathbf{V}_2$. Thus the existence of a nontrivial (i.e., different from the zero and unit operator) idempotent operator defines a decomposition of the space into a direct sum.

Given two representations $(\mathbf{V}, T)$ and $(\mathbf{V}', T')$ of a group $\mathcal{G}$, their *direct sum* $(\mathbf{V}' \oplus \mathbf{V}'', T' \oplus T'')$ is defined by assigning $g \mapsto T_g' \oplus T_g''$, where the latter operators on $\mathbf{V}' \oplus \mathbf{V}''$ are defined by

$$\left(T_g' \oplus T_g''\right)(v' \oplus v'') := T_g' v' \oplus T_g'' v''. \tag{6.6.7}$$

The product rule (6.5.7), and therefore the representation property of the assignment, is easily verified.

Conversely, if a representation $(\mathbf{V}, T)$ admits two complementary invariant subspaces $\mathbf{V}'$ and $\mathbf{V}''$, then the representation T decomposes as a direct sum $T' \oplus T''$ of the *subrepresentations induced in the invariant subspaces*. (Formally, the T_g', T_g'' are defined by $T_g' v = T_g v$ resp. $T_g'' v = T_g v$ for $v \in \mathbf{V}'$ resp. $\in \mathbf{V}''$, i.e., by restricting T to the subspaces.) In this case T commutes with the projection operators P', P'' on $\mathbf{V}'$, $\mathbf{V}''$:

$$T_g\, P' = P'\, T_g, \qquad\qquad T_g\, P'' = P''\, T_g. \tag{6.6.8}$$

The reducibility of the representation T in the presence of the operator P' commuting with T illustrates Schur's lemma; but here we have decomposability, corresponding to the fact that the eigenspaces of P' do span V (as a direct sum): $\mathbf{V}'$ belongs to the eigenvalue 1, $\mathbf{V}''$ belongs to the eigenvalue 0. (From idempotency, there are no other eigenvalues, and $P^2 - P = 0$ is the 'minimal equation' satisfied by P, thus having simple roots: this is known to be a general criterion for the eigenspaces to span the whole space.)

In general, there is no converse to Schur II: if the commutant of a complex representation is trivial $(= \mathbf{C} \cdot \mathrm{id})$, it only follows that the representation does not decompose; however it may be reducible—as stressed above, a single invariant subspace does not define a projection. An example for this is given in exercise 11. However, there are many groups for which all finite-dimensional representations are completely reducible, but this goes much beyond the role of the group as an indexing set for the representing operators. Also, we have complete reducibility in all *unitary* representations (cf. sect. 7.5).

We now illustrate these general considerations by looking at the simplest tensor representations of the Lorentz group. As was to be shown in exercise 5 of sect. 6.5 the (defining) 4-vector representation is irreducible, even after complexification; so we turn to the space of 4-tensors of degree 2. Here the subspaces of symmetric and of antisymmetric tensors are Lorentz invariant. Hence the tensor product representation

$g \mapsto L(g) \otimes L(g)$ (see eq. (6.4.12) but interpret actively) is reducible. Projection operators P_S, P_A onto these subspaces are obtained from the decomposition

$$D^{ik} = \frac{1}{2}\left(D^{ik} + D^{ki}\right) + \frac{1}{2}\left(D^{ik} - D^{ki}\right) =: T^{ik} + F^{ik} = \left(P_S{}^{ik}{}_{mn} + P_A{}^{ik}{}_{mn}\right) D^{mn} \quad (6.6.9)$$

as the invariant tensors

$$P_S{}^{ik}{}_{mn} := \frac{1}{2}\left(\delta^i{}_m \delta^k{}_n + \delta^i{}_n \delta^k{}_m\right), \quad P_A{}^{ik}{}_{mn} := \frac{1}{2}\left(\delta^i{}_m \delta^k{}_n - \delta^i{}_n \delta^k{}_m\right), \quad (6.6.10, 11)$$

which as linear operators in the tensor space commute with $L(g) \otimes L(g)$. In the symmetric (antisymmetric) subspace, P_S (P_A) acts as the identity, while it annihilates antisymmetric (symmetric) tensors; eq. (6.6.9) says that $P_S + P_A = \mathrm{id}$.

In the tensor spaces of higher degree, the operators (5.5.3,5) of *total* (anti)symmetrization are idempotent again, cf. exercise 1 of sect. 5.5; but their sum is not the identity in these tensor spaces, and besides total symmetry and total antisymmetry there are other symmetry types—see, e.g., Boerner (1955) or Fulton and Harris (1991).

Consider now the subspace of *symmetric tensors*. Here the scalar multiples of the (invariant!) metric tensor form a 1-dimensional subspace onto which the Lorentz invariant operator P

$$P^{ik}{}_{mn} = \frac{1}{4}\eta^{ik}\eta_{mn}, \qquad\qquad P^2 = P, \qquad\qquad (6.6.12)$$

projects: we have

$$P^{ik}{}_{mn} T^{mn} = \frac{1}{4}\eta^{ik} \tau, \qquad\qquad \tau := \eta_{mn} T^{mn}. \qquad\qquad (6.6.13)$$

The complementary projector $\mathrm{id} - P$ projects the tensors onto their trace-free parts:

$$\left(\delta^i{}_m \delta^k{}_n - \frac{1}{4}\eta^{ik}\eta_{mn}\right) T^{mn} = T^{ik} - \frac{1}{4}\eta^{ik} \tau;$$

$$\eta_{ik}\left(T^{ik} - \frac{1}{4}\eta^{ik}\tau\right) = 0. \qquad\qquad (6.6.14)$$

In these subspaces—consisting of multiples of the metric tensor or of trace-free tensors—we have as subrepresentations the trivial one or a 9-dimensional one, respectively, whose irreducibility will emerge later.

Frequently the situation presents itself in a slightly different way, which we still illustrate by the example just considered. From a given representation $(\mathbf{V}, T)$ (like the one by symmetric tensors) we derive by some invariant linear operation (like taking the trace) an intertwiner π to another representation $(\mathbf{V}_1, T_1)$ (like the trivial one). Let $(\mathbf{V}'', T'')$ be the subrepresentation induced on $\mathbf{V}'' := \ker\pi$ and assume π to be a *surjection*, so that $\mathbf{V}_1 \cong \mathbf{V}/\mathbf{V}''$—this will in particular be the case if T_1 is irreducible. One then tries to find an intertwining *injection* $\iota\colon \mathbf{V}_1 \to \mathbf{V}$ which is a right inverse to π ($\iota\colon \alpha \mapsto \frac{1}{4}\alpha \cdot \eta^{ik}$ in the example): $\pi \circ \iota = \mathrm{id}_1$. This must be possible in case of full reducibility. (Without this, the original representation is not determined by T_1, T'' alone, see Kirillov (1976) for the structures arising in this situation.) If ι has been found (which may be tricky), then $P' := \iota \circ \pi$ is idempotent: $P'^2 = \iota \circ \pi \circ \iota \circ \pi = \iota \circ \pi = P'$, and projects to an invariant complement $\mathbf{V}' \cong \mathbf{V}_1$ for $\mathbf{V}''$.

In the subspace of *antisymmetric tensors* (sixtors) a further decomposition into subspaces invariant under the full Lorentz group $\mathcal{L}$ is not possible (see exercise 4 of sect. 6.5). The Kronecker square $L(g) \otimes L(g)$ of the 4-vector representation of the full Lorentz group $\mathcal{L}$ therefore decomposes into three irreducible constituents,

$$[4] \otimes [4] = [1] \oplus [9] \oplus [6], \tag{6.6.15}$$

where the tensor representations were symbolized by putting their dimensions in square brackets.

If we restrict to the group $\mathcal{L}_+$, however, the operator S given by the *-operation treated in sect. 5.5,

$$S^{ik}{}_{mn} F^{mn} := *F^{ik} = \frac{1}{2}\epsilon^{ik}{}_{mn} F^{mn}, \tag{6.6.16}$$

commutes with all transformations $L(g) \otimes L(g)$, $g \in \mathcal{L}_+$, since ϵ_{iklm} is an invariant tensor under $\mathcal{L}_+$. By Schur's lemma, this representation is reducible in the space of complexified sixtors. It emerges from the proof of the lemma that invariant subspaces are obtained as eigenspaces of S; i.e., we have to look for eigentensors of the *-operation. From eq. (5.5.6) we have $S^2 = -$id and thus for the eigenvalues $s^2 = -1$, $s = \pm i$ and for the tensors in question

$$*F = iF, \qquad \text{or} \qquad *F = -iF. \tag{6.6.17}$$

These *complex-selfdual* or *-antiselfdual* sixtors form two subspaces invariant under $\mathcal{L}_+$ (but getting interchanged under $\mathcal{L}_-$ since ϵ_{iklm} here changes sign). A physical interpretation of such sixtors was given in exercise 7 to sect. 5.5. From $S^2 = -$id it follows that the operators $1/2(\text{id} \mp iS)$ are idempotent, thus giving the projectors onto the invariant subspaces:

$$F = \frac{1}{2}(F - i *F) + \frac{1}{2}(F + i *F). \tag{6.6.18}$$
$$\text{selfdual} \qquad\qquad \text{antiselfdual}$$

(This nomenclature involves a convention.) The projections just correspond to the complex combinations $-(\mathbf{E}+i\mathbf{B})/2$ and $-(\mathbf{E}-i\mathbf{B})/2$; the considerations made earlier show that the resulting subrepresentations are irreducible and inequivalent.

Let us put together our results obtained about tensors of degree 2 under $\mathcal{L}_+$: The Kronecker product $L(g) \otimes L(g)$ of two (irreducible!) 4-vector representations $L(g)$ of $\mathcal{L}_+$ is reducible, the *Clebsch-Gordan decomposition* into irreducible parts being given by

$$[4] \otimes [4] = [1] \oplus [9] \oplus [3] \oplus [3^*]. \tag{6.6.19}$$

([3*] is to indicate the representation complex-conjugate to [3].)

This ends our introductory considerations on representations of the Lorentz group. A systematic derivation of all finite-dimensional irreducible complex representations will be given in chap. 8. It will make use of 6 simpler theory of representations of the rotation group, to be developed in chap. 7.

Let us add here some notation on tensor products. The direct sum of vector spaces, $\mathbf{V} \oplus \mathbf{W}$, and of representations, $T \oplus D$, was abstractly defined in this section. On the other hand, the tensor product of vectors $v \in \mathbf{V}$ with vectors $w \in \mathbf{W}$ was so far only described using components $v^i w^\alpha$, and we do not give the abstract definition here (see Appendix B.8 for one version of it). However, we introduce the abstract notation: $v \otimes w$ is a vector of a linear space $\mathbf{V} \otimes \mathbf{W}$, whose components are $v^i w^\alpha$ when referred to a basis $\{e_i\}$ for $\mathbf{V}$ and a basis $\{f_\alpha\}$ for $\mathbf{W}$; and $T \otimes D$ is a linear operator in $\mathbf{V} \otimes \mathbf{W}$ with matrix $T^i{}_k D^\alpha{}_\beta$. The multiplication laws were already written down before.

Exercises[1]

1. Assume that a representation space $\mathbf{V}$ decomposes as a direct sum of invariant subspaces $\mathbf{V}_\mu$, $\mu = 1, 2, \ldots$. Let $\mathbf{V}' \subset \mathbf{V}$ be an invariant irreducible subspace. Then either $\mathbf{V}' \subset \mathbf{V}_\mu$ for one value of μ, or some $\mathbf{V}_\mu$ each contain a subrepresentation equivalent to the one in $\mathbf{V}'$. Proof?

   Hint: The parallel projectors $P_\mu : \mathbf{V} \to \mathbf{V}_\mu$ define linear maps $S_\mu : \mathbf{V}' \to \mathbf{V}_\mu$. At least one of them must be injective; now distinguish the case where precisely one of them is injective and the case where several are injective.

2. A completely reducible representation is called *multiplicity free* if a decomposition into irreducible representations results in pairwise inequivalent ones. Show that in this case any irreducible subspace has to agree with one of those occurring in this decomposition, which is therefore essentially unique; every invariant subspace is a direct sum of some of the irreducible subspaces from the decomposition.

   Hint: Use the theorem proved in exercise 1!

   While the theorems expressed in exercises 1 and 2 hold over an arbitrary ground field, the latter will be assumed to be $\mathbf{C}$ (or algebraically closed) in all exercises that follow, together with finite dimensionality of representations.

3. Prove Schur II as a consequence of Schur I: from $T_g S = S T_g$ it follows $T_g (S - s \cdot \mathrm{id}_\mathbf{V}) = (S - s \cdot \mathrm{id}_\mathbf{V}) T_g$ for all $s \in \mathbf{C}$. Now choose s such as to make $S - s \cdot \mathrm{id}_\mathbf{V}$ singular.

4. Show that for two given equivalent irreducible representations the intertwiner is unique up to a scalar factor. Note that this remains true if the intertwiner is replaced by an *antilinear* map satisfying eq. (6.6.2).

   Hint: Let S, S' be two possible intertwiners, S nonsingular. Now consider $S^{-1}S'$ and use Schur II.

---

[1]These exercises are of a more abstract nature and are intended to demonstrate the applicability of Schur's lemma; otherwise, the theorems formulated could be just accepted as results from mathematics when they are used in the sequel.

5. Assume that an irreducible representation T is equivalent to its complex-conjugate T^*. Show for the equivalence map C that the product CC^* is a *real* multiple of the identity and may be made equal to $\pm$id by changing C by a scalar factor. What is the freedom left in C? Note that composing C with complex conjugation one obtains an *antilinear operator $\mathcal{C}$ commuting with T*.

   Hint: Write the condition of equivalence, take its complex-conjugate and apply the theorem expressed in exercise 4!

   Remark: This result yields a *classification of irreducible complex representations* into 3 classes:

   1. *Complex type*: C does not exist.

   2. *Real type* (also called potentially real or virtually real): $CC^* = +$id can be reached; there is then a basis $\{e_i\}$ with $Ce_i = e_i^*$, the T_g having real matrices with respect to it.

   3. *Quaternionic type* (also called pseudo-real or antireal): $CC^* = -$id can be reached; as we do not explain in any detail, the dimension of the representation may be halved by using quaternions instead of complex numbers here.

6. Let a completely reducible representation be a *multiple* of some irreducible representation $(\mathbf{V}_0, {}_0T)$ (also called *isotypic representation* or *factor representation* of type ${}_0T$): this means that in some decomposition into irreducible subrepresentations, $\mathbf{V} = \mathbf{V}_1 \oplus \mathbf{V}_2 \oplus \ldots \oplus \mathbf{V}_h$, all $\mathbf{V}_\mu$ turn out to be equivalent to one fixed irreducible $\mathbf{V}_0$; let $A_\mu : \mathbf{V}_0 \to \mathbf{V}_\mu$ be the corresponding equivalence maps. The natural number h is called the *multiplicity* of ${}_0T$ in T. Show the following:
   i. To each choice of ratios $a^1 : a^2 : \ldots \neq 0 : 0 : \ldots$ there is an invariant irreducible subspace $\mathbf{V}' := A'\mathbf{V}_0 \subset \mathbf{V}$ equivalent to $\mathbf{V}_0$ with intertwining injection A' given by

$$A'v_0 = a^\rho A_\rho v_0 \quad \text{for} \quad v_0 \in \mathbf{V}_0. \tag{6.6.20}$$

   ii. Every irreducible invariant subspace $\mathbf{V}' \subset \mathbf{V}$ can be obtained this way with uniquely determined ratios $a^1 : a^2 : \ldots$ and is, therefore, equivalent to $\mathbf{V}_0$.
   iii. Every decomposition into irreducible subspaces is of the form

$$\begin{aligned} \mathbf{V} &= \mathbf{V}_1' \oplus \mathbf{V}_2' \oplus \ldots \oplus \mathbf{V}_h' \\ \mathbf{V}_\mu' &:= A_\mu' \mathbf{V}_0, \qquad A_\mu' v_0 = a^\rho{}_\mu A_\rho v_0 \quad \text{for} \quad v_0 \in \mathbf{V}_0, \end{aligned} \tag{6.6.21}$$

   where $a^\rho{}_\mu$ is a nonsingular $h \times h$ matrix; conversely every such matrix yields a decomposition according to eq. (6.6.21).
   Hints: Ad i. A' must possess an inverse, since $A'v_0 = o$ for $v_0 \neq o$ would yield a nontrivial decomposition of the zero vector with respect to the $\mathbf{V}_\mu$.

   Ad ii. Writing S_μ as in the hint to exercise 1, one can apply the theorem of exercise 4 to the maps $A_1^{-1}S_1$, $A_2^{-1}S_2$, $\ldots$.

   Ad iii. For a given decomposition $\mathbf{V}_1' \oplus \mathbf{V}_2' \oplus \ldots$, form the matrix a according to ii; if it were singular, the linear dependence of its columns would immediately give a nontrivial decomposition of the zero vector with respect to the $\mathbf{V}'_\mu$. For the converse, observe $\mathbf{V}_\mu = \{v = (a^{-1})^\rho{}_\mu A'_\rho v_0 \mid v_0 \in \mathbf{V}_0\}$ and the dimensions.

7. Determine the form of the elements of the commutant, $A : \mathbf{V} \to \mathbf{V}$, in case $\mathbf{V}$ is completely reducible and (a) multiplicity free, or (b) isotypic!
   Solution: Since A carries irreducible subspaces into irreducible ones or annihilates them, one obtains, using the above notation

$$\begin{aligned}
\text{(a)} \quad & A = \lambda_1 \operatorname{id}_{\mathbf{V}_1} \oplus \lambda_2 \operatorname{id}_{\mathbf{V}_2} \oplus \ldots = \lambda_1 P_1 + \lambda_2 P_2 + \ldots, \quad \lambda_\mu \in \mathbf{C} \\
\text{(b)} \quad & A = A'_1 A_1^{-1} \oplus A'_2 A_2^{-1} \oplus \ldots = A'_1 A_1^{-1} P_1 + \ldots.
\end{aligned} \qquad (6.6.22)$$

8. Another useful description of the invariant subspaces and self-intertwiners of isotypic representations is the following.
   i. $\mathbf{V}$ may be thought of as a tensor product $\mathbf{V}_h \otimes \mathbf{V}_0$, where $\mathbf{V}_h$ is an auxiliary h-dimensional vector space; invariant subspaces $\mathbf{V}'$ of $\mathbf{V}$ are of form $\mathbf{V}'_h \otimes \mathbf{V}_0$, where $\mathbf{V}'_h \subset \mathbf{V}_h$ is 1-dimensional iff $\mathbf{V}'$ is irreducible. (Hence the auxiliary space can be considered as the linear space of intertwiners A', eq. (6.6.20).)
   ii. The representing operators T_g have the form $\operatorname{id}_{\mathbf{V}_h} \otimes {}_0T_g$.
   iii. The self-intertwiners $A : \mathbf{V} \to \mathbf{V}$ have the form $A_h \otimes \operatorname{id}_{\mathbf{V}_0}$, where A_h is an arbitrary linear map of $\mathbf{V}_h$ into itself.
   Prove this reformulation by choosing a basis in $\mathbf{V}_0$.

   Hints: Let $\{b_1, b_2, \ldots\}$ be a basis in $\mathbf{V}_0$; then $\{b_{\mu i}\} := \{A_\mu b_i\}$ is a basis in $\mathbf{V}_\mu$ and $\{b_{11}, \ldots, \ldots, b_{h1}, \ldots\}$ is a basis in $\mathbf{V}$ with respect to which $v \in \mathbf{V}$ has components $v^{\mu i}$. If ${}_0T_g b_k = t^i{}_k(g)\, b_i$, then we also have $T_g b_{\mu k} = t^i{}_k(g)\, b_{\mu i}$ and further $A\, b_{\mu k} = a^\nu{}_\mu\, b_{\nu k}$, thus $(T_g v)^{\mu i} = t^i{}_k(g)\, v^{\mu k} = \delta^\mu{}_\nu\, t^i{}_k(g)\, v^{\nu k}$, $(Av)^{\mu i} = a^\mu{}_\nu\, v^{\nu i} = a^\mu{}_\nu\, \delta^i{}_k\, v^{\nu k}$.

9. Prove Burnside's Lemma: Let $(\mathbf{V}, T)$ be a finite-dimensional irreducible complex representation of the group $\mathcal{G}$. Then for $W \in \mathrm{L}(\mathbf{V},\mathbf{V})$ (the linear space of linear maps $\mathbf{V} \to \mathbf{V}$) the condition $\operatorname{Tr}(T_g W) = 0$ for all $g \in \mathcal{G}$ implies $W = 0$.

   Hints: The possible solutions W form a subspace $\mathbf{W}$ of $\mathrm{L}(\mathbf{V},\mathbf{V}) = \mathbf{V} \otimes \widetilde{\mathbf{V}}$ (see Appendix B.8). Replace g by gh in the condition for W to see that $\mathbf{W}$ is invariant under the isotypic representation $(\mathbf{V} \otimes \widetilde{\mathbf{V}}, T \otimes \widetilde{\operatorname{id}})$ of $\mathcal{G}$. Now conclude from the previous exercise that $\mathbf{W} = \mathbf{V} \otimes \mathbf{W}'$ for some subspace $\mathbf{W}' \subset \widetilde{\mathbf{V}}$ and is spanned by irreducible subspaces $\mathbf{V} \otimes \mathbf{C}\tilde{w}$, $\tilde{w} \in \mathbf{W}'$. Thus the possible W are arbitrary linear combinations of those of the form $W^i_j = v^i \tilde{w}_j$, where $v \in \mathbf{V}$ is arbitrary. The starting condition then implies $T_g^\top \tilde{w} = 0$; but the T_g are invertible.

10. What can you say from exercises 2, 6, 7, 8 about invariant subspaces, uniqueness of irreducible decompositions and the commutant of arbitrary completely reducible finite-dimensional complex representations?

    Hint: Perform an *isotypic decomposition*!

11. The matrices of form $\begin{pmatrix} a & b \\ 0 & 1 \end{pmatrix}$ with $a \neq 0$ form a group and at the same time a reducible representation of it. Show that the commutant is trivial. For the

subgroup where $a = 1$ the commutant is nontrivial but the representation is not decomposable.

12. Investigate the commutant of *real irreducible representations* by studying first the commutant of its complexification (cf. exercise 8 of sect. 6.5).
    Hint: If the complexification is irreducible the real commutant is obviously $\mathbf{R}\,\mathrm{id}_\mathbf{V}$ (*real type*). However, if it decomposes into two complex-conjugate subrepresentations in the subspaces $\mathbf{W}$, $\mathbf{W}^*$ with projectors P, P^*, there are two cases. (a) The subrepresentations are inequivalent: then the real commutant is $\mathbf{R}\,\mathrm{id}_\mathbf{V} + \mathbf{R}I$, where I is determined by its complexification $I^c := i(P - P^*)$ and satisfies $I^2 \equiv -\mathrm{id}_\mathbf{V}$ (*complex type*). (b) The subrepresentations are equivalent: let S be a nonzero intertwiner as in exercise 5. Then the real commutant is given by $\mathbf{R}\,\mathrm{id}_\mathbf{V} + \mathbf{R}I + \mathbf{R}J + \mathbf{R}K$, where I, J, K are determined by their complexifications $I^c := i(P - P^*)$, $J^c := PS^*P^* + P^*SP$, $K^c := i(PS^*P^* - P^*SP)$ and satisfy $I^2 \equiv -\mathrm{id}_\mathbf{V}$, $IJ + JI \equiv 0$, $IJ \equiv K$, $J^2 \equiv K^2 \equiv \pm\mathrm{id}_\mathbf{V}$ if $S^*S = \pm\mathrm{id}_\mathbf{W}$; the upper sign is excluded, however, otherwise J, K would have real invariant eigenspaces (*quaternionic type*).

13. Consider the space of totally symmetric 4-tensors of degree 4 and the Lorentz invariant map π to symmetric tensors of degree 2 obtained by a single contraction, $T^{ijk\ell} \mapsto T_k^{ijk}$. Find a Lorentz invariant injection ι of symmetric tensors of degree 2 into symmetric tensors of degree 4 giving a right inverse to π.

    Hint: A multiple of $T^{(ij}\eta^{k\ell)}$ does not work, but you can correct by a multiple of $\eta^{(ij}\eta^{k\ell)}$!

7 Representation Theory of the Rotation Group

Before looking for all (finite-dimensional) irreducible representations of the Lorentz group we treat the same problem for the rotation group SO(3,$\mathbf{R}$). There are four reasons for this.

- The general methods are easy to demonstrate here.

- The isomorphism between $\mathcal{L}_+^\uparrow$ and the complex rotation group SO(3,$\mathbf{C}$) mentioned in sect. 6.5 leaves us with the expectation that some analytic continuation of the representations of SO(3,$\mathbf{R}$) will lead to representations of the Lorentz group. (It will turn out that we do not get all representations this way, but the remaining ones are then easily found.)

- The *unitary* representations of the rotation group play an important role in the quantum mechanics of angular momentum, so that connections between the abstractly treated problems and physical applications are easily established.

- The unitary irreducible representations of SO(3,$\mathbf{R}$) will be directly required in chap. 9 for the representation theory of the Poincaré group.

The finite-dimensional irreducible representations of the rotation group SO(3,$\mathbf{R}$) may be classified and constructed by elementary means; one can also prove *full* reducibility for reducible representations and carry out the reduction; finally one can extend the results from SO(3,$\mathbf{R}$) to SO(3,$\mathbf{C}$) and thus to the restricted Lorentz group $\mathcal{L}_+^\uparrow$. This route to the finite-dimensional representations of the rotation and Lorentz group is described, e.g., in Cartan (1966).

However, the way how the principle of relativity is realized in *quantum mechanics* requires the construction of representations of the Poincaré group in a space of quantum states, i.e., in a Hilbert space (see sect. 9.2), which in general will be of infinite dimension. For a mathematically rigorous treatment of this, deeper considerations from functional analysis together with the theory of integration on groups would be necessary. It would be impossible within the bounds of this book even to define all concepts precisely, let alone to prove the fundamental theorems. We shall therefore simply quote some of these theorems and work, as far as infinite-dimensional representations are concerned, with formal analogies to the finite-dimensional case, whose precise meaning can be given only by constructions from functional analysis.

For the rotation group—just as for any other *compact* topological group—the general theory tells us that all continuous representations in a Hilbert space are equivalent to unitary ones and thus are completely reducible, and that the irreducible representations are all finite-dimensional. We shall therefore introduce the concept of *unitary representation* and construct the irreducible unitary representations. Methodologically, we will go beyond previous chapters by making systematic use of group elements 'infinitely' close to the unit element, assigned to which are 'infinitesimal' transformations. This way of proceeding is not necessary; the irreducible representations may be constructed globally; a concept of completeness for the system of irreducible representations may be formulated on the global level as well, and the completeness of the system of representations found may be proved. However, the use of infinitesimal transformations in physics and geometry is useful and necessary. When they are used, new objects emerge beyond the tensors that so far made up our examples of representation spaces: the *spinors*. Although they get eliminated again upon subsequent global considerations, they not only remain important because of the quantum mechanical version of the representation problem but also for tensorial questions alone.

7.1   The Rotation Group SO(3,R)

By the rotation group we mean the group of homogeneous linear transformations

$$\mathbf{x}' = R\,\mathbf{x} \tag{7.1.1}$$

of a Euclidean 3-dimensional vector space into itself which preserve lengths and orientation. We can read eq. (7.1.1) in three ways:

a. as an abstract transformation, leaving invariant some positive-definite quadratic form defined on the space, $\mathbf{x}^2 = (R\,\mathbf{x})^2$, together with some determinant function;

b. as a matrix equation for this transformation, where the latter is carried out actively; $\mathbf{x}$ and $\mathbf{x}'$ then symbolize the columns of components, with respect to a fixed orthonormal basis of the space, of the original and the rotated vector, and R is the orthogonal matrix with $\det R = +1$ connecting them;

c. as a matrix equation for the transformation carried out passively, in which only the basis gets rotated but vectors remain fixed. The $\mathbf{x}$ and $\mathbf{x}'$ again are columns, this time formed from the components of the same vector, once referred to the original frame and once to the rotated one; R is the matrix relating them.

We shall not make a notational distinction for the three ways of reading eq. (7.1.1); but it has to be observed that the matrices R appearing in the versions b and c are *inverses* of each other if it is the *same* rotation that acts in b on all vectors, the basis remaining fixed, while it acts in c on just that basis alone, the unchanged vector getting referred to the new basis. (Using indices one could make a notational distinction between the two ways b and c, if required, as follows:

$$x'^{\mu} = R^{\mu}{}_{\nu}\, x^{\nu} \qquad \text{(active transformation)} \tag{7.1.2}$$

$$x^{\mu'} = R^{\mu'}{}_{\nu}\, x^{\nu} \qquad \text{(passive transformation).)} \tag{7.1.3}$$

In this chapter we will have to deal with objects related to Euclidean 3-space only; we shall use greek indices μ, ν, $\ldots = 1, 2, 3$ that will be lowered and raised using the Euclidean metric and inverse metric tensors $\delta_{\mu\nu}$ and $\delta^{\mu\nu}$, respectively, each numerically given by the Kronecker symbol, as we shall stick to orthonormal frames.

An active rotation about the axis $\boldsymbol{\alpha}$ through the angle $\alpha = |\boldsymbol{\alpha}|$ in the sense of the usual right-hand rule is given by

$$\mathbf{x}' = \mathbf{x}\cos\alpha + \frac{\boldsymbol{\alpha}\,\mathbf{x}}{\alpha^2}\,\boldsymbol{\alpha}(1-\cos\alpha) + \frac{\boldsymbol{\alpha}\times\mathbf{x}}{\alpha}\sin\alpha, \tag{7.1.4}$$

$$x'^{\mu} = x^{\mu}\cos\alpha + \frac{\alpha_\nu x^\nu}{\alpha^2}\,\alpha^\mu(1-\cos\alpha) - \frac{\epsilon^{\mu\nu\lambda}\alpha^\lambda x^\nu}{\alpha}\sin\alpha, \tag{7.1.5}$$

the difference in sign as compared to eqs. (1.3.1,2) arising from the change to active instead of passive execution of the rotation. We can read off the matrix $R^\mu{}_\nu$ as

$$R^\mu{}_\nu = \delta^\mu{}_\nu\cos\alpha + \frac{\alpha^\mu\alpha_\nu}{\alpha^2}\,(1-\cos\alpha) - \frac{\epsilon^{\mu\nu\lambda}\alpha^\lambda}{\alpha}\sin\alpha. \tag{7.1.6}$$

The *trace* of R—which is well known to be independent of the basis to which the matrix refers—yields the angle in terms of R as

$$\operatorname{Tr} R = 1 + 2\cos\alpha. \tag{7.1.7}$$

Since a rotation through α about the axis $\mathbf{n} = \boldsymbol{\alpha}/\alpha$ and a rotation through $2\pi - \alpha$ about the axis $-\mathbf{n}$ lead to the same result, it is necessary to restrict the angle to the interval $0 \le \alpha < \pi$ in order to get a 1:1 assignment between rotations and *rotation vectors*. To get all rotations we must certainly add the value $\alpha = \pi$, however, but here we cannot avoid that the *same* rotation is given by $\boldsymbol{\alpha}$ *and* $-\boldsymbol{\alpha}$. Now conversely every proper-orthogonal matrix R may be written as in eq. (7.1.6) and is a rotation around some axis: we can calculate a rotation angle α from eq. (7.1.7) and (when $\alpha \ne 0$) direction cosines α^μ/α for an axis from

$$\frac{\alpha^\mu}{\alpha}\sin\alpha = -\frac{1}{2}\,\epsilon^{\mu\nu\lambda}\,R^\nu{}_\lambda, \tag{7.1.8}$$

consistency of these equations and reality of α being secured by $R^{\top}R = \mathbf{1}$, det R=1. This is unique except when $\alpha = \pi$ where eq. (7.1.8) breaks down. In the latter case one can use the fact that $\boldsymbol{\alpha}$ is the only independent eigenvector of $R = R(\boldsymbol{\alpha}) \ne \mathbf{1}$ for the only real eigenvalue 1.

If the space is complexified there are two more eigenvalues $\exp(\pm i\alpha)$ and corresponding eigendirections which are in the plane orthogonal to $\boldsymbol{\alpha}$ and are called 'isotropic' since they do not change under the rotation without lying in the axis; when the scalar product is extended bilinearly to the complex domain, their scalar squares are zero. On first sight it seems unphysical to consider them, but it will turn out to be of advantage to use them together with $\boldsymbol{\alpha}/\alpha$ as a basis. Group theoretically, their occurrence illustrates Schur's lemma: the rotations about a fixed axis $\boldsymbol{\alpha}$ form a commutative group, the matrices $R(\boldsymbol{\alpha})$ form a representation of it which after complexification of the representation space becomes decomposable into three 1-dimensional representations. (The irreducible complex representations of any commutative group must be 1-dimensional by Schur II.) Note the exceptional cases $\alpha = 0, \pi$.

We thus have a bijection between rotations and points of a solid sphere $0 \leq |\boldsymbol{\alpha}| \leq \pi$ whose antipodal surface points are identified so that each diameter topologically becomes a circle representing the subgroup of rotations about a fixed axis, as indicated in Fig. 7.1.

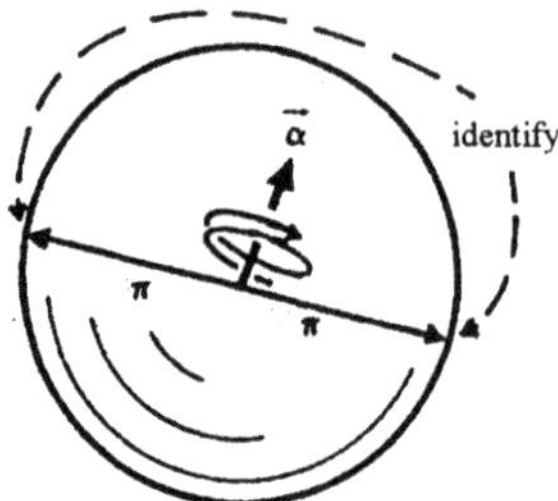

Fig. 7.1. Picturing the rotation group in parameter space $\{\boldsymbol{\alpha} \in \mathbf{R}^3 : |\boldsymbol{\alpha}| \leq \pi\}$.

The abstract group $SO(3,\mathbf{R})$ is therefore a 3-dimensional compact connected Lie group. The identification just described causes an interesting topological complication which will turn out finally to be responsible for the occurrence of spinors and half integer values of spin in quantum mechanics.

On the group manifold, for which we now have a model, one can use coordinates other than the $\boldsymbol{\alpha}$. A possibility that is used very much is given by the *Euler angles*. They are defined as follows (Fig. 7.2):

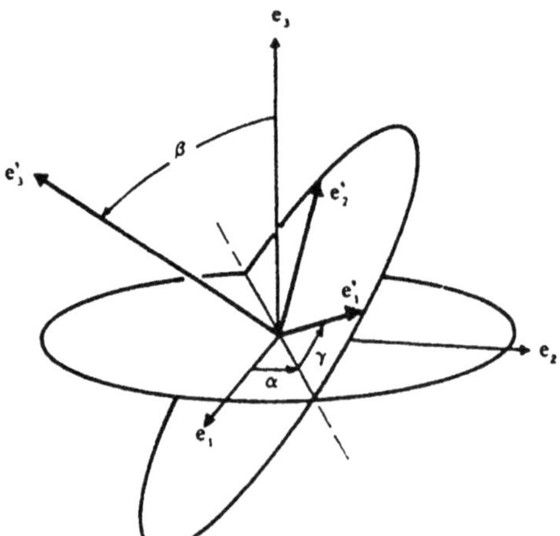

Fig. 7.2. Euler angles

Let $\{\mathbf{e}_\mu\}$ be a right-handed orthonormal frame and $\{\mathbf{e}_\mu'\}$ a rotated one. The intersection of the 1,2-plane with the $1', 2'$-plane is the *nodal line*; it is oriented as

$\mathbf{e}_3 \times \mathbf{e}'_3$. Now we carry $\{\mathbf{e}_\mu\}$ into $\{\mathbf{e}'_\mu\}$ by three successive *positive* rotations (i.e., forming right-hand screws when combined with translations along the positively oriented axes): one rotation about $\mathbf{e}_3$ through the angle α ($0 \leq \alpha < 2\pi$) which takes $\mathbf{e}_1$ into the positive nodal line; one rotation about the nodal line through the angle β; and finally a rotation about $\mathbf{e}'_3$ through the angle γ ($0 \leq \gamma < 2\pi$) which takes the nodal line into $\mathbf{e}'_1$. Formally, we have ($0 \leq \beta < 2\pi$)

$$
\begin{pmatrix} \mathbf{e}'_1 \\ \mathbf{e}'_2 \\ \mathbf{e}'_3 \end{pmatrix} = \underbrace{\begin{pmatrix} \cos\gamma & \sin\gamma & 0 \\ -\sin\gamma & \cos\gamma & 0 \\ 0 & 0 & 1 \end{pmatrix} \begin{pmatrix} 1 & 0 & 0 \\ 0 & \cos\beta & \sin\beta \\ 0 & -\sin\beta & \cos\beta \end{pmatrix} \begin{pmatrix} \cos\alpha & \sin\alpha & 0 \\ -\sin\alpha & \cos\alpha & 0 \\ 0 & 0 & 1 \end{pmatrix}}_{=:\, \mathrm{R}(\alpha,\beta,\gamma)\,.} \begin{pmatrix} \mathbf{e}_1 \\ \mathbf{e}_2 \\ \mathbf{e}_3 \end{pmatrix} \tag{7.1.9}
$$

This parametrization of the rotation group is 1:1 except for the cases where $\beta = 0$ or $\beta = \pi$, in which the nodal line is indeterminate.

The inverse matrix for $\mathrm{R}(\alpha, \beta, \gamma)$ is easy to get from eq. (7.1.9); however, the angles then occurring do not fall into the domains given. One can verify that

$$
\mathrm{R}^{-1}(\alpha, \beta, \gamma) = \mathrm{R}(\pi - \gamma, \beta, \pi - \alpha) \tag{7.1.10}
$$

satisfies all conditions.

A further parametrization which also gives a simple description of the multiplication table will come from the spinor representation.

Exercise

Find the relation between $\boldsymbol{\alpha}$ and the Euler angles!

7.2 Infinitesimal Transformations

A characteristic feature of many investigations in Lie groups is the use of infinitesimal transformations. One considers group elements that deviate only little from the unit element; i.e., elements whose parameters differ only by small amounts from the parameters of the unit element. Other group elements are then reached by composing such 'small' elements. One thus combines the manifold aspect (that allows to do calculus[1] on the group) and the group aspect.

Let us write a small rotation R as

$$
\mathrm{R} = \mathbf{1} + \Omega, \tag{7.2.1}
$$

where the elements of the matrix Ω are small quantities of first order, so that their squares and products can be neglected. To this accuracy, the orthogonality condition

$$
\mathrm{R}\,\mathrm{R}^\top = \mathbf{1} \tag{7.2.2}
$$

---

[1] We will use here as far as possible the 'physicist's version' where one writes infinitesimal quantities instead of considering limits.

requires

$$\Omega + \Omega^\mathsf{T} = 0. \tag{7.2.3}$$

Thus Ω is an antisymmetric matrix and may be written as

$$\Omega = \begin{pmatrix} 0 & -\alpha_3 & \alpha_2 \\ \alpha_3 & 0 & -\alpha_1 \\ -\alpha_2 & \alpha_1 & 0 \end{pmatrix} = \boldsymbol{\alpha}\,\boldsymbol{\Lambda}, \tag{7.2.4}$$

where $\boldsymbol{\alpha} = (\alpha_1, \alpha_2, \alpha_3)$ and $\boldsymbol{\Lambda}$ indicates a triple of matrices:

$$\Lambda_1 := \begin{pmatrix} 0 & 0 & 0 \\ 0 & 0 & -1 \\ 0 & 1 & 0 \end{pmatrix}, \quad \Lambda_2 := \begin{pmatrix} 0 & 0 & 1 \\ 0 & 0 & 0 \\ -1 & 0 & 0 \end{pmatrix}, \quad \Lambda_3 := \begin{pmatrix} 0 & -1 & 0 \\ 1 & 0 & 0 \\ 0 & 0 & 0 \end{pmatrix}. \tag{7.2.5}$$

For the $\mu\nu$-element of the matrix Λ_λ we read off

$$\Lambda_{\lambda\mu\nu} = -\epsilon_{\lambda\mu\nu}. \tag{7.2.6}$$

With eqs. (7.2.1,4), the transformation $\mathbf{x}' = \mathrm{R}\,\mathbf{x}$ reads $x'^\mu = x^\mu + \epsilon^{\mu\lambda\nu}\alpha^\lambda x^\nu$ or $\mathbf{x}' = \mathbf{x} + \boldsymbol{\alpha} \times \mathbf{x}$, which is the form taken by eq. (7.1.5) for small α. Thus the $\boldsymbol{\alpha}$ in eq. (7.2.4) are just the components of the rotation vector of the infinitesimal rotation.

Heuristically, we can obtain the relation between infinitesimal and finite rotations as follows: write the finite rotation $\mathrm{R}(\boldsymbol{\alpha})$ as

$$\mathrm{R}(\boldsymbol{\alpha}) = \mathrm{R}\left(\frac{\boldsymbol{\alpha}}{2}\right) \mathrm{R}\left(\frac{\boldsymbol{\alpha}}{2}\right) = \ldots = \left[\mathrm{R}\left(\frac{\boldsymbol{\alpha}}{N}\right)\right]^N; \tag{7.2.7}$$

for sufficiently large N, $\boldsymbol{\alpha}/N$ becomes sufficiently small, so that we can put $\mathrm{R}\left(\frac{\boldsymbol{\alpha}}{N}\right) \approx \mathbf{1} + \boldsymbol{\alpha}\boldsymbol{\Lambda}/N$, and $N \to \infty$ gives

$$\mathrm{R}(\boldsymbol{\alpha}) = \exp(\boldsymbol{\alpha}\boldsymbol{\Lambda}) = (\exp(\boldsymbol{\alpha}\boldsymbol{\Lambda}/N))^N. \tag{7.2.8}$$

In this way we can *generate* any rotation from an 'infinitesimal' one. One can check that summing the power series for $\exp(\boldsymbol{\alpha}\boldsymbol{\Lambda})$ leads back to eq. (7.1.5) (cf. exercise 1).

For fixed $\boldsymbol{\alpha}_0$ but variable τ the rotations $\mathrm{R}(\tau\boldsymbol{\alpha}_0) = \exp(\tau\boldsymbol{\alpha}_0\boldsymbol{\Lambda})$ form a 1-parameter subgroup, $\tau = 0$ giving the unit element and $\tau = 1$ giving the rotation $\mathrm{R}(\boldsymbol{\alpha}_0)$. Every matrix of the form $\boldsymbol{\alpha}_0\boldsymbol{\Lambda}$ is, in this sense, the *generator of a 1-parameter subgroup*. Sums and (real) multiples of generators are obviously again generators, which therefore constitute a real 3-dimensional vector space, in which the generators for rotations about the 1-, 2-, 3-axis, i.e., $\Lambda_1, \Lambda_2, \Lambda_3$, form a basis. On the other hand, multiplying two matrices of that form would lead out of this vector space, since the product of two antisymmetric matrices is no more antisymmetric in general. However, the *commutator*

$$[\mathrm{A}, \mathrm{B}] := \mathrm{A}\mathrm{B} - \mathrm{B}\mathrm{A} = -[\mathrm{B}, \mathrm{A}] \tag{7.2.9}$$

of two antisymmetric matrices is again antisymmetric:

$$[\mathrm{A}, \mathrm{B}]^\mathsf{T} = [\mathrm{B}^\mathsf{T}, \mathrm{A}^\mathsf{T}] = -[\mathrm{A}^\mathsf{T}, \mathrm{B}^\mathsf{T}] = -[\mathrm{A}, \mathrm{B}]. \tag{7.2.10}$$

The commutator of two generators thus turns out to be a generator again and therefore can be written in the form $\alpha\Lambda$. Writing $A = \mathbf{m}\Lambda$, $B = \mathbf{n}\Lambda$, α must, as a bilinear concomitant of $\mathbf{m}$, $\mathbf{n}$ of axial vector type, have the form $const \cdot \mathbf{m} \times \mathbf{n}$; comparing coefficients yields $const = 1$:

$$[\mathbf{m}\Lambda, \mathbf{n}\Lambda] = (\mathbf{m} \times \mathbf{n})\Lambda \tag{7.2.11}$$

or

$$[\Lambda_\mu, \Lambda_\nu] = \epsilon_{\mu\nu\lambda}\Lambda_\lambda. \tag{7.2.12}$$

These are the fundamental *commutation relations for the generators of the rotation group*, on which the derivation of all representations will be based.

There is a new abstract structure here: In the (real) vector space of generators A, B, C, ... we have a bilinear product defined by $A \circ B := [A, B]$ which does not lead out of the space and which, because of eq. (7.2.9) and the *Jacobi identity*

$$[[A, B], C] + [[C, A], B] + [[B, C], A] \equiv 0 \tag{7.2.13}$$

(valid for all commutators), satisfies the usual distributive laws and the relations

$$A \circ B = -B \circ A, \quad (A \circ B) \circ C + (C \circ A) \circ B + (B \circ C) \circ A = 0. \tag{7.2.14}$$

An abstract vector space with a bilinear multiplication $\circ$ satisfying the formal requirements expressed in eqs. (7.2.14) is called a *Lie algebra*. The latter conditions show that a Lie algebra is neither associative nor commutative.

When the vector space is finite-dimensional (as in our case), one may choose a basis $\{X_A\}$. Then, because of bilinearity, it suffices to know the products $X_A \circ X_B$, and these in turn may be fixed by giving their components with respect to the basis chosen:

$$X_A \circ X_B = C^D{}_{AB} X_D. \tag{7.2.15}$$

The *structure constants* $C^D{}_{AB}$—components of the *structure tensor*, since they are basis-dependent in the appropriate way—determine the structure of the algebra.

To define an n-dimensional Lie algebra, these constants cannot be chosen as arbitrary n^3 numbers; rather, they have to satisfy the relations

$$C^D{}_{AB} = -C^D{}_{BA} \tag{7.2.16}$$

$$C^D{}_{AB}\, C^E{}_{CD} + C^D{}_{CA}\, C^E{}_{BD} + C^D{}_{BC}\, C^E{}_{AD} = 0 \tag{7.2.17}$$

implied by eqs. (7.2.14). Abstract Lie algebras over arbitrary fields form a subject for their own in algebra, cf. Jacobson (1962).

We can state therefore that the generators of (the defining representation of) the rotation group $SO(3,\mathbf{R})$ form a 3-dimensional Lie algebra over the field of *real* numbers. (The latter is because in $\alpha\Lambda$ the α has to be real, otherwise $\exp(\alpha\Lambda)$ is not a real rotation matrix; this must be emphasized since we are going to work in complex representation spaces most of the time, and the complex rotation group $SO(3,\mathbf{C})$, already mentioned in sect. 6.5, will become important later as well.) The structure constants of the Lie algebra of the rotation group, which is sometimes written so$(3,\mathbf{R})$, were derived above to be $\epsilon_{\mu\nu\lambda}$.

In the next section we investigate the consequences that the Lie algebra structure has on representations.

Exercises

1. Show that the matrix $\mathbf{n}\Lambda$ satisfies the relations $(\mathbf{n}\Lambda)^2 = \mathbf{n}\,\mathbf{n}^\mathsf{T} - \mathbf{1}$, $(\mathbf{n}\Lambda)^3 = -\mathbf{n}\Lambda$, where $\mathbf{n} = \boldsymbol{\alpha}/\alpha$. Use them to sum the power series $\mathrm{R}(\boldsymbol{\alpha}) = \exp(\boldsymbol{\alpha}\Lambda) = \sum_{k=0}^{\infty} \frac{1}{k!}(\boldsymbol{\alpha}\Lambda)^k$. Compare with the geometrically derived eq. (7.1.4)!

2. Show, at the same low level of mathematical rigor as in our derivation of eq. (7.2.8): If $\exp(\Omega) = \mathrm{R}$, then $\det \mathrm{R} = \exp(\mathrm{Tr}\,\Omega)$; if $\Omega = -\Omega^\mathsf{T}$, then $\mathrm{R}^\mathsf{T}\mathrm{R} = \mathbf{1}$, $\det \mathrm{R} = +1$.

3. Verify eq. (7.2.12) directly.

4. Verify eq. (7.2.13).

5. Show that the vector space $\mathbf{R}^3$ with the usual cross product ($\times$) as the multiplication $\circ$ is a Lie algebra: compare its structure constants to those of so(3,$\mathbf{R}$)!

6. Construct all 2-dimensional Lie algebras, simplifying the structure constants as much as possible by a suitable choice of basis.

   Remark: The same problem in 3 dimensions is already more complicated, but has application in certain cosmological investigations ('Bianchi models').

7.3   Lie Algebra and Representations of SO(3)

We now regard the proper orthogonal transformations of $\mathbf{R}^3$ as a special representation $g \mapsto R_g$ of the abstract Lie group[1] SO(3)—its defining representation[2]. It is irreducible—there are no subspaces in $\mathbf{R}^3$ invariant under rotations, since there are no distinguished directions and, orthogonal to directions, no distinguished planes. Our task of finding *all* irreducible complex representations of SO(3) will be facilitated by the following considerations.

Let $(\mathbf{V}, T)$ be a representation of SO(3); then to a 1-parameter subgroup $g(\tau)$ (where τ varies over some interval beginning at 0, and where $g(0) = e$ is the unit element) there corresponds, under the representation homomorphism, a 1-parameter group of transformations or matrices $T_{g(\tau)}$ in $\mathbf{V}$. For small τ we have

$$T_{g(\tau)} \approx \mathrm{id}_{\mathbf{V}} + \tau t, \tag{7.3.1}$$

where

$$t := \frac{\partial}{\partial \tau} T_{g(\tau)} \bigg|_{\tau = 0} \tag{7.3.2}$$

is called the *generator* of the subgroup in the representation considered. We want to show that the generators of all 1-parameter subgroups taken together form a real

---

[1] In this chapter we shall write SO(3) instead of SO(3,$\mathbf{R}$).

[2] There exist Lie groups in the abstract sense for which there is no defining representation, in the sense that they do not possess any faithful *finite-dimensional* representation. Lie groups that do are called *linear Lie groups*.

vector space, in which the generators of the rotations about the three coordinate axes in $\mathbf{R}^3$ form a basis and satisfy the commutation relations

$$[t_\mu, t_\nu] = \epsilon_{\mu\nu\lambda} t_\lambda. \tag{7.3.3}$$

Therefore also the generators in an arbitrary representation form a Lie algebra, whose structure is isomorphic to the Lie algebra (7.2.11) except for the trivial representation, in which all generators vanish.

The problem of finding all irreducible representations (of the infinitely many elements) of the Lie group SO(3) *is thus reduced to the determination of the three generators of the Lie algebra (7.3.3).* This problem will be solved in sect. 7.5.

To prove this basic claim, we choose $S \in SO(3)$ and start from the relation

$$S\,R(\boldsymbol{\alpha})\,S^{-1} = R(S\boldsymbol{\alpha}) \tag{7.3.4}$$

which can be verified from eq. (7.1.4).

Its significance is the following: consider an active rotation, given with respect to the positively oriented orthonormal basis $\{e_\mu\}$ by the orthogonal matrix $R(\boldsymbol{\alpha})$, eq. (7.1.6). With respect to the basis $\{\bar{e}_\mu\}$, obtained from $\{e_\mu\}$ by applying the rotation $\bar{e}_\mu = S_{\mu\nu} e_\nu$, the rotation considered before is given by the matrix $S\,R(\boldsymbol{\alpha})\,S^{-1}$. Since $\boldsymbol{\alpha}$ in the new basis becomes $\bar{\boldsymbol{\alpha}} = S\boldsymbol{\alpha}$, we must have eq. (7.3.4).

Regarding now $R(\boldsymbol{\alpha}) =: S_{g(\boldsymbol{\alpha})}$ and $S =: S_h$ as representing matrices of the abstract group elements $g(\boldsymbol{\alpha})$ and h in the defining representation, eq. (7.3.4) just tells us that

$$h\,g(\boldsymbol{\alpha})\,h^{-1} = g\left(S_h\boldsymbol{\alpha}\right). \tag{7.3.5}$$

Passing now from the abstract group to an arbitrary representation T, we must have there, homomorphically,

$$T_h\,T_{g(\boldsymbol{\alpha})}\,T_{h^{-1}} = T_{g(S_h\boldsymbol{\alpha})} \tag{7.3.6}$$

We now replace $\boldsymbol{\alpha}$ by $\tau\boldsymbol{\alpha}$ and go to small τ:

$$T_{g(\tau\boldsymbol{\alpha})} \cong \mathrm{id}_\mathbf{V} + \tau t, \qquad t := \left.\frac{\partial}{\partial\tau} T_{g(\tau\boldsymbol{\alpha})}\right|_{\tau=0}. \tag{7.3.7}$$

By the chain rule we have

$$t = \alpha_\mu t_\mu = \boldsymbol{\alpha}\,\mathbf{t}, \qquad \mathbf{t} := (t_1, t_2, t_3), \tag{7.3.8}$$

where

$$t_\mu := \left.\frac{\partial}{\partial\alpha_\mu}\,T_{g(\boldsymbol{\alpha})}\right|_{\boldsymbol{\alpha}=0} \tag{7.3.9}$$

are the generators of rotations about the coordinate axes of $\mathbf{R}^3$ in the representation considered.

Equations (7.3.7–9) show that the generators t form a (real) vector space spanned by t_1, t_2, t_3, as claimed. This vector space is 3-dimensional in all faithful representations; since SO(3)—as will be shown later—is *simple*, i.e., has only trivial invariant subgroups, only the trivial representation is not faithful. To arrive at the last part of our claim we insert eqs. (7.3.7–9) into eq. (7.3.6) and replace $\boldsymbol{\alpha} \to \tau\boldsymbol{\alpha}$, $\tau \ll 1$, to obtain

$$T_h\,\boldsymbol{\alpha}\,\mathbf{t}\,T_h^{-1} = (S_h\boldsymbol{\alpha})\,\mathbf{t}. \tag{7.3.10}$$

If we now also take h close to unity, i.e., if we write it as $h(\tau\boldsymbol{\beta})$ with $\tau \ll 1$, we have

$$T_h \cong \mathrm{id}_\mathbf{V} + \tau\boldsymbol{\beta}\,\mathbf{t}, \qquad\qquad T_h^{-1} \cong \mathrm{id}_\mathbf{V} - \tau\boldsymbol{\beta}\,\mathbf{t}, \tag{7.3.11}$$

and from the behavior of vectors under infinitesimal rotations

$$S_h\,\boldsymbol{\alpha} \cong \boldsymbol{\alpha} + \tau\boldsymbol{\beta} \times \boldsymbol{\alpha}. \tag{7.3.12}$$

Thus (7.3.10) becomes

$$[\boldsymbol{\beta}\,\mathbf{t},\, \boldsymbol{\alpha}\,\mathbf{t}] = (\boldsymbol{\beta} \times \boldsymbol{\alpha})\,\mathbf{t} \tag{7.3.13}$$

or

$$[t_\mu, t_\nu] = \epsilon_{\mu\nu\lambda}t_\lambda. \tag{7.3.14}$$

The commutation relations (7.3.14) are therefore valid in *arbitrary* representations, which proves the claim made above.

The decisive relation (7.3.10) can be rewritten to produce additional insight. Since $\boldsymbol{\alpha}$ in eq. (7.3.10) is arbitrary, we also have

$$T_h\,t_\mu\,T_h^{-1} = (S_h)_{\nu\mu}\,t_\nu, \tag{7.3.15}$$

or, replacing $h \to h^{-1}$ and taking into account that $\left(S_h^{-1}\right)_{\nu\mu} = (S_h)_{\mu\nu}$,

$$T_h^{-1}\,\mathbf{t}\,T_h = S_h\,\mathbf{t}. \tag{7.3.16}$$

Generally, a triple $\mathbf{v}$ of operators on $\mathbf{V}$ satisfying the relations

$$T_h^{-1}\,\mathbf{v}\,T_h = S_h\mathbf{v} \tag{7.3.17}$$

is called a *vector operator* on $\mathbf{V}$. When the infinitesimal version (7.3.11) is inserted here we get

$$[\mathbf{v}, \boldsymbol{\beta}\,\mathbf{t}] = \boldsymbol{\beta} \times \mathbf{v}. \tag{7.3.18}$$

The *square* $\mathbf{v}^2 := v_\mu v_\mu$ *of a vector operator is invariant under the representation*, since from eq. (7.3.17) it follows that

$$\mathbf{v}^2 = (S_h\mathbf{v})^2 = T_h^{-1}\,\mathbf{v}\,T_h\,T_h^{-1}\,\mathbf{v}\,T_h = T_h^{-1}\,\mathbf{v}^2\,T_h$$
$$T_h\,\mathbf{v}^2 = \mathbf{v}^2\,T_h \tag{7.3.19}$$

i.e., $\mathbf{v}^2$ commutes with all operators T_h of the representation. When the representation considered is complex-irreducible, Schur II tells us that $\mathbf{v}^2$ must be a multiple of the unit operator $\mathrm{id}_\mathbf{V}$.

In particular, if we take $\mathbf{v} = \mathbf{t}$ we obtain the *Casimir operator*

$$\mathbf{C} := \mathbf{t}^2, \tag{7.3.20}$$

which commutes with all operators of the representation.

Exercises

1. Show that the infinitesimal version of eq. (7.3.4) is $S \Lambda_\mu S^{-1} = S_{\mu\nu} \Lambda_\nu$. Prove the latter equation from the definition of the Λ_μ and the properties $S^\top S = S S^\top = 1$, $\det S = +1$ without using the geometric significance of $\boldsymbol{\alpha}$.

2. Verify the commutator relation

$$[A, BC] = [A, B]C + B[A, C] \tag{7.3.21}$$

and use it to deduce $[\mathbf{v}^2, \boldsymbol{\beta}\, t] = 0$ from eq. (7.3.18).

7.4 Lie Algebras of Lie Groups

In view of later applications, we now parallel the considerations above as far as possible for an arbitrary Lie group $\mathcal{G}$ and some representation $(\mathbf{V}, T)$ of it. The group elements g are specified by n parameters $\beta_1 \ldots \beta_n$: $g(\beta_1, \ldots, \beta_n) =: g(\beta_A)$, where we always require $g(0) = e$. To a curve $\beta_A = \beta_A(\tau)$ in parameter space there corresponds a curve $g(\beta_A(\tau)) = g(\tau)$ in the group manifold. We consider curves through the unit element, so that $g(\beta_A(0)) = e$; then near e we have in the representation T

$$T_{g(\tau)} \approx \mathrm{id}_\mathbf{V} + \tau t, \tag{7.4.1}$$

where

$$t := \frac{\partial}{\partial \tau} T_{g(\tau)} \bigg|_{\tau = 0} = \frac{\partial \beta_A}{\partial \tau} \bigg|_{\tau = 0} \frac{\partial}{\partial \beta_A} T_{g(\beta)} \bigg|_{\beta = 0}. \tag{7.4.2}$$

t is the *generator* of a 1-parameter subgroup in the representation considered. The finite transformations of this subgroup are given by $\exp(\tau t)$, where multiplication in the subgroup is given, in the representation, by $\exp(\tau_1 t) \exp(\tau_2 t) = \exp((\tau_1 + \tau_2)t)$.

It has to be noted that an arbitrary curve through e will in general not be a 1-parameter subgroup, since the latter is fixed already by the generator, i.e., by the values $\partial \beta_A / \partial \tau$ at $\tau = 0$. $\exp(\tau t)$ is the representation of that 1-parameter subgroup which touches the original curve $g(\tau)$ at e, as indicated in Fig. 7.3.

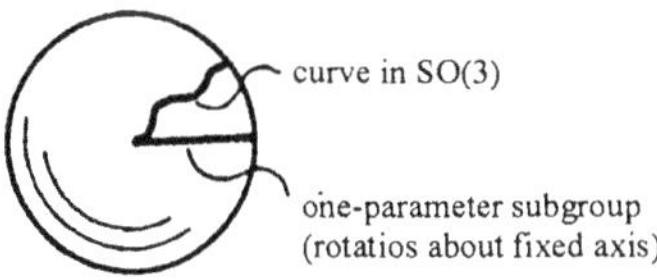

Fig. 7.3. A curve and a 1-parameter subgroup in the model of SO(3), Fig. 7.1

Let $g(\tau)$, $g_1(\tau)$ be two curves through e and c a real number; then the products $g(c\tau) g_1(\tau)$ also lie on a curve through e. In the representation T we get for infinitesimal τ

$$g(c\tau) g_1(\tau) \rightarrow T_{g(c\tau)} T_{g_1(\tau)} \approx \mathrm{id}_\mathbf{V} + \tau(ct + t_1), \tag{7.4.3}$$

where t_1 is defined in analogy to eqs. (7.4.1,2). This shows that the generators form a real vector space $\mathbf{L_V}$ spanned by

$$t_A := \left.\frac{\partial}{\partial\beta_A}T_{g(\beta)}\right|_{\beta\,=\,0}, \qquad\qquad A = 1\dots n. \qquad\qquad (7.4.4)$$

In a faithful representation $\mathbf{L_V}$ would be n-dimensional. Note that even if both curves are actually 1-parameter subgroups, the product curve is, in general, *not* a subgroup but only a curve through e. This is reflected in representations: in general one has $\exp(\tau t)\exp(\tau t_1) \neq \exp(\tau t + \tau t_1)$, equality emerging only if the generators t, t_1 commute.

To show that the generators also form a Lie algebra under commutators we form, from a given curve $g(\tau)$ through e, by conjugation with $h \in \mathcal{G}$ another curve $h\,g(\tau)\,h^{-1}$ through e. (This step is analogous to eq. (7.3.5).) For small τ we then have from eq. (7.4.1)

$$T_{hg(\tau)h^{-1}} = T_h\,T_{g(\tau)}\,T_h^{-1} \approx \mathrm{id}_{\mathbf{V}} + \tau T_h\,t\,T_h^{-1}, \qquad\qquad (7.4.5)$$

i.e., for $t \in \mathbf{L_V}$ also $T_h\,t\,T_h^{-1} \in \mathbf{L_V}$ is a generator. If in this last statement we replace h by $g_1(\tau)$ and consider small τ, so that

$$T_{g_1(\tau)} \approx \mathrm{id}_{\mathbf{V}} + \tau t_1, \qquad\qquad T_{g_1(\tau)}^{-1} \approx \mathrm{id}_{\mathbf{V}} - \tau t_1, \qquad\qquad (7.4.6)$$

then we see that

$$\mathbf{L_V} \ni \left(T_h\,t\,T_h^{-1} - t\right)/\tau \approx [t_1, t]; \qquad\qquad (7.4.7)$$

thus indeed $\mathbf{L_V}$ is a Lie algebra. (This is somewhat cavalier as concerns the limit involved if the representation is infinite-dimensional, but we cannot go into any details here, although we definitely must include infinite-dimensional representations for later purposes.)

We note that the assignment $t \mapsto T_h\,t\,T_h^{-1}$ is, for every $h \in \mathcal{G}$, a bijective linear map of $\mathbf{L_V}$ to itself which is called the *adjoint action* of h on $\mathbf{L_V}$; so, associated to a representation T on $\mathbf{V}$ is a representation on the space $\mathbf{L_V}$ by the adjoint action. Equation (7.4.7) says that the generator of this action corresponding to the curve $g_1(\tau)$ is $[t_1, \cdot]$.

As we remarked above, only in *faithful* representations the vector space $\mathbf{L_V}$ has the same dimension as the group $\mathcal{G}$ itself. Now in most physical examples the group $\mathcal{G}$ is either given as a matrix group anyway, or a finite-dimensional faithful representation can easily be found. The general definition of a Lie group also includes, for good reasons, the case where a finite-dimensional faithful representation does not exist— so-called nonlinear Lie groups. But an infinite-dimensional faithful representation in a function space, with linear differential operators of first order as generators, always exists and is indeed canonically associated with the group. It is called its *regular representation*; its construction will be indicated in sect. 7.7—for the moment only its existence is needed.

So, assuming now that the representation $(\mathbf{V}, T)$ is faithful, the operators of the adjoint action are written Ad_h; the assignment $h \mapsto \mathrm{Ad}_h$ in the n-dimensional space $\mathbf{L_V}$ is called the *adjoint representation* of the group. In the case of the rotation group this representation happens to agree with the defining representation (7.1.6), as follows from eq. (7.3.16).

To every $t \in \mathbf{L_V}$ of the faithful representation T there belongs, conversely, a one-parameter subgroup and thus a generator $t'_t \in \mathbf{L_{V'}}$ for every other—not necessarily faithful—representation $(T', \mathbf{V}')$; and from the representation property we have

$$T'_h\, t'_t\, T'_h{}^{-1} = t'_{T_h\, t\, T_h^{-1}} = t'_{\mathrm{Ad}_h\, t}. \tag{7.4.8}$$

When we put $h = g_1(\tau)$ here as well, use primed analogs of eq. (7.4.6) with $t'_1 = t'_{t_1}$ and also eq. (7.4.7), we get

$$[t'_{t_1}, t'_t] = t'_{[t_1, t]}; \tag{7.4.9}$$

this means that $t \mapsto t'_t$ is a homomorphism of Lie algebras $\mathbf{L_V} \to \mathbf{L_{V'}}$. For the adjoint representation we have $\mathbf{V}' = \mathbf{L_V}$, $T'_h = \mathrm{Ad}_h$, $t'_{t_1} =: \mathrm{ad}_{t_1}$, where $\mathrm{ad}_{t_1} t = [t_1, t]$ (exercise).

To summarize: the Lie algebras $\mathbf{L_V}$ belonging to faithful representations—such always exist—are all isomorphic, the underlying abstract n-dimensional Lie algebra will be written $\mathbf{L}(\mathcal{G})$ and is called *Lie algebra of the Lie group* $\mathcal{G}$; the generators in any representation form a Lie algebra of linear operators which is a *homomorphic image* (a representation) of $\mathbf{L}(\mathcal{G})$.

The theory of Lie groups shows that the component $\mathcal{G}_e$ of unity of a Lie group $\mathcal{G}$ is locally determined fully by its Lie algebra, i.e., by its structure tensor; we shall, however, not try to make precise here the meaning of the restriction 'locally'—it will be illustrated later by way of examples furnished by the rotation and Lorentz groups. It is also true that every finite-dimensional Lie algebra over the field of real numbers is the Lie algebra of some Lie group.

An important step in the search for representations of a connected Lie group is the search for representations of its abstract Lie algebra, where to the product $\circ$ in eqs. (7.2.14,15) there corresponds the commutator of the operators assigned to the elements of the abstract algebra. The representation of the 'finite' group elements is then found by first composing group elements from ones that can be reached from the unit element along one-parameter subgroups and then representing the latter by operators $\exp(\tau t)$ (cf. eq. (7.2.8)), where t is a generator. This is always possible, but in noncompact groups it may happen indeed that a given element is not reached directly by a one-parameter subgroup—see exercise 8 of sect. 8.2 for an example.

The general problem of classifying (real or complex) Lie algebras is unsolved: while it is possible to classify, over $\mathbf{R}$ and $\mathbf{C}$, symmetric tensors S_{AB} (by rank and signature), antisymmetric tensors F_{AB} (by rank), and mixed tensors $T^A{}_B$ (by elementary divisors), and to give normal forms for them in each dimension, it has not been possible to do so for tensors $C^D{}_{AB}$ satisfying eqs. (7.2.16,17) when the dimension exceeds four. (The classification of 3- and 4-dimensional algebras plays a role in the study of gravitational fields with symmetries in General Relativity—cf. Petrov (1969).)

It is only for the so-called *semisimple* Lie algebras that one has a complete classification. These belong to *semisimple* Lie groups, i.e., groups where all Abelian invariant subgroups are discrete. In this case one can easily form analogs to the Casimir operator (7.3.20), the recipe being the following. From the structure tensor one defines the *Killing-Cartan tensor*

$$g_{AB} := C^C{}_{DA}\, C^D{}_{CB} = g_{BA} = \mathrm{Tr}\left(\mathrm{ad}_{X_A}\, \mathrm{ad}_{X_B}\right), \tag{7.4.10}$$

which is invariant under the transformations of the adjoint representation, since already the structure tensor possesses this property (see exercise). According to a theorem of Cartan, semisimplicity is characterized by $\det g_{AB} \neq 0$ and thus by the existence of an inverse tensor g^{AB}. For every representation $X_A \mapsto t_A$ of the algebra it then follows that the *Casimir operator*

$$C := g^{AB}\, t_A\, t_B \tag{7.4.11}$$

is invariant under the adjoint action, i.e., $T_g^{-1} C T_g = C$. If we further define

$$X^A := g^{AB} X_B, \quad t^A := g^{AB} t_B \tag{7.4.12}$$

(the latter constituting a vector operator under the adjoint action), we have $C = g_{AB} t^A t^B$, and the similarly constructed operators

$$\mathrm{Tr}\,(\mathrm{ad}_{X_A}\mathrm{ad}_{X_B}\ldots\mathrm{ad}_{X_C})\,t^A t^B \ldots t^C = C^E{}_{DA}\,C^F{}_{EB}\ldots C^D{}_{GC}\,t^A\,t^B\ldots t^C \tag{7.4.13}$$

are also invariant, i.e., commute with the representation operators T_g. In a (complex) irreducible representation they must be multiples of the unit operator; their eigenvalues serve to classify these representations.

The classification of semisimple Lie algebras is effected by a detailed study of their adjoint representation, the first step being the selection of a maximal set of linearly independent commuting elements. The number of these, the rank of the algebra, is the principal classifying parameter. We cannot go into this theory here, but we note that the classification together with the related construction of their irreducible representations plays a fundamental role for the 'inner symmetries' of elementary particle physics, as can be seen from any modern text on this subject. (One of the earliest systematic presentations of the initial stages of this development is found in Urban (1964).)

The first step in the classification and representation theory just mentioned is easy for the rotation group. As remarked above, the adjoint representation is equivalent to the defining one here, and the generators Λ_μ satisfy eq. (7.2.12), so that obviously only one generator and its multiples can be diagonalized simultaneously. It is customary to diagonalize Λ_3. From the third of eqs. (7.2.5) one finds the eigenvectors

$$\begin{pmatrix} 0 \\ 0 \\ 1 \end{pmatrix} = \mathbf{e}_3, \qquad\qquad \begin{pmatrix} 1 \\ \pm i \\ 0 \end{pmatrix} = \mathbf{e}_1 \pm i\mathbf{e}_2 \tag{7.4.14}$$

$$\Lambda_3\,\mathbf{e}_3 = 0, \qquad\qquad \Lambda_3(\mathbf{e}_1 \pm i\mathbf{e}_2) = \mp i(\mathbf{e}_1 \pm i\mathbf{e}_2), \tag{7.4.15}$$

i.e., the vector $\mathbf{e}_3$ of the rotation axis together with the 'isotropic' vectors $\mathbf{e}_1 \pm i\mathbf{e}_2$ (note $(\mathbf{e}_1\pm i\mathbf{e}_2)^2 = 0$!) contained in the plane orthogonal to $\mathbf{e}_3$. Note that our originally real 3-space has been complexified now, according to the strategy explained in sect. 6.5; otherwise we could not have achieved diagonalization. Therefore, the *Cartan-Weyl basis* of the Lie algebra we are looking for is, in the defining representation, $\Lambda_1 \pm i\Lambda_2$, Λ_3, and is $t_1 \pm it_2$, t_3 in other representations. With this basis we shall continue to work in the next section. Forming complex combinations of the real generators means that we are actually considering now the compexified Lie algebra.

Exercises

1. Show that $\mathrm{ad}_{t_1} t = [t_1, t]$ and that eq. (7.4.9) then becomes the Jacobi identity.

2. Show that the generators of a subgroup form a subalgebra and the generators of an invariant subgroup form an *ideal* (i.e., an ad-invariant subspace) of the Lie algebra.

3. How does the adjoint representation for an Abelian Lie group look like? Show that the adjoint representation of a semisimple Lie group is faithful, and that for a simple group it is irreducible. What about the converse?

Hint: As in the text, base your arguments on a faithful representation. Then the exponentials of generators annihilated by ad generate a central subgroup.

Remark: Observe that although we announced to consider mainly complex representations, the Lie algebra and with it the adjoint representation (for a real Lie group) are real 'by birth'. Thus the irreducibility of the adjoint representation for simple groups is in the sense of real representations and might get lost upon complexification. As we shall see, this is the case for the Lorentz group!

4. Show the Ad-invariance of the structure tensor and of the more general tensors $\phi_{AB...C} := \mathrm{Tr}(t_A t_B ... t_C)$ defined by any representation.

5. Let X_A be the elements of a basis for a Lie algebra and let C^C_{AB} be the structure constants with respect to that basis. Find the matrices for the maps ad_{X_A} and prove the equality of the two versions of g_{AB} given in eq. (7.4.10).

7.5 Unitary Irreducible Representations of SO(3)

a. Generalities on unitary representations
As has been said already, the formulation of physical laws in the regime of quantum theory requires Hilbert spaces, so that the description of relativistic symmetry in that regime involves group representations in Hilbert space and therefore concepts and tools from functional analysis. There are two basic theorems simplifying the representation problem for SO(3) considerably, which we quote, without proof, at the beginning of this section[1]:

1. *Every continuous representation of a compact Lie group in a Hilbert space is equivalent to a unitary representation.*

2. *Every continuous irreducible representation of a compact Lie group in a Hilbert space is finite-dimensional; every continuous unitary representation of a compact Lie group is a direct orthogonal sum of irreducible (and thus finite-dimensional) subrepresentations.*

We begin with a few definitions. A *Hilbert space* **H** is a complex vector space (of infinite dimension, in general) in which there is a *scalar*, or *inner*, *product* associating to each pair of vectors $x \in$ **H**, $y \in$ **H** a complex number $\langle x \,|\, y \rangle \in$ **C**, such that ($\alpha, \beta \in$ **C**, * means complex conjugation)

$$\langle x, \alpha y_1 + \beta y_2 \rangle = \alpha \langle x, y_1 \rangle + \beta \langle x, y_2 \rangle \tag{7.5.1a}$$

$$\langle x, y \rangle = \langle y, x \rangle^* \tag{7.5.1b}$$

$$\| x \|^2 := \langle x, x \rangle > 0 \qquad \text{for} \qquad x \neq 0. \tag{7.5.1c}$$

---

[1]See, e.g., Naimark (1960) or Dieudonné (1977); 'Lie group' may be replaced by 'topological group' here, but we shall not need this generality.

In the infinite-dimensional case, the definition also involves the postulate that $\mathbf{H}$ be complete in the sense of the metric topology defined by the norm $\| x \|$—see the references cited in the preceding footnote, where all necessary conceptual refinements from functional analysis can be found, such as closedness of subspaces, domains of definition for operators, Hermitian and self-adjoint, isometric and unitary operators, etc., and spectral theory.

Vectors x, y with $\langle x, y \rangle = 0$ are called orthogonal to each other. The vectors orthogonal to all vectors of some subset form a linear subspace of $\mathbf{H}$, called the *orthogonal subspace* of that subset. If the subset is a subspace $\mathbf{H}_1 \subset \mathbf{H}$, the orthogonal subspace forms the *orthogonal complement* $\mathbf{H}_2$ of $\mathbf{H}_1$, meaning that we have $\mathbf{H}_1 \cap \mathbf{H}_2 = \{0\}$ and that every vector $x \in \mathbf{H}$ has a unique decomposition $x = x_1 + x_2$, where $x_1 \in \mathbf{H}_1$, $x_2 \in \mathbf{H}_2$ (x_1 is the vector from $\mathbf{H}_1$ for which $\| x - x_1 \|$ is minimum). If $y \in \mathbf{H}$ is decomposed in the same way, we have from the orthogonality between $\mathbf{H}_1$ and $\mathbf{H}_2$:

$$\langle x, y \rangle = \langle x_1, y_1 \rangle + \langle x_2, y_2 \rangle. \tag{7.5.2}$$

In this situation one writes $\mathbf{H} = \mathbf{H}_1 \oplus \mathbf{H}_2$ and calls it *orthogonal direct sum* of $\mathbf{H}_1$ and $\mathbf{H}_2$. (Again this is the 'internal' version, whereas the external direct sum of Hilbert spaces is defined as in sect. 6.5; it is made into a Hilbert space by defining a scalar product as in eq. (7.5.2).

Note that the availability of the orthogonality concept including property (7.5.1c) now enables *orthogonal projections* $x_1 =: P_1 x$, $x_2 =: P_2 x$ to be well-defined by giving one subspace $\mathbf{H}_1$ alone, whereas the *parallel projections* defined in sect. 6.6 required two subspaces for their definition. A direct sum decomposition effected by an idempotent linear operator P as in sect. 6.6 is an orthogonal direct sum iff P is Hermitian (see definition below).

A representation $g \mapsto T_g$ of a group in a Hilbert space is called *unitary* if for all $g \in \mathcal{G}$ and all $x, y \in \mathbf{H}$ we have

$$\langle T_g x, T_g y \rangle = \langle x, y \rangle; \tag{7.5.3}$$

this means that the operators T_g leave scalar products invariant, and such operators, if invertible, are called *unitary*. The unitary operators T_g corresponding to infinitesimal elements of a Lie group are given by

$$T_{g(\tau)} \approx \mathrm{id}_\mathbf{H} + \tau t, \tag{7.5.4}$$

where for the operators t it follows from eq. (7.5.3) that

$$\langle tx, y \rangle + \langle x, ty \rangle = 0. \tag{7.5.5}$$

Thus the generators of unitary representations are *anti-Hermitian*, and multiplication with $\pm i$, by eq. (7.5.1a), makes them *Hermitian*:

$$\langle \pm itx, y \rangle = \langle x, \pm itx \rangle. \tag{7.5.6}$$

The latter are called the *Hermitian generators* of the corresponding one-parameter unitary subgroups of the representation. (Unfortunately, the sign $\pm$ is subject to varying convention.)

For completeness we mention that the *adjoint* (or *Hermitian conjugate*) $A^\dagger$ of a linear operator A is defined by

$$\langle A^\dagger x, y \rangle = \langle x, Ay \rangle. \tag{7.5.7}$$

Hermitian operators are selfadjoint, $A^\dagger = A$, anti-Hermitian ones satisfy $A^\dagger = -A$, unitary ones $A^\dagger = A^{-1}$. Hermitian operators have real eigenvalues, anti-Hermitian operators have pure imaginary eigenvalues (including zero), while the eigenvalues of unitary operators are phase factors (absolute value 1). All these operators, and more generally, operators commuting with their adjoints possess a complete orthonormal system of eigenvectors—i.e., a system spanning all of $\mathbf{H}$. Note that these definitions and statements have to be refined considerably in the infinite-dimensional case. For instance, the isometric property (7.5.3) alone would guarantee invertibility automatically in a finite-dimensional space (write out in matrix form and take determinants!); however, this is not the case in infinite dimension, as simple examples show. Again, the reader is referred to Naimark (1960), Dieudonné (1977), Reed and Simon (1972).

A reducible unitary representation decomposes as a direct orthogonal sum of two unitary subrepresentations, since the orthogonal complement of an invariant subspace is also invariant. Unitary (finite-dimensional) representations are therefore fully reducible.

For *finite* groups, the equivalence expressed in theorem 1 may be shown by a simple device: if $\langle\,,\,\rangle_\circ$ is any scalar product to start with, then

$$\langle x, y \rangle := \sum_{g'} \langle T_{g'} x, T_{g'} y \rangle_\circ \tag{7.5.8}$$

(the sum being extended over all elements of the group) is a new scalar product which is *invariant*; i.e., it satisfies condition (7.5.3) in addition to conditions (7.5.1):

$$\langle T_g x, T_g y \rangle = \sum_{g'} \langle T_{g'} T_g x, T_{g'} T_g y \rangle_\circ = \sum_{g'} \langle T_{g'g} x, T_{g'g} y \rangle_\circ = \sum_{g''} \langle T_{g''} x, T_{g''} y \rangle_\circ = \langle x, y \rangle, \tag{7.5.9}$$

since $g'' = g'g$ runs over $\mathcal{G}$ exactly once if g' does. For Lie groups, the sum appearing in eq. (7.5.8) has to be replaced by an integral over the parameters; the volume element in parameter space has to be chosen such that it is invariant against the (right) translation $g' \mapsto g'g$. Such a *right invariant integral* indeed exists in every Lie group (in fact in every locally compact topological group) and is unique up to a multiplicative constant. The easiest example is the additive group of real numbers with the invariant 'volume element' dx. Since the integration analogous to the summation in eq. (7.5.8) has to be extended over the whole group, the latter should better be compact for the integral to be finite. For the rotation group as parametrized by the Euler angles the (in fact also left) invariant integral is given by

$$\int_0^{2\pi} d\alpha \int_0^{\pi} d\beta \int_0^{2\pi} d\gamma \sin\beta \ldots \tag{7.5.10}$$

(cf. sect. 7.6). Invariant integration is an important tool in the theory of groups and representations (see the references cited above).

Let us add some remarks here about the objects that have been termed *scalar* or *inner product* on a real or complex vector space $\mathbf{V}$ so far. Writing them generally as $\langle\,,\,\rangle$, they all share the requirements

$$\langle x, \alpha y_1 + \beta y_2 \rangle = \alpha \langle x, y_1 \rangle + \beta \langle x, y_2 \rangle \tag{7.5.11}$$

$$\langle x, y \rangle = 0 \quad \text{for all } x \text{ implies } y = 0 \qquad \text{nondegeneracy} \tag{7.5.12}$$

$$\langle \alpha x_1 + \beta x_2, y \rangle = \begin{array}{ll} \alpha \langle x_1, y \rangle + \beta \langle x_2, y \rangle & \text{bilinearity} \qquad (7.5.13a) \\ \qquad\qquad or & \\ \alpha^* \langle x_1, y \rangle + \beta^* \langle x_2, y \rangle & \text{sesquilinearity} \qquad (7.5.13b) \end{array}$$

Orthogonality between two vectors x and y is defined by $\langle x, y \rangle = 0$; and since one wants orthogonality to be a symmetric concept, one requires the last equation always to imply $\langle y, x \rangle = 0$. Over the reals one then must have $\langle x, y \rangle = c\langle y, x \rangle$, while over the complex ground field there is also the possibility $\langle x, y \rangle = c\langle y, x \rangle^*$, where c is real or complex, respectively.

The first possibility implies bilinearity (7.5.11,13a) together with $c^2 = 1$; the resulting postulates

$$\langle x, y \rangle = \langle y, x \rangle \quad \text{or} \quad \langle x, y \rangle = -\langle y, x \rangle \qquad (7.5.14a, b)$$

of symmetry or antisymmetry define a *(pseudo-)Euclidean* or a *symplectic structure* in $\mathbf{V}$, respectively. The real symmetric case has a final subdivision according to the *signature* of the quadratic form $\langle x, x \rangle$: if it is definite, we have the usual proper Euclidean structure, with the standard meaning of scalar product in the positive case. (There is a tendency in mathematics to use the term scalar product on a real vector space only in this sense, but that is ignored by 'relativists'.) Recall that the signature is given by the maximum dimension of a subspace on which the quadratic form is positive or negative, or alternatively by the number of positive and of negative squares that arise when the form is diagonalized: $\langle x, x \rangle = \sum \pm x_i x_i$. Note that in the complex symmetric case there is no notion of signature, one simply has complex-Euclidean structure. Also note that the real proper Euclidean case is distinguished by the fact that a self-orthogonal vector (also called isotropic, or null) coincides with the zero vector, i.e., one has condition (7.5.1c) in the positive case.

The second possibility implies sesquilinearity (7.5.11,13b) and $|c|^2 = 1$, $c = e^{i\gamma}$, $\gamma \in \mathbf{R}$. Redefining $\langle \, , \, \rangle$ by absorbing a factor $e^{-i\gamma/2}$, we arrive at the *Hermiticity* condition

$$\langle x, y \rangle = \langle y, x \rangle^* \qquad (7.5.15)$$

characterizing a *(pseudo-)unitary structure* in $\mathbf{V}$. The final subdivision here is again by the signature of $\langle x, x \rangle$, defined again by the maximum dimension of a subspace of positive or negative definiteness, or by the numbers of positive and negative absolute squares in a diagonalization $\langle x, x \rangle = \sum \pm |x_i|^2$. The definite case is proper unitary structure, without nonzero null vectors, i.e., we have property (7.5.1c) in the positive-definite case (Hilbert space). (Again, there is a tendency in mathematics to restrict the term scalar product to this case only, and this time even relativists do not object.)

Many of the different types of 'scalar product' occur in physics: we mention the usual scalar product in Euclidean 3-space, the scalar product in the Hilbert spaces of quantum mechanics; we have the Lorentz invariant but indefinite scalar product of signature (1,3) or (3,1), as the convention may be; a real scalar product of symplectic type is at the basis of Hamiltonian mechanics; a complex symplectic one will appear in the spinor formalism of chap. 8; a Lorentz invariant symmetric complex-bilinear scalar product was encountered in eq. (6.5.12).

It therefore has to be stressed again that the definition of unitary representations of a group always involves an invariant scalar product of the kind (7.5.1). Thus the 4-vector representation of the Lorentz group is *not* unitary despite the existence of the invariant 4-scalar product $x_i y^i$. The latter is not definite, there are self-orthogonal (=lightlike) vectors, so that the orthogonal space of a lightlike direction contains the latter and a direct sum decomposition does not always arise from a single subspace. Similarly, the representation of $\mathcal{L}_+$ in the complex space of selfdual antisymmetric tensors is *not* unitary despite the invariant scalar square $(\mathbf{E} \pm i\mathbf{B})^2$, since the latter is in the sense of complex-Euclidean rather than unitary geometry.

b. Classification of irreducible representations of SO(3)

We now come to the classification of the irreducible representations of the rotation group. By the theorems quoted at the beginning, we can restrict ourselves to finite-dimensional unitary representations. Our strategy will be to assume, in this subsection, the existence of a representation and derive a number of necessary conditions for it which turn out to characterize it up to equivalence. In the next subsection we then convince ourselves that the equivalence classes are not empty.

Thus consider the generators t_μ of such a representation: they are anti-Hermitian operators on a finite-dimensional Hilbert space $\mathbf{H}$ that satisfy eq. (7.7.3). The *Her-*

mitian generators

$$J_\mu := it_\mu = J_\mu^\dagger, \tag{7.5.16}$$

whose relation to the angular momentum operators of quantum mechanics will be discussed in sect. 7.7, satisfy

$$[J_\mu, J_\nu] = i\,\epsilon_{\mu\nu\lambda}\, J_\lambda \tag{7.5.17}$$

and further, for all $x \in \mathbf{H}$, $\mu = 1, 2, 3$,

$$\langle\, x, J_\mu^2\, x\,\rangle = \langle\, J_\mu x, J_\mu x\,\rangle \geq 0. \tag{7.5.18}$$

By eq. (7.3.20), $\mathbf{J}^2$ is a multiple of the unit operator by irreducibility,

$$\mathbf{J}^2 = \lambda\,\mathrm{id}_\mathbf{H}\,, \tag{7.5.19}$$

where from eq. (7.5.18) we have $\lambda \geq 0$.

The key to the solution of the problem now lies in the analysis of the (real) eigenvalue spectrum of a Hermitian generator, J_3 say. According to a remark made at the end of sect. 7.4, it is advantageous to use the combinations

$$J_\pm := J_1 \pm i\,J_2 = J_\mp^\dagger \quad \text{and} \quad J_3, \tag{7.5.20}$$

which satisfy

$$[J_+, J_-] = 2J_3, \qquad\qquad [J_3, J_\pm] = \pm J_\pm, \tag{7.5.21}$$

$$\mathbf{J}^2 = J_\pm J_\mp \mp J_3 + J_3^2. \tag{7.5.22}$$

Consider the eigenvectors of J_3: they form a complete orthonormal system in $\mathbf{H}$. Let x_m be a normalized eigenvector for the eigenvalue m (called a weight vector for the *weight m*),

$$J_3\, x_m = m\, x_m, \qquad\qquad \|\, x_m\,\| = 1; \tag{7.5.23}$$

then we have from eqs. (7.5.21,22)

$$J_3\, J_\pm\, x_m = (m \pm 1)\, J_\pm\, x_m \tag{7.5.24}$$

$$\langle\, J_\pm\, x_m, J_\pm\, x_m\,\rangle = \langle\, x_m, J_\mp\, J_\pm\, x_m\,\rangle = \lambda \mp m - m^2. \tag{7.5.25}$$

From eq. (7.5.24) it follows that either $J_\pm\, x_m$ is the zero vector, or $m \pm 1$ is also an eigenvalue of J_3. Since the representation is finite-dimensional, there are only finitely many eigenvalues; let j be the largest among them. For an eigenvector to this eigenvalue we must then have, according to eqs. (7.5.24,25),

$$J_+\, x_j = 0, \qquad\qquad \lambda = j + j^2. \tag{7.5.26}$$

We now form the sequence $J_-\, x_j$, $(J_-)^2\, x_j$, ... of (not necessarily normalized) eigenvectors, belonging to the eigenvalues $j - 1$, $j - 2$, ... After a finite number $N - 1 \geq 0$ of applications of J_- a smallest eigenvalue j' will be reached, so that N applications annihilate x_j:

$$(J_-)^{N-1} x_j \neq 0, \qquad J_3 (J_-)^{N-1} x_j = j' (J_-)^{N-1} x_j, \tag{7.5.27}$$

but

$$J_-(J_-)^{N-1}x_j = 0. \tag{7.5.28}$$

Equations (7.5.25,27) then give

$$\lambda = j^2 + j = j'^2 - j', \qquad\qquad j - (N-1) = j', \tag{7.5.29}$$

or $(j+j')(j-j'+1) = 0$, thus $j' = -j$ and hence $2j+1 = N = $ natural number. As possible values for j and λ we therefore obtain

$$j = 0,\ 1/2,\ 1,\ 3/2,\ 2,\ \dots \tag{7.5.30}$$

$$\lambda = j(j+1) = 0,\ 3/4,\ 2,\ 15/4,\ 6,\ \dots \tag{7.5.31}$$

Consider the ladder of eigenvalues and pertinent eigenvectors

$$m = j,\ j-1,\ \dots,\ -j+1,\ -j \tag{7.5.32}$$

$$x_j,\ J_-\,x_j,\ \dots,\ (J_-)^{2j}x_j \tag{7.5.33}$$

for J_3 obtained so far: they are orthogonal and thus linearly independent and span a $(2j+1)$-dimensional subspace of $\mathbf{H}$ which is invariant under the action of J_3 and J_-. We now show that it is also invariant under J_+. Indeed, from eq. (7.5.22) we deduce

$$J_+(J_-)^p = J_+J_-(J_-)^{p-1} = (\mathbf{J}^2 + J_3 - J_3^2)(J_-)^{p-1};$$

applying this to x_j, taking into account eqs. (7.5.19,23), then yields

$$J_+(J_-)^p\,x_j \propto (J_-)^{p-1}\,x_j, \tag{7.5.34}$$

demonstrating that J_+ simply goes up the ladder. The subspace just constructed is thus invariant under $\mathbf{J}$ and also under $\exp(i\boldsymbol{\alpha}\mathbf{J})$. Since we are interested in irreducible representations, the subspace must agree with $\mathbf{H}$.

The eigenvalue j—and with it, all the others—cannot be degenerate in an irreducible representation because if there were a further eigenvector, not proportional to x_j, we could construct a whole new ladder and a corresponding invariant subspace. The eigenvectors for each of the possible eigenvalues m from the list (7.5.32) are complex multiples of one of them, x_m, which we can still restrict by normalization, $\|x_m\| = 1$; it is then unique up to a *phase factor*. From eqs. (7.5.23,24) it follows that

$$J_3\,x_m = m\,x_m$$
$$\tag{7.5.35}$$
$$J_\pm\,x_m = \rho_\pm(m)\,x_{m\pm 1},$$

where we have from eq. (7.5.25)

$$|\rho_\pm(m)|^2 = j(j+1) \mp m - m^2. \tag{7.5.36}$$

The phases of $\rho_\pm$ are still restricted by

$$\rho_\pm(m) = \langle\, x_{m\pm 1}, J_\pm\,x_m\,\rangle = \langle\, J_\mp\,x_{m\pm 1}, x_m\,\rangle = \langle\, x_m, J_\mp\,x_{m\pm 1}\,\rangle^* = \rho_\mp^*(m\pm 1),$$

which follows from $J_\pm^\dagger = J_\mp$ and the orthogonality of the x_m. Since eq. (7.5.36) is consistent with this restriction, a possible choice of phases is

$$\rho_\pm(m) := {}_+\sqrt{j(j+1) \mp m - m^2};\qquad\qquad(7.5.37)$$

The $\{x_m\}$ so restricted will be called a *canonical basis* for the representation which is thus seen to be uniquely determined, up to equivalence (=choice of basis) by the maximal eigenvalue j of the Hermitian generator J_3, called the *highest weight of the representation*—or by its dimension $2j+1$, or by the eigenvalue $j(j+1)$ of the Casimir operator $\mathbf{J}^2$; the possible values of j are among $0, 1/2, 1, 3/2,\ldots$

c. Existence of representations for the highest weights j
So far we have been exploiting the assumed existence of an irreducible representation with maximal eigenvalue j for J_3. This led us to the construction of a canonical basis satisfying eqs. (7.5.35). Everything got fixed to the extent that we can write down matrices for the operators $J_\pm$, J_3 with respect to the canonical basis $\{x_m, m = j, j-1, \ldots, -j+1\}$: we simply multiply the equation $J_\mu x_m = (J_\mu)_{nm} x_n$ scalarly by x_n to obtain the following matrices for J_3, $J_\pm$, $J_1 = 1/2(J_+ + J_-)$, $J_2 = 1/2i(J_+ - J_-)$, and $\mathbf{J}^2$:

$$(J_{3\,nm}) = \begin{pmatrix} j & & & & \\ & j-1 & & 0 & \\ & & \ddots & & \\ & 0 & & -j+1 & \\ & & & & -j \end{pmatrix}, \quad (\mathbf{J}^2{}_{nm}) = \begin{pmatrix} 1 & & & & \\ & 1 & & 0 & \\ & & \ddots & & \\ & 0 & & 1 & \\ & & & & 1 \end{pmatrix} \cdot j(j+1),$$

$$(J_{+\,nm}) = \begin{pmatrix} 0 & \rho_+(j-1) & & & \\ & 0 & \rho_+(j-2) & & 0 \\ & & 0 & \ddots & \\ & 0 & & \ddots & \rho_+(-j) \\ & & & & 0 \end{pmatrix},$$

$$(J_{-\,nm}) = \begin{pmatrix} 0 & & & & \\ \rho_-(j) & 0 & & 0 & \\ & \rho_-(j-1) & 0 & & \\ & & \ddots & \ddots & \\ & 0 & & \rho_-(-j+1) & 0 \end{pmatrix}. \qquad (7.5.38)$$

If we now can check that these matrices indeed do satisfy our starting relations (7.5.17) or (7.5.21), we have also shown the *existence* of representations, for all admissible values of j, of the Lie *algebra* of the rotation group SO(3). We leave this basically straightforward checking to the reader, however. More subtle is the question whether all these representations of the Lie algebra exponentiate to yield representations of the *group* itself—remember that our infinitesimal considerations again so far led us to *necessary* conditions. We postpone this question to sect. 7.6 and continue the present subsection with the simplest examples.

$j = 0$ ... 1-dimensional representation; $\mathbf{J} = 0$, i.e., this is the trivial representation, and there are no other 1-dimensional representations.

$j = 1$ ... 3-dimensional representation; eqs. (7.5.38,20) give

$$J_3 = \begin{pmatrix} 1 & 0 & 0 \\ 0 & 0 & 0 \\ 0 & 0 & -1 \end{pmatrix} \qquad J_1 = \frac{1}{\sqrt{2}} \begin{pmatrix} 0 & 1 & 0 \\ 1 & 0 & 1 \\ 0 & 1 & 0 \end{pmatrix} \qquad J_2 = \frac{1}{\sqrt{2}} \begin{pmatrix} 0 & 1 & 0 \\ -1 & 0 & 1 \\ 0 & -1 & 0 \end{pmatrix}$$

$$\mathbf{J}^2 = \begin{pmatrix} 2 & 0 & 0 \\ 0 & 2 & 0 \\ 0 & 0 & 2 \end{pmatrix} \tag{7.5.39}$$

This representation is equivalent to the defining or adjoint representation: J_3 is the diagonalized form of $i\Lambda_3$ from eq. (7.2.5). This must be so also on general grounds—there is at most one equivalence class of 3-dimensional irreducible representations of SO(3).

$j = 2$ ... a 5-dimensional representation; it turns out to be equivalent to the representation of SO(3) in the space of symmetric trace-free tensors $T^{\mu\nu}$ (not tensor fields!). Such tensors form an invariant subspace under the product representation $g \mapsto R_g \otimes R_g$. The generators $\mathbf{t}$ of the representation $R_g \otimes R_g$ can be read off from

$$(1 + \tau\boldsymbol{\alpha}\Lambda) \otimes (1 + \tau\boldsymbol{\alpha}\Lambda) \approx 1 \otimes 1 + \tau\boldsymbol{\alpha}\left(1 \otimes \Lambda + \Lambda \otimes 1\right) = 1 \otimes 1 + \tau\boldsymbol{\alpha}\,\mathbf{t}.$$

We therefore have

$$\mathbf{t}^2 = (1 \otimes \Lambda_\mu + \Lambda_\mu \otimes 1)(1 \otimes \Lambda_\mu + \Lambda_\mu \otimes 1) = \Lambda^2 \otimes 1 + 1 \otimes \Lambda^2 + 2\Lambda_\mu \otimes \Lambda_\mu. \tag{7.5.40}$$

The defining representation $g \mapsto R_g$ has $j(j+1) = 2$, hence $\Lambda^2 = -2 \cdot 1$, so that only $\Lambda_\mu \otimes \Lambda_\mu$ remains to be computed:

$$(\Lambda_\mu \otimes \Lambda_\mu)^{\alpha\beta}{}_{\rho\sigma} = \epsilon_{\mu\alpha\rho}\,\epsilon_{\mu\beta\sigma} = \delta_{\alpha\beta}\delta_{\rho\sigma} - \delta_{\alpha\sigma}\delta_{\beta\rho}. \tag{7.5.41}$$

Therefore we get for a symmetric trace-free tensor $T^{\mu\nu}$

$$(\mathbf{t}^2)^{\alpha\beta}{}_{\mu\nu}\,T^{\mu\nu} = -4T^{\alpha\beta} + 2\delta^{\alpha\beta}\,T^\mu{}_\mu - 2T^{\beta\alpha} = -6T^{\alpha\beta},$$

or symbolically

$$\mathbf{t}^2 \cdot T = -2(2+1)T = -j(j+1)T.$$

In the 5-dimensional space of trace-free symmetric tensors, $\mathbf{t}^2$ is thus indeed a multiple of the unit operator, with eigenvalue corresponding to $j = 2$; this representation is therefore irreducible with highest weight $j = 2$.

Generally it turns out that the irreducible representations of *integer weight* j are just the ones that obtain from reducing tensor representations, i.e., from forming totally symmetric and trace-free tensors of degree j. This will result later from spinor algebra. It will also turn out that these tensor representations are the only irreducible ones of the Lie algebra that survive transition to the group level.

$j = 1/2 \ldots$ 2-dimensional representation; the matrices for the generators in a canonical basis are $\mathbf{J} = 1/2\,\boldsymbol{\sigma}$, where the matrices

$$\sigma_3 = \begin{pmatrix} 1 & 0 \\ 0 & -1 \end{pmatrix}, \qquad \sigma_1 = \begin{pmatrix} 0 & 1 \\ 1 & 0 \end{pmatrix}, \qquad \sigma_2 = \begin{pmatrix} 0 & -i \\ i & 0 \end{pmatrix} \qquad (7.5.42)$$

are the *Pauli spin matrices*.

Beyond the commutation relations which they satisfy from our construction,

$$[\sigma_\mu, \sigma_\nu] = 2i\epsilon_{\mu\nu\lambda}\sigma_\lambda, \qquad (7.5.43)$$

they also satisfy the *anticommutator relations (Clifford algebra relations for the Euclidean 3-metric $\delta_{\mu\nu}$, cf eq. (9.1.17))*

$$\{\sigma_\mu, \sigma_\nu\} := \sigma_\mu\sigma_\nu + \sigma_\nu\sigma_\mu = 2\delta_{\mu\nu}\,\mathbf{1}, \qquad (7.5.44)$$

from which, together with eq. (7.5.43), we get

$$\sigma_\mu\sigma_\nu = \delta_{\mu\nu}\,\mathrm{id} + i\epsilon_{\mu\nu\lambda}\sigma_\lambda. \qquad (7.5.45)$$

Equation (7.5.45) comprises the multiplication table of Hamilton's *quaternion units* $\mathbf{1}, -i\sigma_\mu$.

This representation is called *spinor representation* of SO(3)—or rather, of its Lie *algebra* so(3). Just as all representations with half integer weights, it *cannot* be obtained by reducing tensor representations, and is thus a genuinely new result of representation theory, first obtained—for arbitrary rotation groups SO(n,$\mathbf{C}$)—by E. Cartan in 1913. It was independently rediscovered by W. Pauli in the process of setting up the quantum mechanical theory of the spinning nonrelativistic electron; hence the name (coined by P. Ehrenfest). It will be treated in more detail in the next section where we shall see that although it does *not* lead to a representation, in the strict sense, of the rotation *group*, it is very useful from the mathematical point of view. The reason why it may be—and in fact is—necessary in physics will be given in sect. 9.2.

d. Reduction of reducible representations

We finally consider *reducible representations* and their *reduction*. Let $\mathbf{H}$ be a Hilbert space on which SO(3) acts by a possibly reducible representation T. How can we find the invariant irreducible subspaces and reduce the representation in a systematic fashion? We are looking for a decomposition

$$\mathbf{H} = \sum_{j\alpha} \oplus \mathbf{H}_{j\alpha}$$

of $\mathbf{H}$ as an orthogonal direct sum of irreducible invariant subspaces $\mathbf{H}_{j\alpha}$, indexed by the highest weight j which we know characterizes irreducible representations up to equivalence, and by a further index α which may be necessary if the same irreducible representation occurs several times in equivalent guise. If we carry out, for a given j, the sum over α we obtain an *isotypic* subspace $\mathbf{H}_j$ of type j (cf. sect. 6.6, exercise 6).

The solution of this problem may be effected again by considering the eigenvalues and eigenvectors of the generator J_3. Notice first that in a reducible representation the Casimir operator $\mathbf{J}^2$ is not necessarily a multiple of $\mathrm{id}_{\mathbf{H}}$; rather, its eigenspaces are just the isotypic subspaces. Let us stress the parallelism of our considerations to some about angular momentum in quantum mechanics and use the *Dirac ket notation* $|\ldots\rangle$ for the vectors in $\mathbf{H}$, putting symbols j, m, α, $\ldots$ inside to further characterize the vectors. Then the vectors of $\mathbf{H}_j$ are just those satisfying

$$\mathbf{J}^2 |j, \ldots\rangle = j(j+1)|j, \ldots\rangle, \tag{7.5.46}$$

where the possible values of j are contained in the set $0, 1/2, 1, \ldots$ Within each $\mathbf{H}_j$ we have eigenvectors of J_3:

$$J_3 |j, m, \ldots\rangle = m |j, m, \ldots\rangle, \tag{7.5.47}$$

where the dots indicate a possible multiplicity of the weight m. We now pick one of them, taking, e.g., $m = j$: i.e., we pick a solution $|j, j, 1\rangle$ of

$$J_3 |j, j, \ldots\rangle = j |j, j, \ldots\rangle, \qquad\qquad J_+ |j, j, \ldots\rangle = 0 \tag{7.5.48}$$

and apply to it the ladder operator J_- to construct recursively the vectors

$$|j, m-1, 1\rangle = \frac{J_-}{\rho_-(m)} |j, m, 1\rangle, \qquad m = j, j-1, \ldots, -j; \tag{7.5.49}$$

they form the canonical basis for an invariant irreducible subspace $\mathbf{H}_{j1}$. If there is a second independent eigenvector $|j, j, 2\rangle$ solving eqs. (7.5.48), we can again construct vectors $|j, m, 2\rangle$ spanning $\mathbf{H}_{j2}$, and so on. (We could have started equally well from some other weight m, but note that eqs. (7.5.48) imply eq. (7.5.46).) The subspaces $\mathbf{H}_{j\alpha}$ are not uniquely determined but depend on the actual choice of the independent vectors $|j, j, 1\rangle$, $|j, j, 2\rangle$, $\ldots$ —one frequently chooses them to be orthogonal to each other and/or possibly as eigenvectors of some further operator commuting with J_3 and $\mathbf{J}^2$ that might be available in a concrete situation.

Consider, as an example, the representation of SO(3) in the space of tensors $T^{\mu\nu}$. In analogy to the considerations for the Lorentz group we can form the projectors P_A and P_S onto the subspaces of antisymmetric and symmetric tensors, respectively:

$$P_A{}^{\alpha\beta}{}_{\mu\nu} := \frac{1}{2}\left(\delta^{\alpha}{}_{\mu}\delta^{\beta}{}_{\nu} - \delta^{\alpha}{}_{\nu}\delta^{\beta}{}_{\mu}\right), \qquad P_S{}^{\alpha\beta}{}_{\mu\nu} := \frac{1}{2}\left(\delta^{\alpha}{}_{\mu}\delta^{\beta}{}_{\nu} + \delta^{\alpha}{}_{\nu}\delta^{\beta}{}_{\mu}\right) \tag{7.5.50}$$

as well as the projector to the subspace of multiples of the Euclidean metric tensor $\delta^{\mu\nu}$:

$$P_\delta{}^{\alpha\beta}{}_{\mu\nu} := \frac{1}{3}\delta^{\alpha\beta}\delta_{\mu\nu}. \tag{7.5.51}$$

Then the projector to the symmetric traceless tensors is $P_{SS} := P_S - P_\delta$ (because of $P_S P_\delta = P_\delta P_S = P_\delta$ this is idempotent, $P_{SS}^2 = P_{SS}$). By eqs. (7.5.40,41) we now have

$$\mathbf{J}^2 = -\mathbf{t}^2 = 4 \cdot \mathbf{1} \otimes \mathbf{1} - 2\Lambda_\mu \otimes \Lambda_\mu = 4(P_S + P_A) - 2(3P_\delta - (P_S - P_A)) =$$
$$= 6(P_S - P_\delta) + 2P_A = 2 \cdot 3P_{SS} + 1 \cdot 2P_A + 0 \cdot 1P_\delta. \tag{7.5.52}$$

Since $\mathbf{1} \otimes \mathbf{1} = P_{SS} + P_A + P_\delta$, we have achieved a complete reduction, the operators P_{SS}, P_A, P_δ projecting onto irreducible subspaces belonging to the highest weights 2, 1, 0: none of these can occur more than once on dimensional grounds ($9 = 5 + 3 + 1$). The equivalence of the antisymmetric tensor representation to the vector representation ($j = 1$) is of course given, in the tensorial setting, by the *-operation associated to $\epsilon_{\mu\nu\lambda}$. One can construct now the canonical basis in each of the irreducible subspaces, but we refrain from doing so here.

Exercises

1. Show that the direct sum decomposition effected by a linear idempotent operator $P = P^2$ is orthogonal iff P is Hermitian: $P^\dagger = P$.

2. Show that the quadratic form $\phi_{AB}\xi^A\xi^B$ associated with the symmetric tensor $\phi_{AB} := \mathrm{Tr}(t_A t_B)$ belonging to a *unitary* representation is always negative (the ξ^A being assumed real).

   Hint: Use your knowledge about eigenvalues of anti-Hermitian operators.

3. Show, for a compact simple group, that all Ad-invariant symmetric tensors ϕ_{AB} over the Lie algebra are multiples of the Killing-Cartan tensor g_{AB} of eq. (7.4.10).

   Hint: Consider $t_B^A = g^{AC}t_{CB}$ as (the matrix of) an operator on the Lie algebra. Since according to the previous exercise the quadratic form associated with the Killing-Cartan tensor is negative definite (negative and nondegenerate), the operator t_B^A has real Ad-invariant eigenspaces. Now you can use the result of exercise 3 of sect. 7.4 and Schur II.

4. (a) Show that the existence of an invariant nondegenerate *bi*linear form B for a (finite-dimensional, complex) representation $(\mathbf{V}, T)$—in eqs. (7.5.11,12,13a,14) we simply wrote $\langle x, y \rangle$ instead of $B(x,y)$— is the same thing as saying that the representation T and the contragredient representation $\widetilde{T}$ (on the dual vector space $\widetilde{\mathbf{V}}$) are equivalent. (You have seen an example of this in eq. (6.4.9), hence a hint is to use a component-matrix notation: $B(x,y) = \mathrm{x}^\top B\mathrm{y}$. Note also that if B is degenerate it still gives an intertwiner.) (b) By transposing the matrix version of the invariance condition on B, conclude that also the transposed bilinear form $B^\top$ is invariant and does the same job. (c) Assuming *irreducibility* now, conclude that B and $B^\top$, if nonzero, are automatically nondegenerate and proportional to each other (cf. exercise 4 of sect. 6.6). Taking transposes again, conclude that B *must* be *symmetric* or *antisymmetric*, eq. (7.5.14), making the representation (complex-)orthogonal or symplectic, respectively.

5. (a) Show, similarly, that the existence of an invariant nondegenerate *sesqui*linear form A is the same thing as saying that the representation is equivalent to the complex-conjugate of the contragredient representation, a degenerate A still giving an intertwiner. (b) By Hermitian conjugation of the matrix version of the invariance condition on A, conclude that also the Hermitian conjugate $A^\dagger$ is invariant and does the same job. (c) Assuming *irreducibility* now, conclude that A and $A^\dagger$, if nonzero, are automatically nondegenerate and proportional to each other. Taking Hermitian conjugates again, conclude that the factor of proportionality is a phase factor $\exp(i\alpha)$, so that $\exp(i\alpha/2)A$ is a *Hermitian* sesquilinear form, eq. (7.5.15), unique up to a *real* factor, making the representation pseudo-unitary.

6. (The 'ABC Theorem'.) Assume that for an irreducible (finite-dimensional complex) representation A and B from the previous exercises exist both. Conclude that this puts you into the situation of exercise 5, sect. 6.6, and that A and B^*C must be proportional. From the properties $A = A^\dagger$, $B^\top = \beta B$, $CC^* = \gamma\mathrm{id}$—where $\beta^2 = \gamma^2 = 1$—and the freedom left in the choice of A, B, C conclude that for a given choice of B one can arrange for $B = A^\top C$. Finally show that with $y := (Cx)^*$ one has, from all these relations,

$$A(y, y) = \beta\gamma A(x, x). \tag{7.5.53}$$

This equation couples the signature of A and sign $\beta\gamma$, in that definiteness of A implies $\beta\gamma = +1$, while $\beta\gamma = -1$ implies 'neutral' signature for A (i.e., equal numbers of $+1$ and -1 in its diagonalized form).

7. From the canonical unit vectors $\mathbf{e}_1$, $\mathbf{e}_2$, $\mathbf{e}_3$ of the defining representation in $\mathbf{R}^3$, construct a canonical basis, starting with $x_0 = \mathbf{e}_3$ and using all conventions made.

   Remark: Clearly, the complex vectors of eq. (7.4.14) reappear: they are null and nonorthogonal in the sense of the Euclidean scalar product if the latter is extended *bilinearly* into the complex domain; they are orthogonal and normalizable, however, in the sense of the *sesquilinearly* extended scalar product that makes $\mathbf{C}^3 = $ complexified $\mathbf{R}^3$ into a Hilbert space.

8. Consider the representation of SO(3) furnished by totally symmetric tensors of degree p. What is its dimension? What is the dimension if we additionally impose tracelessness (i.e., the vanishing of all contractions)? What are the eigenvalues of the Casimir operator in the latter space? Conclude irreducibility! Illustrate the ABC theorem in this example.

   Hint: Use a computation similar to eqs. (7.5.40,41).

9. In an isotypic representation of type j, two weight vectors belonging to the same weight m are not necessarily orthogonal. Show that the scalar products $\langle j\,m\,1\,|\,j\,m\,2\rangle$ do not depend on m.

7.6 SU(2), Spinors, and Representation of Finite Rotations

a. SU(2) and its homomorphism to SO(3)

Having constructed all possible representation spaces and canonical bases therein for the unitary irreducible representations, we still have to find representing matrices for *finite* rotations from those for the Hermitian generators. We already know the method for doing this in principle: to represent the group element $g(\boldsymbol{\alpha})$, we consider the 1-parameter subgroup $g(\tau\boldsymbol{\alpha})$ containing it. If $t = \boldsymbol{\alpha}\,\mathbf{t}$ is the generator of that subgroup, then the representing operator we are looking for is

$$g(\boldsymbol{\alpha}) \mapsto T_{g(\boldsymbol{\alpha})} = \exp(\boldsymbol{\alpha}\,\mathbf{t}) = \exp(-i\boldsymbol{\alpha}\mathbf{J}). \tag{7.6.1}$$

The exponential has to be evaluated for the matrices $\mathbf{J}$ found in sect. 7.5c. In principle, this may be done using the Sylvester formula (cf. Smirnov 1964) for functions of a matrix, since the eigenvalues of $-i\boldsymbol{\alpha}\,\mathbf{J}$ are known: by rotational covariance, they differ from those of J_3 only by the factor $-i|\boldsymbol{\alpha}|$. This route is practical only[1] for the lowest values $j = 0, 1/2, 1$, and we shall find a different one later.

The case $j = 0$ is trivial; the case $j = 1$ was treated in exercise 1 of sect. 7.2, leading to the defining representation R_g. For $j = 1/2$ we have

$$\boldsymbol{\alpha}\,\mathbf{J} = \frac{1}{2}\,\boldsymbol{\alpha}\,\boldsymbol{\sigma} = \frac{1}{2}\,\alpha_\mu\sigma_\mu,$$

$$(\boldsymbol{\alpha}\boldsymbol{\sigma})^2 = \alpha_\mu\alpha_\nu\sigma_\mu\sigma_\nu = \frac{1}{2}\,\alpha_\mu\alpha_\nu\,(\sigma_\mu\sigma_\nu + \sigma_\nu\sigma_\mu) = \boldsymbol{\alpha}^2\cdot\mathbf{1}. \tag{7.6.2}$$

Putting $\boldsymbol{\alpha} = \alpha\mathbf{n}$, where $\mathbf{n}^2 = 1$, we obtain

$$(-i\boldsymbol{\alpha}\,\mathbf{J})^2 = -\left(\frac{\alpha}{2}\right)^2\cdot\mathbf{1}, \qquad (-i\boldsymbol{\alpha}\,\mathbf{J})^3 = +\left(\frac{\alpha}{2}\right)^3\mathbf{n}\,\boldsymbol{\sigma},$$

and so on. Therefore the series expansion for the exponential function gives

$$\exp(-i\boldsymbol{\alpha}\,\mathbf{J}) = \mathbf{1}\cdot\cos\frac{\alpha}{2} - i\,\mathbf{n}\,\boldsymbol{\sigma}\sin\frac{\alpha}{2} =: \mathrm{U}(\boldsymbol{\alpha}) \tag{7.6.3}$$

$$\mathrm{U}(\boldsymbol{\alpha}) = \begin{pmatrix} \cos\dfrac{\alpha}{2} - in_3\sin\dfrac{\alpha}{2} & -i(n_1 - in_2)\sin\dfrac{\alpha}{2} \\[2ex] -i(n_1 + in_2)\sin\dfrac{\alpha}{2} & \cos\dfrac{\alpha}{2} + in_3\sin\dfrac{\alpha}{2} \end{pmatrix}. \tag{7.6.4}$$

Thus the matrices representing finite rotations in the spinor representation have been found. By construction, they are unitary, and in addition they are *unimodular*: $\det\mathrm{U}(\boldsymbol{\alpha}) = \det\exp(-i\boldsymbol{\alpha}\boldsymbol{\sigma}/2) = \exp(\mathrm{Tr}(-i\boldsymbol{\alpha}\boldsymbol{\sigma}/2)) = 1$, as $\mathrm{Tr}\,\sigma_\mu = 0$. (Unitarity alone, $\mathrm{U}^\dagger\mathrm{U} = \mathbf{1}$, only implies $|\det\mathrm{U}| = 1$).

There is something peculiar about the 'representation' $\mathrm{U}(\boldsymbol{\alpha})$, however. For instance, when we compose two rotations through the angle π about an axis $\mathbf{n}$ we get a rotation through 2π, which is the unit element of the group SO(3); by contrast,

$$\mathrm{U}(\pi\mathbf{n})\,\mathrm{U}(\pi\mathbf{n}) = \mathrm{U}(2\pi\mathbf{n}) = -\mathbf{1}. \tag{7.6.5}$$

---

[1]See, however, A. Torruella, J. Math. Phys. *16*, 1637 (1975).

Generally, while the $U(\boldsymbol{\alpha})$ do have representation properties when finite but sufficiently small rotations are composed, this is not true when too large rotations are involved. The set of matrices $U(\boldsymbol{\alpha})$ forms a group only if the domain $0 \le |\boldsymbol{\alpha}| \le \pi$ is extended to become $0 \le |\boldsymbol{\alpha}| \le 2\pi$, so that the set of rotations SO(3) gets *doubly covered* since we have $U(-(2\pi - \alpha)\mathbf{n}) = -U(\alpha\mathbf{n})$. Note, in particular, that the identification $R(\pi\mathbf{n}) = R(-\pi\mathbf{n})$ is undone here: $U(\pi\mathbf{n}) = -U(-\pi\mathbf{n})$. To each rotation thus there correspond *two* unitary unimodular matrices, $U(\boldsymbol{\alpha})$ and $-U(\boldsymbol{\alpha})$. This situation is, strictly speaking, not included in our definition of representation, and is sometimes referred to as *two-valued representation*, or representation up to a sign. While we would have to exclude $j = 1/2$, and similarly all the other half integers, from our list of possible values if it were only for representations of SO(3) in the strict sense, these 'representations' are important both from a mathematical and from a physical point of view. We shall see the former immediately and give the reasons for physical relevance in sect. 9.2, where we discuss the principle of relativity in the quantum context. Generalities on many-valued representations will be given in sect. 7.10.

Our situation here is similar to one in complex analysis where, e.g., the function $w = z^{1/2}$ is single-valued and continuous either only locally or if we introduce for its domain of definition a Riemann surface covering the complex plane twice. Modern mathematics tends to dislike expressions such as 'many-valued functions', 'many-valued representations', replacing them by 'functions on a covering space', 'representation of a covering group' (or of some other *extension* of the group (cf. Appendix A)), 'projective representation', 'ray representation' (cf. sects. 7.10 and 9.2).

Now, letting $\boldsymbol{\alpha}$ range over $0 \le |\boldsymbol{\alpha}| \le 2\pi$, the $U(\boldsymbol{\alpha})$ range over the whole group SU(2) of all *unitary unimodular matrices*. To see that indeed none of those is omitted, write for a complex 2×2 matrix

$$U = \begin{pmatrix} a & b \\ c & d \end{pmatrix};$$

then unitarity requires $c = -\lambda b^*$, $d = \lambda a^*$, $|a|^2 + |b|^2 = 1$, for some complex λ satisfying $|\lambda| = 1$, while unimodularity narrows this down to $\lambda = 1$, so that

$$U = \begin{pmatrix} a & b \\ -b^* & a^* \end{pmatrix} \quad \text{whith} \quad |a|^2 + |b|^2 = 1. \tag{7.6.6}$$

This implies $|\operatorname{Re} a| \le 1$, and we can find exactly one α, $0 \le \alpha \le 2\pi$, having $\operatorname{Re} a = \cos \alpha/2$. A unit vector $\mathbf{n}$ is then determined from $\operatorname{Im} a = -n_3 \sin \alpha/2$, $\operatorname{Re} b = -n_2 \sin \alpha/2$, $\operatorname{Im} b = -n_1 \sin \alpha/2$: it is unique except for $U = \pm\mathbf{1}$ where it is arbitrary.

The last mentioned circumstance already shows that the group manifold of SU(2) is that of a 3-sphere $\mathbf{S}_3$. This becomes more evident when $\operatorname{Re} a$, $\operatorname{Im} a$, $\operatorname{Re} b$, $\operatorname{Im} b$ are interpreted as Cartesian coordinates in $\mathbf{R}^4$: then eq. (7.6.6) tells us that we are on the unit sphere. Since U and $-U$ belong to the same rotation, we can look at the group manifold of SO(3) as $\mathbf{S}_3$ with antipodal points identified (which is also the same as real projective 3-space). Going back again by restricting to values $0 \le |\boldsymbol{\alpha}| \le \pi$ we can forget the identification except for the points of the boundary 2-sphere $|\boldsymbol{\alpha}| = \pi$ (the equator of $\mathbf{S}_3$; see Fig. 7.4). Stereographic projection of the 3-hemisphere gives back, topologically speaking, the earlier model of Fig. 7.1.

This picture allows us to study the topological situation in some detail. The continuous curve $g(\tau)$ on $\mathbf{S}_3 = \mathrm{SU}(2)$ that leads from $\alpha = 0$ to $\alpha = 2\pi$ becomes closed up, i.e., becomes a *loop* when

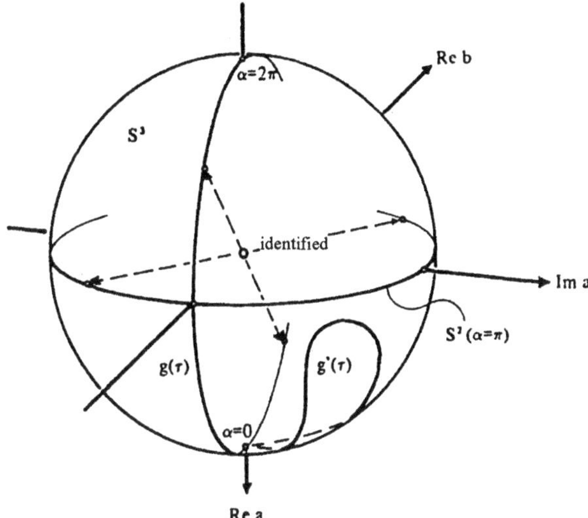

Fig. 7.4. $SU(2) = S_3$ and SO(3). The coordinate Im b has been omitted

the identifications of antipodal points are made, and one that *cannot* continuously be deformed (shrunk) to the point e, in contrast to the loop $g'(\tau)$. If such nonshrinkable loops exist, the manifold is called *multiply connected*. In SO(3) there are two classes of loops, where the members of a class are continuously deformable into each other inside SO(3): one class is of the (homotopy) type $g'(\tau)$, continuously deformable to the trivial loop consisting of e alone; the other class is of the type $g(\tau)$; one says that SO(3) is doubly connected. (See Boerner (1955) for more details: omitting one dimension allows for visualization, but topological properties often depend sensitively on dimension!) By going to the *universal covering group* $S_3 = SU(2)$ of SO(3), undoing the identifications, we achieved that within SU(2) all loops are contractible to a point, the ones formerly uncontractible having been opened up: one says that SU(2) is *simply connected*. Note that in addition to being a topologically simpler manifold, SU(2) is a Lie group; and instead of saying that we have a 2-valued representation of SO(3) by SU(2), we note that the above antipodal identification also has a group theoretical aspect: U and $-$U together form a coset in SU(2) with respect to the discrete invariant (in fact, central) subgroup $\{1, -1\} = \mathcal{Z}_2$, so that $SO(3) \cong SU(2)/\mathcal{Z}_2$ is a homomorphic image of SU(2). We will describe this homomorphism explicitly below.

Generally, by a covering group of a Lie group $\mathcal{G}$ one means a Lie group $\mathcal{G}'$ together with a continuous covering homomorphism $\mathcal{G}' \to \mathcal{G}$ such that for every $g \in \mathcal{G}$ the inverse image is discrete. If $\mathcal{G} = \mathcal{G}_e$ is connected, then among the *connected* covering groups of it there exists one (and, up to isomorphism, only one) which is simply connected—its *universal covering group* $\widetilde{\mathcal{G}} = \widetilde{\mathcal{G}}_e$. (See Dieudonné (1972), Chevalley (1946), or Pontryagin (1966) for proof.) In our example $\mathcal{G} = SO(3)$ we gave $\widetilde{\mathcal{G}}$ in a concrete fashion as the matrix group SU(2). It should be pointed out, however, that there exist linear (i.e., matrix) Lie groups for which the universal covering group, while being a Lie group in the abstract sense, is not a linear Lie group, i.e., has no faithful finite-dimensional representation. Also, if $\mathcal{G} = \mathcal{G}_e$ is compact, the universal covering group need not be compact, the inverse image of each element then being an infinite set. However, it is a theorem of H. Weyl (see Helgason 1962) that the universal covering group of a connected compact *semisimple* Lie group *is* compact. Another theorem is that every compact Lie group has a faithful finite-dimensional representation. Our example of SO(3) illustrates both theorems.

It also illustrates the fact that a Lie group and any one of its covering groups possess the same Lie algebra and are isomorphic in sufficiently small neighborhoods of the unit elements ('local isomorphism'). In the large, one has a homomorphism $\widetilde{\mathcal{G}} \to \mathcal{G}$ whose kernel is a discrete central subgroup of $\widetilde{\mathcal{G}}$ and is isomorphic to the *fundamental*, or *first homotopy group* of $\mathcal{G}$, which is made up of the homotopy (=continuous deformation) classes of (continuous) loops through the unit element in $\mathcal{G}$. In our example we already described the two classes of loops and associated them with the elements $\mathbf{1}, -\mathbf{1}$ of SU(2). Generally, the Lie algebra determines a (connected) Lie group uniquely if the latter is required to be simply connected in addition. Other candidates are obtained from the simply connected one by quotienting with respect to a discrete central subgroup, which introduces identifications and nontrivial classes of loops.

The uniqueness of $\widetilde{\mathcal{G}}$ also says that a connected and simply connected group has no connected covering group other than itself. This implies that it has no (locally continuous) *discretely* multi-valued representations if the set of representing operators is to be connected. There are, however, examples of multiply connected Lie groups without multivalued *finite*-dimensional representations. (See Cartan 1966; the argument given there indeed breaks down in the infinite-dimensional case.)

The spinor representation $g(\boldsymbol{\alpha}) \mapsto (\pm)\mathrm{U}(\boldsymbol{\alpha})$ yields the most compact version of the *multiplication table of the rotation group*: from

$$\left(\cos\frac{\alpha_1}{2}\,\mathbf{1} - i\sin\frac{\alpha_1}{2}\,\mathbf{n}_1\boldsymbol{\sigma}\right)\left(\cos\frac{\alpha_2}{2}\,\mathbf{1} - i\sin\frac{\alpha_2}{2}\,\mathbf{n}_2\boldsymbol{\sigma}\right) = \cos\frac{\alpha_3}{2}\,\mathbf{1} - i\sin\frac{\alpha_3}{2}\,\mathbf{n}_3\boldsymbol{\sigma} \qquad (7.6.7)$$

one can read off $\boldsymbol{\alpha}_3 = \alpha_3\mathbf{n}_3$ once the left hand-side has been multiplied out and rearranged into the form of the right-hand side with the help of eq. (7.5.45). Without having a matrix representation, Hamilton in 1843 (and before him, Gauss in 1819) discovered the *quaternions* $\mathbf{1}, (-i\boldsymbol{\sigma})$ when looking for 'numbers' whose multiplication would correspond to the composition of spatial rotations in the same way that the multiplication of complex numbers corresponds to the composition of rotations in the plane. Starting from an ansatz $a \cdot \mathbf{1} - i\,\mathbf{a}\,\boldsymbol{\sigma}$ for the 'hypercomplex numbers' in question he arrived at the rules (7.5.45). However, he was not able to fully settle the matter, and it required contributions from O. Rodrigues and A. Cayley to do so—see Altmann (1986) for the complete history; $(\cos\alpha/2, \sin\alpha/2\,n_\mu)$ are sometimes called the Euler-Rodrigues parameters, and the a, b in eq. (7.6.6) are called Cayley-Klein parameters.

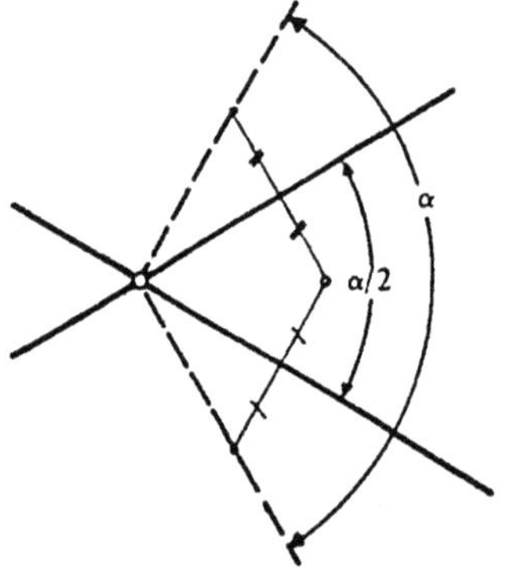

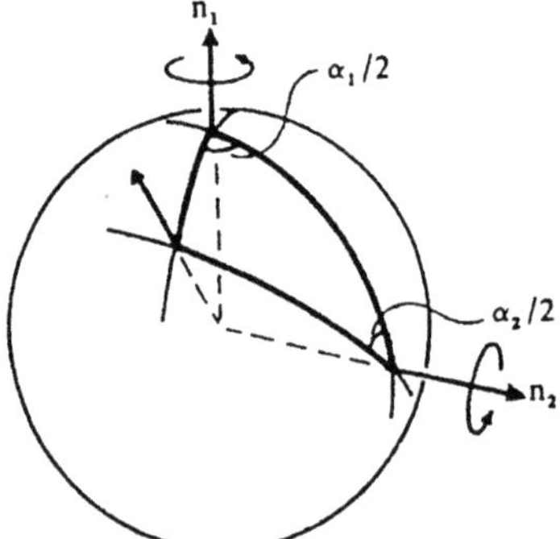

a) Replacing a rotation by two reflections   b) Composition of two rotations

Fig. 7.5. Rotations as composed from reflections

The occurrence of half-angles can be understood geometrically according to Fig. 7.5a: every rotation through the angle α around the axis $\mathbf{n}$ may be replaced by the succession of two plane reflections, the planes intersecting along the axis and enclosing the angle $\alpha/2$ but being arbitrary otherwise. To compose two rotations with axes $\mathbf{n}_1, \mathbf{n}_2$ and angles α_1, α_2, one replaces each of them by two reflections, choosing as one of the planes in each case the plane spanned by the axes. Then upon composition the two reflections in the plane of $\mathbf{n}_1$, $\mathbf{n}_2$ cancel and there remain two reflections, i.e., one rotation. The resulting axis is the intersection of the remaining planes. In Fig. 7.5b the traces of these planes on the unit sphere are shown; the formulae for α_3, $\mathbf{n}_3$ that result from eqs. (7.6.7), (7.5.45) reflect the spherical trigonometry of the figure.

We can use the composition law of SO(3) just described geometrically together with eq. (7.3.4) to give the promised proof that the group SO(3) is *simple*, i.e., has no nontrivial invariant subgroup: from eq. (7.2.12) it follows that there is no nontrivial ideal in the Lie algebra (cf. exercise 2 of sect. 7.4 for the relation between invariant subgroups and ad-invariant subalgebras (ideals)), but this only implies that the connected component of an invariant subgroup consists of the unit element alone—while we here show that there is not even a discrete invariant subgroup. So let $\mathcal{N}$ be a nontrivial invariant subgroup: it contains a rotation $R(\alpha) \neq \mathbf{1}$. Since all powers of the latter are also in $\mathcal{N}$, we may assume $2\alpha \geq \pi$. From eq. (7.3.4) we see that $\mathcal{N}$ contains all rotations $R(\beta)$ with $\beta = \alpha$ as well. To see that we get not only all directions of axes but also all angles, we now form $R(\beta)R(\alpha) = R(\gamma) \in \mathcal{N}$ and vary the angle between α and β from π to 0: by continuity, γ then takes all values from 0 to $2\alpha \geq \pi$. But this says that $\mathcal{N} = $ SO(3), the other trivial possibility.

Because of SU(2) $= \mathbf{S}_3$ we can read the equation $U(\alpha_1)\,U(\alpha_2) = U(\alpha_3)$ also as follows: the rotation $g(\alpha_1)$ transforms, via $U(\alpha_1)$, the *point* $U(\alpha_2) \in \mathbf{S}_3$ into the point $U(\alpha_3)$. Since the Cartesian coordinates $\operatorname{Re} a_2$, $\operatorname{Im} a_2, \ldots$ get transformed *linearly* into the Cartesian coordinates $\operatorname{Re} a_3$, $\operatorname{Im} a_3, \ldots$ this transformation is a (special kind of) *rotation* of the 4-dimensional space. If in the whole 4-space we put

$$x_4 = r\operatorname{Re} a, \qquad x_3 = r\operatorname{Im} a, \qquad x_1 = r\operatorname{Re} b, \qquad x_2 = r\operatorname{Im} b \tag{7.6.8}$$

and write down the rotationally invariant 4-volume element $dx_4\,dx_1\,dx_2\,dx_3 = r^3\,dr\,d\mathcal{V}$, then $d\mathcal{V}$ is the surface element of $\mathbf{S}_3$, also invariant under all 4-rotations. A fortiori, $d\mathcal{V}$ then is the *invariant volume element* on SU(2) and SO(3) mentioned in sect. 7.5a. Indeed, if a, b are expressed by the Euler angles α, β, γ instead of α, one arrives, after some calculation, at eq. (7.5.10). Since $U(\alpha_1)$ was written above to the left of $U(\alpha_2)$, our argument shows the left invariance of eq. (7.5.10). We can, however, simply interchange the roles of α_1, α_2 and obtain right invariance. (It is a general theorem that for a compact group right (left) invariant integrals are also left (right) invariant. Compactness is only sufficient but not necessary for this to happen; however, there are easy examples where it does *not* happen.)

The *right translations* $U \mapsto U \cdot U(\alpha)$ and *left translations* $U \mapsto U(\alpha) \cdot U$ of $\mathbf{S}_3$ are by no means the most general rotations in four dimensions. Rather, the latter form the 6-parameter Lie group SO(4), while the former each form a 3-parameter group only. However, if we consider the set of transformations $U \mapsto U(\alpha) \cdot U \cdot U^{-1}(\beta)$, where α, β each varies independently over $0 \leq |\alpha| \leq 2\pi, 0 \leq |\beta| \leq 2\pi$, we obtain a 6-parameter group of transformations whose elements may be assigned to the pairs $(g(\alpha), g(\beta))$ of the direct product SU(2) $\times$ SU(2) such that the representation property holds. We thus get a homomorphism of SU(2) $\times$ SU(2) $\to$ SO(4) which in fact is onto, both groups being connected. The identity of SO(4), $U \mapsto U$, results only from the pairs $(\mathbf{1}, \mathbf{1})$ and $(-\mathbf{1}, -\mathbf{1})$. So we again have a local isomorphism, both groups have isomorphic Lie algebras; in the large, they are different since SO(4) is doubly connected like SO(3), while SU(2) $\times$ SU(2) is simply connected and is thus the universal covering group of SO(4). (To contract a loop in the product, just project it onto the factors and contract the projections: the product of the contracting projections gives a contraction of the original loop.) The local isomorphism is the isomorphism SO(4) $\cong$ SU(2) $\times$ SU(2)$/\mathcal{Z}_2$, with the discrete subgroup $\mathcal{Z}_2 = \{(\mathbf{1}, \mathbf{1}), (-\mathbf{1}, -\mathbf{1})\}$. Inside SU(2) $\times$ SU(2) there are further discrete central subgroups: $\mathcal{Z}_2' = \{(\mathbf{1}, \mathbf{1}), (-\mathbf{1}, \mathbf{1})\}$, $\mathcal{Z}_2'' = \{(\mathbf{1}, \mathbf{1}), (\mathbf{1}, -\mathbf{1})\}$ and $\mathcal{V}_4 = \{(\mathbf{1}, \mathbf{1}), (-\mathbf{1}, \mathbf{1}), (\mathbf{1}, -\mathbf{1}), (-\mathbf{1}, -\mathbf{1})\}$; with them, the quotients SU(2) $\times$ SU(2)$/\mathcal{Z}_2' \cong$ SO(3) $\times$ SU(2), SU(2) $\times$ SU(2)$/\mathcal{Z}_2'' \cong$ SU(2) $\times$ SO(3) and SU(2) $\times$ SU(2)$/\mathcal{V}_4 \cong$ SO(3) $\times$ SO(3) $\cong$ SO(4)$/\{E, -E\}$ may be formed (E is the 4×4 unit matrix). Locally, all these groups are isomorphic.

In many of the applications of group theory to particle physics, only the Lie algebras of the groups involved play a role. However, in sect. 9.4b we shall have occasion to use a global argument to obtain the correct spectrum of *helicities*. As we pointed out in sect. 6.1, there are situations where one uses the group manifold itself in a direct physical sense.

We have seen that instead of a genuine representation $SO(3) \to SU(2)$ we have a genuine homomorphism in the opposite direction, $SU(2) \to SO(3)$. The latter can be made explicit in another way, effectively eliminating the trigonometric functions between eqs. (7.6.3,4,6) and (7.1.6). This will be useful in the development of a systematic spinor algebra in sect. 8.3. Here we give just a matrix version. Thus, with every 3-vector $\mathbf{x}$ we associate the 2×2 matrix

$$X = \mathbf{x}\,\boldsymbol{\sigma}. \tag{7.6.9}$$

Since $\mathbf{x}$ is real, while the σ_μ are Hermitian and traceless, we have

$$X = X^\dagger, \qquad\qquad\qquad \mathrm{Tr}\,X = 0. \tag{7.6.10}$$

Conversely, every Hermitian and traceless matrix X may be written as in eq. (7.6.9) with a unique real $\mathbf{x}$ which can be calculated from X via

$$\mathbf{x} = \frac{1}{2}\,\mathrm{Tr}\,X\,\boldsymbol{\sigma}, \tag{7.6.11}$$

since from eq. (7.5.45) we have

$$\mathrm{Tr}\,\sigma_\mu\,\sigma_\nu = 2\,\delta_{\mu\nu}. \tag{7.6.12}$$

X further satisfies (cf. eq. (7.6.2))

$$X^2 = \mathbf{x}^2 \cdot \mathbf{1}, \qquad\qquad\qquad \det X = -\mathbf{x}^2. \tag{7.6.13}$$

We now pick $U \in SU(2)$ and form

$$X' = U\,X\,U^{-1} = U\,X\,U^\dagger. \tag{7.6.14}$$

The matrix X' is again Hermitian and traceless:

$$X'^\dagger = (U\,X\,U^\dagger)^\dagger = U\,X^\dagger\,U^\dagger = U\,X\,U^\dagger$$

$$\mathrm{Tr}\,X' = \mathrm{Tr}\,U\,X\,U^{-1} = \mathrm{Tr}\,U^{-1}\,U\,X = \mathrm{Tr}\,X = 0$$

and defines, according to (7.6.9,11), a linear transformation $\mathbf{x} \mapsto \mathbf{x}' = R\mathbf{x}$ which, because of

$$\mathbf{x}'^2 \cdot \mathbf{1} = X'^2 = U\,X\,U^{-1}\,U\,X\,U^{-1} = U\,X^2\,U^{-1} = \mathbf{x}^2 \cdot \mathbf{1}$$

or

$$-\mathbf{x}'^2 = \det X' = \det U \det X \det U^\dagger = \det X = -\mathbf{x}^2,$$

must be orthogonal. The homomorphism property of this assignment is easily seen. If we write

$$x'_\mu = R_{\mu\nu}\,x_\nu \tag{7.6.15}$$

and compare $X' = x'_\mu \sigma_\mu = R_{\mu\nu}\,x_\nu\,\sigma_\mu$ with $U\,x_\nu\,\sigma_\nu\,U^\dagger$ we see that

$$R_{\mu\nu}\,\sigma_\mu = U\,\sigma_\nu\,U^\dagger = U\,\sigma_\nu\,U^{-1}. \tag{7.6.16}$$

From this we get $R_{\mu\nu}$ explicitly, using eq. (7.6.12):

$$R_{\mu\nu} = \frac{1}{2}\,\mathrm{Tr}\,\sigma_\mu\,\mathrm{U}\,\sigma_\nu\,\mathrm{U}^\dagger = \frac{1}{2}\,\mathrm{Tr}\,\sigma_\mu\,\mathrm{U}\,\sigma_\nu\,\mathrm{U}^{-1}. \tag{7.6.17}$$

It is obvious that our assignment is a continuous map from the connected 3-dimensional Lie group SU(2) to the 3-dimensional Lie group O(3); so on general grounds the image must be the connected component SO(3). (We could use the parametrization by eq. (7.6.3) and (7.1.6) to see this explicitly; the property $\det \mathrm{R} = 1$ can also be seen from the more tensorial approach described in sect. 8.4 which says that R is essentially $\mathrm{U}\otimes\tilde{\mathrm{U}}$: one then just applies a well-known determinant formula.)

It is possible to express, conversely, U in terms of R explicitly. Since for every 2×2 matrix M the following identity holds (exercise)

$$\sigma_\nu\,\mathrm{M}\,\sigma_\nu = 2\,\mathrm{Tr}\,\mathrm{M} \cdot \mathbf{1} - \mathrm{M}, \tag{7.6.18}$$

we get from eq. (7.6.16) on multiplying by σ_ν: $R_{\mu\nu}\,\sigma_\mu\,\sigma_\nu = \mathrm{U}\,(2\,\mathrm{Tr}\,\mathrm{U}^\dagger \cdot \mathbf{1} - \mathrm{U}^\dagger) = (2\,\mathrm{Tr}\,\mathrm{U})\,\mathrm{U} - \mathbf{1}$. Taking traces yields $2(\mathrm{Tr}\,\mathrm{U})^2 = 2(1 + \mathrm{Tr}\,\mathrm{R})$, and therefore

$$\mathrm{U} = \pm\frac{\mathbf{1} + R_{\mu\nu}\,\sigma_\mu\,\sigma_\nu}{2\sqrt{1 + \mathrm{Tr}\,\mathrm{R}}}. \tag{7.6.19}$$

It is not hard to check that this agrees with the earlier form $\mathrm{U}(\alpha)$; observe that the formula breaks down for rotations through 180°, which is necessary on topological grounds—there cannot be a consistent continuous distinction between the $+$ and $-$ sign, otherwise SU(2) would have to consist of two disconnected topological copies of SO(3)!

b. Spinors of SO(3)

The vectors of the 2-dimensional representation space on which the $\mathrm{U}(\alpha)$ act are called *spinors*. By definition, under a rotation $\mathrm{R}(\alpha)$ a spinor u transforms as

$$u \mapsto u' = \mathrm{U}(\alpha)\,u. \tag{7.6.20}$$

The question of which sign for U has to be taken here does not arise: for quantum mechanical purposes, a phase factor will be open anyway,[1] and for mathematical purposes we will now take SU(2) as the primary object. Observe: composing a representation (in the strict sense) of SO(3) with the homomorphism SU(2)→SO(3) gives a representation of SU(2), but the latter has more representations. We will show that to every j from our list there is an irreducible representation of the whole group SU(2) in the strict sense. Only those among them where the 'disturbing' kernel $\mathscr{Z}_2$ is represented trivially will give strict representations of SO(3). We also point out that when direct sums are formed, we must not mix integer and half-integer j if we want to get objects whose transformation law leaves open a phase factor only.

The scalar product in the Hilbert space sense, invariant under the transformations (7.6.20), is

$$\langle\, u, v \,\rangle = u_1^* v_1 + u_2^* v_2, \tag{7.6.21}$$

if u is specified by components $u_A = (u_1, u_2)^\top$ relative to a canonical basis (7.5.48,49). The $\mathrm{U}(\alpha)$ being unimodular ($\det\mathrm{U} = 1$), there is an analog to the ϵ- tensor introduced

---

[1] Of course, the *same* choice must be made for all spinors $u, v, \ldots$, so that sums $u + v$ transform in the same way as their summands!

in eqs. (5.5.6,7), namely an ϵ-spinor of degree 2 (since spinor space is 2-dimensional); it thus defines an invariant *bi*linear form (cf. eq. (7.5.14b))

$$\epsilon^{AB} u_A u_B = u_1 v_2 - u_2 v_1 \qquad (7.6.22)$$

for any two spinors u, v. Note that, contrary to this, the scalar product (7.6.21) is *sesqui*linear!

In sects. 5.4 and 5.5 we developed the tensor algebra over an arbitrary vector space. We can now apply this to spinor space, form higher-degree spinors and investigate their transformation behavior, i.e., form Kronecker products $U(\boldsymbol{\alpha}) \otimes U(\boldsymbol{\alpha}) \otimes \ldots$ and study their reduction.

Let us illustrate this first quite explicitly by the simplest example, the reduction of the representation $g(\boldsymbol{\alpha}) \mapsto U(\boldsymbol{\alpha}) \otimes U(\boldsymbol{\alpha})$. Let the spinors u, v be given by the components $(u_1, u_2)^\top$, $(v_1, v_2)^\top$; then $(u_1 v_1, u_1 v_2, u_2 v_1, u_2 v_2)^\top$ are the components of $u \otimes v$. If further $u' = U u$, $v' = U v$, then

$$
\begin{pmatrix} u_1' v_1' \\ u_1' v_2' \\ u_2' v_1' \\ u_2' v_2' \end{pmatrix}
=
\begin{pmatrix}
a^2 & ab & ab & b^2 \\
-ab^* & |a|^2 & -|b|^2 & a^*b \\
-ab^* & -|b|^2 & |a|^2 & a^*b \\
b^{*2} & -a^*b^* & -a^*b^* & a^{*2}
\end{pmatrix}
\begin{pmatrix} u_1 v_1 \\ u_1 v_2 \\ u_2 v_1 \\ u_2 v_2 \end{pmatrix},
\qquad (7.6.23)
$$

where U is specified by eq. (7.6.6). For the antisymmetric (cf. eq. (5.5.3)) part $u_{[A} v_{B]}$ we read off

$$u_1' v_2' - u_2' v_1' = (|a|^2 + |b|^2)(u_1 v_2 - u_2 v_1) = u_1 v_2 - u_2 v_1, \qquad (7.6.24)$$

as claimed by eq. (7.6.22). The antisymmetric spinors of degree 2 thus transform according to the trivial representation of SU(2). In the subspace of symmetric spinors (cf. eq. (5.5.5)) lies the part $u_{(A} v_{B)}$. When we choose the basis such that its components become $(u_1 v_1, (u_1 v_2 + u_2 v_1)/\sqrt{2}, u_2 v_2)^\top$, then eq. (7.6.23) becomes

$$
\begin{pmatrix} (u_1' v_2' - u_2' v_1')/\sqrt{2} \\ u_1' v_1' \\ (u_1' v_2' + u_2' v_1')/\sqrt{2} \\ u_2' v_2' \end{pmatrix}
=
\begin{pmatrix}
1 & 0 & 0 & 0 \\
0 & a^2 & \sqrt{2}ab & b^2 \\
0 & -\sqrt{2}ab^* & |a|^2 - |b|^2 & \sqrt{2}a^*b \\
0 & b^{*2} & -\sqrt{2}a^*b^* & a^{*2}
\end{pmatrix}
\begin{pmatrix} (u_1 v_2 - u_2 v_1)/\sqrt{2} \\ u_1 v_1 \\ (u_1 v_2 + u_2 v_1)/\sqrt{2} \\ u_2 v_2 \end{pmatrix}.
$$

$$(7.6.25)$$

Full reduction is already achieved: one checks that for an infinitesimal rotation around the 3-axis ($b \approx 0$, $a \approx 1 - i\alpha/2$) the generator J_3 takes the form $\mathrm{diag}\,(1, 0, -1)$ within the 3-dimensional subrepresentation, which characterizes an irreducible representation of highest weight $j = 1$. The appearance of the operators $J_\pm$ shows, in addition, that $(u_1 v_1, (u_1 v_2 + u_2 v_1)/\sqrt{2}, u_2 v_2)^\top$ refers to a canonical basis.

In complete analogy one can form totally symmetric spinors of higher degree. The removal of trace parts as in the tensorial case (cf. exercise 5 of sect. 7.5) is not possible here: there is no invariant spin metric that would give a nonzero result on symmetric spinors—the scalar product (7.6.21) is sesquilinear instead of bilinear, while the bilinear ϵ-spinor (7.6.22) is antisymmetric. Indeed, the space of totally symmetric spinors of any given degree p is *irreducible* and of dimension $p+1$, corresponding to $j = p/2$, as we are going to show now.

The dimension is found by counting the number of independent components of a totally symmetric spinor of degree p. Since the order of the indices is irrelevant here, we may take as independent components those where the first p_1 indices are equal to 1 and the remaining $p_2 = p - p_1$ ones are equal to 2. Since we have the possibilities $p_1 = 0, 1, \ldots, p$ we have $p + 1$ independent components. We now investigate the spectrum of the generator J_3 in this space. Generally, in the space of all degree p spinors, an infinitesimal rotation about the μ-axis has the form

$$\mathrm{U}(\tau\mathbf{e}_\mu) \otimes \ldots \otimes \mathrm{U}(\tau\mathbf{e}_\mu) \approx \mathbf{1} \otimes \ldots \otimes \mathbf{1} - i\tau J_\mu,$$

where

$$J_\mu = \frac{1}{2}(\sigma_\mu \otimes \ldots \otimes \mathbf{1} + \ldots + \mathbf{1} \otimes \ldots \otimes \sigma_\mu). \tag{7.6.26}$$

Now let $u^\pm$ be the eigenspinor of J_3 in 2-spinor space (i.e., the eigenspinor of $\sigma_3/2$) for the eigenvalue $\pm 1/2$. Then $u^\pm \otimes u^\pm \otimes \ldots \otimes u^\pm$ belongs to the subspace of totally symmetric spinors of degree p and verifies to be an eigenspinor of J_3 for the eigenvalue $\pm p/2$, annihilated by $J_\pm$. From the general procedure of sect. 7.5b we then know that the representation space must contain an irreducible subrepresentation of highest weight $p/2$, which is $p + 1$-dimensional. There is no room left, therefore, for anything else in our space. Note that for even $p = 2j$ the sign ambiguity $\pm\mathrm{U}$ drops out in the transformation law and we get a representation of SO(3) in the strict sense; while such is not the case when p is odd, where the kernel $\mathcal{Z}_2$ of the homomorphism SU(2)$\to$SO(3) is represented nontrivially.

c. Representation matrices for finite rotations

We now finally use the above realization of the irreducible representations of highest weight j to obtain an explicit form of the representing matrices for finite rotations— or finite elements of SU(2)—for *all* values of j. (Up to now we wrote them down only for $j = 0, 1/2, 1$.)

A symmetric spinor of degree p transforms as the pth tensorial power $u_A u_B \ldots u_C$ of a 2-component spinor $u = u_1 u^+ + u_2 u^-$; its independent components are the $p + 1$ monomials

$$(u_1)^p, \ (u_1)^{p-1}u_2, \ \ldots, \ u_1(u_2)^{p-1}, \ (u_2)^p,$$

if the basis in the symmetric tensor product space is taken as consisting of all

$$\binom{p}{p_1} u^+_{(A} u^+_B \ldots u^-_{C)}$$

with p_1 factors u^+ and p_2 factors u^-, where, as above, $p_1 + p_2 = p$ and $p_1 = 0, 1, \ldots, p$. One can see that these basis spinors are eigenspinors of J_3, eq. (7.6.26), for the eigenvalues $m = (p_1 - p_2)/2$. Under rotation, the monomials get transformed into the corresponding monomials $(u_1')^p, \ldots, (u_2')^p$ formed from the transformed components $u' = \mathrm{U}(\boldsymbol{\alpha})\,u$,

$$u_1' = a\,u_1 + b\,u_2$$
$$u_2' = -b^*\,u_1 + a^*\,u_2 \;,$$

so that simply carrying out all multiplications and comparing coefficients will yield the matrix elements we are looking for. To get the matrices in explicitly unitary form and conforming to our phase conventions of sect. 7.5, we observe that the basis vectors above turn out to almost constitute a *canonical* basis in the symmetric spinor space. More precisely, if we take the spinor $(p!)^{1/2}u_A^+ u_B^+ \ldots u_C^+$ as our $|\,j\,j\,\rangle$, it is not hard to show by induction that the basis vectors $|\,j\,m\,\rangle$ constructed from it according to eq. (7.5.49), with J_- formed from eq. (7.6.26), are

$$\binom{p}{p_1}^{1/2} u_{(A}^+ u_B^+ \ldots u_{C)}^-. \tag{7.6.27}$$

It follows that the correct normalization, *including phases*, of the above monomials is given by

$$\binom{p}{p_1}^{1/2} (u_1)^{p_1} (u_2)^{p_2}. \tag{7.6.28}$$

Our choice for the normalization of $|\,j\,j\,\rangle$ corresponds to the choice of

$$\frac{1}{p!}\,u_A^*\,u_B^* \ldots u_C^*\,v_A\,v_B \ldots v_C = \frac{1}{p!}\,(u_A^*\,v_A)^p = \frac{1}{p!}\langle u, v \rangle^p \tag{7.6.29}$$

for the (obviously invariant) scalar product between $u_A\,u_B \ldots u_C$ and $v_A\,v_B \ldots v_C$ in our tensor space (cf. also Appendix B.11). Using the binomial theorem, we can check this directly:

$$(u_1^*\,v_1 + u_2^*\,v_2)^p = \sum_{p_1=0}^{p} \binom{p}{p_1} (u_1^*\,v_1)^{p_1} (u_2^*\,v_2)^{p_2} =$$

$$= \sum_{p_1} \binom{p}{p_1}^{1/2} (u_1^*)^{p_1} (u_2^*)^{p_2} \binom{p}{p_1}^{1/2} (v_1)^{p_1} (v_2)^{p_2}.$$

The elements of the representing matrices are now obtained by expanding

$$\binom{p}{p_1}^{1/2} (u_1')^{p_1} (u_2')^{p_2} = \binom{p}{p_1}^{1/2} (a\,u_1 + b\,u_2)^{p_1} (-b^*\,u_1 + a^*\,u_2)^{p_2}$$

and reading off the coefficients of $\sqrt{p!}(u_1)^{q_1} (u_2)^{q_2}/\sqrt{q_1!\,q_2!}$. We still reinstate j, m by $p_1 = j + m$, $p_2 = j - m$ and write, similarly, $q_1 = j + n$, $q_2 = j - n$; then we get the matrix elements $(m, n = -j, \ldots, +j)$:

$$\mathrm{D}_{mn}^{(j)}(\boldsymbol{\alpha}) = \sum_{\ell}(-1)^{\ell}\,\frac{\sqrt{(j+m)!(j-m)!(j+n)!(j-n)!}}{(j-m-\ell)!(j+n-\ell)!(m-n+\ell)!\ell!}\,a^{j+n-\ell}a^{*j-m-\ell}b^{m-n+\ell}b^{*\ell}.$$

$$\tag{7.6.30}$$

Here ℓ is an integer to be summed from 0 up to $j - m$, but all its values that would lead to factorials of negative integers are to be omitted. The index (j) affixed to the matrix elements $D^{(j)}_{mn}(\alpha)$ indicates the irreducible representation of highest weight j.

We will use the symbol $D^{(j)}$ not only for the matrix with elements (7.6.30) but also for the equivalence class of irreducible representations of highest weight j. It is easy to convince oneself that eq. (7.6.30) reproduces the earlier results for $j=1/2, 1$. When a given representation $(\mathbf{H}, T)$ is reduced as described in sect. 7.5d, $\mathbf{H} = \sum \oplus \mathbf{H}_{j\alpha}$, and in each irreducible subspace $\mathbf{H}_{j\alpha}$ a canonical basis $\{\,|\,j\,m\,\alpha\,\rangle\,\}$ is constructed, then these vectors transform according to

$$T_g\,|\,j\,m\,\alpha\,\rangle = \sum_n D^{(j)}_{nm}(g)\,|\,j\,n\,\alpha\,\rangle. \tag{7.6.31}$$

Summing up, the 2-dimensional spinor representation permitted us to get all irreducible representations of SU(2) by reducing its Kronecker, or tensorial, powers. A representation of this kind is called a *fundamental representation*.

Exercises

1. Express the spinor representation of a rotation by Euler angles, decomposing it into three rotations, taking their spinor representations and multiplying together. (When you compare with eq. (7.6.4) this must be consistent with your solution to the exercise of sect. 7.1.)

2. With the result of the foregoing exercise; form the expressions of eqs. (7.6.8) and calculate dV in terms of Euler angles. (Cf. eq. (7.5.10).)

   Hint: For the solution of exercise 1 one obtains $a = \exp(-i(\alpha + \gamma)/2)\cos \beta/2$, $b = -i\exp(-i(\alpha - \gamma)/2)\sin \beta/2$; the computation is then simplified by using the calculus of differential forms and the relation $|a|^2 + |b|^2 = 1$. One gets
   $dx_4\,dx_1\,dx_2\,dx_3 = -1/4\,d(ra)\,d(ra^*)\,d(rb)\,d(rb^*) = r^3\,dr\,dV$,
   $dV = -1/2(a\,da^*\,d(b\,db^*) + b\,db^*\,d(a\,da^*)) = \ldots = 1/8\sin\beta\,d\alpha\,d\beta\,d\gamma$.

3. Show that the sets of transformations $U \mapsto U(\alpha)\,U$ and of transformations $U \mapsto U\,U(\beta)$ intersect only in the discrete set $U \mapsto \pm U$. The union of both sets thus forms a transformation group with not less that 6 parameters.
   Hint: First take $U = \mathbf{1}$, and then use Schur.

4. Show that the transformation $U \mapsto U(\alpha)\,U\,U^{-1}(\beta)$ results in the identity only if $U(\alpha) = \mathbf{1} = U(\beta)$ or $U(\alpha) = -\mathbf{1} = U(\beta)$.

5. Which of the transformations $U \mapsto U(\alpha)\,U\,U^{-1}(\beta)$ leave a given point $U \in$ SU(2) $= \mathbf{S}_3$ fixed? Show that these transformations form a group isomorphic to SO(3), as should be clear geometrically.

6. Verify the identity $\sigma_\nu\,M\,\sigma_\nu = 2\,\mathrm{Tr}\,M \cdot \mathbf{1} - M$ for every 2×2 matrix M.

7. We shall generalize the transformation (7.6.14) in two ways. One will appear in sect. 8.2; consider here the transformation $X \mapsto U X U^{-1}$ without the restriction on U to be unitary. Then traceless X are carried into traceless X', and also $X^2 = \mathbf{x}^2 \cdot \mathbf{1}$ remains unchanged, while the Hermiticity $X = X^\dagger$ would get lost and thus is not assumed here from the outset. The transformation thus gives *complex rotations* of complex $\mathbf{x}$, explicitly by the second version of eq. (7.6.17). The identity results only if $U = \lambda \cdot \mathbf{1}$, and by restricting to $\det U = 1$ the ambiguity is reduced to $\lambda = \pm 1$. With this restriction on U to belong to the group $SL(2,\mathbf{C})$ of all complex unimodular matrices we also have $\mathrm{Tr}\, U^{-1} = \mathrm{Tr}\, U$ and thus the inversion formula (7.6.19). All in all, this leads to an isomorphism $SL(2, \mathbf{C})/\mathcal{Z}_2 \cong SO(3, \mathbf{C})$. Study all details of these considerations!

8. When in the considerations of the previous exercise the U are restricted to be real unimodular matrices and the $\mathbf{x}$ are restricted to $x_1 = $ real, $x_3 = $ real, $x_2 = $ pure imaginary, X and X' are real. Show that in this way an isomorphism $SL(2, \mathbf{R})/\mathcal{Z}_2 \cong SO_e(2, 1)$ is obtained.

9. Find all invariant subgroups of $SU(2)$.

   Hint: Use our solution of the analogous problem for $SO(3)$.

10. Verify that the spinors (7.6.27) satisfy eqs. (7.5.35,37).

7.7   Representations on Function Spaces

Relativistic electrodynamics has shown that besides tensors also *tensor fields* are necessary for the formulation of covariant laws of nature. In this section we begin to analyze the relation between fields on space-time and representation theory by starting with the simplest case, *scalar* fields on Euclidean 3-space and their behavior under rotations. The treatment of spinor and tensor fields will follow in the next section.

The behavior of a scalar field under rotations $\Phi(\mathbf{x})$ is completely analogous to the one for scalar fields on Minkowski space as described in sect. 3.4:

$$\Phi'(\mathbf{x}') = \Phi(\mathbf{x}) \qquad \text{or} \qquad \Phi'(\mathbf{x}) = \Phi(R^{-1}\mathbf{x}). \qquad (7.7.1)$$

Here $\Phi'(\mathbf{x}')$ may be regarded as the same field as $\Phi(\mathbf{x})$ but referred to new, rotated coordinates $\mathbf{x}'$ (passive transformation), or Φ' defines a new scalar field taking the same value at $\mathbf{x}$ which Φ takes at $R^{-1}\mathbf{x}$ (active transformation).

The complex-valued scalar fields on $\mathbf{R}^3$ form an infinite-dimensional complex vector space $\mathbf{H} = \mathbf{H}(\mathbf{R}^3)$ if addition and multiplication with complex scalars are defined pointwise; i.e., the field $\alpha\Phi + \beta\Psi$ is defined by

$$(\alpha\,\Phi + \beta\,\Psi)(\mathbf{x}) = \alpha\,\Phi(\mathbf{x}) + \beta\,\Psi(\mathbf{x}) \qquad \alpha, \beta \in \mathbf{C}. \qquad (7.7.2)$$

Now every rotation assigns to Φ the field Φ' given by eq. (7.7.1). This assignment constitutes a linear transformation of $\mathbf{H}$:

$$\Phi' = T_g\Phi, \qquad\qquad (T_g\Phi)(\mathbf{x}) = \Phi(R_g^{-1}\mathbf{x})$$
$$T_g(\alpha\,\Phi + \beta\,\Psi) = \alpha\,T_g\,\Phi + \beta\,T_g\,\Psi. \qquad (7.7.3)$$

Thus to every $g \in \mathrm{SO}(3)$ there is a linear operator T_g, and it should be clear from the geometric picture that the assignment $g \mapsto T_g$ is a representation of $\mathrm{SO}(3)$ in $\mathbf{H}$. Explicitly, we have

$$(T_{gh}\Phi)(\mathbf{x}) = \Phi(\mathrm{R}_{gh}^{-1}\mathbf{x}) = \Phi(\mathrm{R}_h^{-1}\,\mathrm{R}_g^{-1}\,\mathbf{x}) = (T_h\Phi)(\mathrm{R}_g^{-1}\,\mathbf{x}) = (T_g(T_h\Phi))(\mathbf{x}) \qquad (7.7.4)$$

$$(T_e\,\Phi)(\mathbf{x}) = \Phi(\mathbf{x}), \qquad (7.7.5)$$

and so $T_g\,T_h = T_{gh}$, $T_e = \mathrm{id}_{\mathbf{H}}$. (The purpose of using R^{-1} instead of R in eq. (7.7.3) was just to get the order of factors correctly here.) The representation $(\mathbf{H}, T)$ becomes *unitary* when $\mathbf{H}$ is made into a Hilbert space on using the scalar product

$$\langle\, \Phi, \Psi \,\rangle := \int d^3x\, \Phi^*(\mathbf{x})\, \Psi(\mathbf{x}), \qquad (7.7.6)$$

well known from wave mechanics. (To guarantee the existence of these integrals one admits square integrable fields Φ only: $\int d^3x\, |\Phi^2| < \infty$; however, we promised not to go into any details from functional analysis.) This is checked by

$$\langle\, T_g\Phi, T_g\Psi \,\rangle = \int d^3x\, \Phi^*(\mathrm{R}_g^{-1}\mathbf{x})\, \Psi(\mathrm{R}_g^{-1}\mathbf{x}) = \int d^3y\, \Phi^*(\mathbf{y})\, \Psi(\mathbf{y}) = \langle\, \Phi, \Psi \,\rangle, \qquad (7.7.7)$$

where a new integration variable $\mathrm{R}_g^{-1}\mathbf{x} = \mathbf{y}$ was introduced, the essential ingredient being the *rotational invariance of the Euclidean volume element* $d^3x = d^3y$.

Being infinite-dimensional, our representation must be reducible. The reduction procedure described in sect. 7.5 requires the knowledge of the Hermitian generators $\mathbf{J}$ of the representation. We have

$$(T_{g(\tau\boldsymbol{\alpha})}\Phi)(\mathbf{x}) = \Phi(\mathrm{R}^{-1}(\tau\boldsymbol{\alpha})\,\mathbf{x}) \approx \Phi(\mathbf{x} - \tau\boldsymbol{\alpha} \times \mathbf{x}) \approx \Phi(\mathbf{x}) - \tau(\boldsymbol{\alpha} \times \mathbf{x})\, \boldsymbol{\nabla}\Phi\big|_{\mathbf{x}} =$$

$$= \Phi(\mathbf{x}) - \tau\boldsymbol{\alpha}(\mathbf{x} \times \boldsymbol{\nabla})\Phi\big|_{\mathbf{x}} = ((\mathrm{id}_{\mathbf{H}} - i\tau\boldsymbol{\alpha}\,\mathbf{J})\Phi)(\mathbf{x}).$$

The generators $\mathbf{J}$ are therefore given by the *first-order homogeneous linear differential operators*

$$\mathbf{L} := \mathbf{x} \times \frac{1}{i}\,\boldsymbol{\nabla}, \qquad (7.7.8)$$

which up to a factor $\hbar$ agree with the operators of *orbital angular momentum* of wave mechanics. Note that here $\mathbf{x}$ and $\boldsymbol{\nabla}$ constitute the first examples of vector operators (defined in eq. (7.3.17)) that are different from the generators themselves (exercise 1).

The equations (7.5.46–48) serving to determine the irreducible subspaces $\mathbf{H}_{j\alpha}$ thus become *homogeneous linear differential equations*. In particular, the equation $\mathbf{J}^2\Phi(\mathbf{x}) = j(j+1)\Phi(\mathbf{x})$ is a rotationally invariant differential equation—and quite generally, invariant homogeneous linear differential equations on a function space define invariant subspaces of the function space. To solve these differential equations it is useful to go over to polar coordinates r, θ, φ, since rotations do not change the value of $r = |\mathbf{x}|$, so that the $\mathbf{J}$ do not involve the variable r.

Therefore, let us then first consider $\mathbf{H}(\mathbf{S}_2)$, the space of functions $\Phi = \Phi(\theta, \varphi)$ defined on the unit sphere. A rotationally invariant scalar product on it results from $d^3x = r^2\,dr\,\sin\theta\,d\theta\,d\varphi$ as

$$\langle\, \Phi, \Psi \,\rangle_{\mathbf{S}_2} = \int_{\theta=0}^{\pi} \int_{\varphi=0}^{2\pi} \sin\theta\,d\theta\,d\varphi\, \Phi^*\,\Psi =: \int d\Omega\, \Phi^*\,\Psi. \qquad (7.7.9)$$

A simple change of variables now gives

$$L_3 = \frac{1}{i}\frac{\partial}{\partial\varphi}, \qquad\qquad L_\pm = e^{\pm i\varphi}\left(\pm\frac{\partial}{\partial\theta} + i\,\mathrm{ctg}\,\theta\,\frac{\partial}{\partial\varphi}\right),$$

$$\mathbf{L}^2 = -\left(\frac{1}{\sin\theta}\frac{\partial}{\partial\theta}\sin\theta\frac{\partial}{\partial\theta} + \frac{1}{\sin^2\theta}\frac{\partial^2}{\partial\varphi^2}\right). \tag{7.7.10}$$

$\mathbf{L}^2$ is nothing but $-r^2 \times$ (angular part of the Laplace operator $\Delta := \partial_\mu\,\partial_\mu$). The solutions of $L_3\Phi = m\Phi$ have the form $f(\theta)\exp(im\varphi)$, where m has to be an integer for a single-valued function on the sphere to result. The solutions of $L_+\Phi = 0$, $L_3\Phi = j\Phi$ are $const\cdot(\sin\theta)^j\exp(ij\varphi)$, where the highest weight $j = \ell = 0, 1, 2, \ldots$ has to be an *integer*. Up to a phase, the constant follows from normalization with respect to the scalar product (7.7.9):

$$1 = |const.|^2\, 2\pi \int_0^\pi (\sin\theta)^{2\ell}\sin\theta\,d\theta = |const.|^2\, 4\pi\,\frac{2.4.6\ldots 2\ell}{1.3.5\ldots(2\ell+1)}.$$

We thus find that in our representation every integer value ℓ appears exactly once as a highest weight, the canonical basis in the irreducible subspace $\mathbf{H}_\ell(\mathbf{S}_2)$ being given recursively by the $2\ell + 1$ functions

$$\mathrm{Y}_{\ell\ell}(\theta,\varphi) = \frac{(-1)^\ell}{\sqrt{4\pi}}\sqrt{\frac{1.3.5\ldots(2\ell+1)}{2.4.6\ldots 2\ell}}\,(\sin\theta)^\ell e^{i\ell\varphi}$$

$$\mathrm{Y}_{\ell,m-1} = \frac{L_-}{\sqrt{\ell(\ell+1)+m-m^2}}\,\mathrm{Y}_{\ell m} \qquad (m = \ell,\,\ell-1,\,\ldots,\,-\ell+1) \tag{7.7.11}$$

The functions $\mathrm{Y}_{\ell m}(\theta,\varphi)$—where $(-1)^\ell$ is a conventional over-all phase factor—are nothing but the usual *spherical harmonics*, since $-\mathbf{L}^2$ also is the Laplacian on the unit sphere $\mathbf{S}_2$. As the full set of eigenfunctions of the commuting Hermitian operators L_3, $\mathbf{L}^2$ they form a complete orthonormal system on $\mathbf{H}(\mathbf{S}_2)$:

$$\langle\, \mathrm{Y}_{\ell m}, \mathrm{Y}_{\ell'm'}\,\rangle = \int d\Omega\, \mathrm{Y}^*_{\ell m}\,\mathrm{Y}_{\ell'm'} = \delta_{\ell\ell'}\,\delta_{mm'}. \tag{7.7.12}$$

To the decomposition

$$\mathbf{H}(\mathbf{S}_2) = \sum_{\ell=0}^\infty \oplus\mathbf{H}_\ell(\mathbf{S}_2) \tag{7.7.13}$$

there corresponds the unique expansion

$$\Phi(\theta,\varphi) = \sum_{\ell=0}^\infty \Phi_\ell(\theta,\varphi), \qquad \Phi_\ell(\theta,\varphi) := \sum_{m=-\ell}^\ell c_{\ell m}\mathrm{Y}_{\ell m}(\theta,\varphi) \in \mathbf{H}_\ell(\mathbf{S}_2) \tag{7.7.14}$$

of Φ into spherical harmonics. The components $c_{\ell m}$ of Φ with respect to the basis $\{\mathrm{Y}_{\ell m}\}$ are obtained from the orthogonality relation (7.7.12) as $c_{\ell m} = \langle\,\mathrm{Y}_{\ell m},\Phi\,\rangle$.

The projection operator P_ℓ to the *finite-dimensional* subspace $\mathbf{H}_\ell(\mathbf{S}_2)$, defined by $P_\ell\Phi = \Phi_\ell$, has the explicit form

$$P_\ell\Phi = \sum_{m=-\ell}^{\ell} \langle Y_{\ell m}, \Phi \rangle Y_{\ell m}$$

$$(P_\ell\Phi)(\theta, \varphi) = \int d\Omega' \left[\sum_{m=-\ell}^{\ell} Y_{\ell m}^*(\theta', \varphi') \, Y_{\ell m}(\theta, \varphi) \right] \Phi(\theta', \varphi'),$$

$$(7.7.15)$$

i.e., is a linear integral operator of finite rank.

This opens up the wide area of relations between group theory and the theory of 'special functions of mathematical physics' (see, e.g., Talman and Wigner 1968, Dieudonné 1980). Let us indicate here only a derivation of the *addition theorem of spherical harmonics* which is very natural from a group theoretical point of view. Let us write $\mathbf{n} = (\sin\theta\cos\varphi, \sin\theta\sin\varphi, \cos\theta)$ instead of (θ, φ) as the argument of the $Y_{\ell m}$; then by eq. (7.6.31) we have

$$Y_{\ell m}(\mathrm{R}_g^{-1}\mathbf{n}) = (T_g Y_{\ell m})(\mathbf{n}) = \mathrm{D}_{nm}^{(\ell)}(g) Y_{\ell n}(\mathbf{n}).$$

$$(7.7.16)$$

Now a more detailed study of the $\mathrm{D}_{nm}^{(\ell)}$ (see, e.g., Edmonds (1960), who uses the passive interpretation, however) shows their relation to the so-called Jacobi polynomials, which for $m = 0$ reduce to Legendre polynomials, so that $\mathrm{D}_{n0}^{(\ell)}$ may be expressed by spherical harmonics, leading to the addition theorem. The following argument is closer to our point of view here. From unitarity and eq. (7.7.16) one verifies the relation $(\mathbf{n}' = \mathrm{R}_g^{-1}\mathbf{n})$

$$\sum_{m=-\ell}^{\ell} Y_{\ell m}^*(\mathbf{n}_1') \, Y_{\ell m}(\mathbf{n}_2') = \sum_{m=-\ell}^{\ell} Y_{\ell m}^*(\mathbf{n}_1) \, Y_{\ell m}(\mathbf{n}_2)$$

$$(7.7.17)$$

whose geometric content is clear: the kernel function, with respect to the invariant measure $d\Omega$, of the projection operator to $\mathbf{H}_\ell(\mathbf{S}_2)$ must not depend on the special orthonormal system used in $\mathbf{H}_\ell(\mathbf{S}_2)$ for its construction. Given $\mathbf{n}_1$, $\mathbf{n}_2$, we now choose g such as to make $\theta_1' = 0$ for $\mathbf{n}_1'$; this is done because from eqs. (7.7.10,11) we see that $Y_{\ell m}(0, \varphi) \propto \delta_{0m}$, so that the sum on the left-hand side reduces to one term, depending only on θ_2' which now equals the angle between $\mathbf{n}_1$, $\mathbf{n}_2$.

Let us now come back to our original space $\mathbf{H} = \mathbf{H}(\mathbf{R}^3)$ of fields defined on $\mathbf{R}^3$. Contrary to the space $\mathbf{H}(\mathbf{S}_2)$—where we had a unique decomposition into pairwise inequivalent irreducible representations, eq. (7.7.13)—such is not the case for $\mathbf{H} = \mathbf{H}(\mathbf{R}^3)$. Rather, we here encounter the more general situation

$$\mathbf{H}(\mathbf{R}^3) = \sum_{\ell\alpha} \oplus \mathbf{H}_{\ell\alpha}(\mathbf{R}^3)$$

envisaged in sect. 7.5d. This is because if $f(r)$ is any function on the half line $0 \leq r < \infty$ (square integrable with respect to the measure $r^2 dr$), then for any fixed ℓ the functions $f(r)Y_{\ell m}(\theta, \varphi)$ span an irreducible invariant subspace of $\mathbf{H} = \mathbf{H}(\mathbf{R}^3)$ which is isomorphic to $\mathbf{H}_\ell(\mathbf{S}_2)$, carrying the representation $\mathrm{D}^{(\ell)}$. Now the space of all such $f(r)$ may be decomposed into 1-dimensional subspaces in many ways—the theory of the rotation group has nothing to say here. The specific way how the isotypic components

$$\mathbf{H}_\ell(\mathbf{R}^3) = \sum_{\alpha} \oplus \mathbf{H}_{\ell\alpha}(\mathbf{R}^3)$$

$$(7.7.18)$$

of type $D^{(\ell)}$, onto which the projection operators (7.7.15) constructed above project, are to be decomposed into irreducible ones depends on other criteria, and the sum over α may even be an integral (*direct integral* of representations; see, e.g., Naimark 1960, Reed and Simon 1978). Examples for the various choices are encountered, e.g., in wave mechanics, where one diagonalizes L_3 and $\mathbf{L}^2$ together with some rotationally invariant Hamiltonian: it then depends on the potential what the 'radial quantum number' α and the radial functions will be that are used to span the space of radial functions, and whether $\mathbf{H}_\ell(\mathbf{R}^3)$ appears decomposed as a direct sum, as a direct integral, or as a mixture of both. (E.g., free particle–spherical Bessel functions, direct integral; spherical harmonic oscillator–generalized Laguerre functions, direct sum; Coulomb potential–Laguerre functions, mixed case; etc. It is well known that some potentials are related to larger groups containing SO(3)—e.g., the free particle is related to the full group of Euclidean rotations *and* translations—but the generic rotationally symmetric potential is not.) Note that the theorem we quoted about full reducibility of representations of compact groups in the sense of a direct sum does not exclude, in the infinite-dimensional case, the alternative appearance of the representation space as a direct integral.

Let us isolate from the considerations above some 'abstract nonsense' part. We started from a set $\mathbf{M}$ on which a group $\mathcal{G}$ acted as a group of transformations, i.e., there was a homomorphism of the group into the self-bijections (permutations) of $\mathbf{M}$. (In our example, the latter was $\mathbf{R}^3$ or the unit sphere $\mathbf{S}_2$, thus not necessarily a linear space, while the group was SO(3).) We then considered the set $\mathbf{H(M)}$ of functions defined on $\mathbf{M}$ with values in the field $\mathbf{C}$. This is a complex vector space by a definition identical to eq. (7.7.2). For each $g \in \mathcal{G}$ we then define a linear operator T_g on $\mathbf{H(M)}$ by an equation analogous to eq. (7.7.3). The assignment $g \mapsto T_g$ is a representation of $\mathcal{G}$ by a reasoning analogous to the one in eqs. (7.7.4,5)—no additional structure is needed up to this point. For unitarity, one needs a *measure* on $\mathbf{M}$ which is *invariant* under the transformations by the $g \in \mathcal{G}$; then a scalar product can be defined in analogy to eq. (7.7.6). If $\mathbf{M}$ is a manifold and $\mathcal{G}$ is a Lie group with differentiable action on $\mathbf{M}$ one can define *generators* by a procedure similar to the one that led to eq. (7.7.8), i.e., the generators will be *linear first-order differential operators* on $\mathbf{M}$.

Of particular importance is the case where $\mathcal{G}$ acts *transitively* on $\mathbf{M}$, meaning that any point can be transformed into any other point (as is the case for the action of the rotation group on the sphere $\mathbf{S}_2$ but not on $\mathbf{R}^3$) by some group element g. Thus all points are on equal footing. But now we select one of them at will, $p_0 \in \mathbf{M}$ (e.g., the 'north pole' of the sphere) and consider all $g \in \mathcal{G}$ leaving p_0 fixed. They form a subgroup $\mathcal{H}_{p_0}$ (called the *isotropy group* or *stabilizer* of p_0). (Stabilizers of other points are conjugate to it; in the example of the sphere, the stabilizer of the north pole is a subgroup isomorphic to the group SO(2) of rotations in a plane.) Now all the other points $p \in \mathbf{M}$ may be put in bijection with the left cosets $g\mathcal{H}_{p_0}$ where g is some element that transforms p_0 into p—the assumption of transitivity guaranteeing its existence. The set of all left cosets is denoted by $\mathcal{G}/\mathcal{H}_{p_0}$; but remember that this set has no group structure in general—the exception being the case where the subgroup is an invariant one. Thus the information on the possible sets $\mathbf{M}$ on which $\mathcal{G}$ can act transitively is already contained in the group itself. Such sets are also called *homogeneous spaces* of the group. In the case of a Lie group acting differentiably on a manifold one just requires the subgroup to be a closed Lie subgroup. In our example above we thus have $\mathbf{S}_2 \leftrightarrow$ SO(3)/SO(2); and similarly $\mathbf{S}_3 \leftrightarrow$ SO(4)/SO(3) etc.

It should be clear that $\mathcal{G}$ is a homogeneous space of itself: on $\mathbf{M} = \mathcal{G}$, $\mathcal{G}$ can act as a transitive group of transformations by left or by right multiplication: if $p \in \mathcal{G} = \mathbf{M}$, put $g(p) = gp$ or $= pg^{-1}$, respectively. This action of the group on itself is *simply transitive* (also termed *free* and transitive, meaning that the isotropy group of every point consists just of the unit element). (Note that there is a conceptual difference between a space $\mathbf{M}$ on which $\mathcal{G}$ acts simply transitively and $\mathcal{G}$ itself, in that the two are in bijection in many ways, depending on which point $p_0 \in \mathbf{M}$ we single out, while

$\mathcal{G}$ has the unit element as a distinguished point. Our main example here is the set $\mathcal{I}$ of all inertial frames: after singling out any of them as I_0, all the other $I \in \mathcal{I}$ are reached from it by exactly one Poincaré transformation, which gives a bijection between $\mathcal{I}$ and the Poincaré group $\mathcal{P}$. But it is conceptually clear that a frame is not the same thing as a transformation!) When we apply the construction described above to the situation $\mathbf{M} = \mathcal{G}$, we obtain a *faithful* representation of the group canonically associated with it, called the left or right *regular representation* on $\mathbf{H}(\mathcal{G})$. For a Lie group, this allows to define the Lie algebra as the Lie algebra formed by the generators of the regular representation with respect to commutators. For compact groups, the regular representation may be made unitary by using the left = right invariant measure mentioned in sect. 7.5a.

For the group SO(3)—or even better for SU(2), because of its relation to the sphere $\mathbf{S}_3$—this is nicely illustrated. The transformation $p \mapsto gp$ or pg^{-1} is then just the left or the right translation of the sphere $\mathbf{S}_3$ considered in sect. 7.6a. If the group is parametrized by Euler angles, the (left = right) invariant measure is given by eq. (7.5.10) (cf. also exercise 2 of sect. 7.6). Functions on the group are then functions $f(\alpha, \beta, \gamma)$ which for SU(2) and SO(3) differ in their periodicity properties. Generators of the regular representation can be obtained from the 'multiplication table' (7.6.7). This allows the regular representation to be reduced according to the usual procedure.

Concerning the reduction of the regular representation, we can make here an additional general remark. Consider the matrix elements $D_{mn}(g)$ of a finite-dimensional representation of a group $\mathcal{G}$ as functions on $\mathcal{G}$, and let $g \mapsto T_g$ be the right regular representation. Then

$$(T_{g_1} D_{mn})(g) = D_{mn}(gg_1) = D_{mk}(g) D_{kn}(g_1)$$
$$T_{g_1} D_{mn} = D_{kn}(g_1) D_{mk}.$$

$$(7.7.19)$$

This says that for every fixed m the functions $D_{mn}(g)$ span an invariant subspace of $\mathbf{H}(\mathcal{G})$. If the representation D is irreducible, then (according to a lemma of Burnside) the functions D_{mn}, taken for all n and m, are linearly independent. It follows that, for each fixed m, in the invariant subspace just constructed the regular representation acts in equivalence with the representation D and that the representation D occurs in $\mathbf{H}(\mathcal{G})$ a number of times at least equal to its dimension. (Note the twin role of the D_{mn}: they occur as representing matrices and also as basis vectors of invariant subspaces of the regular representation! It is therefore possible to find the $D_{mn}^{(j)}$ for the rotation group in analogy to the way we constructed the spherical harmonics $Y_{\ell m}$ by solving differential equations—now in the variables α, β, γ: this method is described, e.g., in Gelfand, Minlos, and Shapiro (1963)).

For *compact* groups one can say more about the number of times an irreducible representation occurs in the regular representation: the multiplicity *equals* the dimension, and if D runs through the system of irreducible representations, the functions $D_{mn}(g)$ form a complete system in $\mathbf{H}(\mathcal{G})$; i.e., every $f \in \mathbf{H}(\mathcal{G})$ possesses a unique expansion in terms of them. (This is one—crude—version of the celebrated theorem of Peter and Weyl from 1927. From it follows the complete reducibility of the regular representation as well as the finite dimensionality of all irreducible representations. Compactness is an essential assumption here.)

Exercises

1. Fill in the details leading to the addition theorem for the spherical harmonics.

2. Let x_μ be the components of the position vector in 3-space; then the tensor components x_μ, $x_\mu x_\nu, \ldots$ are at the same time functions that get transformed into each other by rotations. In the cases written, find their relations to the spherical harmonics.

3. Show that the wave mechanical *position vector* $\mathbf{X}$, defined by $(\mathbf{X}\Phi)(\mathbf{x}) = \mathbf{x}\,\Phi(\mathbf{x})$, and the *momentum vector* $\mathbf{P}$, defined by $(\mathbf{P}\,\Phi)(\mathbf{x}) = \hbar/i\,\nabla\Phi(\mathbf{x})$ or $(\exp(-i\mathbf{a}\mathbf{P}/\hbar)\Phi)(\mathbf{x}) = \Phi(\mathbf{x} - \mathbf{a})$, are vector operators.

4. Calculate the generators of the left and right regular representation of the rotation group in terms of Euler angles.

5. Convince yourself that the version of Burnside's lemma needed to continue with eq. (7.7.19) is equivalent to the one given in exercise 9 of sect. 6.6!

6. For the generators (7.7.8) verify by direct computation the theorem that the (Hermitian) generators of any (unitary) representation of SO(3) have to satisfy eq. (7.5.15). Also verify the expressions (7.7.10).

7.8   Description of Particles with Spin

Besides scalar fields, in physics one needs vector and tensor *fields*, as we saw already in chap. 5; in addition, in quantum mechanics also spinor fields are needed for the description of (Fermi) particles. Tensor and spinor fields are given by multicomponent wave functions describing inner angular momentum (spin, angular momentum in the rest system) or polarization degrees of freedom. In this section we investigate the group theoretical aspects of the transformation behavior of such fields under space rotations.

Let us begin with the example of *vector fields*, since they are easiest to visualize. A vector field associates a vector $\mathbf{v}(\mathbf{x})$ to each point $\mathbf{x}$ of Euclidean 3-space. In analogy to the well-known flux lines and lines of force, this field may be illustrated by its *field lines*. Under an active rotation the pattern of field lines is rotated rigidly, as shown in Fig. 7.6, producing a new vector field that may be compared with the old one. When the two patterns coincide after the first has been rotated, the vector field is called *rotation invariant* (cf. Fig. 7.8). (Note that although no nonzero vector in 3-space is invariant under all rotations, a vector field may have that property.)

Formally, the behavior of vector and tensor fields was already given in eqs. (3.4.10)

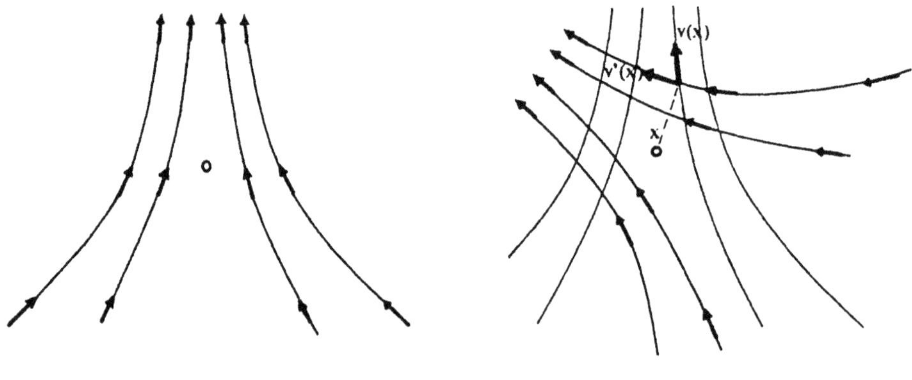

a) original vector field   b) field and rotated field

Fig. 7.6. Rotating a vector field

and (5.6.2), we just have to lower the number of space dimensions by one:

$$\bar{\mathbf{v}}(\mathbf{x}) = \mathbf{R}\,\mathbf{v}\,(\mathbf{R}^{-1}\mathbf{x}). \tag{7.8.1}$$

Again we can interpret this passively, as we did earlier, or actively, as we will be doing further on.

More generally, we can pick some finite-dimensional representation $(\mathbf{V}, D)$ of the rotation group and define a *field of type D* as a map v associating to every point $\mathbf{x}$ of 3-space an element $v(\mathbf{x}) \in \mathbf{V}$. ($\mathbf{V}$ may be a space of tensors, of spinors, ...) A rotation R carries v into the field v' given by

$$v'(\mathbf{x}) = D_g(v\,(\mathrm{R}_g^{-1}\mathbf{x})), \tag{7.8.2}$$

which behavior is also illustrated by the commuting diagram of Fig. 7.7.

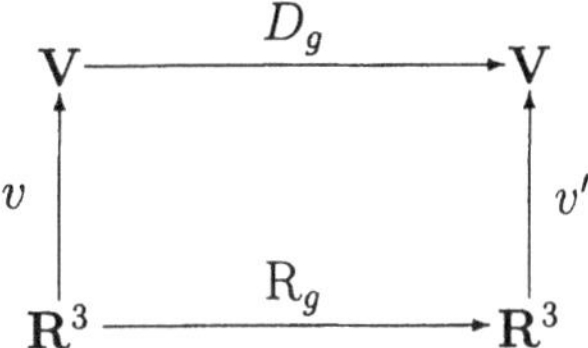

Fig. 7.7. Commuting diagram for the transformation behavior of fields

The fields of type D form an infinite-dimensional vector space $\mathbf{H}(\mathbf{V}, \mathbf{R}^3)$ if addition and multiplication by scalars is defined pointwise as in the case of scalar fields. The assignment $v \mapsto v' =: {}_\mathrm{D}T_g v$ defines a linear operator ${}_\mathrm{D}T_g$ on $\mathbf{H}$ for each g, and the assignment $g \mapsto {}_\mathrm{D}T_g$ is a representation of SO(3) in the space $\mathbf{H}$, which becomes unitary upon introduction of the scalar product

$$\langle\, v, w\,\rangle := \int d^3x \langle\, v(\mathbf{x}), w(\mathbf{x})\,\rangle_\mathbf{V} \tag{7.8.3}$$

(where $\langle\ \rangle_\mathbf{V}$ indicates the invariant scalar product in $\mathbf{V}$). In particular, if D is an irreducible representation $\mathrm{D}^{(s)}$ (*fields for spin s*), and if the fields v, w are specified by their components v_σ, w_σ with respect to a canonical basis in $\mathbf{V}$, we have

$$\langle\, v, w\,\rangle = \int d^3x \sum_{\sigma=-s}^{s} v_\sigma^*(\mathbf{x})\, w_\sigma(\mathbf{x}). \tag{7.8.4}$$

Interpreting $\mathbf{x}$ as a 'continuous index', it is at least plausible that the representation ${}_\mathrm{D}T$ is just the tensor product of the representation $(\mathbf{V}, D)$ and the representation $(\mathbf{H}(\mathbf{R}^3), T)$ in the space of scalar fields considered in the last section:

$$_\mathrm{D}T_g = D_g \otimes T_g, \qquad\qquad \mathbf{H}(\mathbf{V}, \mathbf{R}^3) = \mathbf{V} \otimes \mathbf{H}(\mathbf{R}^3). \tag{7.8.5}$$

To reduce this representation, we determine its generators $\mathbf{J}$. For this, write $\mathbf{S}$ for the Hermitian generators of $(\mathbf{V}, D)$ and invoke the $\mathbf{L}$ of eq. (7.7.8); then

$$_D T_{g(\tau\boldsymbol{\alpha})} \approx (\mathrm{id}_{\mathbf{V}} - i\tau\boldsymbol{\alpha}\mathbf{S}) \otimes (\mathrm{id}_{\mathbf{H(R^3)}} - i\tau\boldsymbol{\alpha}\mathbf{L}) \approx (\mathrm{id}_{\mathbf{H(V,R^3)}} - i\tau\boldsymbol{\alpha}\mathbf{J}),$$

$$\mathbf{J} := \mathrm{id}_{\mathbf{V}} \otimes \mathbf{L} + \mathbf{S} \otimes \mathrm{id}_{\mathbf{H(R^3)}},$$

$$(7.8.6)$$

which is usually written $\mathbf{J} = \mathbf{L} + \mathbf{S}$, accompanied by the words 'L acts on $\mathbf{x}$, $\mathbf{S}$ acts on the discrete index'. Quantum mechanically, $\mathbf{J}$ is the operator of *total angular momentum*, additively composed of *orbital angular momentum* $\mathbf{L}$ and *spin* $\mathbf{S}$. From the more precise form $\mathrm{id}_{\mathbf{V}} \otimes \mathbf{L} + \mathbf{S} \otimes \mathrm{id}_{\mathbf{H(R^3)}}$ there follows at once

$$[\mathbf{L}, \mathbf{S}] = 0$$

$$[L_\mu, L_\nu] = i\epsilon_{\mu\nu\lambda} L_\lambda, \qquad\qquad [S_\mu, S_\nu] = i\epsilon_{\mu\nu\lambda} S_\lambda.$$

$$(7.8.7)$$

The distributive property (6.5.8) of the tensor product now permits, for the purpose of reducing the tensor product, first to reduce each factor, D and T. For the latter, let us confine to the space of functions on the unit sphere, $\mathbf{H(S_2)}$, so that we have $T_g = \sum \oplus D_g^{(\ell)}$. From D we pick any of the irreducible components $D_g^{(s)}$ contained in it. The problem that remains to be solved is to decompose the tensor product $D^{(j)} \otimes D^{(j')}$ of two irreducible representations into irreducible parts. (For slightly greater generality we replace, for the moment, the integer weight ℓ by the arbitrary weight $j' = 0, 1/2, 1 \ldots$.) The solution is given by the *Clebsch-Gordan series*[1]

$$D^{(j)} \otimes D^{(j')} = D^{(j+j')} \oplus D^{(j+j'-1)} \oplus \ldots \oplus D^{(|j-j'|+1)} \oplus D^{(|j-j'|)},\qquad (7.8.8)$$

as will be shown in sect. 8.3, using spinor algebra. Therefore we have

$$D^{(s)} \otimes T = D^{(s)} \otimes \left(D^{(0)} \oplus D^{(1)} \oplus D^{(2)} \oplus \ldots \right) =$$

$$= D^{(s)} \oplus D^{(s-1)} \oplus D^{(s)} \oplus D^{(s+1)} \oplus D^{(s-2)} \oplus D^{(s-1)} \oplus \ldots$$

$$(7.8.9)$$

The highest weights have multiple occurrence here; more precisely, the highest weights $j \geq s + m$ appear $(2s + 1)$ times ($m \geq 0$ integer); the highest weights $j = s - m$ appear $(2j + 1)$ times ($0 < m \leq s$, m integer).

To construct a *canonical basis* in each of the irreducible subspaces we have to solve eqs. (7.5.48) with $\mathbf{J} = \mathbf{L} + \mathbf{S}$. This problem is known in quantum mechanics as *addition of angular momenta*. Since a highest weight j occurs several times in the decomposition (7.8.9), the eigenvectors $|j, j, \ldots\rangle$ form a space of a dimension given above, and one selects a basis in each by requiring them to be eigenvectors of the operators $\mathbf{L}^2$ and $\mathbf{S}^2$ also—being squares of vector operators, they commute with $\mathbf{J}$. This choice of eigenvectors $|j, j, \ell, s\rangle$ precisely removes the degeneracy of j, since it just states from which of the products $D^{(s)} \otimes D^{(\ell)}$ the $D^{(j)}$ under consideration stems.

---

[1] A. Clebsch and P. Gordan were leading figures in the branch of mathematics called 'theory of invariants'; cf. Weitzenböck (1923). The Clebsch-Gordan series determines the structure of the representation ring.

In the last section we saw that in the representation T the canonical basis $|\,\ell\, m\,\rangle$ for the irreducible part $\mathrm{D}^{(\ell)}$ is given by the spherical harmonics $\{Y_{\ell m}\}$. For $\mathrm{D}^{(s)}$ we as well assume given a canonical basis $\{v_\sigma,\ \sigma=-s,\ \ldots,s\}$. Then the space of $\mathrm{D}^{(s)}\otimes\mathrm{D}^{(\ell)}$ is spanned by the tensor products $v_\sigma\otimes Y_{\ell\lambda}$, which are eigenvectors of $\mathbf{L}^2$, L_3, $\mathbf{S}^2$, S_3 and are usually written in Dirac notation as $|\,\ell\,\lambda\,s\,\sigma\,\rangle$; they form an orthonormal system. From them, the canonical basis vectors $|\,j\,m\,\ell\,s\,\rangle$ for each $\mathrm{D}^{(j)}$ contained in $\mathrm{D}^{(s)}\otimes\mathrm{D}^{(\ell)}$ in accordance with eq. (7.7.8) are to be constructed:

$$|\,j\,m\,\ell\,s\,\rangle = \sum_{\lambda\sigma} c_{\lambda\sigma}\,|\,\ell\,\lambda\,s\,\sigma\,\rangle \qquad (j=\ell+s,\ \ell+s-1,\ldots,|\ell-s|). \tag{7.8.10}$$

Because of $J_3=L_3+S_3$ we must always have $m=\lambda+\sigma$. The coefficients $c_{\lambda\sigma}$, which because of the orthonormal property of the $|\,\ell\,\lambda\,s\,\sigma\,\rangle$ are just the scalar products

$$c_{\lambda\sigma} = \langle\,\ell\,\lambda\,s\,\sigma\,|\,j\,m\,\ell\,s\,\rangle, \tag{7.8.11}$$

are called *Clebsch-Gordan coefficients*. There are several methods for computing them explicitly—cf. the comprehensive treatment in Biedenharn and Louck (1981). An elementary derivation and tables are found in Edmonds (1960).

The canonical basis vectors for the irreducible constituents $\mathrm{D}^{(j)}$ of the space of fields of type $\mathrm{D}^{(s)}$ thus are

$$|\,j\,m\,\ell\,s\,\rangle = \sum_{\lambda}\langle\,\ell,\lambda,s,m-\lambda\,|\,j\,m\,\ell\,s\,\rangle\,v_\sigma\otimes Y_{\ell\lambda}. \tag{7.8.12}$$

As a concrete example, let us consider the space of *vector fields*; i.e., as the finite-dimensional representation D we take the irreducible representation $\mathrm{D}^{(1)}$. In this case, the basis vectors $|\,j\,m\,\ell\,1\,\rangle$ are called *vector spherical harmonics*. They are written $\mathbf{Y}_{j\ell m}(\theta,\varphi)$ $(j=\ell+1,\ \ell,\ \ell-1)$, and by eq. (7.8.12) we have

$$\mathbf{Y}_{j\ell m}(\theta,\varphi) = \sum_{\lambda}\langle\,\ell,\lambda,1,m-\lambda\,|\,j\,m\,\ell\,1\,\rangle\,\mathbf{e}_\sigma Y_{\ell\lambda}(\theta,\varphi), \tag{7.8.13}$$

where $\mathbf{e}_\sigma$ $(\sigma=-1,\,0,\,1)$ are the canonical basis vectors

$$\mathbf{e}_{\pm1} = -(\pm\mathbf{e}_x + i\mathbf{e}_y)/\sqrt{2}, \qquad\qquad \mathbf{e}_0 = \mathbf{e}_z \tag{7.8.14}$$

resulting as the solution of exercise 7 of sect. 7.5. (A common phase factor was chosen conventionally; vector and tensor components with respect to this basis are often called *spherical components*.)

The vector spherical harmonics form a complete orthonormal system for vector fields $\mathbf{v}(\theta,\varphi)$ defined on the unit sphere, i.e., every $\mathbf{v}(\theta,\varphi)$ has a unique expansion

$$\mathbf{v}(\theta,\varphi) = \sum_{j=0}^{\infty}\mathbf{v}_j(\theta,\varphi)$$

$$\mathbf{v}_j(\theta,\varphi) = \sum_{\ell=j-1}^{j+1}\mathbf{v}_{j\ell}(\theta,\varphi)$$

$$\mathbf{v}_{j\ell}(\theta,\varphi) = \sum_{m=-j}^{j} c_{j\ell m}\mathbf{Y}_{j\ell m}(\theta,\varphi) \tag{7.8.15}$$

$$c_{j\ell m} - \int d\Omega\, \mathbf{Y}^*_{j\ell m}(\theta,\varphi)\,\mathbf{v}(\theta,\varphi).$$

(If fields on 3-space are considered by adding an r-dependence, the $c_{j\ell m}$ become functions of r which as before in the case of scalar fields can be decomposed in many ways about which the theory of the rotation group alone has nothing to say. Rather, it will depend on the specific problem at hand which decomposition is appropriate. In typical applications, some rotationally invariant system of field equation is to be solved, and the decomposition above takes care of the angular dependence, leaving behind some 'radial equation(s)'.)

Note in particular that for $j = 0$ there is only one vector spherical harmonic $\mathbf{Y}_{010}$, transforming according to the trivial representation, i.e., giving an *invariant vector field*. It is geometrically clear (cf. Fig. 7.8) that (in 3 and higher dimensions) such a vector field has to be of the form $\mathbf{v}(\mathbf{x}) = f(r)\,\mathbf{x}/r$, and this of course does result from eq. (7.8.15). So if one does not make a *notational* distinction between the vector $\mathbf{x}$ and the position vector *field* $\mathbf{x}$ one has to keep in *mind* that it transforms according to $\mathrm{D}^{(1)}$ for the first interpretation but according to $\mathrm{D}^{(0)}$ for the second!

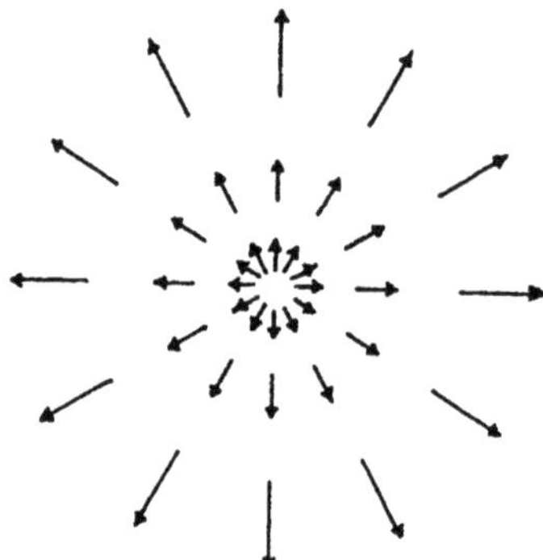

Fig. 7.8. Invariant vector field

Vector spherical harmonics are used to separate vectorial field equations such as eqs. (5.2.1,2) in spherical coordinates, just as one separates scalar field equations like $\Delta\Phi = 4\pi\rho$ using an expansion into the $Y_{\ell m}$. For this, it becomes necessary to expand expressions like $r\nabla\, \mathbf{Y}_{j\ell m}$, $\mathbf{x}/r\, \mathbf{Y}_{j\ell m}$ and $r\nabla\, Y_{\ell m}$, $\mathbf{x}/r\, Y_{\ell m}$, $r\nabla \times \mathbf{Y}_{j\ell m}$ into scalar and vector spherical harmonics, respectively, in order to be able to compare coefficients. Eq. (7.8.13) shows that this task may be reduced to calculating the scalar products ('matrix elements') $\langle Y_{\ell'm'}, \mathbf{v}\, Y_{\ell m}\rangle$, where $\mathbf{v} = v_\sigma\, \mathbf{e}_\sigma$ is a vector operator such as $\mathbf{x}/r$, $r\nabla$ etc. The task is facilitated by the *Wigner-Eckart theorem*, according to which the entire m, m', σ dependence of such matrix elements enters through a Clebsch-Gordan coefficient only, so that the specific nature of the vector operator has to be taken into account only in the calculation for one single set of values (often $\sigma = 0$, $m = m' = 0$ or $1/2$ is the easiest choice). One can understand this theorem by first concluding from eq. (7.3.17) that $\mathbf{v}\,|\,\ell m\rangle$ transforms according to $\mathrm{D}^{(1)} \otimes \mathrm{D}^{(\ell)}$ and thus may be decomposed via Clebsch-Gordan into vectors $|\,\ell',m'',1\rangle$ (where $(\ell' = \ell+1, \ell, \ell-1)$ and where 1 is a further index to characterize these vectors depending on the special nature of the vector operator $\mathbf{v}$) and by then using the lemma formulated as exercise 6 of sect. 7.5. See Edmonds (1960) for details and Jackson (1999) or Blatt and Weisskopf (1952) for an application to electromagnetic multipole radiation.

One can give an analysis of spinor and tensor fields in complete analogy to the above, defining appropriate spinor and tensor spherical harmonics. (The latter appear in a multipole expansion of gravitational waves—the formalism is developed, e.g., in F. Zerilli, J. Math. Phys. *11*, 2203 (1970).)

Let us finally come back to the general ideas displayed at the end of the last section. One can generalize them directly by considering functions on **M** with values in a vector space **V** on which $\mathcal{G}$ acts via a representation D. It is useful to consider an even more general situation where we associate to every point $p \in$ **M** an isomorphic copy $\mathbf{V}_p$ of **V**. This situation is referred to as a *vector bundle* over **M** with *standard fiber* **V**; the copies $\mathbf{V}_p$ are called the *fibers*. Instead of a function on **M** with values in **V** we consider an assignment of an element $v_p \in \mathbf{V}_p$ to each $p \in$ **M**, called a *cross section* of the vector bundle. These cross sections form a linear space under pointwise addition and multiplication by scalars, and one can define a scalar product between cross sections, given a scalar product in each fiber as well as an invariant measure on **M**. So far there seems to be no essential difference between the **V**-valued functions and cross sections of a vector bundle. The difference comes in when topological and continuity properties are added—which we cannot go into here—and when group actions are considered. Namely, we consider the situation where the group $\mathcal{G}$ acts on **M**: $p \mapsto gp$, and at the same time maps the fiber over p linearly and invertibly to the fiber over gp. If we choose a basis in each fiber, these linear maps will be given by matrices that depend not only on the group elements g as before but also on the point p, in contrast to the situation of **V**-valued functions. Despite this, we nevertheless get a representation of $\mathcal{G}$ in the space of cross sections by a definition similar to eq. (7.8.2), with the representing operator D_g being replaced by a linear map $D(g, p)$. When the g are restricted to the stabilizer subgroup $\mathcal{H}_{p_0}$ of a point p_0 the $D(g, p_0)$ furnish an ordinary representation of that subgroup on the vector space $\mathbf{V}_{p_0}$. For unitarity one again needs a $\mathcal{G}$-invariant measure on **M**.

The important situation where $\mathcal{G}$ acts transitively on **M** is referred to as a *homogeneous vector bundle*. Here the whole representation on the space of sections of the bundle is already determined by the representation D_0 of a stabilizer subgroup just mentioned and is called *induced* by the latter. We shall see this fact in sect. 9.4 for the example where the group is the Lorentz group and the stabilizer subgroup is, in one case, the rotation group; but it will be obvious that the argument generalizes. A more abstract argument, not using bases as in sect. 9.4, is given in modern treatments like Hermann (1966) (but beware of misprints!), Mackey (1968), Loebl (1968). Such induced representations are—as may transpire from the examples above—by no means irreducible. The easiest case is the one where D_0 is an irreducible representation of the subgroup. (This is not the case in our example of vector fields $\mathbf{v}(\theta, \varphi)$: here the subgroup is SO(2), and D_0 decomposes into 3 irreducible parts, corresponding to the normal component and two tangential components of **v** with respect to the sphere.) Then the question of the reduction of the induced representation is answered— under suitable general assumptions, e.g., for compact groups—by the *Frobenius reciprocity theorem*. Let D_{irr} be an irreducible representation of $\mathcal{G}$: when restricted to the subgroup it will in general become reducible, and the irreducible representation D_0 of the subgroup will occur there a certain number of times. The theorem says that this multiplicity is the same as the multiplicity of the occurrence of D_{irr} in the representation induced by D_0. (See also Shaw (1983) for a proof.) When D_0 is not irreducible one has just to decompose it.

The possibility of obtaining representations of groups by the inducing construction is used extensively in representation theory. In particular, for noncompact groups big progress was achieved by inducing with the help of maximal compact subgroups. The inducing construction was used by O. Nachtmann (Commun. Math. Phys. *6*, 1 (1967)) to construct a theory of free quantum fields on homogeneous spaces of groups.

It turns out to be useful even in the case of the rotation group to consider representations induced from *irreducible* representations of its (only, up to conjugation, connected) subgroup SO(2); remember **M** $=$ SO(3)/SO(2) $=$ **S**$_2$. Being commutative, its irreducible representations are 1-dimensional by Schur II and are easily seen to be given by $\alpha \mapsto e^{is\alpha}$ (where α is the rotation angle and s is integer for single-valued representations and half-integer for double-valued ones). The fibers of the corresponding vector bundles over **S**$_2$ are thus (complex) 1-dimensional and its cross sections are described, after a choice of basis in each fiber, by complex-valued functions which, however, transform differently as compared to the scalar fields considered earlier: e.g., under a rotation through α around the 3-axis they pick up, in addition to eq. (7.7.1), a factor $e^{is\alpha}$.

Although in sect. 9.4 we shall stick to this description of sections using bases in the fibers, this is not useful in some ways, in particular when the homogeneous space is topologically noncontractible.

It is then advisible to use the description given in the works quoted above, and although we cannot go into any details here we should like to make a little propaganda for it by giving a few indications in the case of the rotation group. In this alternative description, cross sections v are replaced by functions $\tilde{v}$ on $\mathcal{G}$ itself that take values in $\mathbf{V}_{p_0}$: $\tilde{v}(g)$ is obtained from $v(gp_0)$ applying the linear transformation associated with g^{-1}. These functions verify to behave as $\tilde{v}(gh) = D_{0h}^{-1}\tilde{v}(g)$ as one moves within the cosets, whereas the representation of $\mathcal{G}$ is by operators defined in the same way as in the left regular representation mentioned in sect. 7.7. For SO(3) this means considering functions $f(\alpha,\beta,\gamma)$ of the Euler angles which are eigenfunctions of the operators of right translations (cf. sect. 7.6) by elements of the subgroup SO(2), $\gamma \mapsto \gamma + \tau$: $f(\alpha,\beta,\gamma) = e^{i\gamma s}f(\alpha,\beta)$. The eigenvalue of the generator J_3^{right} is thus s—in order to include half integer values one has to replace SO(3) by SU(2) and SO(2) by the unitary subgroup U(1) covering it. Since one can interpret the remaining Euler angles α, β as polar coordinates on a sphere (and γ as a directional angle for an oriented orthonormal tangent frame at the position (α,β) on the sphere, thus identifying SO(3) with the bundle of all such frames, similar to what was said about the Lorentz group in the appendix to sect. 4.1), our functions are essentially given by functions on $\mathbf{S}_2$. Occasionally they are referred to as *spin-weighted functions* (see, e.g., J. Goldberg et al., J. Math. Phys. *8*, 2155 (1967), Penrose and Rindler (1984); for a geometrical visualization see also Gelfand, Minlos, and Shapiro (1963), p. 101) and s is referred to as their *spin weight*. If the space of spin-weight s functions is decomposed into irreducible subspaces, Frobenius reciprocity tells us that the representation $\mathrm{D}^{(j)}$ will occur (precisely once) iff $j \geq |s|$. Those functions on $\mathbf{S}_2$ that give a canonical basis for the irreducible subspace corresponding to $\mathrm{D}^{(j)}$ within the space of spin weight s are called *spin-weighted spherical harmonics*. They are also known as *monopole spherical harmonics*, since they occur in the quantum mechanical description of the motion of a charged particle in the field of a (hypothetical) spherically symmetric magnetic monopole (s is then related to the Dirac quantum number mentioned in sect. 5.7; see, e.g., Biedenharn and Louck 1981). They are to be kept strictly apart from the spinor spherical harmonics that would be constructed in analogy to vector spherical harmonics via Clebsch-Gordan; they are obtained as usual by diagonalizing $\mathbf{J}^{2\text{left}}$, J_3^{left} formed from the generators of left translations (which commute with the right translations, so in particular with J_3^{right}). They are also related to the $\mathrm{D}_{mn}^{(j)}$ of eq. (7.6.30), as results from a consideration analogous to eq. (7.7.19) for left translations. What is remarkable in this construction and makes it useful in practical calculations is the fact that the ladder operators $J_{\pm}^{\text{right}}$ for the right action raise/lower the eigenvalue of J_3^{right} while commuting with $\mathbf{J}^{\text{left}}$. This gives two s-dependent operators on functions on the sphere to raise and lower the spin weight (Goldberg et al., loc. cit.). By continued application of these operators one can obtain functions with arbitrary integer spin weight from ordinary functions or, conversely, construct scalar 'potentials' for functions of integer weight. There results a formalism for the separation of vectorial and tensorial field equations in polar coordinates using radial and tangential components (*not* to be confused with what were called spherical components following eq. (7.8.14) where the basis vectors are position-independent!) which is simpler than the formalism using vectorial and tensorial spherical harmonics: already the formulas to obtain the $\mathbf{Y}_{j\ell m}(\theta,\varphi)$ from the $\mathbf{Y}_{\ell m}$ by applying the operators $\mathbf{L}$, $\mathbf{x}/r$, $r\boldsymbol{\nabla}$ (cf. Edmonds 1960) and also their relation to the Debye potentials for vector fields (see Born and Wolf 1970) are more complicated, and when it comes to higher-degree tensor spherical harmonics, the situation is much more involved in the sense that many more applications of Clebsch-Gordanology become necessary; cf. Zerilli, loc. cit., and, by contrast, M. Carmeli, J. Math. Phys. *10*, 1699 (1969). The simplifications correspond to those achieved in the analysis of the scattering matrix upon use of the 'helicity basis' (M. Jacob, G.C. Wick, Ann. Phys. (N.Y.) *7*, 404 (1959); Halpern (1968), Appendix 2). One can even turn things completely around and give a derivation of the Clebsch-Gordan coefficients using the spin-weight formalism!

We must end here our spin-weight propaganda and refer to the quoted literature.

7.9　The Full Orthogonal Group O(3)

In this section we consider the full orthogonal group O(3) generated by proper orthogonal transformations (=rotations) and improper ones (=reversals). Rotations have determinant $+1$, while reversals have determinant -1. The latter may be uniquely written as a product of a rotation $R \in \mathrm{SO}(3)$ and a special reversal, the *space reversal* or *parity operation* or *reflection in the origin*,[1] P,

$$P\mathbf{x} = -\mathbf{x}, \qquad P^2 = \mathbf{1}, \qquad PR = RP. \tag{7.9.1}$$

Since the determinant is a continuous function on the group, the latter cannot be connected: it consists of two connected components: $\mathrm{SO}(3)$ and $P \cdot \mathrm{SO}(3)$, which constitute an invariant subgroup and a single coset. It is then clear that $\mathrm{O}(3)$ is compact.

$\mathbf{1}$ and P form the cyclic subgroup $\mathcal{Z}_2 = \{\mathbf{1}, \mathrm{P}\}$, which is an invariant subgroup of $\mathrm{O}(3)$ as is $\mathrm{SO}(3)$. We can easily establish an isomorphism between the (outer) direct product (cf. exercise 6 of sect. 3.1) group $\mathrm{SO}(3) \times \mathcal{Z}_2$ and $\mathrm{O}(3)$ by $(R, \mathbf{1}) \leftrightarrow R$, $(R, P) \leftrightarrow RP$: one says that $\mathrm{O}(3)$ is the (inner) direct product of those subgroups.

While the infinitesimal methods used so far to classify and reduce representations are of no direct help in the case of nonconnected groups, a direct product structure is, due to the following theorem, whose first part is actually a statement about equivalence classes:

One obtains all (single-valued, finite-dimensional) irreducible representations of the direct product $\mathcal{G}$ of two groups $\mathcal{G}_1$ and $\mathcal{G}_2$ by taking all tensor products of some irreducible representation of $\mathcal{G}_1$ and some of $\mathcal{G}_2$, equivalence among product representations implying and being implied by equivalence between the corresponding representations of both factors. Moreover, all (finite-dimensional) representations of $\mathcal{G}$ are fully reducible if the same is true for both $\mathcal{G}_1$ and $\mathcal{G}_2$.

We shall convince ourselves at this place only about the representation property of the tensor product representations, leaving the proof of the theorem to an exercise with hints, or referring to Cartan (1966), Shaw (1983). If (g_1, g_2) and (h_1, h_2) are two elements from $\mathcal{G} = \mathcal{G}_1 \times \mathcal{G}_2$, their product is $(g_1 h_1, g_2 h_2)$; let $_1T$ and $_2T$ be representations of $\mathcal{G}_1$ and $\mathcal{G}_2$, respectively; then we have

$$(g_1, g_2) \mapsto {_1T}_{g_1} \otimes {_2T}_{g_2} \qquad\qquad (h_1, h_2) \mapsto {_1T}_{h_1} \otimes {_2T}_{h_2}$$

and further, by the multiplication rule for tensor products

$$\left({_1T}_{g_1} \otimes {_2T}_{g_2}\right)\left({_1T}_{h_1} \otimes {_2T}_{h_2}\right) = {_1T}_{g_1}\,{_1T}_{h_1} \otimes {_2T}_{g_2}\,{_2T}_{h_2} = {_1T}_{g_1 h_1} \otimes {_2T}_{g_2 h_2},$$

proving the representation property. Note that when we apply equivalence maps S_1, S_2 to the factors, the tensor product undergoes an equivalence transformation by the map $S_1 \otimes S_2$, as is easily seen.

For the problem at hand, what is still lacking is the set of irreducible representations of $\mathcal{Z}_2$. But these are easy to find: since the group is Abelian, they are 1-dimensional; to P is assigned a complex number whose square is 1 because of $P^2 = \mathbf{1}$. We thus get only two possibilities, $\mathbf{1} \mapsto 1$, $P \mapsto 1$ and $\mathbf{1} \mapsto 1$, $P \mapsto -1$. This leads to the irreducible representations

---

[1] Observe that this operation is improper only in an odd number of space dimensions; for a reflection (=involutive orthogonal transformation, i.e., squaring to the identity) that is always improper in any number of dimensions one can take a reflection in a hyperplane.

$$R \mapsto D^{(\ell)}(R), \qquad\qquad R\,P \mapsto D^{(\ell)}(R); \qquad\qquad (7.9.2a)$$

$$R \mapsto D^{(\ell)}(R), \qquad\qquad R\,P \mapsto -D^{(\ell)}(R). \qquad\qquad (7.9.2b)$$

of O(3). (We are considering only integer weights ℓ, since for two-valued representation the situation is more complicated: see below and sect. 7.10.) The representations defined by eq. (7.9.2a) and (7.9.2b) will be symbolized $D^{(\ell,+)}$ and $D^{(\ell,-)}$ (*positive* and *negative parity*), respectively.

Reducible representations of O(3) are decomposed as before, but in addition the operator assigned to P, which commutes with the generators of SO(3), has also to be diagonalized, the possible eigenvalues being ± 1 because of $P^2 = \mathbf{1}$. If both eigenvalues do indeed occur, the representation space contains vectors with *undefined parity*.

Let us now look at *tensor* representations. Since tensors furnish representations even for the full linear group (cf. sect. 5.4 as specialized to dimension $n = 3$), a well-defined behavior under reversals is automatically implied. If $M^\mu{}_\nu$ is an orthogonal matrix (proper *or* improper), we have for the transformation behavior of *proper tensors*

$$T'^{\mu\nu\cdots} = M^\mu{}_\alpha\, M^\nu{}_\beta \ldots T^{\alpha\beta\cdots}. \qquad\qquad (7.9.3a)$$

Thus a proper tensor of degree p has the definite parity $(-1)^p$, the representation decomposes, for p even or odd, into irreducible parts $D^{(\ell,+)}$ or $D^{(\ell,-)}$, respectively.

Pseudotensors by definition transform as

$$T'^{\mu\nu\cdots} = \det \mathrm{M} \cdot M^\mu{}_\alpha\, M^\nu{}_\beta \ldots T^{\alpha\beta\cdots}, \qquad\qquad (7.9.3b)$$

a pseudotensor of degree p has thus parity $(-1)^{p+1}$. Upon decomposition one obtains, for p even or odd, irreducible parts $D^{(\ell,-)}$ or $D^{(\ell,+)}$.

The proper tensors of degree 1 are called *polar vectors* (representation $D^{(1,-)}$); the pseudotensors of degree 1 are called *axial vectors* (representation $D^{(1,+)}$). We already made use of this distinction in chap. 1. The ε-tensor[1] $\varepsilon_{\mu\nu\lambda}$ is an invariant pseudotensor of degree 3 and thus transforms as $D^{(0,+)}$; transvections with it do not change parity. For instance, if we form the tensor product of two polar vectors $\mathbf{x}$, $\mathbf{y}$ we get the proper tensor $x^\mu y^\nu$ of degree 2; transvection with $\varepsilon_{\mu\nu\lambda}$ gives the pseudo(=axial) vector $\mathbf{z}$:

$$z_\lambda = \varepsilon_{\mu\nu\lambda}\, x^\mu\, y^\nu,$$

which is nothing but the vector product $\mathbf{z} = \mathbf{x} \times \mathbf{y}$, transforming according to $D^{(1,+)}$.

Let us now consider the spaces $\mathbf{H}(\mathbf{R}^3)$ and $\mathbf{H}(\mathbf{S}_2)$ of scalar fields studied in sect. 7.7. We can define an action of P in the same way as we defined one for SO(3), viz., by the unitary transformation $\Phi \mapsto \Phi'$, where $\Phi'(\mathbf{x}) = \Phi(-\mathbf{x})$ (*proper* scalar field). Its eigenfunctions are the even or odd functions:

$$\Phi(-\mathbf{x}) = \Phi(\mathbf{x}) \qquad \text{or} \qquad \Phi(-\mathbf{x}) = -\Phi(\mathbf{x}).$$

Thus only even or odd functions have a well-defined parity. The whole space decomposes directly according to the decomposition

$$\Phi(\mathbf{x}) = \frac{1}{2}\,[\Phi(\mathbf{x}) + \Phi(-\mathbf{x})] + \frac{1}{2}\,[\Phi(\mathbf{x}) - \Phi(-\mathbf{x})] \qquad\qquad (7.9.4)$$

---

[1] Observe the remarks made in this context after eq. (5.5.14)!

into two eigenspaces of P.

On the sphere $\mathbf{S}_2$ the space reflection $\mathbf{x} \mapsto -\mathbf{x}$ gives the transformation

$$\theta \mapsto \pi - \theta, \qquad\qquad \varphi \mapsto \varphi + \pi \qquad\qquad (7.9.5)$$

of the polar coordinates. The decomposition of the representation into the $D^{(\ell,\pm)}$ then follows from the behavior of the spherical harmonics under space reflection. Since all generators and thus also the ladder operators $L_\pm$ commute with it, it suffices to consider, following eq. (7.7.11), the behavior of $Y_{\ell\ell}$. We immediately get

$$Y_{\ell m}(\pi - \theta, \varphi + \pi) = (-1)^\ell \, Y_{\ell m}(\theta, \varphi), \qquad\qquad (7.9.6)$$

i.e., the $Y_{\ell m}$ transform according to $D^{(\ell,+)}$ or $D^{(\ell,-)}$ for ℓ even or odd. Therefore, when an even or odd function is expanded into spherical harmonics the terms with ℓ odd or even are absent.

One can analyze representations in spaces of other transformation character relative to parity. For an application to simplify the formalism of vector spherical harmonics in electromagnetic multipole radiation, see Blatt and Weisskopf (1952) or Jackson (1999).

We now come to the discussion of two-valued representations. As we have stressed several times, there are mathematical and physical reasons for looking at them for SO(3). It turns out in the case of O(3) that the mathematical and physical reasons do not lead directly to the same results. We therefore postpone the general discussion about this to sect. 7.10 and present here one specific *covering group of* O(3) which is associated with some geometric ideas on the behavior of spinors under reflections (Cartan 1966), so that its defining representation can be called fundamental in much the same sense as the spinor representation was fundamental for SO(3) (see the end of sect. 7.6). This covering group covers the subgroup SO(3) by SU(2) as before. Since $P^2 = \mathbf{1}$ and since to the identity rotation there correspond the matrices $\pm\mathbf{1}$ of SU(2), one possibility of representing P is $P \mapsto \pm i \cdot \mathbf{1}$. The 2×2 matrix $i \cdot \mathbf{1}$ is unitary with determinant -1, and every unitary 2×2 matrix with determinant -1 may be written as a product iU, where $U \in$ SU(2). Thus, a possible covering group of O(3) is given by the group $S_\pm U(2)$ of all unitary 2×2 matrices with determinant ± 1. It is compact and consists of the two connected components SU(2) and iSU(2). We have $O(3) \cong S_\pm U(2)/\mathcal{Z}_2$, the cyclic subgroup $\{\mathbf{1}, P\} \cong \mathcal{Z}_2$ being covered by the cyclic subgroup $\{\mathbf{1}, -\mathbf{1}, i\mathbf{1}, -i\mathbf{1}\} \cong \mathcal{Z}_4$, which is *not* a direct product.

As for its representations, let us first just write down some. Besides the defining representation $A \mapsto A$ for $A \in S_\pm U(2)$, the assignment $A \mapsto \det A \cdot A$ gives an inequivalent one (exercise 2) which is nothing but the tensor product of the defining representation and the (only) nontrivial 1-dimensional representation $A \mapsto \det A$ furnished by pseudoscalars. There are thus *two kinds of spinors* as regards their behavior under reversals. Further representations are obtained by forming tensor products of these two, and as in the SU(2) case one gets irreducible ones by symmetrization. For each of the representations $D^{(j)}$ of SU(2) one obtains two inequivalent ones for $S_\pm U(2)$:

$$\mathrm{U} \mapsto \mathrm{D}^{(j)}(\mathrm{U}), \qquad\qquad\qquad i\mathrm{U} \mapsto i^{2j}\,\mathrm{D}^{(j)}(\mathrm{U}) \qquad\qquad (7.9.7a)$$

$$\mathrm{U} \mapsto \mathrm{D}^{(j)}(\mathrm{U}), \qquad\qquad\qquad i\mathrm{U} \mapsto -i^{2j}\,\mathrm{D}^{(j)}(\mathrm{U}). \qquad\qquad (7.9.7b)$$

If j is an even or an odd integer, eq. (7.9.7a) gives the representation denoted earlier as $\mathrm{D}^{(j,+)}$ or $\mathrm{D}^{(j,-)}$, while for eq. (7.9.7b) the situation is reversed. The invariant scalar products introduced before remain invariant for the extended representations also, making them unitary.

Up to equivalence, eqs. (7.9.7) turn out to give all irreducible representations of the covering group $\mathrm{S}_{\pm}\mathrm{U}(2)$. However, as this group does *not* have a direct product structure, we cannot invoke here the theorem quoted at the beginning of this section. We shall therefore quote two more theorems which are applicable in the new situation and which will also be useful later when we discuss representations of the full Lorentz group. They refer to a situation where a group $\mathcal{G}$ consists of an (invariant) subgroup $\mathcal{G}_1$ with one single coset $\mathcal{G}_2$: $\mathcal{G} = \mathcal{G}_1 \cup \mathcal{G}_2$. The theorems are:

1. *If an irreducible representation of $\mathcal{G}$ remains irreducible upon restriction to $\mathcal{G}_1$, then there is exactly one more inequivalent irreducible representation of $\mathcal{G}$ whose restriction to $\mathcal{G}_1$ is the same as the former one. (If in the first representation one has $g \mapsto T_g$ for $g \in \mathcal{G}_2$, then in the second one has $g \mapsto -T_g$.)*

From this theorem it follows that the representations (7.9.7) are the only ones that give $\mathrm{D}^{(j)}$ upon restriction from $\mathrm{S}_{\pm}\mathrm{U}(2)$ to $\mathrm{SU}(2)$.

2. *If an irreducible representation of $\mathcal{G}$ becomes reducible upon restriction to $\mathcal{G}_1$, then the restricted representation decomposes into two inequivalent irreducible representations of $\mathcal{G}_1$ whose dimensions are equal to each other and which determine the representation of $\mathcal{G}$ uniquely up to equivalence.*

In our case the situation of theorem 2 cannot occur, since for $\mathcal{G}_1 = \mathrm{SU}(2)$ the irreducible representations are characterized uniquely up to equivalence by their dimensions. This shows that indeed we already have all irreducible representations of $\mathrm{S}_{\pm}\mathrm{U}(2)$. For the proof of the above theorems see exercise 6 and sect. 8.5, or Cartan (1966).

Naturally, the group just considered is compact, so its reducible representations are direct sums of irreducible ones. In this way we obtain certain multivalued representations of $\mathrm{O}(3)$. It is, however, neither clear whether these give all multivalued representations of it, in a sense yet to be defined, nor whether they are all physically relevant. This will be discussed in the next section.

Exercises

1. Find the behavior of the vectorial spherical harmonics $\mathbf{Y}_{j\ell m}(\theta,\varphi)$ under P!

2. Using Schur's lemma, demonstrate the inequivalence of the defining representation of the group $\mathrm{S}_{\pm}\mathrm{U}(2)$ and the one given by $\mathrm{A} \mapsto \det \mathrm{A} \cdot \mathrm{A}$.

3. Investigate the behavior of the bilinear form (7.6.22) under reversals.

4. *Inner direct product of subgroups.* Consider a group $\mathcal{G}$ containing two subgroups $\mathcal{G}_1$, $\mathcal{G}_2$ with the following properties:
   1. $\mathcal{G}_1 \cap \mathcal{G}_2 =$ unit element
   2. $g_1 g_2 = g_2 g_1$ for $g_i \in \mathcal{G}_i$
   3. Every $g \in \mathcal{G}$ may be written $g = g_1 g_2$, where $g_i \in \mathcal{G}_i$.
   In such a situation $\mathcal{G}$ is called the inner direct product of the two subgroups. Show that $\mathcal{G}_1$, $\mathcal{G}_2$ are invariant; in $g = g_1 g_2$ the g_i are uniquely determined; $\mathcal{G}$ is isomorphic to the outer direct product $\mathcal{G}_1 \times \mathcal{G}_2$.

5. Let $\mathcal{G}$ be the direct product of two groups $\mathcal{G}_1$, $\mathcal{G}_2$. Prove: (1) Every (finite-dimensional complex) irreducible representation of $\mathcal{G}$ is (equivalent to) the tensor product of an irreducible representation of $\mathcal{G}_1$ with an irreducible representation of $\mathcal{G}_2$, and all such tensor products are irreducible. (Make sure that this is really a statement about equivalence classes!) (2) If all representations of $\mathcal{G}_1$, $\mathcal{G}_2$ are fully reducible, so are the representations of $\mathcal{G}$.

   Hints (relying heavily on the results of exercises 6, 7, 8 of sect. 6.6!): Ad 1. Let $(\mathbf{V}, T)$ be an irreducible representation of $\mathcal{G}$: in general, it will be reducible upon restriction to the subgroup $\mathcal{G}_2$ alone. Pick a subspace $\mathbf{V}_2$ which is irreducible under $\mathcal{G}_2$ and form the subspaces $T_g \mathbf{V}_2$, where $g \in \mathcal{G}_1$. They all carry representations of $\mathcal{G}_2$ which are equivalent to the one in $\mathbf{V}_2$, and $\mathbf{V}$ must be an isotypic direct sum of some of them. Therefore $\mathbf{V}$ has the structure $\mathbf{V}_1 \otimes \mathbf{V}_2$, where T_{g_2} is given by $\mathrm{id}_{\mathbf{V}_1} \otimes {}_2T_{g_2}$, in which ${}_2T_{g_2}$ is the irreducible representation of $\mathcal{G}_2$ in $\mathbf{V}_2$. Since they commute with the T_{g_2}, the T_{g_1} act in it as ${}_1T_{g_1} \otimes \mathrm{id}_{\mathbf{V}_2}$, where the ${}_1T_{g_1}$ are linear operators on $\mathbf{V}_1$, obviously forming a representation of $\mathcal{G}_1$. Irreducibility of ${}_1T_{g_1}$ is necessary and also sufficient to guarantee irreducibility of $\mathbf{V}_1 \otimes \mathbf{V}_2$ under $\mathcal{G}$. Writing $g = g_1 g_2$, we have $T_g = T_{g_1} T_{g_2}$, and this acts in $\mathbf{V}_1 \otimes \mathbf{V}_2$ as $({}_1T_{g_1} \otimes \mathrm{id}_{\mathbf{V}_2})(\mathrm{id}_{\mathbf{V}_1} \otimes {}_2T_{g_2}) = {}_1T_{g_1} \otimes {}_2T_{g_2}$.—Ad 2. If $(\mathbf{V}, T)$ is reducible, decompose, inside each isotypic component $\mathbf{V}_1 \otimes \mathbf{V}_2$ for $\mathcal{G}_2$, the space $\mathbf{V}_1$ into irreducible parts for $\mathcal{G}_1$.

6. Prove Theorem 1.
   Hints: Let $(\mathbf{V}, T)$ and $(\mathbf{V}, D)$ be two irreducible representations of $\mathcal{G}$ whose restrictions to $\mathcal{G}_1$ are identical and irreducible: $D_{g_1} = T_{g_1}$ for $g_1 \in \mathcal{G}_1$. Let $g_2 \in \mathcal{G}_2$, then $\bar{g}_1 := g_2^{-1} g_1 g_2 \in \mathcal{G}_1$ and therefore $D_{\bar{g}_1} = T_{\bar{g}_1}$. Conclude from this that $D_{g_2} T_{g_2}^{-1}$ commutes with all T_{g_1}, so by Schur $D_{g_2} = \lambda T_{g_2}$. Replacing g_2 by $h_2 = h_1 g_2$ with $h_1 \in \mathcal{G}_1$ we can see that λ is independent of the special $g_2 \in \mathcal{G}_2$ chosen; replacing it by $g_2^{-1} \in \mathcal{G}_2$, we thus see that $\lambda = 1/\lambda$ or $\lambda = \pm 1$. Both possibilities are consistent with the representation property, and they are inequivalent since $D_g A = A T_g$ for $g \in \mathcal{G}_1$ already implies $A \propto \mathrm{id}_{\mathbf{V}}$, which, however, gives a contradiction when $g \in \mathcal{G}_2$.

7. Prove the following *supplements to theorems 1 and 2*, for whose formulation we first give some notation. Let $g_1 \mapsto T_{g_1}$ be an irreducible representation of the subgroup $\mathcal{G}_1 \subset \mathcal{G}$; choose again a fixed $g_2 \in \mathcal{G}_2$ and observe that $g_2^{-1} g_1 g_2$ and also $g_2^2 =: g_0$ belongs to $\mathcal{G}_1$. Then the assignment $g_1 \mapsto T'_{g_1} := T_{g_2^{-1} g_1 g_2}$ is again a

representation of $\mathcal{G}_1$ (called *conjugate* to T with respect to $\mathcal{G}$). If $g_1 \mapsto T_{g_1}$ may be obtained by restriction to $\mathcal{G}_1$ of a representation of $\mathcal{G}$, the representations T_{g_1} and T'_{g_1} of $\mathcal{G}_1$ are of course equivalent. Now assume, conversely, that T_{g_1} and T'_{g_1} are equivalent, $T'_{g_1} = S^{-1}T_{g_1}S$; then it follows that $(S^2)^{-1}T_{g_1}S^2 = T_{g_0^{-1}g_1g_0} = T_{g_0}^{-1}T_{g_1}T_{g_0}$; by Schur, S^2 and T_{g_0} are proportional, where the free factor in S is determined up to sign by requiring $S^2 = T_{g_0}$. Now the supplements to the above theorems are:

a. If T_{g_1} and $T'_{g_1} = S^{-1}T_{g_1}S$ are equivalent and if we choose $S^2 = T_{g_0}$, then the assignments $g_1 \mapsto T_{g_1}$, $g_2 \mapsto \pm S$ may each be extended to give a representation of $\mathcal{G}$ on the same space.

b. If T'_{g_1} and T_{g_1} are inequivalent, the assignment

$$g_1 \mapsto T_{g_1} \oplus T'_{g_1}, \qquad g_2 \mapsto \begin{pmatrix} 0 & T_{g_0} \\ 1 & 0 \end{pmatrix}$$

may be extended to give an irreducible representation of $\mathcal{G}$.

Hint for b: What can be said about $\mathcal{G}$-invariant subspaces when the result of exercise 2 of sect. 6.6 is used with respect to $\mathcal{G}_1$?

7.10   On Multivalued and Ray Representations

It is time now to come to grips with multivalued representations. From a purely formal point of view, one could think of a situation where to each group element $g \in \mathcal{G}$ there is assigned a whole bunch of operators such that any of those for g times any of those for h gives some of those for gh. But this is too naive in that the simplest algebraic operations with such 'representations', like forming direct sums, lead to troubles, as the reader can try out. The best sense one can make out of the idea seems to be to consider ordinary representations of some larger group $\tilde{\mathcal{G}}$ of which $\mathcal{G}$ is a homomorphic but not isomorphic image. Then any representation of $\mathcal{G}$ gives one for $\tilde{\mathcal{G}}$ by composing the representation homomorphism and the homomorphism between the groups; but in general, a representation of $\tilde{\mathcal{G}}$ gives only a multivalued one for $\mathcal{G}$, associating to $g \in \mathcal{G}$ all the operators assigned by the representation to the elements of the coset in $\tilde{\mathcal{G}}$ that is mapped to g by the homomorphism (cf. Appendix A). The group $\tilde{\mathcal{G}}$ is called an *extension* of $\mathcal{G}$ and is highly nonunique.

The precise kind of multivalued representation resulting from the needs of group covariance in quantum theory is called *multiplier representation*. For its detailed physical motivation we again refer to sect. 9.2, but its mathematical description will be given now, up to the omission of one formal point (the inclusion of antilinear operators).

By a multiplier representation we will understand the situation where to each group element $g \in \mathcal{G}$ there is assigned not one single operator T_g on the space $\mathbf{V}$ but a set of multiples αT_g, where α is allowed to run through the set $\mathbf{C}^\times$ of nonzero complex numbers or a multiplicative subgroup $\mathcal{A}$ thereof. (Up to now, we had $\mathcal{A} = \{1,\text{-}1\}$;

if our operators are required to be unitary, as happens in quantum theory, we must have $\mathcal{A} \subset \mathrm{U}(1)$, the unit circle in $\mathbf{C}$). These sets are called *operator rays*, and the representation property is here weakened to require that any operator from the ray assigned to g times any one from the ray for h give some operator from the ray for gh. Having chosen one operator T_g from the ray for each g, our requirement comes down to postulate

$$T_g\, T_h = \omega(g,h)\, T_{gh} \tag{7.10.1}$$

for some $\omega(g,h) \in \mathcal{A}$. Other names for such multivalued representations are *ray representations* or, if $\mathcal{A} = \mathbf{C}^\times$ is allowed, *projective representations*.

From the associativity of the ordinary operator product and from the normalizing choice $T_e = \mathrm{id}_\mathbf{V}$ we obtain the following restrictions on $\omega(\,.\,,\,.\,)$:

$$\omega(g,h)\,\omega(gh,k) = \omega(g,hk)\,\omega(h,k) \qquad (\text{2-cocycle condition}), \tag{7.10.2a}$$

$$\omega(e,g) = \omega(g,e) = 1, \tag{7.10.2b}$$

as may be verified as an exercise. One can further verify now (exercise) that the Cartesian product set $\mathcal{G} \times \mathcal{A}$ by the multiplication rule

$$(g,\alpha)\,\textcircled{\tiny$\cdot$}\,(h,\beta) := (gh, \omega(g,h)\,\alpha\beta) \tag{7.10.3}$$

becomes a group, in which the elements (e,α) form a *central*[1] (invariant) subgroup, so that the quotient group is isomorphic to $\mathcal{G}$, but where $\mathcal{G} \times \mathcal{A}$ in general does *not* contain a subgroup isomorphic to $\mathcal{G}$ as would be the case in a semidirect product. This group is written $\mathcal{G} \times_\omega \mathcal{A}$ and is called *central extension* of $\mathcal{G}$ by $\mathcal{A}$ with *extension cocycle ω*.

The purpose of this artificially looking construction—which would even work if $\mathcal{A}$ were *any* Abelian group in which ω takes values satisfying eq. (7.10.2)—is the following: the assignment $(g,\alpha) \mapsto \alpha T_g$ is now an *ordinary* representation of the centrally extended group, and conversely, any ordinary representation of the extended group which *represents the central subgroup by multiples of the unit operator* gives a multivalued representation of $\mathcal{G}$ in the sense above (exercise). One says that the multivalued representation of $\mathcal{G}$ has been *lifted* to a representation of $\mathcal{G} \times_\omega \mathcal{A}$ which is thus a special example of an extension group $\tilde{\mathcal{G}}$ in the sense of the beginning of this section.

The concept of *equivalence* must be generalized to read

$$T'_g = \lambda_g\, S\, T_g\, S^{-1}, \tag{7.10.4}$$

for some $\lambda_g \in \mathcal{A}$ for each $g \in \mathcal{G}$. To the equivalent representation belongs an extension cocycle ω' which is related to ω by

$$\omega'(g,h) = \lambda_g\, \lambda_h\, \lambda_{gh}^{-1}\, \omega(g,h) \tag{7.10.5}$$

(exercise); ω, ω' are called equivalent, or *cohomologous*, cocycles. Two extensions $\mathcal{G} \times_\omega \mathcal{A}$, $\mathcal{G} \times_{\omega'} \mathcal{A}$ are isomorphic as extensions iff eq. (7.10.5) holds for some choice of λ_g for each g (exercise).

---

[1]This means that these elements commute with all others, they belong to the center of the group.

Sometimes one can achieve by equivalence that ω' takes its values in a genuine subgroup $\mathcal{A}' \subset \mathcal{A}$. One tries to make $\mathcal{A}'$ as small as possible, since one can then use the smaller extension group $\mathcal{G} \times_{\omega'} \mathcal{A}'$ for the same purpose. Extensions equivalent to one having $\omega' \equiv 1$ (direct product) are called trivial. To find all multivalued representations, one has thus first to find all equivalence classes of solutions of eq. (7.10.2), taking into account the prescribed domain $\mathcal{A}$ for the values of the λ_g. As we shall see, it will be important to realize that if ω, ω' both take values in $\mathcal{A}' \subset \mathcal{A}$, they may be inequivalent as cocycles with values in $\mathcal{A}'$ but equivalent as cocycles with values in $\mathcal{A}$, because in the latter case the λ are allowed to take values in the bigger group $\mathcal{A}$, so it is easier for them to satisfy eq. (7.10.5) than if *their* values were restricted to $\mathcal{A}'$ *a priori.*

As an example, consider the 4-group $\mathcal{V}_4 \cong \{E, P, T, PT\} \subset \mathcal{L}$. For this group, the assignment $E \mapsto \mathbf{1}$, $P \mapsto \sigma_1$, $T \mapsto \sigma_2$, $PT \mapsto \sigma_3$ is a 2-dimensional ray representation because of relations (7.5.45), the cocycle taking values in $\{1, i, -1, -i\} \cong \mathcal{Z}_4$. We know that the σ act irreducibly on $\mathbf{C}^2$, so that it is impossible to achieve $\omega' \equiv 1$ by our generalized kind of equivalence transformations, because $\omega' \equiv 1$ would mean that we have an ordinary complex irreducible 2-dimensional representation of an Abelian group, contradicting Schur II. However, when we take $T \mapsto i\sigma_2$ instead, the cocycle takes values in $\mathcal{A}' = \{1, -1\}$, which is a genuine subset of $\mathcal{Z}_4$.

When $\mathcal{G}$ is a Lie group, continuity requirements are added, which in ordinary representations simply say that the T_g depend continuously on g (from which one then proves even analyticity). For multivalued representations it would be too restrictive to postulate that the T_g as well as the ω should be continuous on the whole of $\mathcal{G}$, because it is possible to pass from continuous to discontinuous $T_g, \omega(g, h)$ by discontinuous equivalences (7.10.5)—i.e., a discontinuous choice of λ. Rather, the appropriate postulate here is that the T_g, ω be continuous on open subsets of $\mathcal{G}$ which together cover the group and where one has continuous equivalences (7.10.5) on all intersections of those open sets.

As an example, consider $\mathcal{G} = $ SO(3), $\mathcal{A} = \mathcal{Z}_2 = \{1, -1\}$: if one postulated $\omega(\mathbf{1}, \mathbf{1}) = 1$ and continuity on the whole group one would obtain $\omega \equiv 1$, excluding the central extension SU(2).—It is possible to characterize the extension necessary for lifting a projective representation without such a patching construction; cf. Simms (1968), Kirillov (1976), Varadarajan (1985); M. S. Raghunathan, Rev. Math. Phys. *6*, 207 (1994).

Let us now look first at the situation when the group is connected, $\mathcal{G} = \mathcal{G}_e$. As already mentioned, in this case there exists an essentially unique connected and simply connected Lie group $\widetilde{\mathcal{G}}_e$—its universal covering group—from which all connected covering groups (=extensions of $\mathcal{G}_e$ by discrete $\mathcal{A}$) may be obtained by quotienting out some discrete central subgroup. The first result here is that every (continuous) complex or unitary *finite-dimensional* ray representation ($\mathcal{A} = \mathbf{C}^\times$ or U(1)) of a connected and simply connected Lie group is equivalent to an ordinary representation in the same space $\mathbf{V}$, the equivalence being given by

$$\lambda_g := (\det T_g)^{-1/\dim \mathbf{V}}, \tag{7.10.6}$$

as we can see by taking determinants in eq. (7.10.1)—the simple connectedness of the group avoids the possibility of getting into a tangle of different values of the root appearing in eq. (7.10.6). This result shows why it makes sense to consider not only representations of an original group but also those of its universal covering group. For the rotation group SO(3) the universal covering group is SU(2): so we are indeed in the possession of its finite-dimensional irreducible ray representations.

In the case of infinite-dimensional unitary ray representations, where $\mathcal{A} = \mathrm{U}(1)$, V. Bargmann (Ann. Math. **59**, 1 (1954)) has shown that for *compact* connected groups one can always lift to an ordinary representation of a compact connected covering group. This takes care of the infinite-dimensional unitary ray representations of SO(3) and shows that one can narrow down again from $\mathcal{A} = \mathrm{U}(1)$ to $\mathcal{A}' \cong \mathcal{Z}_2$, i.e., to two-valued representations. (We shall sketch another argument for this in sect. 9.2 which also works in the case of the Poincaré group.)

Let us now come back to the nonconnected group O(3)! Let $R, S \in$ SO(3), let P be the space reversal as before, and consider a multiplier representation with $\mathcal{A} = \mathbf{C}^\times$ or $\mathcal{A} = \mathrm{U}(1) \subset \mathbf{C}^\times$. We first show that the values $\omega(R,S)$, $\omega(P,R)$, $\omega(P,P)$ already determine the cocycle $\omega(.\,,.)$ on all of O(3), i.e., determine the values $\omega(R,P)$, $\omega(PR,S)$, $\omega(S,PR)$ and $\omega(PR,PS)$. Let us work, for easier manipulation of the cocycle condition, with the representing operators and their associativity. Then from $PRP^{-1} = R$ it follows with some $\gamma(R) \in \mathcal{A}$:

$$T_P T_R T_P^{-1} = \gamma(R) T_R. \tag{7.10.7}$$

Evaluating now the product

$$(T_P T_R T_P^{-1})(T_P T_S T_P^{-1}) = T_P(T_R(T_P^{-1} T_P)T_S)T_P^{-1}$$

in the sense of both bracketings, we obtain after cancelling the factor $\omega(R,S)T_{RS}$:

$$\gamma(R)\,\gamma(S) = \gamma(RS).$$

This shows that the assignment $R \mapsto \gamma(R)$ is a 1-dimensional representation of SO(3)—and the only one there exists is $\gamma(R) = 1$. Equation (7.10.7) goes over into

$$\omega(P,R)T_{PR} = T_P T_R = T_R T_P = \omega(R,P)T_{PR}$$

$$\omega(P,R) = \omega(R,P).$$

Multiplying with $T_P T_S$ we get, because of $P^2 = \mathbf{1}$—and hence $T_P^2 = \omega(P,P)\mathrm{id}_{\mathbf{V}}$—:

$$\omega(PR,PS) = \frac{\omega(P,P)\,\omega(R,S)}{\omega(P,R)\,\omega(P,S)}.$$

Finally, the cocycle relations belonging to $T_P T_R T_S$ and $T_R T_S T_P$ permit the calculation of the cocycle values

$$\omega(PR,S) = \frac{\omega(R,S)\,\omega(P,RS)}{\omega(P,R)}$$

$$\omega(R,PS) = \frac{\omega(R,S)\,\omega(P,RS)}{\omega(P,S)}.$$

Having proved our claim we investigate what can be achieved by the equivalence (7.10.5). For $\omega(R,S)$ we know that its range can be narrowed down to $\{1,-1\}$, the remaining freedom of the λ_R being restricted to that domain. Taking for λ_P one of the values of $(\omega(P,P))^{-1/2}$ we achieve $\omega'(P,P) = 1$, and putting $\lambda_{PR} = \lambda_P\,\omega(P,R)$ also achieves $\omega'(P,R) = 1$. (Note that in this step it was essential that in the groups $\mathcal{A} = \mathbf{C}^{\times}$ or $\mathcal{A} = \mathrm{U}(1)$ one can do square roots!) Now depending on whether the $\omega(R,S)$ can still be brought to the value 1 or not, we obtain the group $\mathrm{SO}(3) \times \{\mathbf{1},\mathrm{P}\} \cong \mathrm{O}(3)$ itself or $\mathrm{SU}(2) \times \{\mathbf{1},\mathrm{P}\}$ as a relevant central extension. For both groups we know, from the theorem at the beginning of sect. 7.9, how to construct the ordinary representations. This gives us the irreducible ray representations of $\mathrm{O}(3)$.

Let us underline here the conceptual difference between the ray representations just constructed and *a priori two-valued* representations of $\mathrm{O}(3)$, where we have $\mathcal{A} \cong \mathcal{Z}_2 = \{1,-1\}$ and thus *also* $\lambda_P \in \{1,-1\}$ *to start with*. Here we have the two possibilities $\omega(P,P) = +1$ or -1 from the beginning, but in the latter case we *cannot* get $\omega'(P,P) = +1$, since $\sqrt{-1} \notin \{1,-1\}$! $\omega'(P,R) = 1$ may be achieved, however. The two possibilities obtained so far, together with the two possibilities for the $\omega(R,S)$, give four inequivalent central extensions of $\mathrm{O}(3)$ by $\mathcal{Z}_2$. Two of them are the groups obtained in the preceding paragraph; one is isomorphic to the group $S_{\pm}U(2)$ considered in the last section. It and the group $\mathrm{SU}(2) \times \{\mathbf{1},\mathrm{P}\}$ are the only covering groups of $\mathrm{O}(3)$ in which the component of unity, $\mathrm{SO}(3)$, is covered by a *connected* subgroup. Although these two covering groups are not isomorphic, they are, as we have seen, equally good for the purposes of quantum mechanics, giving isomorphic central extensions of $\mathrm{O}(3)$ by $\mathrm{U}(1)$.

Exercises

1. From eq. (7.10.1), deduce the cocycle condition (7.10.2).

2. Verify the group axioms for the multiplication law (7.10.3).

3. Verify that the assignment $(g,\alpha) \mapsto \alpha T_g$ gives a representation of the group defined by eq. (7.10.3).

4. Deduce eq. (7.10.5) from $T'_g T'_h = \omega'(g,h)\,T'_{gh}$ and eq. (7.10.4).

5. Show that the assignment $(g,\alpha) \mapsto (g,\lambda_g^{-1}\alpha)$ is an isomorphism between the extensions defined by ω, ω' if eq. (7.10.5) holds.

6. Verify in detail that the two covering groups of $\mathrm{O}(3)$ that cover $\mathrm{SO}(3)$ by $\mathrm{SU}(2)$ are isomorphic to $\mathrm{SU}(2) \times \mathcal{Z}_2$ and $S_{\pm}U(2)$; show that the remaining nontrivial extension of $\mathrm{O}(3)$ by $\mathcal{Z}_2$ is isomorphic to $\mathrm{SO}(3) \times \mathcal{Z}_4$ (where, as before, $\mathcal{Z}_4$ is a cyclic group with 4 elements).

7. Study the behavior of ray representations and their cocycles upon (a) passage to the contragredient representation, (b) passage to the complex-conjugate representation, (c) formation of direct sums, (d) formation of tensor products, (e) passage to a homomorphic group!

8 Representation Theory of the Lorentz Group

We now come to fulfill the program formulated in chap. 6: to find and classify all quantities that behave linearly under Lorentz transformations just as tensors do—or in other words, to construct all finite-dimensional representations of the Lorentz group. From the commutation relations one reads off the adjoint representation, which happens to be identical with the representation in the space of sixtors (antisymmetric tensors of degree two) considered in sect. 6.5. From it one deduces that its Lie algebra is semisimple in the sense of the definition given in sect. 7.4. (The point here is the semisimplicity of its *complexification*: for the real Lorentz group, we already demonstrated even simplicity on the group level in appendix 2 to sect. 6.3.) It is an important theorem of H. Weyl that the finite-dimensional representations of semisimple Lie groups are fully reducible,[1] so that for their classification it suffices to find all irreducible representations. There result two fundamental representations, from which all others may be obtained by reducing tensor products: they are 2-dimensional and 2-valued and are again called *spinor representations*. From them, we develop some spinor algebra and give the relation to tensors. Finally we consider representations of the full Lorentz group.

It will turn out that apart from multiples of the trivial representation there are *no finite-dimensional unitary representations* of $\mathcal{L}_+^\uparrow$. Its infinite-dimensional irreducible unitary representations are found, e.g., in Naimark (1964). In this chapter we will not consider unitary representations, since in relativistic quantum theory one rather needs unitary representations of the *Poincaré group*, which we discuss in sect. 9; of course, unitary representations of the Lorentz group are obtained by restriction, but their irreducible components have not found significant applications so far.[2]

8.1 Lie Algebra and Representations of $\mathcal{L}_+^\uparrow$

To determine the representations of $\mathcal{L}_+^\uparrow$ we first consider—in line with the general theory indicated in sect. 7.4—its Lie algebra. The commutation relations can be taken from the defining representation: infinitesimal Lorentz transformations $L(\mathbf{v}, \boldsymbol{\alpha})$ may be composed, according to sect. 1.5, from infinitesimal rotations and infinitesimal boosts, so by eq. (1.5.13) we have for infinitesimal $\mathbf{v}$, $\boldsymbol{\alpha}$

$$L(\mathbf{v}, \boldsymbol{\alpha}) \approx L(\mathbf{0}, \boldsymbol{\alpha})\, L(\mathbf{v}, \mathbf{0}) \approx E + \boldsymbol{\alpha}\, \mathbf{M} + \mathbf{v}\, \mathbf{N}. \tag{8.1.1}$$

Here E is the 4×4 unit matrix, and

$$M_\mu := \begin{pmatrix} 0 & \mathbf{0}^\top \\ \mathbf{0} & \Lambda_\mu \end{pmatrix}, \qquad N_\mu := \begin{pmatrix} 0 & -\mathbf{e}_\mu^\top \\ -\mathbf{e}_\mu & 0 \end{pmatrix}, \tag{8.1.2}$$

---

[1] See, e.g., Samelson (1990), who gives a general proof as well as one for the covering group SL(2,C) of the Lorentz group.

[2] Cf. the pertinent remarks in H. Joos, Fortschr. Phys. *10*, 65 (1962).

where Λ_μ is defined by eq. (7.2.5) and where $\mathbf{e}_\mu$ are the usual Cartesian unit vectors.

From eqs. (8.1.2) one can verify the commutation relations

$$[M_\mu, M_\nu] = \epsilon_{\mu\nu\lambda}\, M_\lambda \tag{8.1.3a}$$

$$[N_\mu, N_\nu] = -\epsilon_{\mu\nu\lambda}\, M_\lambda \tag{8.1.3b}$$

$$[N_\mu, M_\nu] = \epsilon_{\mu\nu\lambda}\, N_\lambda, \tag{8.1.3c}$$

which define the structure of the Lie algebra $\mathbf{L} = \mathrm{so}(1,3)$ of the Lorentz group. Comparing eqs. (8.1.3c) and (7.3.18) we see that $\mathbf{N}$ is a vector operator under rotations— which is a consequence of the fact that $\mathbf{v}$ are vector components. Finally, eq. (8.1.3b) is the infinitesimal algebraic relation corresponding to the Thomas rotation. Now a more suitable (in the sense of the general remarks near the end of sect. 7.4) choice of basis in this Lie algebra, or rather its complexification, is given by the complex linear combinations

$$\mathbf{M}^{\pm} = \frac{1}{2}(\mathbf{M} \pm i\mathbf{N}), \tag{8.1.4}$$

satisfying the commutation relations

$$[M_\mu^{\pm}, M_\nu^{\pm}] = \epsilon_{\mu\nu\lambda}\, M_\lambda^{\pm}, \qquad\qquad [M_\mu^{+}, M_\nu^{-}] = 0. \tag{8.1.5}$$

The complexified Lie algebra $\mathbf{L}^c$ therefore decomposes as the *direct sum* of two complex 3-dimensional Lie algebras $\mathbf{L}^{+}$, $\mathbf{L}^{-}$, spanned by $\mathbf{M}^{+}$ and $\mathbf{M}^{-}$ (meaning $\mathbf{L}^c = \mathbf{L}^{+} \oplus \mathbf{L}^{-}$ as a vector space, while the elements of $\mathbf{L}^{+}$ commute with those of $\mathbf{L}^{-}$). Both, $\mathbf{L}^{+}$ as well as $\mathbf{L}^{-}$, have the structure of the complexified Lie algbra of the rotation group; i.e., the real linear combinations of the $\mathbf{M}^{+}$ and $\mathbf{M}^{-}$ each give a real Lie algebra isomorphic to the algebra of $\mathrm{SO}(3)$.

Having explored the structure of the complexified Lie algebra $\mathbf{L}^c$ we now put this to use in finding the irreducible representations. Given a complex irreducible representation of $\mathbf{L}$, we can extend it to an irreducible representation of $\mathbf{L}^c$ by simply considering complex linear combinations of generators instead of real ones. Note that irreducibility is not touched by this step since we started from a complex representation anyway!

Also note that in this way we at the same time introduced the concept of *representation of a complex Lie algebra* as a *complex-linear* map of the algebra into the algebra of linear operators on a complex vector space sending Lie algebra products to commutators. It must be observed now that the concept of the complex-conjugate representation gets modified: if $X \mapsto t_X$ is a representation in our sense, $X \mapsto (t_X)^*$ is not, as it is antilinear in X. Rather, the $(t_X)^*$ furnish a representation of the *complex-conjugate algebra*, formed by the complex-conjugates X^* (see Appendix B.3 for the vector space aspect of this, and define the Lie product of X^* and Y^* as $(X \circ Y)^*$), by assigning $X^* \mapsto (t_X)^*$.

Now $\mathbf{L}^c$ is a direct sum of the complex Lie algebras $\mathbf{L}^{+}$ and $\mathbf{L}^{-}$, and (by an argument entirely analogous to the one in the hints given for exercise 5 of sect. 7.9) it follows that every irreducible representation of $\mathbf{L}^c$ is just given by the tensor product of some irreducible representation of $\mathbf{L}^{+}$ with some of $\mathbf{L}^{-}$, each determined uniquely up to equivalence. Conversely, given an irreducible representation of $\mathbf{L}^c$, by restriction we get an irreducible complex representation of $\mathbf{L}$. Similarly, the irreducible representations of $\mathbf{L}^\pm$ are in bijective correspondence with the complex irreducible representations of the real rotation group. This solves our classification problem: the (equivalence classes of) complex irreducible representations of $\mathbf{L}$ are of the form $\mathrm{D}^{(j,j')} := \mathrm{D}^{(j)} \otimes \mathrm{D}^{(j')}$, where j, j' are highest weights of irreducible representations of the rotation group. The dimension of the product representation is then $(2j+1)(2j'+1)$. From the Casimir operators $(\mathbf{M}^\pm)^2 = (\mathbf{M}^2 - \mathbf{N}^2 \pm 2i\mathbf{MN})/4$ of the rotation groups with the values $-j(j+1)$, $-j'(j'+1)$ one can find the values of the Casimir operators

$$\frac{1}{2}\,(\mathbf{M}^2 - \mathbf{N}^2), \qquad\qquad \mathbf{M}\,\mathbf{N} \qquad\qquad (8.1.6)$$

for the Lorentz group $\mathcal{L}_+^\uparrow$ in the representations $\mathrm{D}^{(j,j')}$.

The concept of direct sum of Lie algebras introduced above is, of course, related to the concept of direct product of Lie groups. Using the regular representation, it is not hard to see that the Lie algebra of a direct product of Lie groups is the direct sum of the Lie algebras of the factors; the converse is true in the sense of *local* isomorphism. So what follows from our finding above on the group level is that the complex Lorentz group SO(4,C) (which is the same as the complex rotation group in 4 dimensions, since signature makes no sense in the complex domain) is locally isomorphic to the direct product of the complex rotation group in 3 dimensions with itself (or rather its complex conjugate, if isomorphism is to be understood in the sense of the 'category of complex Lie groups', i.e., as a holomorphic mapping). Indeed, in sect. 8.2 we shall find the global relation between these groups as well as the relation between the various real groups contained in the complex group. Here we just remind the reader of the local product structure of the real rotation group SO(4) contained in SO(4,C), a consequence of the global product structure of its (universal) covering group SU(2)×SU(2) which was already discussed in sect. 7.6a. The invariant measure on SU(2) obtained there yields one on the product group and its quotient SO(4). Invariant integration over a compact group allowed to prove unitarity of representations of compact groups (cf. sect. 7.5a). The ensuing *full reducibility* carries over, in the finite-dimensional case, to representations of the complexification SO(4,C) and its other real forms, and thus to $\mathcal{L}_+^\uparrow$.

Having found the classification scheme, we now want to find the representations more explicitly. For a real infinitesimal Lorentz transformation we get from eqs. (8.1.1,4)

$$L(\mathbf{v}, \boldsymbol{\alpha}) \approx E + \boldsymbol{\alpha}(\mathbf{M}^+ + \mathbf{M}^-) - i\mathbf{v}(\mathbf{M}^+ - \mathbf{M}^-) = E + (\boldsymbol{\alpha} - i\mathbf{v})\mathbf{M}^+ + (\boldsymbol{\alpha} + i\mathbf{v})\mathbf{M}^-,$$
$$(8.1.7)$$

saying that the coefficients of $\mathbf{M}^\pm$ for real Lorentz transformations are just complex-conjugates of each other. The procedure to construct the operators for real infinitesimal Lorentz transformations in the representation $\mathrm{D}^{(j,j')}$ is, therefore, the following: let $\mathrm{D}^{(j)}(\boldsymbol{\alpha})$, $\mathrm{D}^{(j')}(\boldsymbol{\alpha}')$ be operators corresponding to infinitesimal real rotations and replace $\boldsymbol{\alpha}$ by the complex parameter $\boldsymbol{\alpha} - i\mathbf{v}$, $\boldsymbol{\alpha}'$ by the complex-conjugate parameter $\boldsymbol{\alpha} + i\mathbf{v}$; then the representation is given by

$$L(\mathbf{v}, \boldsymbol{\alpha}) \mapsto \mathrm{D}^{(j,j')}(\mathbf{v}, \boldsymbol{\alpha}) = \mathrm{D}^{(j)}(\boldsymbol{\alpha} - i\mathbf{v}) \otimes \mathrm{D}^{(j')}(\boldsymbol{\alpha} + i\mathbf{v}). \qquad (8.1.8)$$

When we want to pass to *finite* Lorentz transformations by exponentiation, some care is necessary, since for finite $\boldsymbol{\alpha}$, $\mathbf{v}$ we have $L(\mathbf{v}, \boldsymbol{\alpha}) \neq \exp\{E + (\boldsymbol{\alpha} - i\mathbf{v})\mathbf{M}^| +$

$(\alpha + i\mathbf{v})\mathbf{M}^-\}$: the 1-parameter subgroup connecting $L(\mathbf{v}, \alpha)$ with the unit element is not given in parameter space by the curve $(\mathbf{v}(\tau), \alpha(\tau)) = (\tau\mathbf{v}, \tau\alpha)$. The reason is twofold. First, boosts and rotations do not commute (except for $\alpha \propto \mathbf{v}$). Second, for a given direction of $\mathbf{v}$ its length $|\mathbf{v}| = v$ is *not* an additive parameter, at variance with the situation for the rotation angle. This is of course a consequence of relativistic velocity addition, and we have seen in sect. 2.1—cf. eq. (2.1.8)—that the quantity $\operatorname{ar\,tanh} v$ is additive instead. (In the theory of Lie groups, an additive parameter for a 1-parameter subgroup is also called a *canonical parameter*.) For these reasons, if $(\mathbf{v}, \alpha)$ is finite, the matrix $\mathrm{D}^{(j)}(\alpha \pm i\mathbf{v})$ will represent some Lorentz transformation, but not the one specified by $L(\mathbf{v}, \alpha)$! To find the latter, we use the decomposition (1.5.13) and the additive parameter $\operatorname{ar\,tanh} v$ to obtain

$$\mathrm{D}^{(j,j')}(\mathbf{v}, \alpha) = \mathrm{D}^{(j)}(\alpha)\,\mathrm{D}^{(j)}(-i\mathbf{u}) \otimes \mathrm{D}^{(j')}(\alpha)\,\mathrm{D}^{(j')}(i\mathbf{u})$$

$$\mathbf{u} := \operatorname{ar\,tanh} v \cdot \frac{\mathbf{v}}{v} \tag{8.1.9}$$

($-i|\mathbf{u}|$ is just the imaginary angle φ of eq. (2.1.6).)

When the representations of $\mathcal{L}_+^\uparrow$ so obtained are restricted to a subgroup they may become reducible. In particular, when restricted to the rotation subgroup SO(3) they decompose as

$$\mathrm{D}^{(j,j')}(\mathbf{0}, \alpha) = \mathrm{D}^{(j)}(\alpha) \otimes \mathrm{D}^{(j')}(\alpha) = \mathrm{D}^{(j+j')}(\alpha) \oplus \ldots \oplus \mathrm{D}^{(|j-j'|)}(\alpha), \tag{8.1.10}$$

corresponding to eq. (7.8.8). They thus remain irreducible only if $j' = 0$ or $j = 0$.

The simplest nontrivial irreducible representations are those having $j = 1/2$, $j' = 0$ and $j = 0$, $j' = 1/2$. These are two inequivalent two-valued 2-dimensional representations which we will study in more detail in sect. 8.2. In particular, from what we learned in sect. 7.6b it follows directly that these *spinor representations of the Lorentz group* form a system of *fundamental representations*: every irreducible representation may be obtained by reducing (i.e., symmetrizing—see sect. 8.3 for detail) suitable tensor products of the fundamental representations. This fact automatically implies a relation between spinors and 4-tensors, which will be elaborated in sect. 8.4. Alongside with this important mathematical aspect of spinors comes the physical one that arises from the needs of relativistic invariance in the quantum domain.

Let us consider here the representations characterized by $j = 1$, $j' = 0$ and $j = 0$, $j' = 1$. When $\mathbf{v} = \mathbf{0}$ they both go over into the defining representation of the rotation group, which is real-orthogonal. Going to $\mathbf{v} \neq \mathbf{0}$ is an analytic continuation, so the representations remain orthogonal in the complex sense (see eqs. (7.5.13a,14a)) but not unitary; they are single-valued and faithful, corresponding to the isomorphism $\mathcal{L}_+^\uparrow \cong \mathrm{SO}(3, \mathbf{C})$ mentioned in sect. 6.5. The representations of the Lorentz group $\mathcal{L}_+^\uparrow$ may therefore be viewed as representations of the complex rotation group SO(3,**C**). If the latter is parametrized by complex rotation vectors $\alpha + i\mathbf{v}$, then the form (8.1.8) of the representations remains valid also for finite values of $\alpha, \mathbf{v}$.

The derivation of the classification of irreducible representations for $\mathcal{L}_+^\uparrow$ mixes two mathematical strategies: passage to the complexified Lie algebra, and recognition of a complex structure in the original real Lie algebra. Let us point out here first a special feature of complex Lie groups, whose general definition is of course analogous to the one given around eqs. (6.1.9) but with the additional requirement that the group may be parametrized by complex parameters instead of real ones such that the composition functions f and the parameters of the inverse are holomorphic functions. Examples encountered so far are the groups $SO(3,\mathbf{C})$, $SO(4,\mathbf{C})$; note that, on the other hand, $SU(2)$ is not a complex Lie group, although consisting, by definition, of matrices with complex entries: its defining relation (7.6.6) is not a holomorphic restriction to the complex variables a, b, and it is indeed 3-dimensional, while a complex Lie group depends on an even number of real parameters. Clearly, the Lie algebra of a complex Lie group is a complex Lie algebra. The (continuous finite-dimensional) representations of real Lie groups are *real-analytic* in suitable real parameters, and thus give complex-analytic ($=holomorphic$) representations of the complexified group. This is because the representing matrices for 1-parameter subgroups are expressed in terms of their generators t as $\exp(\tau t)$ (see sect. 7.4). Thus, e.g., the assignment $\boldsymbol{\alpha}+i\mathbf{v} \mapsto D^{(j)}(\boldsymbol{\alpha}+i\mathbf{v})$ is analytic in the real parameters $\boldsymbol{\alpha}$, $\mathbf{v}$; but it is also holomorphic in the complex parameters $\boldsymbol{\alpha}+i\mathbf{v}$. It is clear, however, that the representation $\boldsymbol{\alpha}+i\mathbf{v} \mapsto D^{(j)}(\boldsymbol{\alpha}-i\mathbf{v})$ of $SO(3,\mathbf{C})$, albeit continuous in the complex parameters $\boldsymbol{\alpha}+i\mathbf{v}$, and real-analytic in their real and imaginary parts, is *not* holomorphic, since $\boldsymbol{\alpha}-i\mathbf{v}$ is not a holomorphic function of $\boldsymbol{\alpha}+i\mathbf{v}$ (it is *anti-holomorphic*). Now all continuous finite-dimensional representations of a complex Lie group $\mathcal{G}$, being real-analytic representations of its '*realification*' (the same group when viewed as a real Lie group, i.e., when the real and imaginary parts of its complex parameters are viewed as its real parameters) may be analytically continued to give holomorphic representations of the complexification of that real group. This complexification is locally isomorphic to $\mathcal{G} \times \mathcal{G}$— where again the second factor should be $\mathcal{G}^*$ if isomorphism of complex Lie groups is to include holomorphy, in line with the 'categorical' thinking of modern mathematics—cf. Cartan (1966), and Samelson (1990) for a clearer (by modern terminology) argument on the Lie algebra level. (See also exercise 7 below.) This allows the theorem on the finite-dimensional irreducible representations of direct products of groups (or of direct sums of Lie algebras) to be applied. From the way the original complex group (or algebra) is imbedded in the complexification of its realification one then sees that the essentially new thing here is the occurrence of anti-holomorphic representations. These are complex-conjugates to holomorphic ones, as in the example above—note, however, that in the more general case where the complex group $\mathcal{G}$ does not possess a real form, so that there is no compatible complex conjugation *in* the group, one must take $(T_g)^*$ for the complex-conjugate representations, T running through the holomorphic ones; we shall verify by eq. (8.2.15) that in our example of $SO(3,\mathbf{C})$ both methods—conjugating the group element or conjugating the representation matrix—give equivalent results for every single representation, not only for the list of irreducible anti-holomorphic representations as a whole.

The method used above for $\mathcal{L}_+^\uparrow$—to pass to the complexified Lie algebra and from there to another real form of it which belongs to a compact Lie group—may be applied to all semisimple Lie groups: all complex semisimple Lie groups possess a compact real form. From the unitary nature of the representations of the compact form one concludes the full reducibility, which property is, in finite-dimensional representations, preserved under the passage back to the complexified group and its other real forms. (This method is known as the 'unitary trick' of H. Weyl.) Although the concepts of complexification, realification, complex structure, real (or reality) structure are considered elementary by mathematicians, physics readers may find them confusing on first sight and are advised to disentangle them using the modern abstract formulation, to be found in many— but not all—texts on abstract linear algebra. (See also Appendix B and the exercises to the present section.) What is added here is their interplay with the Lie algebra structure, which is much less trivial.

As should transpire from eq. (8.1.9), the representations found are double-valued iff $j + j' =$ half-integer. We shall find in the next section that they are single-valued representations of the universal covering group, so that by eq. (7.10.6) we also found all irreducible continuous multivalued representations of $\mathcal{L}_+^\uparrow$, up to equivalence.

Exercises[1]

1. Show that the adjoint representation of $\mathcal{L}_+^\uparrow$ agrees with the one in the space of antisymmetric tensors $F_{ik} = (\mathbf{E}, \mathbf{B})$ ('sixtors') considered in sect. 6.5, and that the decomposition (8.1.4,5) corresponds to the reduction carried out there. Also demonstrate the semisimplicity of the Lie algebra.

2. The equation of motion (4.1.10), (5.3.2) of a charged particle in a *constant* electromagnetic field $(F^i{}_k) = F$ possesses the first integral

$$u(s) = \exp\left(\frac{e}{m} F s\right) u(0).$$

   Show that $\exp(\frac{e}{m} F s)$ is the matrix of a Lorentz transformation.

3. The structure of the complex rotation group SO(3,$\mathbf{C}$) is given by the same commutation relations as for SO(3,$\mathbf{R}$), viz., eq. (7.2.12). The difference is that now the Lie algebra consists of all *complex* linear combinations of the Λ_μ. In the *realification* of this algebra, $\Lambda_\mu =: M'_\mu$ and $i\Lambda_\mu =: N'_\mu$ are to be considered as linearly independent over $\mathbf{R}$. Show that $\mathbf{M'}, \mathbf{N'}$ satisfy the same eqs. (8.1.3) as do the $\mathbf{M}, \mathbf{N}$ defined in eqs. (8.1.2).

4. Show, conversely, that if a linear map $J\colon \mathbf{L} \to \mathbf{L}$ of the real Lorentz algebra $\mathbf{L}$ into itself is defined by $J\mathbf{M} = \mathbf{N}$, $J\mathbf{N} = -\mathbf{M}$, then

   (i) $J^2 = -\mathrm{id}_\mathbf{L}$,   (ii) J commutes with ad_X for every $X \in \mathbf{L}$.

   One can make $\mathbf{L}$ into a complex Lie algebra by defining, for every $z \in \mathbf{C}$, $zX := (\mathrm{Re}z)X + (\mathrm{Im}z)JX$: verify the axioms of a complex vector space and complex bilinearity of the algebra multiplication!

   Remark: For any real Lie algebra $\mathbf{L}$, a linear map J satisfying i and ii is called a *complex structure* for $\mathbf{L}$; in the way just given, the pair $\mathbf{L}$, J is then a complex Lie algebra; 'forgetting' about J gives back the original real Lie algebra: this is an alternative way of describing the realification process. $-J$ is then also a complex structure, complex-conjugate to J.

5. A real(ity) structure for a complex Lie algebra $\mathbf{L}$ is a real structure for the underlying vector space, i.e., (cf. Appendix B.6) an antilinear map $C\colon \mathbf{L} \to \mathbf{L}$, satisfying

   (i) $C^2 = \mathrm{id}_\mathbf{L}$,   (ii) $[CX, CY] = C[X, Y]$,

   i.e., it is an anti-involution of first kind leaving the structure tensor invariant. The elements of $\mathbf{L}$ left invariant by (and called real in the sense of) C then form a real Lie algebra, called the *real form* of $\mathbf{L}$ determined by C.

   a. Show that with respect to a real basis the structure constants are then real.

---

[1]From exercise 3 on, these exercises are in part of a more abstract nature and are intended to getting used to complex or real structures in real or complex Lie algebras as well as to the processes of realification and complexification.

b. Show that if $t: X \mapsto t_X$ is a representation of $\mathbf{L}$ then $X^* \mapsto t_{CX}$ is a representation of the complex-conjugate algebra $\mathbf{L}^*$.

c. Find the operator $\mathcal{C}$ for the real forms so(3) and so(1,2) (Lorentz algebra in 3-dimensional space-time) of the complex algebra so(3,$\mathbf{C}$), and similarly for the real forms so(4) and so(2,2) of the complex algebra so(4,$\mathbf{C}$)!

Remark: The algebra so(3,$\mathbf{C}$) is made up of complex linear combinations of 3 basis elements X_μ satisfying $[X_\mu, X_\nu] = \epsilon_{\mu\nu\lambda} X_\lambda$. If $\mathcal{C}$ is defined by $\mathcal{C}X_\mu = X_\mu$ and antilinearity, the elements invariant by $\mathcal{C}$ are just the real linear combinations of the X_μ, constituting the algebra of so(3) as a real form of so(3,$\mathbf{C}$). If $\mathcal{C}'$ is defined by $\mathcal{C}'X_1 = -X_1$, $\mathcal{C}'X_2 = -X_2$, $\mathcal{C}'X_3 = X_3$ and antilinearity, then the real combinations of $X_1' := iX_1$, $X_2' := -iX_2$, $X_3' := X_3$ are $\mathcal{C}'$-invariant. The structure constants with respect to the primed basis are real again and correspond to the algebra of so(1,2); no *real* change of this basis can lead to a basis with structure constants as in so(3). Declaring elements to be real in the sense of some real structure for the complex algebra is not to be confused with what happens in defining matrix representations! For instance, the matrices Λ_μ of eq. (7.2.5) give a faithful (defining) matrix representation of the X_μ, and since they have real matrix elements, this fits in with real structure constants. However, to the X_μ', real in the sense of $\mathcal{C}'$, correspond the matrices $i\Lambda_1, -i\Lambda_2, \Lambda_3$ which are *not* all real in the sense of having real matrix elements. Using a complex equivalence transformation $S = \mathrm{diag}(1, -i, -i)$ one can transform them into real matrices (analogs of the ones of eq. (8.1.2)), but this is not the essential point here!

6. So far we considered the complexification of two real simple Lie algebras, namely so(3) and so(1,3). In the first case, the complexification so(3,$\mathbf{C}$) was simple again, in the second case the complexification so(4,$\mathbf{C}$) decomposed as a direct sum of two simple complex-conjugate algebras, while the original real algebra had a complex structure. Show that these two situations exhaust all possibilities for any real simple Lie algebra in view of the results expressed in exercise 3 of sect. 7.4, exercise 8 of sect. 6.5 and exercise 12 of sect. 6.6!

7. Show that the complexification of the realification of a complex Lie algebra is isomorphic, in the sense of complex Lie algebras, to the (outer) direct sum of the original algebra and its complex-conjugate.

Hints: To simplify notation, write $\mathbf{L}$ for the realification, so that the original algebra is $(\mathbf{L}, J)$ with some complex structure J as described above. When the real-linear operator J is now extended complex-linearly to the complexification $\mathbf{L}^c$, it has eigenvalues $\pm i$ there and a corresponding eigenspace decomposition of $\mathbf{L}^c$. Show that this gives a (inner) direct sum decomposition in the sense of Lie algebras and that the restriction to $\mathbf{L}$ of the projection operators onto the eigenspaces give isomorphisms which are complex-linear in the sense of $(\mathbf{L}, \pm J)$, so that their direct sum yields the required isomorphism.

8.2   The Spinor Representation

We now investigate the spinor representation $j = 1/2$, $j' = 0$ in more detail. The generators are the trace-free 2×2 matrices $-i(\boldsymbol{\alpha} - i\mathbf{v})\,\boldsymbol{\sigma}/2$, which for $\mathbf{v} = \mathbf{0}$—i.e., for pure rotations—are anti-Hermitian, while for boosts ($\boldsymbol{\alpha} = \mathbf{0}$) they are Hermitian. Therefore their exponentials $\exp[-i(\boldsymbol{\alpha} - i\mathbf{v})\,\boldsymbol{\sigma}/2]$ are all *unimodular* (determinant $= 1$), but are unitary only for $\mathbf{v} = \mathbf{0}$; for boosts, $\boldsymbol{\alpha} = \mathbf{0}$, they are *Hermitian positive-definite*. Our exponentials thus all belong to the group of all complex unimodular 2×2 matrices, which is denoted by SL(2,C). We stress again that $\exp[-i(\boldsymbol{\alpha} - i\mathbf{v})\,\boldsymbol{\sigma}/2]$ does represent a Lorentz tranformation—but not one where the vectors $\mathbf{v}$, $\boldsymbol{\alpha}$ have their usual significance. For the latter, we rather have, according to eq. (8.1.9),

$$ \mathrm{D}^{(1/2,0)}(\mathbf{v}, \boldsymbol{\alpha}) = \exp(-i\boldsymbol{\alpha}\boldsymbol{\sigma}/2)\,\exp(-\mathbf{u}\boldsymbol{\sigma}/2) \qquad (\neq \exp[-i(\boldsymbol{\alpha} - i\mathbf{v})\boldsymbol{\sigma}/2])\,, \quad (8.2.1) $$

where $\mathbf{u} := (\mathrm{ar\,tanh}\,v)\,\mathbf{v}/v$ and the exponential is to be evaluated as in eqs. (7.6.1,2,3).

As in the case of the rotation group, there is an alternative description of the spinor representation, bringing out the fact that we again here have a *covering homomorphism* $\mathrm{SL(2,C)} \to \mathcal{L}_+^\uparrow$. From a 4-vector x^i form the 2×2 matrix

$$ \mathrm{X} := x^0 \cdot \mathbf{1} + \mathbf{x}\,\boldsymbol{\sigma} = x^i \sigma_i = \begin{pmatrix} x^0 + x^3 & x^1 - ix^2 \\ x^1 + ix^2 & x^0 - x^3 \end{pmatrix} \tag{8.2.2} $$

(where $\{\sigma_i\} = \{\mathbf{1}, \sigma_1, \sigma_2, \sigma_3\}$), which is Hermitian precisely for real x^i. But now X is not trace-free; rather we have $\mathrm{Tr\,X} = 2\,x^0$. If besides the σ_i we formally introduce matrices $\tilde{\sigma}^i$ by

$$ \tilde{\sigma}^i := \sigma_i \tag{8.2.3} $$

($\tilde{\sigma}^i$ has to be distinguished from $\sigma^i := \eta^{ik}\sigma_k$!), we have

$$ \mathrm{X} = x^i \sigma_i \leftrightarrow x^i = \frac{1}{2}\,\mathrm{Tr\,X}\,\tilde{\sigma}^i. \tag{8.2.4} $$

Only the second of eqs. (7.6.13) generalizes to the present case:

$$ \det \mathrm{X} = (x^0)^2 - \mathbf{x}^2 = x^i x_i. \tag{8.2.5} $$

With an *arbitrary complex unimodular* 2×2 matrix A we now form the matrix

$$ \mathrm{X}' = \mathrm{A\,X\,A}^\dagger. \tag{8.2.6} $$

If X is Hermitian, so is X'. The 4-vector components formed from it according to $x'^i = \frac{1}{2}\,\mathrm{Tr\,X}'\,\tilde{\sigma}^i$ depend linearly on x^i, and the 4-square satisfies, because of $\det \mathrm{A} = 1$,

$$ x'^i x'_i = \det \mathrm{X}' = \det \mathrm{X} = x^i x_i. \tag{8.2.7} $$

Therefore, eq. (8.2.6) defines a Lorentz transformation whose coefficients $L^i{}_k$ are given from eq. (8.2.4) as

$$ L^i{}_k = \frac{1}{2}\,\mathrm{Tr\,A}\,\sigma_k\,\mathrm{A}^\dagger\,\tilde{\sigma}^i. \tag{8.2.8} $$

From $L^0{}_0 = \frac{1}{2}\mathrm{Tr}\,A\,A^\dagger > 0$ we see that only orthochronous Lorentz transformations result in this way. It is also not hard to see (exercise) that only proper Lorentz transformations can result. As with the rotation group in sect. 7.6, we thus constructed a homomorphism $\mathrm{SL}(2,\mathbf{C}) \to \mathcal{L}^\uparrow_+$ which by eq. (8.2.1) is onto and again 2:1, only A and $-$A leading to the same transformation $X \mapsto X'$. (This may be shown from eq. (8.2.6) as an exercise, using Schur II. It is also possible to find an explicit formula for $\pm$A expressing it by the $L^i{}_k$, similar to eq. (7.6.19).)

The factorization (8.2.1) is, by the way, the special 2×2 version of the well-known matrix analog of the polar decomposition $z = |z|\exp(i\arg z)$ of a complex number, i.e., the fact that an arbitrary complex nonsingular square matrix A may be uniquely written as a product

$$A = U\,H \tag{8.2.9}$$

of a Hermitian positive-definite matrix H and a unitary matrix U. (From $A^\dagger A = H^2$, H may be constructed here quite explicitly by solving the Cayley-Hamilton equation $H^2 - H\,\mathrm{Tr}\,H + \det H\,\mathbf{1} = 0$ and its trace for H and Tr H; U is then defined as $A\,H^{-1}$.) When $\det A = 1$ it follows that $U \in \mathrm{SU}(2)$ and $\det H = 1$. If we assign to H a real 4-vector by $h^i = \frac{1}{2}\mathrm{Tr}\,\tilde{\sigma}^i H$, it follows from $\det H = 1$ that $h^i h_i = 1$, and $h^0 > 1$ by positive definiteness. Thus h^i lies on the sheet $h^0 = {}_+\sqrt{1+\mathbf{h}^2}$ of a hyperboloid in 4-vector space (cf. the hyperboloid of 4-velocities considered in the appendix of sect. 4.1.). This sheet has topology $\mathbf{R}^3$, so that for $\mathrm{SL}(2,\mathbf{C})$ we get the topology $\mathbf{R}^3 \times \mathrm{SU}(2) = \mathbf{R}^3 \times \mathbf{S}_3$, due to uniqueness and continuity of the decomposition (8.2.9). In particular, as the topological—not group theoretical!—product of two simply connected manifolds, $\mathrm{SL}(2,\mathbf{C})$ is simply connected and is therefore the *universal covering group* of the Lorentz group $\mathcal{L}^\uparrow_+$. The latter is thus doubly connected, the complications coming, of course, from the rotation subgroup. All multivalued (=two-valued) representations of $\mathcal{L}^\uparrow_+$ are therefore single-valued representations of the covering group $\mathrm{SL}(2,\mathbf{C})$. From the fact that $\mathrm{SL}(2,\mathbf{C})$ is connected and that the $L^i{}_k$ are continuous functions of A it follows again that the homomorphism given by eq. (8.2.6,8) can only be onto the connected component $\mathcal{L}^\uparrow_+ \cong \mathrm{SL}(2,\mathbf{C})/\mathcal{Z}_2, \;\; \mathcal{Z}_2 := \{1, -1\}$.

We now remind the reader about the group isomorphisms obtained in sect. 7.6; also noting that for $x^0, x^1, x^3 =$ real, $x^2 =$ imaginary the matrix X is real, the signature of the quadratic form (8.2.5) becoming $(+-+-)$, we can give the following overview of group isomorphisms. The equation $X' = A\,X\,B^\dagger$ defines a linear transformation $x^i \mapsto x'^i$ which is a

complex Lorentz transformation	for	$(A, B) \in \mathrm{SL}(2, \mathbf{C}) \times \mathrm{SL}(2, \mathbf{C})$
real Lorentz transformation	for	$B = A \in \mathrm{SL}(2, \mathbf{C})$
complex 3-dimensional rotation	for	$B^\dagger = A^{-1} \in \mathrm{SL}(2, \mathbf{C})$
real 4-dimensional rotation	for	$(A, B) \in \mathrm{SU}(2) \times \mathrm{SU}(2)$
transformation $\in \mathrm{SO}_e(2, 2)$	for	$(A, B) \in \mathrm{SL}(2, \mathbf{R}) \times \mathrm{SL}(2, \mathbf{R})$
real 3-dimensional rotation	for	$B = A \in \mathrm{SU}(2)$
transformation $\in \mathrm{SO}_e(1, 2)$	for	$B^\dagger = A^{-1} \in \mathrm{SL}(2, \mathbf{R})$.

$$\tag{8.2.10}$$

From this derive the following isomorphisms:

$$\mathrm{SO}(3) \cong \mathrm{SU}(2)/\mathcal{Z}_2, \qquad \mathrm{SO}_e(1,2) \cong \mathrm{SL}(2,\mathbf{R})/\mathcal{Z}_2, \qquad \mathrm{SO}(3,\mathbf{C}) \cong \mathcal{L}_+^\uparrow \cong \mathrm{SL}(2,\mathbf{C})/\mathcal{Z}_2,$$

$$\mathrm{SO}(4,\mathbf{C}) \cong (\mathrm{SL}(2,\mathbf{C}) \times \mathrm{SL}(2,\mathbf{C}))/\{(\mathbf{1},\mathbf{1}),(-\mathbf{1},-\mathbf{1})\}, \qquad \mathrm{SO}_e(2,2) \cong \text{idem with } \mathbf{C} \to \mathbf{R},$$

$$(\mathrm{SL}(2,\mathbf{C}) \times \mathrm{SL}(2,\mathbf{C}))/\mathcal{V}_4 \cong \mathrm{SO}(3,\mathbf{C}) \times \mathrm{SO}(3,\mathbf{C}) \cong \mathrm{SO}(4,\mathbf{C})/\{E,-E\},$$

$$(\mathrm{SL}(2,\mathbf{R}) \times \mathrm{SL}(2,\mathbf{R}))/\mathcal{V}_4 \cong \mathrm{SO}_e(1,2) \times \mathrm{SO}_e(1,2) \cong \mathrm{SO}_e(2,2)/\{E,-E\},$$

$$(8.2.11)$$

where $\mathcal{V}_4 = \{(\mathbf{1},\mathbf{1}),(-\mathbf{1},\mathbf{1}),(\mathbf{1},-\mathbf{1}),(-\mathbf{1},-\mathbf{1})\}$ is the Kleinian four-group and where the subscript e indicates the component of unity. In addition, there are the isomorphisms written in sect. 7.6.

We now use the relation between $\mathcal{L}_+^\uparrow$ and the complex Lie groups $\mathrm{SO}(3,\mathbf{C})$ or $\mathrm{SL}(2,\mathbf{C})$ to show that every *unitary* finite-dimensional representation of $\mathcal{L}_+^\uparrow$ is a multiple of the trivial representation.[1] Since we can invariantly declare the irreducible constituents of a direct sum representation to be orthogonal, by full reducibility it suffices to show that a unitary irreducible representation must be trivial. We saw before that an irreducible representation is the tensor, or Kronecker, product of some holomorphic and some antiholomorphic representation. It is clear now that a holomorphic representation even cannot be pseudo-unitary, since this would mean equivalence to the complex conjugate of the contragredient representation (cf. exercise 5 of sect. 7.5), but equivalence and contragredience preserve holomorphicity, while complex conjugation does not. Similarly, an antiholomorphic representation cannot be pseudo-unitary. Now look at the general representation $\mathrm{D}^{(j,j')}$. We claim that the complex conjugate contragredient representation is just $\mathrm{D}^{(j',j)}$, so that for pseudo-unitarity we must have $j = j'$. In sect. 8.4 these representations will be seen to be complexifications of real 4-tensor representations, for which we have an invariant scalar product

$$T_{ij\ldots}\,T'^{ij\ldots} \tag{8.2.12}$$

which in turn can be extended invariantly to the complexification in two ways: one is by just copying expression (8.2.12)—this is complex-bilinear (symmetric) and even invariant under complex Lorentz transformations—, and one is sesquilinear (Hermitian) and still invariant under real Lorentz transformations:

$$T_{ij\ldots}^{*}\,T'^{ij\ldots}. \tag{8.2.13}$$

Since in the irreducible case an invariant Hermitian scalar product is unique up to a real factor (again exercise 5 of sect. 7.5), our only candidate representations are pseudo-unitary but not unitary, as this scalar product is obviously not definite when $T\ldots = T'\ldots$. (In the irreducible case, the tensors will be required to be totally symmetric and traceless, so indefiniteness becomes manifest on taking $T^{ij\ldots} = x^i x^j \ldots$, where x is lightlike.)

To verify our claim we first look at the representation having $j = 0$, $j' = 1/2$, whose matrices have the form $\exp[-i(\boldsymbol{\alpha} + i\mathbf{v})\,\boldsymbol{\sigma}/2]$. Because of

$$\sigma_2\,\boldsymbol{\sigma}\,\sigma_2^{-1} = -\boldsymbol{\sigma}^{*} = -\boldsymbol{\sigma}^{\top} \tag{8.2.14}$$

---

[1] Generally, connected noncompact semisimple Lie groups have no faithful finite-dimensional unitary representations.

we have

$$\sigma_2 \exp\left[-\frac{i}{2}(\boldsymbol{\alpha}+i\mathbf{v})\,\boldsymbol{\sigma}\right]\sigma_2^{-1} = \exp\left[\frac{i}{2}(\boldsymbol{\alpha}+i\mathbf{v})\,\boldsymbol{\sigma}^{*}\right] = \left(\exp\left[-\frac{i}{2}(\boldsymbol{\alpha}-i\mathbf{v})\,\boldsymbol{\sigma}\right]\right)^{*},$$
$$(8.2.15)$$

showing that this representation is equivalent to the complex conjugate of the one having $j = 1/2, j' = 0$. On the other hand, the latter is equivalent to its contragredient:

$$\sigma_2 \exp\left[\frac{i}{2}(\boldsymbol{\alpha}-i\mathbf{v})\,\boldsymbol{\sigma}^{\top}\right]\sigma_2^{-1} = \exp\left[-\frac{i}{2}(\boldsymbol{\alpha}-i\mathbf{v})\,\boldsymbol{\sigma}\right]. \qquad (8.2.16)$$

This verifies our claim for the fundamental representations; since the other ones are obtained by forming tensor products of these, the claimed equivalence is obtained using appropriate tensorial powers of σ_2 for the equivalence map, by the usual rules for the composition of tensor products of linear maps (cf. eq. (6.5.5)). The invariant bilinear form that exists on account of the equivalence (8.2.16) and exercise 4 of sect. 7.5 has as its matrix a multiple of σ_2. When the factor is chosen as i, one gets the form (7.6.22), which is of the symplectic kind and will be used in the sequel. Similarly, there is an invariant symplectic form for $j = 0, j' = 1/2$, and from the appropriate tensor products of these one gets invariant bilinear forms for the higher representations, symplectic (or symmetric) for $j + j' = $ odd (or even).

Note that it is only when $\mathbf{v} = \mathbf{0}$ that eq. (8.2.15) gives an equivalence between the spinor representation and its own complex conjugate, or, using eq. (8.2.16) as well, between the representation and its conjugate contragredient one: this is just the case of the subgroup SU(2).

Relation (8.2.14) can also be written as

$$\sigma_2 \,\tilde{\sigma}_i \,\sigma_2^{-1} = \sigma_i^{*} = \sigma_i^{\top}, \qquad (8.2.17)$$

showing that one encounters the complex-conjugate spinor representation when one takes, instead of the 2×2 matrices $X = x^i \sigma_i$, the matrices

$$\tilde{X} = x^i \tilde{\sigma}_i. \qquad (8.2.18)$$

The facts just mentioned will be built into a systematic spinor algebra in the next sections.

Exercises

1. Show that eq. (8.2.6) cannot yield the space reversal.

   Hint: Being basis independent, the determinant of $L^i{}_k$ is also the determinant of the transformation (8.2.6) and thus is equal to the determinant of the Kronecker product $A \otimes A^{*}$, and we have
   $\det(A \otimes A^{*}) = \det((A \otimes \mathbf{1})(\mathbf{1} \otimes A^{*})) = (\det A)^2 (\det A^{*})^2 = |\det A|^4 = +1.$

2. Show that only $-A \in SL(2, \mathbf{C})$ effects the same transformation $X \mapsto X'$, eq. (8.2.6), as does $A \in SL(2, \mathbf{C})$.

3. Conclude that $X \tilde{X} = \tilde{X} X = x^i \, x_i \cdot \mathbf{1}$ by verifying the relations

$$\sigma_{(i}\tilde{\sigma}_{k)} = \tilde{\sigma}_{(i}\sigma_{k)} = \eta_{ik}\cdot\mathbf{1} \tag{8.2.19}$$

$$\frac{1}{2}\operatorname{Tr}\sigma_i\tilde{\sigma}_k = \eta_{ik}. \tag{8.2.20}$$

4. For a given Lorentz transformation, find an explicit formula for $\pm A$ similar to eq. (7.6.19).

   Hints: Insertion of $x'^i = L^i{}_k\, x^k$ into eq. (8.2.6) gives $L^i{}_k\,\sigma_i = A\,\sigma_k\,A^\dagger$. Now eq. (7.6.18) can be rewritten as

$$\sigma_i\, M\,\tilde{\sigma}^i = 2\operatorname{Tr} M\cdot\mathbf{1}, \tag{8.2.21}$$

so that

$$A = \frac{1}{N}\, L^i{}_k\,\sigma_i\,\tilde{\sigma}^k, \tag{8.2.22}$$

where the denominator $N = 2\operatorname{Sp} A^\dagger$ may be determined from the condition $\det A = 1$ as

$$N = \pm\sqrt{\det L^i{}_k\,\sigma_i\,\tilde{\sigma}^k}. \tag{8.2.23}$$

The formula obtained has to break down for some L on topological grounds again—what are these L?

5. Every 2×2 matrix M may be written in the form $M = m^\ell\,\sigma_\ell$, where $m^\ell = \frac{1}{2}\operatorname{Tr} M\,\tilde{\sigma}^\ell$. We shall need decompositions of this kind for the cases $M = \sigma_i\,\tilde{\sigma}_j\,\sigma_k$, $\sigma_i\,\tilde{\sigma}_j\,\sigma_k\,\tilde{\sigma}_m\,\sigma_n,\ldots$, or equivalently, we shall need the traces $\frac{1}{2}\operatorname{Tr}\sigma_i\,\tilde{\sigma}_j\,\sigma_k\,\tilde{\sigma}_\ell,\ldots.$ They may all be reduced recursively to simpler products using

$$\sigma_i\,\tilde{\sigma}_j\,\sigma_k = \eta_{ij}\,\sigma_k + \eta_{jk}\,\sigma_i + \eta_{ik}\,\sigma_j + i\epsilon_{ijk\ell}\,\sigma^\ell. \tag{8.2.24}$$

This equation follows in turn from

$$\frac{1}{2}\operatorname{Tr}\sigma_i\,\tilde{\sigma}_j\,\sigma_k\,\tilde{\sigma}_\ell = \eta_{ij}\,\eta_{k\ell} + \eta_{jk}\,\eta_{i\ell} - \eta_{ik}\,\eta_{j\ell} + i\epsilon_{ijk\ell}. \tag{8.2.25}$$

Prove this last formula in two steps.

   a. For the part symmetric in i, k show that

$$\frac{1}{2}(\sigma_i\,\tilde{\sigma}_j\,\sigma_k\,\tilde{\sigma}_\ell + \sigma_k\,\tilde{\sigma}_j\,\sigma_i\,\tilde{\sigma}_\ell) = \eta_{ij}\,\sigma_k\,\tilde{\sigma}_\ell + \eta_{jk}\,\sigma_i\,\tilde{\sigma}_\ell - \eta_{ik}\,\sigma_j\,\tilde{\sigma}_\ell,$$

   by reshuffling factors of the first term, using eq. (8.2.19) three times, until it takes the form of the second term; the trace is then obtained using eq. (8.2.20).

   b. For the part antisymmetric in i, k, namely $\frac{1}{2}\operatorname{Tr}(\sigma_i\,\tilde{\sigma}_j\,\sigma_k\,\tilde{\sigma}_\ell - \sigma_k\,\tilde{\sigma}_j\,\sigma_i\,\tilde{\sigma}_\ell)$, show its total antisymmetry, and thus proportionality to $\epsilon_{ijk\ell}$, by cyclic permutation under the trace and use of relations (8.2.19,20). Finally, determine the factor of proportionality.

6. As an application, determine the denominator $N = \mathrm{Tr}\, \mathrm{A}^\dagger$ of eq. (8.2.22) in the following manner. Write the analogous equation for $\mathrm{A}^{-1} \mapsto (L^{-1})^i{}_k = L_i{}^k$ and multiply the two; then because of $\mathrm{Tr}\, \mathrm{A}^\dagger = \mathrm{Tr}\, \mathrm{A}^{-1\dagger}$ in the unimodular case one gets

$$N^2 \cdot \mathbf{1} = L^i{}_k \, L_m{}^n \, \sigma_i \, \tilde{\sigma}^k \, \sigma_n \, \tilde{\sigma}^m = L^i{}_k \, L^m{}_n \, \sigma_i \, \tilde{\sigma}^k \, \sigma^n \, \tilde{\sigma}_m$$

$$N^2 = L^i{}_k \, L^m{}_n \cdot \frac{1}{2} \, \mathrm{Tr}\, \sigma_i \, \tilde{\sigma}^k \, \sigma^n \, \tilde{\sigma}_m.$$

7. Find the unimodular matrix H explicitly that belongs to a boost (1.5.6), and show that $\mathrm{H}^2 = u^i \, \tilde{\sigma}_i$, where u^i are the 4-velocity components corresponding to the velocity $\mathbf{v}$.

8.   a. Set up the theory of the finite-dimensional representations for the connected component of the unit element of the pseudo-orthogonal group SO(1,2) (Lorentz group in a space-time with 2 space dimensions only, or rotations in the sense of a metric $(dx_1)^2 - (dx_2)^2 - (dx_3)^2$).

     b. From the spinor representation of this group, deduce its isomorphism to the group $SL(2,\mathbf{R})/\mathcal{Z}_2$ and investigate the topology of the group $SL(2,\mathbf{R})$ of *real* unimodular 2×2 matrices.

     c. Show that an element $\mathrm{A} \in SL(2, \mathbf{R})$ is contained in no 1-parameter subgroup when $\mathrm{Tr}\, \mathrm{A} < -2$, and that the element $-\mathbf{1} \in SL(2, \mathbf{R})$ is contained in infinitely many ones.

Hints: (a) Complexify! (b) Write for a real unimodular 2×2 matrix $\mathrm{A} = \begin{pmatrix} a & b \\ c & d \end{pmatrix} = a_1 \sigma_1 + a_2 \, i\sigma_2 + a_3 \sigma_3 + a_4 \cdot \mathbf{1}$ and consider the real a_i as coordinates in Euclidean $\mathbf{R}^4$. Then $\det \mathrm{A} = 1$ is the equation of a hyperboloid: $(a_2)^2 + (a_4)^2 = 1 + (a_1)^2 + (a_3)^2$. To *each* pair $(a_1, a_3) \in \mathbf{R}^2$ there is a circle $\mathbf{S}_1$. The hyperboloid thus has topology $\mathbf{R}^2 \times \mathbf{S}_1$. (For a sphere $\mathbf{S}_3$: $(a_1)^2 + (a_2)^2 + (a_3)^2 + (a_4)^2 = 1$ or $(a_2)^2 + (a_4)^2 = 1 - (a_1)^2 - (a_3)^2$ one could not draw an analogous conclusion, since to the pairs $(a_1, a_3) \in \mathbf{R}^2$ having $(a_1)^2 + (a_3)^2 = 1$ there is not a circle but a point!) In particular, then, $SL(2,\mathbf{R})$ is infinitely connected. Since $SL(2,\mathbf{R})$ covers the connected component of unity of $SO(1,2)$ twice and since in part a we obtain only single- and two-valued representations, thus single-valued for $SL(2,\mathbf{R})$, we see here an example of a group that has only single-valued representations despite its infinite connectivity. (It must be pointed out that this statement and the argument given for it—which is based on complexification—become *invalid* as soon as infinite-dimensional representations are considered, as has been frequently stressed by Y. Ne'eman.) (c) Exponentiate explicitly an arbitrary Lie algebra element of the defining representation, i.e., a traceless real 2×2 matrix. Try to get a geometric picture!

9. Show that $SL(2, \mathbf{R}) \cong SU(1, 1)$.

10. Demonstrate the following relations between the 4-vector x^i and the associated matrix X:

$$
\begin{array}{lcl}
\text{timelike future-directed} & \Leftrightarrow & \text{positive-definite} \\
\text{timelike past-directed} & \Leftrightarrow & \text{negative-definite} \\
\text{lightlike future-directed} & \Leftrightarrow & \text{positive-semidefinite} \\
\text{lightlike past-directed} & \Leftrightarrow & \text{negative-semidefinite} \\
\text{spacelike} & \Leftrightarrow & \text{indefinite}
\end{array}
$$

8.3 Spinor Algebra[1]

As in the case of the rotation group, the elements of a representation space on which $\mathcal{L}_+^\uparrow$ acts via $\mathrm{D}^{(1/2,0)}$ will be called *spinors* (of degree one). The spinor space $\mathbf{S}$ is therefore complex-2-dimensional, $\dim_\mathbb{C}(\mathbf{S}) = 2$. To every Lorentz transformation $L \in \mathcal{L}_+^\uparrow$ there correspond two unimodular transformations $\pm A$, i.e., after choosing a basis, two SL(2,$\mathbb{C}$) matrices $\pm A$, and a spinor Ψ transforms under it, by definition, as

$$
\Psi \mapsto \Psi' = \mathrm{A}\Psi \qquad \text{or} \qquad \Psi'^J = A^J{}_K \Psi^K \qquad (J,K = 1,2). \tag{8.3.1}
$$

By forming tensor products one constructs spinors of higher degree, transforming correspondingly as

$$
\Psi'^{JK\cdots} = A^J{}_M \, A^K{}_N \ldots \Psi^{MN\cdots}. \tag{8.3.2}
$$

(Of course, there are again the possibilities of active and passive interpretations.) If we call these spinors by convention *contravariant*, then there are also covariant spinors Φ, making up the dual spinor space $\tilde{\mathbf{S}}$, which under L are transformed by $\tilde{\mathrm{A}}$:

$$
\Phi \to \Phi' = \tilde{\mathrm{A}}\,\Phi \quad \text{or} \quad \Phi'_J = A_J{}^K \Phi_K, \quad A_J{}^K A^J{}_L = \delta_L^K.
$$

However, eq. (8.2.16) shows us that there is an equivalence map—given there by the matrix σ_2—between co- and contravariant spinors: if $\Psi' = \mathrm{A}\Psi$, then for $\Phi = \sigma_2\Psi$, $\Phi' = \sigma_2\Psi'$ we have the relation $\Phi' = \tilde{\mathrm{A}}\,\Phi$. We are led, therefore, as in the case of 4-vectors, to identify the corresponding objects and to speak of co- and contravariant components of one and the same spinor only. More precisely, we shall write

$$
\Phi^A := \epsilon^{AB} \Phi_B \quad \text{with} \quad \epsilon = (\epsilon^{AB}) := i\sigma_2 = -\epsilon^T = -\epsilon^{-1} = \begin{pmatrix} 0 & 1 \\ -1 & 0 \end{pmatrix}. \tag{8.3.3}
$$

To explain the factor i here, remember that the equivalence map is unique up to a complex factor and is related either to a symmetric or an antisymmetric bilinear form on general grounds (exercise 4 of sect. 7.5); but in fact we know that it is antisymmetric (symplectic). We can now either think of a given basis and choose the open numerical factor such that eq. (8.3.3) holds with the numerical values written, or we may imagine the map being given and the basis being selected such that its matrix is as in eq. (8.3.3)—sometimes this is called a unimodular, or *spin frame*. In any case, the ϵ^{AB} are components of an invariant antisymmetric spinor of degree 2 that exists due to unimodularity, $\det A = 1$, in complete analogy to the ϵ-tensor of

---

[1]For this and the next section, we highly recommend to the reader the text of Penrose and Rindler (1984)!

sect. 5.5; however, because of its degree, it defines an invariant bilinear form (spinor scalar product)

$$\Phi_A \Psi^A = \epsilon^{AB} \Phi_A \Psi_B = -\epsilon^{BA} \Psi_B \Phi_A = -\Psi_B \Phi^B. \tag{8.3.4}$$

(We encountered this form already in eq. (7.6.22) but now see its Lorentz invariance, whereas the sesquilinear form (7.6.21) is not Lorentz invariant—our representation is not unitary.) In the sense of this invariant 'scalar product', or spinor 'metric', every spinor is orthogonal to itself, and orthogonal spinors must be proportional.

When moving indices with the help of the spinor metric ϵ^{AB} their order has to be observed. The covariant components ϵ_{AB} of the spinor metric have to be chosen in conformity with eq. (8.3.3), i.e., such that $\epsilon^{AB} = \epsilon^{AC} \epsilon^{BD} \epsilon_{CD}$ holds. It follows that

$$\epsilon^{BD} \epsilon_{CD} = \delta^B_C, \qquad (\epsilon_{CD}) = \left(\epsilon^\top\right)^{-1} = \epsilon \tag{8.3.5}$$

(the second of these again involving the use of unimodular bases), so that the inverse of eq. (8.3.3) is

$$\Phi_B = \Phi^A \epsilon_{AB}. \tag{8.3.6}$$

Just as in general tensor algebra, symmetrization and antisymmetrization are invariant processes. However, the dimensionaliy 2 of the spinor space entails that the situation is particularly simple here. Totally antisymmetric spinors of degree higher than 2 vanish identically, while those of degree 2 are multiples of ϵ_{AB}, as they have only one independent component:

$$\Phi_{AB} = -\Phi_{BA} \text{ implies } \Phi_{AB} = \frac{1}{2} \Phi_C{}^C \epsilon_{AB}. \tag{8.3.7}$$

(The factor of proportionality follows by transvecting with ϵ^{AB}). For arbitrary Φ_{AB} we therefore have

$$\Phi_{AB} - \Phi_{BA} = \Phi_C{}^C \epsilon_{AB} = \epsilon_{AB} \epsilon^{CD} \Phi_{CD}. \tag{8.3.8}$$

From this follows the relation

$$\epsilon_{AB} \epsilon^{CD} = \delta_A{}^C \delta_B{}^D - \delta_A{}^D \delta_B{}^C, \tag{8.3.9}$$

which is the analog of eq. (5.5.9e). A further, related simplification is that for totally symmetric spinors all contractions vanish. In fact, we have already seen in the case of the rotation subgroup that total symmetry means irreducibility: such spinors of degree p transform as $D^{(p/2,0)}$.

Spinors of higher degree may be reduced by systematic symmetrization and anti-symmetrization, using eq. (8.3.9). For instance,

$$\Phi_{AB} = \Phi_{(AB)} + \Phi_{[AB]} = \Phi_{(AB)} + \frac{1}{2} \Phi_E{}^E \epsilon_{AB} \tag{8.3.10}$$

yields the reduction of $D^{(1/2)} \otimes D^{(1/2)} = D^{(1)} \oplus D^{(0)}$ for the rotation group, and, correspondingly, of $D^{(1/2,0)} \otimes D^{(1/2,0)}$ for the Lorentz group (ϵ_{AB} as an invariant spinor transforms according to the trivial representation). In the general case we have

$$\Phi_{A_1 \ldots A_p} = \Phi_{(A_1 \ldots A_p)} + \text{remainder}, \tag{8.3.11}$$

where the totally symmetric part transforms as $D^{(p/2,0)}$ and is explicitly given by

$$\Phi_{(A_1 \dots A_p)} = \frac{1}{p!} \sum_{\pi} \Phi_{A_{\pi(1)} \dots A_{\pi(p)}}. \tag{8.3.12}$$

$(\pi(1) \dots \pi(p)$ indicates some permutation of the subindices $1 \dots p$, and the sum over all $p!$ permutations π is to be taken.) The remainder may be written as a sum of $p! - 1$ terms of the form

$$\frac{1}{p!} \left\{ \Phi_{A_1 \dots A_p} - \Phi_{A_{\pi(1)} \dots A_{\pi(p)}} \right\}. \tag{8.3.13}$$

Since every permutation π can be carried out in steps which are simple exchanges of two subindices only, e.g.,

$$\Phi_{ABIJ} - \Phi_{IJAB} = (\Phi_{ABIJ} - \Phi_{IBAJ}) + (\Phi_{IBAJ} - \Phi_{IJAB}), \tag{8.3.14}$$

each of the differences (8.3.13) is, by eq. (8.3.9), a sum of expressions $\Phi_{\dots B \dots J \dots} - \Phi_{\dots J \dots B \dots} = \epsilon_{BJ} \Phi_{\dots E \dots}{}^E{}_{\dots}$. By the invariance of $\epsilon_{\dots}$ the 'effective' degree of the remainder thus has been lowered by 2. Note that, even without symmetries the given $\Phi_{A_1 \dots A_p}$ might possess, there are relations between the spinors $\Phi_{\dots E \dots}{}^E{}_{\dots}$ following from eq. (8.3.9), e.g.,

$$\Phi_A{}^E{}_E + \Phi^E{}_{EA} + \Phi_{EA}{}^E = 0, \dots, \tag{8.3.15}$$

which together with possible symmetries of $\Phi_{A_1 \dots A_p}$ have to be taken into account in discussing the multiplicity of the irreducible representations that occur in the remainder.

Let us write again $D^{(j)}$ instead of $D^{(j,0)}$ to make the notation less clumsy, as long as no other representations come in. As an example, then, let us consider a space of spinors that transform according to $D^{(j_1)} \otimes D^{(j_2)}$, thus having $2j_1 + 2j_2$ indices,

$$\underbrace{\Phi_{A \dots B}}_{2j_1} \underbrace{{}_{I \dots J}}_{2j_2}, \tag{8.3.16}$$

with total symmetry inside the two sets of indices indicated. When we apply the method sketched above, we obtain the part $\Phi_{(A \dots J)}$, transforming as $D^{(j_1+j_2)}$, and remaining terms of the form

$$\epsilon_{AI} \, \Phi_{E \dots B}{}^E{}_{\dots J} \, , \epsilon_{AI} \, \Phi_{E \dots J}{}^E{}_{\dots B} \, , \tag{8.3.17}$$

which, because of the remaining symmetry, transform according to $D^{(j_1-1/2)} \otimes D^{(j_2-1/2)}$. For the reasons given, the terms in the remainder are by no means independent, which restricts the multiplicity of the occurrence of the latter representation. In fact, this multiplicity is just 1, as a simple dimension check shows:

$$(2j_1 + 1)(2j_2 + 1) = 2(j_1 + j_2) + 1 + [2(j_1 - 1/2) + 1]\,[2(j_2 - 1/2) + 1]. \tag{8.3.18}$$

Therefore we get recursively

$$\begin{aligned}
D^{(j_1)} \otimes D^{(j_2)} &= D^{(j_1+j_2)} \oplus \left(D^{(j_1-1/2)} \otimes D^{(j_2-1/2)} \right) = \\
&= D^{(j_1+j_2)} \oplus D^{(j_1+j_2-1)} \oplus \left(D^{(j_1-1)} \otimes D^{(j_2-1)} \right) = \dots \\
&= D^{(j_1+j_2)} \oplus D^{(j_1+j_2-1)} \oplus \dots \oplus D^{(|j_1-j_2|)},
\end{aligned} \tag{8.3.19}$$

proving the *Clebsch-Gordan decomposition* (7.8.8).

For the totally symmetric degree p spinors $\Phi_{A \ldots K}$ there is a (multiplicative) decomposition into *principal spinors* of degree 1, permitting a further classification inside each irreducible representation space (introduced into physics by E. Majorana, Nuovo Cimento *9*, 43 (1932)) and R. Penrose, Ann. Phys. (N.Y.) *10*, 171 (1960)). It is derived as follows: use an auxiliary spinor ξ^A to form the scalar $\Phi_{A \ldots K}\, \xi^A \ldots \xi^K$. This is a homogeneous polynomial of degree p in the variables ξ^1, ξ^2 and thus equal to $(\xi^2)^p$ times a polynomial of degree p in the variable $\xi := \xi^1/\xi^2$. By the fundamental theorem of algebra, the latter decomposes into the product of p linear factors $\xi - \alpha$, $\xi - \beta$, $\ldots$ This gives a decomposition also for $\Phi_{A \ldots K}\, \xi^A \ldots \xi^K$:

$$\Phi_{A \ldots K}\, \xi^A \ldots \xi^K = \left(\alpha_A \xi^A\right) \left(\beta_B \xi^B\right) \ldots \left(\kappa_K \xi^K\right) = \alpha_{(A} \beta_B \ldots \kappa_{K)}\, \xi^A \xi^B \ldots \xi^K,$$

and, since ξ^A is arbitrary:

$$\Phi_{AB \ldots K} = \alpha_{(A} \beta_B \ldots \kappa_{K)}. \tag{8.3.20}$$

Every totally symmetric spinor of degree p can thus be written as a symmetrized product of p principal spinors of degree 1, each unique up to a complex factor. The classification in question now consists in the statement whether and how many principal spinors are proportional. The theory of invariants of binary forms can be invoked to express this in terms of certain concomitants of $\Phi_{A \ldots K}$, at least for low values of the degree p. (See R. Penrose, Ann. Phys. (N.Y.) *10*, 171 (1960), and Penrose and Rindler 1984.) We shall give an application of this decomposition in exercise 8b of sect. 9.4. (The final aim of such classifications would be to separate the orbits of objects inside the irreducible representation spaces under the group considered. It has been achieved only for low degrees.)

We may develop an entirely analogous formalism for the spinors of the complex-conjugate spinor space $\mathbf{S}^*$ that transform according to the complex-conjugate representation $\mathrm{D}^{(0,1/2)}$. It is customary to write (the components of) such spinors with *dotted* or *primed* indices. By definition, a *dotted spinor* of degree 1 transforms under $\mathcal{L}_+^\uparrow$ as

$$\Psi'^{\dot{J}} = A^{\dot{J}}{}_{\dot{K}}\, \Psi^{\dot{K}}, \qquad\qquad A^{\dot{J}}{}_{\dot{K}} := \left(A^J{}_K\right)^*. \tag{8.3.21}$$

We remark here that for an undotted $\mathrm{D}^{(1/2,0)}$-spinor Φ_A the complex-conjugate spinor Φ_A^* transforms according to $\mathrm{D}^{(0,1/2)}$ and thus should be written $\Phi_{\dot{A}}^*$. Note then, however, that for the complexified Lorentz group the representation spaces of $\mathrm{D}^{(1/2,0)}$ and $\mathrm{D}^{(0,1/2)}$ are unrelated, in the sense that complex conjugation and complex Lorentz transformations do not commute, as is clear from the first of relations (8.2.10). (We also remark here that in much of the literature using 2-component spinors, complex conjugation is written using an overbar, and dots are replaced by primes.)

The invariant symplectic 'metric' $\epsilon_{\dot{A}\dot{B}}$ is chosen to be the complex-conjugate of ϵ_{AB}; this means that if we use a unimodular or spin frame and its complex-conjugate, we have the numerical equality $\epsilon_{\dot{A}\dot{B}} = \epsilon_{AB}$ as given in eqs. (8.3.3,5). We now can repeat all considerations made above literally for dotted spinors; in particular, totally symmetric dotted spinors of degree p transform as $\mathrm{D}^{(0,p/2)}$.

For objects transforming according to the irreducible representation $\mathrm{D}^{(j,j')}$ we thus may take tensor products of spinors with $2j$ undotted and $2j'$ dotted indices, i.e., we have general spinors of the form

$$\Phi_{AB \ldots I \dot{X} \dot{Y} \ldots \dot{Z}}, \tag{8.3.22}$$

totally symmetric with respect to A, B, ... , I and also totally symmetric with respect to $\dot{X}$, $\dot{Y}$, ... , $\dot{Z}$. The relative position of dotted and undotted indices is irrelevant since they do not refer to the same space; the basic operations of spinor algebra—symmetrization and contraction—have to operate with indices of the same type only.

Let us finally mention that the components of an irreducible spinor (8.3.22) are sometimes also numbered in the form $\Phi_{\alpha\dot{\beta}}$, where α (or β) is the number of undotted (or dotted) indices that are equal to 1 (say)—a totally symmetric spinor is completely fixed by these data—, α (or β) running from 0 to $2j$ (or $2j'$). Still another numbering is to have α or β running from $-j$ to j or from $-j'$ to j', respectively; here one may also add the normalization relevant for the unitary representations of the rotation group (cf. eqs. (7.6.29,30)).

Appendix: Determination of the Lower Clebsch-Gordan Terms

The explicit calculation of the irreducible parts together with their normalization when referred to canonical bases for the unitary representations of the rotation group is obviously a more complicated combinatorial problem[1] that we do not tackle here in detail: for the Lorentz group, unitarity cannot be obtained anyway, as we saw before; so let us be content with making more explicit the lower terms in the decomposition (8.3.19).

For this purpose, first observe that the remainder in eq. (8.3.11) is, in our case (8.3.16), a sum of terms which are tensor products of ϵ's and partial contractions of Φ, a sum which has, as a whole, the same symmetry as indicated in expression (8.3.16). It must therefore look like $\delta^{(I}_{(A}\Psi^{\cdots\ J)}_{\ \ \ B)}$, where the second set of indices, $I \ldots J$, has been written upstairs for notational ease, and where Ψ is linear in Φ, with symmetry among the $\ldots B$ and among the $\ldots J$. (We write δ^I_A instead of $\epsilon_A{}^I$ in conformity with eq. (8.3.5).) Now the second step in the procedure (8.3.19) is to pick out $\mathrm{D}^{(j_1+j_2-1)}$ by symmetrizing $\Psi_{\ldots B \ldots J}$, which up to a factor is nothing but taking $\Phi_{E(\ldots B}{}^E{}_{\ldots J)}$, since the terms with more ϵ's (more contractions) do not contribute. Let us write $2j_1 = p$, $2j_2 = q$, $2j = p + q - 2r$; it is then clear that, similarly, the $\mathrm{D}^{(j)}$ component will be a multiple of

$$\delta^{(I_1}_{(A_1}\ \cdots\ \delta^{I_r}_{A_r}\ \Psi_{B_{r+1}\ \ldots\ B_p)}{}^{J_{r+1}\ \ldots\ J_q)},\tag{8.3.23}$$

where now

$$\Psi_{\ldots B \ldots J} := \Phi_{E_1\ \ldots\ E_r(B_{r+1}\ \ldots\ B_p}{}^{E_1\ \ldots\ E_r}{}_{J_{r+1}\ \ldots\ J_q)}.\tag{8.3.24}$$

Let us write these operations on Ψ and Φ symbolically as $\iota\,\Psi$ and $\Psi = \pi\,\Phi$; then ι and π are an intertwining injection and an intertwining surjection for $\mathrm{D}^{(j)}$ as described in sect. 6.6. Since $\pi \circ \iota$ is self-intertwining the irreducible representation $\mathrm{D}^{(j)}$, it is a multiple of the identity,

$$\pi \circ \iota = k\,\mathrm{id}_{(j)}.\tag{8.3.25}$$

Of course, the numerical factor k depends on j_1, j_2 and j and is nonzero only if j is in the range given by Clebsch-Gordan. If so, then

$$\iota \circ \pi / k = (\iota \circ \pi / k)^2\tag{8.3.26}$$

is idempotent and gives the *invariant projection* to the irreducible component under consideration. So what is still missing is the numerical factor k, which turns out to be rational and of a purely combinatorial nature. We defer its determination and the result to an exercise.

---

[1] R. Penrose has made an attempt to base far-reaching speculations on the fact that there is a purely combinatorial problem behind quantum mechanical addition of angular momenta—see his article in Klauder (1972).

Exercises

1. Find the decomposition into irreducible parts, analogous to eq. (8.3.10), for a spinor Φ_{ABC} symmetric in A, B!

   Solution: $\Phi_{ABC} = \Phi_{(ABC)} + \frac{2}{3}\Phi_{E(A}{}^{E}\epsilon_{B)C}$.

2. By grouping the permutations of $s + 1$ symbols $E, E_1, E_2, \ldots E_s$ into those beginning with E, with $E_1, \ldots, E_s$, one can rewrite a total symmetrization over $s + 1$ indices $\Omega_{(E\,E_1\,E_2\,\ldots\,E_s)}$ as

$$\frac{1}{s+1}\left[\Omega_{E(E_1\,E_2\,\ldots\,E_s)} + \Omega_{E_1(E\,E_2\,\ldots\,E_s)} + \Omega_{E_2(E_1\,E\,\ldots\,E_s)} + \ldots + \Omega_{E_s(E_1\,E_2\,\ldots\,E)}\right].$$
(8.3.27)

Use this twice in the explicit expression for $\pi\iota\Psi$ that results from the definitions (8.3.25,26), taking into account the vanishing of all contractions of Ψ, to get, with some patience while counting, for the still open combinatorial factor k the recursion

$$p\,q\,k(p,q,r) = r(p + q - r + 1)\,k(p - 1, q - 1, r - 1).$$

This, together with the initial condition $k(*, **, 0) = 1$, gives[1], for $r \leq \min(p,q)$,

$$k = \frac{\binom{p+q-r+1}{r}}{\binom{p}{r}\binom{q}{r}} = \frac{(2j_1)!(2j_2)!(2j+1)!}{(j_1 + j_2 + j + 1)!(-j_1 + j_2 + j)!(j_1 - j_2 + j)!(j_1 + j_2 - j)!}. \quad (8.3.28)$$

8.4 The Relation between Spinors and Tensors

For integer values of $j + j'$, the representations $D^{(j,j')}$ are single-valued—there is an even number of factors A. in the transformation formulae. The spinor representations being fundamental representations and all finite-dimensional representations being completely reducible, it is clear that all 4-tensors may be constructed from spinors; but one will also suspect that the converse is true, namely that all single-valued representations are 4-tensor representations. We are going to show this here, developing a suitable formalism for this purpose.

The simplest case is $D^{(1/2,1/2)}$, the representation carried by spinors with one undotted and one dotted index:

$$X'^{A\dot{Y}} = A^{A}{}_{B}\,A^{*\dot{Y}}{}_{\dot{Z}}\,X^{B\dot{Z}}. \quad (8.4.1)$$

In matrix notation this reads $X' = A\,X\,A^{\dagger}$, so it is identical to eq. (8.2.6), showing that the spinor components $X^{A\dot{B}}$ are linear combinations of the components of a 4-vector whose precise form is given by eq. (8.2.2). To real 4-vectors there correspond Hermitian spinors.

It is sometimes convenient to make eqs. (8.2.4) more symmetric by having the factor $1/2$ shared more symmetrically among the two equations, defining '*soldering quantities*'

$$\sigma_i{}^{A\dot{B}} := \frac{1}{\sqrt{2}}(\sigma_i)^{AB}, \qquad \sigma^{i}{}_{A\dot{B}} = \frac{1}{\sqrt{2}}(\tilde{\sigma}^{iT})_{AB}. \quad (8.4.2)$$

---

[1] Penrose and Rindler (1984)

Then instead of eq. (8.2.4) we have the pair of equations

$$X^{A\dot{Y}} = x^i \sigma_i{}^{A\dot{Y}} \Leftrightarrow x^i = X^{A\dot{Y}} \sigma^i{}_{A\dot{Y}} . \tag{8.4.3}$$

Here it is guaranteed by eq. (8.2.17) that the notation $\sigma_i{}^{A\dot{X}}$, $\sigma^i{}_{A\dot{X}}$ is indeed compatible with the rules (8.3.3,6) (and its dotted versions) for moving spinor indices. (This, and the validity of the formulae (8.4.4–8), while verifiable directly, will appear less miraculous at the end of the section!) Since x^i is arbitrary, we get from eq. (8.4.3)

$$\sigma^i{}_{A\dot{X}} \sigma_k{}^{A\dot{X}} = \delta^i_k, \quad \sigma^i{}_{A\dot{X}} \sigma_i{}^{B\dot{Y}} = \delta^B_A \delta^{\dot{Y}}_{\dot{X}}. \tag{8.4.4a,b}$$

Apparently more general than eq. (8.4.4a) is the formula

$$\sigma^i{}_{A\dot{X}} \sigma^k{}_B{}^{\dot{X}} + \sigma^k{}_{A\dot{X}} \sigma^i{}_B{}^{\dot{X}} = \epsilon_{AB} \eta^{ik} \tag{8.4.5}$$

(and its complex-conjugate), which derives from it by remarking that the left-hand side is, by eq. (8.3.4), antisymmetric in A, B and thus proportional to ϵ_{AB}. One also sees that eq. (8.4.5) is the same as eq. (8.2.19) in the new notation. We shall also need a rewritten version of eqs. (8.3.24,25):

$$\sigma_i{}^{A\dot{X}} \sigma^k{}_{B\dot{X}} \sigma_m{}^{B\dot{Y}} = \frac{1}{2} \left(\delta^k_i \sigma_m{}^{A\dot{Y}} + \delta^k_m \sigma_i{}^{A\dot{Y}} - \eta_{im} \sigma^{kA\dot{Y}} + i \epsilon_i{}^k{}_m{}^n \sigma_n{}^{A\dot{Y}} \right) \tag{8.4.6}$$

$$\sigma_i{}^{A\dot{X}} \sigma^k{}_{B\dot{X}} \sigma_m{}^{B\dot{Y}} \sigma^n{}_{A\dot{Y}} = \frac{1}{2}(\delta^k_i \delta^n_m + \delta^k_m \delta^n_i - \eta_{im} \eta^{kn} + i \epsilon_i{}^k{}_m{}^n). \tag{8.4.7}$$

When eq. (8.4.6) is multiplied by $\sigma^m{}_{C\dot{Y}}$ and eq. (8.4.4b) is used, one gets

$$\sigma_i{}^{A\dot{X}} \sigma_{kC\dot{X}} = \frac{1}{2} \eta_{ik} \delta^A_C - \frac{i}{2} \epsilon_{ik}{}^{mn} \sigma_m{}^{A\dot{X}} \sigma_{nC\dot{X}}. \tag{8.4.8}$$

It is possible (cf. Schmutzer 1968) to deduce from this equation and its complex-conjugate alone all the other relations above without using a special realization of the $\sigma_i{}^{A\dot{X}}$: they have just to be Hermitian solutions of eq. (8.4.8), where indices are moved according to our rules. The symmetric part of eq. (8.4.8) expresses the equivalence between the 4-vector representation and $D^{(1/2,1/2)}$. As we will show now, the antisymmetric part of the relation similarly expresses the equivalence between $D^{(1,0)}$ and the selfdual sixtor representation. As described before, $D^{(1,0)}$ is carried by symmetric spinors $\Phi_{AB} = \Phi_{BA}$; if we use mixed components $\Phi^I{}_J$ (note $\Phi^J{}_J = 0$), their transformation law is

$$\Phi'^I{}_J = A^I{}_K A_J{}^L \Phi^K{}_L, \tag{8.4.9}$$

or, in matrix notation, $\Phi' = A \Phi A^{-1}$ ($\mathrm{Tr}\,\Phi = 0 \Rightarrow \mathrm{Tr}\,\Phi' = 0$). We saw already (cf. eq. (8.2.10) and exercise 7 of sect. 7.6) that this describes complex-orthogonal transformations of the complex vectors $\mathbf{F} = \frac{1}{2} \mathrm{Tr}\,\Phi\,\boldsymbol{\sigma}$. When we now think of $\mathbf{E} := \mathrm{Re}\,\mathbf{F}$, $\mathbf{B} := \mathrm{Im}\,\mathbf{F}$ as the components of a (real) sixtor F_{ik} according to eq. (5.2.18), then using eqs. (5.2.20) and (5.7.1) it can be seen that we have, in 4-dimensional notation,

$$f_{ik} := F_{ik} - i{*}F_{ik} = -\frac{1}{2} \mathrm{Sp}\,\Phi\,\sigma_i \tilde{\sigma}_k, \tag{8.4.10}$$

i.e., Φ determines a selfdual sixtor f_{ik}. Conversely now, one finds, using the antisymmetric part of eq. (8.4.8),

$$\Phi = (E_\lambda + i B_\lambda)\sigma_\lambda = \frac{1}{2} F^{ik} \sigma_i \tilde{\sigma}_k = \frac{-i}{2} {*}F^{ik} \sigma_i \tilde{\sigma}_k = \frac{1}{4} f^{ik} \sigma_i \tilde{\sigma}_k . \tag{8.4.11}$$

Equations (8.4.10,11) are the pair of formulae for $D^{(1,0)}$ analogous to the pair (8.2.4) for $D^{(1/2,1/2)}$; using the notation of eq. (8.4.2) it rewrites

$$\Phi^A{}_B = \frac{1}{2}\, f^{ik}\, \sigma_i{}^{A\dot{X}}\, \sigma_{kB\dot{X}}, \qquad\qquad f_{ik} = \Phi^A{}_B\, \sigma_{iA\dot{X}}\, \sigma_k{}^{B\dot{X}}. \tag{8.4.12}$$

It is seen now that the antisymmetric part of eq. (8.4.8) expresses the selfduality of f_{ik}. A less explicit, more group theoretic, argument will be found below.

In complete analogy to the first of eqs. (8.4.3) we can construct to each 4-tensor an equivalent spinor:

$$T^{ik\cdots} \mapsto T^{A\dot{X}B\dot{Y}\cdots} = T^{ik\cdots}\sigma_i{}^{A\dot{X}}\,\sigma_k{}^{B\dot{Y}}\cdots, \tag{8.4.13}$$

and conversely, to each spinor with equal numbers of dotted and undotted indices we can form an equivalent 4-tensor

$$T^{A\dot{X}B\dot{Y}\cdots} \mapsto T^{ik\cdots} = T^{A\dot{X}B\dot{Y}\cdots}\sigma^i{}_{A\dot{X}}\,\sigma^k{}_{B\dot{Y}}\cdots. \tag{8.4.14}$$

This shows, in particular, that the $D^{(j,j)}$ are equivalent to certain irreducible 4-tensor representations; the total symmetry of the spinors carrying these representations is immediately seen to imply *total symmetry* of the corresponding 4-tensors; but as will be seen by generalizing the example treated below, these tensors are also *trace-free*, and both properties taken together serve to characterize them as irreducible. For reality, see below.

To complete the general situation, consider now the case of spinors with an even total number of indices (even total degree), in particular, spinors carrying the representations $D^{(j,j')}$ with $j + j'$ =integer. The procedure to adopt here suggests itself when we rewrite eq. (8.4.12) as

$$f^{ik} = \Phi^{AB}\, \epsilon^{\dot{X}\dot{Y}}\, \sigma^i{}_{A\dot{X}}\, \sigma^k{}_{B\dot{Y}} :$$

if $\Phi_{A\ldots B\dot{X}\ldots\dot{Y}}$ has a surplus of indices of one kind, fill up the number of indices of the other kind by multiplying with an appropriate number of factors $\epsilon_{CD\ldots}$ or $\epsilon_{\dot{Z}\dot{U}\ldots}$ and apply eq. (8.4.14). Since $\epsilon_{..}$ is invariant, the equivalence class of the representation is not changed.

To reduce a 4-tensor representation, the following method now results: by eq. (8.4.13), translate to an equivalent spinor; reduce the spinor as indicated in sect. 8.3; translate back each irreducible part by itself, using eq. (8.4.14).

As a first example, consider 4-tensors D^{ik} of degree 2: we reduce the equivalent spinor by symmetrization, applying eq. (8.3.10) and its dotted version independently:

$$D^{AB\dot{X}\dot{Y}} = D^{(AB)\dot{X}\dot{Y}} + D^{[AB]\dot{X}\dot{Y}} = D^{(AB)(\dot{X}\dot{Y})} + D^{(AB)[\dot{X}\dot{Y}]} + D^{[AB](\dot{X}\dot{Y})} + D^{[AB][\dot{X}\dot{Y}]} =$$

$$= D^{(AB)(\dot{X}\dot{Y})} + \frac{1}{2} D^{(AB)}{}_{\dot{Z}}{}^{\dot{Z}}\, \epsilon^{\dot{X}\dot{Y}} + \frac{1}{2} D_C{}^{C(\dot{X}\dot{Y})}\, \epsilon^{AB} + \frac{1}{4} D_C{}^{C}{}_{\dot{Z}}{}^{\dot{Z}}\, \epsilon^{AB}\, \epsilon^{\dot{X}\dot{Y}}. \tag{8.4.15}$$

The terms of the last line transform according to $D^{(1,1)}$, $D^{(1,0)}$, $D^{(0,1)}$, $D^{(0,0)}$; therefore (cf. eq. (6.6.19)) we have

$$D^{(1/2,1/2)} \otimes D^{(1/2,1/2)} = D^{(1,1)} \oplus D^{(1,0)} \oplus D^{(0,1)} \oplus D^{(0,0)}. \tag{8.4.16}$$

To translate them back, each one has to be transvected with $\sigma^i{}_{A\dot{X}}\,\sigma^k{}_{B\dot{Y}}$. As for the last one, we have (cf. eq. (8.4.4a))

$$\epsilon^{AB}\,\epsilon^{\dot{X}\dot{Y}}\,\sigma^i{}_{A\dot{X}}\,\sigma^k{}_{B\dot{Y}} = \sigma^i{}_{A\dot{X}}\,\sigma^{kA\dot{X}} = \eta^{ik} \tag{8.4.17}$$

$$D_C{}^{C}{}_{\dot{Z}}{}^{\dot{Z}} = D^{ik}\,\sigma_{iC}{}^{\dot{Z}}\,\sigma_{k}{}^{C}{}_{\dot{Z}} = D^{ik}\,\eta_{ik}. \tag{8.4.18}$$

The first term translates back to a symmetric traceless tensor on which the projection operator

$$\sigma_m{}^{(A}{}_{(\dot{X}}\,\sigma_n{}^{B)}{}_{\dot{Y})}\,\sigma^i{}_{A}{}^{\dot{X}}\,\sigma^k{}_{B}{}^{\dot{Y}} = \frac{1}{2}\left(\delta^i_m\,\delta^k_n + \delta^i_n\,\delta^k_m\right) - \frac{1}{4}\,\eta^{ik}\,\eta_{mn} \tag{8.4.19}$$

projects, as follows from eq. (8.4.7). Similarly we get for the projection operator to $\mathrm{D}^{(1,0)}$:

$$\frac{1}{2}\,\sigma_m{}^{(A}{}_{\dot{Z}}\,\sigma_n{}^{B)\dot{Z}}\,\epsilon^{\dot{X}\dot{Y}}\,\sigma^i{}_{A\dot{X}}\,\sigma^k{}_{B\dot{Y}} = \frac{1}{4}\left(\delta^i_m\,\delta^k_n - \delta^i_n\,\delta^k_m - i\,\epsilon^{ij}{}_{mn}\right), \tag{8.4.20}$$

giving the selfdual sixtor part of D^{ik}. The $\mathrm{D}^{(0,1)}$ part is translated analogously.

We may note at this point that the formulae (8.4.17,19,20), whose content is the same as that of eqs. (8.4.4-8), follow from eqs. (8.4.15,16), our old decomposition (6.6.19), and general theory: the decomposition (8.4.16), being multiplicity-free, is unique up to order (cf. exercise 2 of sect. 6.6); the middle terms correspond to sixtors, due to the antisymmetry of ϵ; the outer terms correspond to symmetric tensors; so the decompositions match precisely, proving the irreducibility of the old decomposition and the equality of the projection operators on the left-hand sides of eqs. (8.4.17,19,20) and the ones formerly calculated explicitly (eqs. (6.6.10,11,12,14) or implicitly (eq. (6.6.18)). Equation (8.4.17) shows at the same time that $\epsilon^{AB}\,\epsilon^{\dot{X}\dot{Y}}$ is the spinor equivalent to the metric tensor. From eq. (8.4.7) one also may deduce the spinor equivalent to the ϵ-tensor:

$$\epsilon_{A\dot{X}B\dot{Y}C\dot{Z}D\dot{U}} = i\left(\epsilon_{AD}\,\epsilon_{BC}\,\epsilon_{\dot{X}\dot{Z}}\,\epsilon_{\dot{Y}\dot{U}} - \epsilon_{AC}\,\epsilon_{BD}\,\epsilon_{\dot{X}\dot{U}}\,\epsilon_{\dot{Y}\dot{Z}}\right). \tag{8.4.21}$$

As was to be expected, both $\eta_{..}$ and $\epsilon_{....}$ are expressible by the ϵ-spinor; we shall indicate below also how spinors determine the time orientation in addition to the total orientation.

When a spinor is translated into an equivalent 4-tensor via eq. (8.4.14), the latter will, in general, not be real. Just as to *real* 4-vectors x^i there belong *Hermitian* matrices $\mathrm{X} = x^i\sigma_i$, the spinors $T^{AB\cdots\dot{X}\dot{Y}\cdots}$ belonging to real 4-tensors $T^{ik\cdots}$ have the generalized Hermitian property

$$T^{AB\cdots\dot{X}\dot{Y}\cdots} = \left(T^{XY\cdots\dot{A}\dot{B}\cdots}\right)^{*}. \tag{8.4.22}$$

Thus the irreducible spinors transforming according to $\mathrm{D}^{(j,j')}$ with even $j + j'$ have a chance to be Hermitian in this sense only if $j = j'$. When $j \neq j'$, only objects transforming according to the *reducible* representation $\mathrm{D}^{(j,j')} \oplus \mathrm{D}^{(j',j)}$ may correspond to real tensors. Conversely, in the reduction of (complexified) real 4-tensor representations the $D^{(j,j')}$ and $D^{(j',j)}$ with $j \neq j'$ always come in pairs, as illustrated by eq. (8.4.15,16).

Let us, as a further example, analyze the quadratic concomitants of the electromagnetic field tensor F_{ik}, i.e., decompose the tensor $A^{ikmn} = F^{ik} F^{mn}$, which transforms according to $[D^{(1,0)} \oplus D^{(0,1)}] \otimes [D^{(1,0)} \oplus D^{(0,1)}]$. Application of the Clebsch-Gordan theorem gives the decomposition

$$D^{(2,0)} \oplus D^{(0,2)} \oplus D^{(1,0)} \oplus D^{(0,1)} \oplus D^{(1,1)} \oplus D^{(1,1)} \oplus D^{(0,0)} \oplus D^{(0,0)}. \qquad (8.4.23)$$

The $D^{(0,0)}$ parts correspond to the invariants of the field tensor. The $D^{(1,1)}$ parts correspond to two symmetric trace-free tensors of degree 2 which in our case coincide (this would not be so had we started from some $F^{ik} G^{mn}$). This part must be of the form

$$F^{ij} F_j^{\ k} - \frac{1}{4} \eta^{ik} F^{hj} F_{jh} = 4\pi T^{ik}, \qquad (8.4.24)$$

thus agreeing with the energy-momentum tensor (5.9.12). The remaining parts are of minor physical importance: instead of considering them we merely look at the further reduction of T^{ik} that takes place if we restrict to the subgroup of space rotations. We then have $D^{(1,1)} = D^{(2)} \oplus D^{(1)} \oplus D^{(0)}$. To $D^{(0)}$ corresponds the energy density T^{00}, to $D^{(1)}$ corresponds the Poynting vector $T^{0\alpha}$, and to $D^{(2)}$ corresponds the shear part of the Maxwell stress tensor.

In view of sect. 7.8 we must point out that in discussing this example we ignored the dependence of the field strengths on the space-time point considered. If this dependence is included as written in eq. (5.6.2), we obtain an infinite-dimensional representation in the space of tensor *fields*. It turns out to make more sense then to analyze this situation from the point of view of representations of the Poincaré group instead of the Lorentz group, as will be done in chap. 9. (Note that 4-tensors and spinors do furnish representations of the Poincaré group also, by assigning the identity transformation to all translations, which is possible since the latter form an invariant subgroup. Fields, however, permit representing the translations nontrivially.)

For conceptual purposes, it is useful to look at the developments of the present as well as of the two preceding sections also from the abstract or geometric rather than matrix point of view. So consider two complex 2-dimensional *spinor spaces* $\mathbf{S}$ and $\dot{\mathbf{S}}$ with antisymmetric (0,2)-spinors ϵ and $\dot{\epsilon}$, on which act transformations A and $\dot{A}$ which leave ϵ and $\dot{\epsilon}$ invariant. $\mathbf{S} \otimes \dot{\mathbf{S}}$ is then complex 4-dimensional with $\epsilon \otimes \dot{\epsilon}$ as a symmetric nondegenerate bilinear form that remains invariant under the transformations $A \otimes \dot{A}$. (This leads to the fourth of the homomorphisms listed in eqs. (8.2.11).) If we now take $\dot{\mathbf{S}} = \mathbf{S}^*$ as the space complex-conjugate to $\mathbf{S}$ (cf. Appendix B) and $\dot{\epsilon} = \epsilon^*$ as well as $\dot{A} = A^*$, then $\mathbf{S} \otimes \dot{\mathbf{S}} = \mathbf{S} \otimes \mathbf{S}^*$ has a reality structure, the real elements being the Hermitian spinors. For them, the bilinear form defined by $\epsilon \otimes \epsilon^*$ is real-valued, the associated quadratic form being of signature $\mathrm{diag}\,(1,-1,-1,-1)$. One may therefore find an invertible linear '*soldering map*' $\tilde{\sigma}$ from $\mathbf{S} \otimes \mathbf{S}^*$ to the Minkowski vector space $\mathbf{V}_4$. Let $\{e_i\}$ be an orthonormal frame for $\mathbf{V}_4$ and $\{\beta_A\}$ a spin frame for $\mathbf{S}$, $\{\beta^*_{\dot{X}}\}$ the complex-conjugate frame for $\mathbf{S}^*$: then the image of $\beta_A \otimes \beta^*_{\dot{X}}$ under $\tilde{\sigma}$ may be expanded as $\tilde{\sigma}(\beta_A \otimes \beta^*_{\dot{X}}) = \sigma^i_{A\dot{X}}\, e_i$. In this way we return to the component version above; but we see that the soldering quantities $\sigma^i_{A\dot{X}}, \ldots$ are the components of a vector-spinor $\tilde{\sigma} \in \mathbf{V}_4 \otimes \tilde{\mathbf{S}} \otimes \widetilde{\mathbf{S}^*}$ which under Lorentz transformations are numerically invariant. It is recommended to reinterpret some of the preceding formulae from this point of view. (One may retain the index notation as 'abstract indices'; a particularly economic version—the 'Rindler convention'—is described at length in Penrose and Rindler (1984).)

Appendix 1: Spinors and Lightlike 4-Vectors

If from an undotted spinor κ^A and a dotted spinor $\dot\kappa^{\dot X}$ we form the product spinor $K^{A\dot X} = \kappa^A \dot\kappa^{\dot X}$, the corresponding (complex) 4-vector $k^i = K^{A\dot X} \sigma^i{}_{A\dot X}$ is null, $k^i k_i = \kappa^A \kappa_A \dot\kappa^{\dot X} \dot\kappa_{\dot X} \equiv 0$. When κ is held fixed and $\dot\kappa$ is varied, k sweeps over a 2-dimensional space of null and pairwise orthogonal vectors (a totally null, or totally isotropic 2-subspace); when $\dot\kappa$ is held fixed and κ is varied a second totally null subspace is obtained; the only common direction of these subspaces is that of k. When $\dot{\mathbf{S}} = \mathbf{S}^*$ and $\dot\kappa^{\dot X} = \gamma \kappa^{*\dot X}$ is taken with real γ, then $K^{A\dot X}$ is Hermitian, k^i is real and lightlike with $k^0 = \gamma(|\kappa^1|^2 + |\kappa^2|^2)/\sqrt{2}$: thus, depending on the sign of γ, k lies on the future or past light cone. The totally null subspaces just described are then complex-conjugate. Conversely, one can find *for each real lightlike 4-vector k* a spinor κ, unique up to a phase factor, such that

$$k^i \sigma_i{}^{A\dot X} = K^{A\dot X} = (\text{sign } k^0)\, \kappa^A \kappa^{*\dot X}. \qquad (8.4.25)$$

This is because from $k^i k_i = 0$ it follows $\det K^{A\dot X} = 0$, thus $K^{A\dot X}$ has rank 1 and can be written as $\kappa^A \dot\kappa^{\dot X}$, where a complex factor in κ remains undetermined. This fixes already the totally null subspaces passing through k. But the Hermiticity of $K^{A\dot X}$ further implies that one can change the normalization of κ such that eq. (8.4.23) holds; a phase factor remains open.

We thus see that the real future-directed lightlike 4-vectors allow to visualize spinors up to a phase factor. It is possible to give a visualization of the information contained in this phase factor, up to a sign, by considering the symmetric spinor $\Phi^{AB} = \kappa^A \kappa^B$ and the real sixtor $\operatorname{Re} f_{ij} = F_{ij}$ corresponding to it by eq. (8.4.12), which satisfies

$$F_{ij} F^{ij} = 0, \qquad F_{ij} *F^{ij} = 0, \qquad F_{ij} k^j = 0, \qquad *F_{ij} k^j = 0, \qquad (8.4.26)$$

since $\Phi_{AB} \Phi^{AB} = \kappa_A \kappa^A \kappa_B \kappa^B \equiv 0$, $K^{A\dot X} \Phi_{AB} = 0$, $K^{A\dot X} \Phi^*_{\dot X \dot Y} \equiv 0$. By exercise 7c of sect. 5.5, F_{ik} determines a real 2-dimensional subspace which contains precisely one lightlike direction, given by k: $F_{ik} = k_{[i} Q_{k]}$. The sixtor F_{ik} is, in this situation, called lightlike, or *null*, and the half-plane through k formed by the positive multiples of all candidates for Q (they differ by multiples of k only) is called a *null flag*. Now when the phase of κ is changed, the sixtor F_{ik} undergoes a *duality rotation*, the null flag getting rotated around the spatial direction of k (for every observer) through twice the phase angle. (This will be interpreted in eq. (9.4.31).) The complex selfdual sixtor f_{ik} also has the form $k_{[i} q_{k]}$, where q is complex ($Q = \operatorname{Re} q$) and $q^2 = qk = 0$; the complex 2-space spanned by k and q is (the 'selfdual') one of the two totally null subspaces through k.

If we start from a *spin frame* containing κ, i.e., add a second spinor λ such that $\kappa_A \lambda^A = 1$ ($\Leftrightarrow 2\kappa_{[A} \lambda_{B]} = \epsilon_{AB}$) we can construct a full unique real orthonormal space- and time-oriented 4-vector basis: form the real null vectors $k^i := \sigma^i{}_{A\dot X} \kappa^A \kappa^{*\dot X}$, $\ell^i := \sigma^i{}_{A\dot X} \lambda^A \lambda^{*\dot X}$ and the complex null vector $m^i := \sigma^i{}_{A\dot X} \kappa^A \lambda^{*\dot X}$; they satisfy $k^2 = \ell^2 = m^2 = km = = \ell m = 0$, $k\ell = -mm^* = 1$. Then $e_0 := (k + \ell)/\sqrt{2}$ is future-oriented and together with $e_3 := (k - \ell)/\sqrt{2}$, $e_1 := (m + m^*)/\sqrt{2}$, $e_2 := (m - m^*)/i\sqrt{2}$ forms a 4-vector basis with the scalar products $e_a e_b = \eta_{ab}$ and the determinant $\epsilon_{ijk\ell} e_0^i e_1^j e_2^k e_3^\ell = +1$. The same basis also arises from the spin frame $-\kappa, -\lambda$ and none more, corresponding to the double-valuedness of the spinor representation. A 1-1 correspondence may thus be obtained only at the expense of homotopy considerations in the set of vector bases ($\cong \mathcal{L}_+^\uparrow$) or in the set of null flags ($\cong$ SO(3)), in analogy with Fig. 7.4.

From the point of view of calculations, it emerges that spinors are of particular advantage when null vectors, null sixtors, … are involved: they are, in the sense of eq. (8.4.25), *square roots* of lightlike 4-vectors; but not only then (cf. exercise 5). The classification of spinors with only undotted, or only dotted, indices given in sect. 8.3 gives a classification of the 4-tensors corresponding to them (if any); to the principal spinors, there correspond *principal null directions* associated to the tensors in a Lorentz invariant fashion. For instance, to the field tensor F^{ik}, to which corresponds a symmetric spinor Φ_{AB}, there belong two principal null directions, which however may coincide in special cases (characterized by the validity of eqs. (8.4.26), where Φ_{AB} has product form $\kappa_A \kappa_B$). The latter is true, e.g., for the field of a plane electromagnetic wave (cf. eq. (5.5.21)) or for the $1/r$ part of the far zone field of a radiating system; it is not true for the Coulomb field.

For a more detailed discussion and applications we again refer to Penrose and Rindler (1984).

Appendix 2: Intrinsic Classification of Lorentz Transformations

As an application of the relation between spinors and 4-vectors, let us treat here the *intrinsic classification and decomposition* of Lorentz transformations which was mentioned in sect. 6.3. Let $L(\neq \mathrm{id}) \in \mathcal{L}_+^\uparrow$ and take $A \neq \pm \mathrm{id}_\mathbf{S}$ to be one of the two unimodular spin transformations corresponding to it. Since $\det A = 1$, its eigenvalues are reciprocals of each other. We distinguish two cases.

a. If the eigenvalues are distinct, there is a spin frame κ, λ such that A has the matrix $A = \mathrm{diag}\,(\alpha, \alpha^{-1})$, i.e., $A\kappa = \alpha\kappa$, $A\lambda = \alpha^{-1}\lambda$. We now form $k = \tilde\sigma(\kappa \otimes \kappa^*)$, $\ell = \tilde\sigma(\lambda \otimes \lambda^*)$, $m = \tilde\sigma(\kappa \otimes \lambda^*)$, $m^* = \tilde\sigma(\lambda \otimes \kappa^*)$ as above. We see then that $Lk = |\alpha|^2 k$, $L\ell = |\alpha|^{-2}\ell$, $Lm = (\alpha/\alpha^*)m$: k, ℓ thus are real lightlike eigenvectors for L belonging to positive reciprocal eigenvalues, spanning an invariant timelike 2-plane; m, m^* are complex-conjugate eigenvectors belonging to reciprocal phase factors for eigenvalues, spanning an invariant real spacelike 2-plane orthogonal to the timelike one. The decomposition $\mathrm{diag}\,(\alpha, \alpha^{-1}) \equiv \mathrm{diag}\,(|\alpha|, |\alpha|^{-1})\,\mathrm{diag}\,(\exp(i\arg\alpha), \exp(-i\arg\alpha))$ corresponds to a decomposition of L as a product of a *timelike rotation* and a *spacelike rotation* which take place in orthogonal 2-planes and which commute. Sometimes a terminology from complex analysis is borrowed where the general case is called loxodromic, the purely timelike case is called hyperbolic, and the purely spacelike case is called elliptic.

b. If the eigenvalues coincide, they have to equal 1 (or -1, in which case we pass to $-A$). We use a spin frame where the matrix A takes on Jordan normal form: $A\kappa = \kappa$, $A\lambda = \lambda + \kappa$. Here we obtain $Lk = k$, $Lm = m + k$, $Lm^* = m^* + k$, $L\ell = \ell + m + m^* + k$, and therefore: k is a real lightlike eigenvector to L for the eigenvalue 1; k and $m + m^*$ span a real lightlike invariant 2-plane; k and $i(m - m^*)$ span an eigenspace orthogonal to it, for the eigenvalue 1; the 3-planes passing through the former plane are also invariant, each being orthogonal to a vector of the latter plane. Such transformations are called *lightlike*, or *null rotations*, or parabolic transformations.

Exercises

1. Prove eq. (8.4.21)!

2. Determine, at every space-time point, the principal null directions of the Coulomb field. (Note that the result follows already from spherical symmetry.)

3. How does the lightlike sixtor $F_{ik} = \mathrm{Re}\,f_{ik}$ associated with a spinor κ according to eq. (8.4.12) with $\Phi^{AB} = \kappa^A \kappa^B$ change when κ is changed by a phase factor? (*Duality rotation.*)

4. Show that a real 4-tensor C_{ikmn} transforming according to $\mathrm{D}^{(2,0)} \oplus \mathrm{D}^{(0,2)}$ is characterized by the following symmetry properties:

$$C_{ik(mn)} = C_{(ik)mn} = C_{i[kmn]} = C^i{}_{kin} = 0. \tag{8.4.27}$$

Remark: A tensor of this kind appears in General Relativity (the *Weyl tensor*).

5. Express the spinor equivalent to the electromagnetic energy-momentum tensor by the symmetric spinor Φ^{AB} associated to the field strength tensor:

$$T^{ik} \leftrightarrow T^{AB\dot X\dot Y} = \frac{1}{2}\,\Phi^{AB}\,\Phi^{*\dot X\dot Y} \tag{8.4.28}$$

and, using this, prove the *Rainich identity*

$$T^{ij}\,T_{jk} = \frac{1}{4}\,\delta^i_k\,T^{\ell j}\,T_{j\ell}. \tag{8.4.29}$$

6. Observe that from the abstract, basis-free point of view the decomposition
(8.2.9) makes no sense without specifying the definite Hermitian form with
respect to which the transformations involved are to be Hermitian or unitary.
This is in line with what we said in appendix 1 to sect. 6.3 about boosts and
rotations: one must specify the observer (by its 4-velocity u) with respect to
whom some Lorentz transformation L of $\mathbf{V}_4$ is a boost or a rotation. Now to
u there is associated $\tilde{U}$ or $U_{I\dot{X}} = u_i\sigma^i{}_{I\dot{X}}$—not to be confused with the unitary
matrix appearing in eq. (8.2.9)!!—which already supplies (sect. 8.2, exercise
10) the Hermitian form needed: Hermiticity, or unitarity, for A with respect
to $\tilde{U}$ means $\tilde{U}A = A^\dagger\tilde{U}$, i.e., $U_{I\dot{X}}A^I{}_J = A^{*\dot{Y}}{}_{\dot{X}}U_{J\dot{Y}}$ using (abstract) indices, or
$A^\dagger\tilde{U}A = \tilde{U}$, i.e. $A^{*\dot{Y}}{}_{\dot{X}}U_{J\dot{Y}}A^J{}_I = U_{I\dot{X}}$. (When bases and component matrices are
used and u is specified as $u^i = \delta^{i0}$, we return to the characterization given at
eq. (8.2.9).) Now use eqs. (8.2.19,20) or (8.4.4,5) in their abstract interpretation
to show the following.

a. If u,u' are two 4-velocities, then the boost that takes u to u' is given, in the
spinor representation, by

$$A = \frac{\mathrm{id_S} + U'\tilde{U}}{\sqrt{2(1+u'u)}}, \quad \text{i.e. } A^I{}_J = \frac{\delta^I{}_J + 2U'^{I\dot{X}}U_{J\dot{X}}}{\sqrt{2(1+u'u)}}. \tag{8.4.30}$$

Hint: Compute A^2, $\mathrm{Tr}A$, show that A is unimodular and Hermitian positive-
definite in the above sense with respect to $\tilde{U}$ and that $AUA^\dagger = U'$ as well as
$AXA^\dagger = X$ when $xu = xu' = 0$; compare to exercise 7 of sect. 8.2.

b. If n,n' are two spacelike unit vectors subtending an angle α ($nn' = -\cos\alpha$),
then with respect to all observers u for which $un = un' = 0$ the spatial rotation
that takes n to n' is given by

$$A = \frac{\mathrm{id_S} - N'\tilde{N}}{2\cos\frac{\alpha}{2}}, \quad \text{i.e., } A^I{}_J = \frac{\delta^I{}_J - 2N'^{I\dot{X}}N_{I\dot{X}}}{2\cos\frac{\alpha}{2}}. \tag{8.4.31}$$

Hint: Show that A is unimodular and unitary with respect to $\tilde{U}$ and that
$ANA^\dagger = N'$, $AUA^\dagger = U$; also calculate

$$\mathrm{Tr}A = 2\cos\frac{\alpha}{2}. \tag{8.4.32}$$

c. Let A be the boost to take u into u', A' to take u' into u'' and A'' to take u''
back to u. Then the product $B = A''A'A$ leaves u fixed and thus should be a
spatial rotation with respect to u. Verify this and calculate the rotation angle
by formula (8.4.32); compare to eq. (2.10.7).

d. If a (real) null 2-plane is given and if κ is a spinor with null flag contained in
it, show that $A^I{}_J = \delta^i{}_J + \kappa^I\kappa_J$ represents a null rotation whose invariant null
2-plane is the given one, the eigenplane being orthogonal to it.

8.5 Representations of the Full Lorentz Group

In this section we discuss the *finite*-dimensional representations of the full Lorentz group $\mathcal{L}$. From a general theorem, to whose proof the reader is guided in exercise 1, it follows that these representations are all completely reducible, so that it is enough to find all irreducible representations. The full group consists of four pieces (cosets; cf. eq. (6.3.3)); we are, therefore, in a new situation as compared to sect. 7.9, and we try to proceed in two steps. One is the transition from $\mathcal{L}_+^\uparrow$ to $\mathcal{L}_+$ or from $\mathcal{L}^\uparrow$ to $\mathcal{L}$, which can be performed as in the case of the full orthogonal group $O(3)$, since $\mathcal{L}_+$ and $\mathcal{L}$ possess direct product structure, as we convinced ourselves in exercise 3 of sect. 6.3:

$$\mathcal{L}_+ \cong \mathcal{L}_+^\uparrow \times \{E, PT\}, \qquad \mathcal{L} \cong \mathcal{L}^\uparrow \times \{E, PT\}. \tag{8.5.1}$$

What requires some new considerations is thus the step from $\mathcal{L}_+^\uparrow$ to $\mathcal{L}^\uparrow$, the orthochronous Lorentz group. $\mathcal{L}^\uparrow = \mathcal{L}_+^\uparrow \cup P\,\mathcal{L}_+^\uparrow$ is *not* a direct product of $\mathcal{L}_+^\uparrow$ and the cyclic group $\{E, P\}$, since by eq. (6.1.10) boosts and space reversal do not commute: we have

$$P\,L(\mathbf{v}, \mathbf{0}) = L(-\mathbf{v}, \mathbf{0})\,P, \tag{8.5.2a}$$

$$P\,L(\mathbf{0}, \boldsymbol{\alpha}) = L(\mathbf{0}, \boldsymbol{\alpha})\,P, \tag{8.5.2b}$$

expressing the polar vector character of $\mathbf{v}$ and the axial nature of rotation vectors. (It is easy to see that the commutation property characterizes the rotation subgroup; see exercise 2!) Thus, while $\mathcal{L}_+^\uparrow$ is an invariant subgroup with quotient $\mathcal{L}^\uparrow/\mathcal{L}_+^\uparrow$ isomorphic to the subgroup $\{E, P\}$, the latter is *not* an invariant subgroup, and $\mathcal{L}^\uparrow$ is only a *semidirect* product of both subgroups (see Appendix A). A similar statement holds for the pair $\mathcal{L}_0$, $\mathcal{L}$.

It follows from these considerations that a nontrivial irreducible representation of $\mathcal{L}_+^\uparrow$ that stays irreducible upon restriction to the rotation subgroup cannot be extended, by adding a representing operator for P, to become a representation of $\mathcal{L}^\uparrow$: that operator would have to commute with the operators representing rotations, so by Schur II must be a multiple of the unit operator, which makes it impossible to represent eq. (8.5.2a).

From the relations (8.5.2) we can deduce much more: combining them with eq. (8.1.9) we can read off the important equivalence

$$L \mapsto D^{(j,j')}(P\,L\,P^{-1}) \cong D^{(j',j)}(L) \tag{8.5.3}$$

for all $L \in \mathcal{L}_+^\uparrow$. We can see this also infinitesimally: eqs. (8.5.2a,b) then read

$$\begin{aligned} P\,\mathbf{N}\,P^{-1} &= -\mathbf{N} \\ P\,\mathbf{M}\,P^{-1} &= \mathbf{M} \end{aligned} \tag{8.5.4}$$

(saying that $\mathbf{M}$ is an axial and $\mathbf{N}$ a polar vector operator in the defining representation); for the complex combinations $\mathbf{M}^\pm$ introduced in eq. (8.1.4) it then follows that

$$P\,\mathbf{M}^\pm\,P^{-1} = \mathbf{M}^\mp. \tag{8.5.5}$$

In the representations $D^{(j,j')}$, $\mathbf{M}^+$ is represented by $D^{(j)}\otimes \mathrm{id}_{j'}$ and $\mathbf{M}^-$ by $\mathrm{id}_j \otimes D^{(j')}$, which thus get exchanged when composed with the automorphism (8.5.4,5) of the Lie algebra.

Relation (8.5.3) can now combined with the general results expressed in Theorems 1 and 2 and their supplements a and b formulated exercise 7 of sect.7.9 to get a complete overview concerning the single-valued finite-dimensional irreducible representations of $\mathcal{L}^\uparrow$ and $\mathcal{L}$. The alternative exposed in those supplements comes down, in our case, to the equivalence or inequivalence between $D^{(j,j')}$ and $D^{(j',j)}$, or to distinguishing between $j = j'$ and $j \neq j'$. In the former case, the representation can be extended in two inequivalent ways from $\mathcal{L}_+^\uparrow$ to $\mathcal{L}^\uparrow$ by adding a representing operator for P, and up to equivalence this gives all the possibilities of irreducible representations of $\mathcal{L}^\uparrow$ which stay irreducible when restricted to $\mathcal{L}_+^\uparrow$.

As we saw in sect. 8.4, this easier case corresponds to certain real irreducible tensor representations of $\mathcal{L}_+^\uparrow$. Let us consider then general tensor representations of $\mathcal{L}$. Tensors transform, by definition, as written in eq. (5.4.5), where we take into account the equivalence between contravariant and covariant that results from the invariance of the Minkowski metric η; but we generalize slightly by forming tensor products of these tensor representations with the four 1-dimensional representations $L \mapsto d(L)$ of $\mathcal{L}$. Thus in this wider sense, tensors transform as

$$T'^{ik\cdots} = d(L)\cdot L^i{}_m\, L^k{}_n \ldots T^{mn\cdots},\tag{8.5.6}$$

where we distinguish the four cases

$$
\begin{aligned}
d(L) &= 1 && \ldots \text{ proper tensors}\\
d(L) &= \operatorname{sign} \det L && \ldots \text{ pseudotensors}\\
d(L) &= \operatorname{sign} L^0{}_0 && \ldots \text{ time-pseudotensors}\\
d(L) &= \operatorname{sign} L^0{}_0 \operatorname{sign} \det L && \ldots \text{ space-pseudotensors.}
\end{aligned}\tag{8.5.7}
$$

When we restrict to $\mathcal{L}^\uparrow$, the third and the fourth class become identical to the first and the second class, respectively. Reduction is achieved by symmetrization, antisymmetrization and contraction with η; the *-operation is not admissible any more, being invariant under $\mathcal{L}_+$ only.

Let us list some examples for tensors from these classes, whose physical discussion has been the subject of exercises to sect. 6.5:

$$
\begin{aligned}
&x^i,\ \partial_i && \ldots \text{ proper 4-vectors}\\
&ds \text{ (proper time)} && \ldots \text{ time-pseudoscalar}\\
&u^i = dx^i/ds \text{ (4-velocity)} && \ldots \text{ time-pseudovector}\\
&b^i = du^i/ds \text{ (4-acceleration)} && \ldots \text{ proper vector}\\
&j^i \text{ (4-current)} && \ldots \text{ time-pseudovector}\\
&A^i \text{ (4-potential in Lorenz gauge)} && \ldots \text{ time-pseudovector}\\
&\varepsilon_{ikmn} && \ldots \text{ pseudotensor}\\
&F^{ik} \text{ (electromagnetic field tensor)} && \ldots \text{ time-pseudotensor}\\
&\star F^{ik} && \ldots \text{ space-pseudotensor.}
\end{aligned}
$$

We point out that the transformation behavior of the tensor of field strengths results from the coupling to its sources. At this point we also would like to remind the reader of the remarks made following eq. (5.5.14).

For tensors transforming under $\mathcal{L}_+^\uparrow$ according to $\mathrm{D}^{(j,j)}$— i.e., for real symmetric tracefree and thus irreducible tensors—the representations of $\mathcal{L}$ given by eqs. (8.5.6,7) are inequivalent and are the only ones obtainable by extending $\mathrm{D}^{(j,j)}$. This follows by applying Theorem 1 of sect. 7.9 twice.

Now consider $\mathcal{L}^\uparrow$-irreducible tensors that are reducible upon restriction to $\mathcal{L}_+^\uparrow$. Here the representation is equivalent to the one carried by the corresponding pseudotensors $(d(L) = \det(L))$, as follows from Theorem 2 of sect. 7.9. As an example, consider the field strength tensor F_{ik}: it transforms as $\mathrm{D}^{(1,0)} \oplus \mathrm{D}^{(0,1)}$ under $\mathcal{L}_+^\uparrow$, the representation of $\mathcal{L}^\uparrow$ on the pseudotensors $\star F_{ik}$ is equivalent to the one on the F_{ik}, the $\star$-operation giving the equivalence map. The $\mathcal{L}_+^\uparrow$-irreducible parts are $\frac{1}{2}(F_{ik} \pm i \star F_{ik})$; they are transformed into each other under space reversal.

Generally, every $\mathcal{L}^\uparrow$-irreducible but $\mathcal{L}_+^\uparrow$-reducible representation is equivalent to a representation determined by the assignment

$$
L \mapsto \begin{pmatrix} \mathrm{D}^{(j,j')}(L) & 0 \\ 0 & \mathrm{D}^{(j',j)}(L) \end{pmatrix}, \qquad P \mapsto \begin{pmatrix} 0 & 1 \\ 1 & 0 \end{pmatrix}, \qquad (8.5.8)
$$

where $L \in \mathcal{L}_+^\uparrow$ and $j \neq j'$. This follows directly from Theorem 2 and Supplement b mentioned above if the equivalence (8.5.3) is made into a matrix equality by using suitable bases: eqs. (8.2.15) and (8.3.3) tell us that this is the case if, e.g., we use spinors with upper indices for $\mathrm{D}^{(j,j')}$ and spinors with lower indices for $\mathrm{D}^{(j',j)}$.

We come to multivalued representations of $\mathcal{L}^\uparrow$. Here a consideration completely analogous[1] to the calculations following eq. (7.10.7) shows that for $L \in \mathcal{L}_+^\uparrow$ we must have

$$
T_P T_L T_P^{-1} = T_{PLP} \qquad (8.5.9)
$$

and that using this relation all values of the cocycle $\omega(. \, , .)$ are expressible in terms of $\omega(L, L')$, $\omega(L, P)$, $\omega(P, P)$, where $L, L' \in \mathcal{L}_+^\uparrow$. Likewise, by changing representatives of the operator rays and cocycles according to eq. (7.10.5) we can achieve $\omega(L, L') = \pm 1$, $\omega(L, P) = 1$.

For $\mathcal{A} = \mathbf{C}^\times$ (projective, or ray representation) we can also achieve $\omega(P, P) = 1$. Unless $\omega(L, L') = +1$ on $\mathcal{L}_+^\uparrow$ can also be achieved, the extension group defined by ω is a (double) covering of $\mathcal{L}^\uparrow$, the subgroup covering the connected component $\mathcal{L}_+^\uparrow$ being connected and isomorphic to SL(2,$\mathbf{C}$). A faithful representation of it is obtained by going from $\mathrm{D}^{(1/2,0)}$ to $\mathrm{D}^{(1/2,0)} \oplus \mathrm{D}^{(0,1/2)}$ and assigning $\pm(8.5.8)$ to the space reversal. The elements of this representation space are called *bispinors* or *Dirac spinors*; we shall encounter them again in sect. 9.1. (Depending on the context, some authors call them just spinors, using one of the terms *semispinors, half-spinors, chiral spinors, Weyl spinors, reduced spinors,* ... for the spinors considered so far.) Higher irreducible representations are to be formed as above.

---

[1] It will be given more explicitly for the full Poincaré group in sect. 9.6!

For *a priori double-valued* representations, where by definition we have from the outset $\mathcal{A} = \{1, -1\}$, we get either $\omega(P, P) = +1$ or $\omega(P, P) = -1$. There are thus precisely two nonisomorphic covering groups to $\mathcal{L}^\uparrow$ that doubly cover $\mathcal{L}_+^\uparrow$ by a connected subgroup ($\cong SL(2, \mathbf{C})$). The possibility corresponding to $\omega(P, P) = -1$ may be faithfully represented in bispinor space by taking for $L \in \mathcal{L}_+^\uparrow$ the same as before and for P

$$P \mapsto \pm i \begin{pmatrix} 0 & 1 \\ 1 & 0 \end{pmatrix}; \qquad\qquad (8.5.10)$$

higher representations are constructed similarly. In this sense, then, there are two kinds of bispinors as far as their space reversal behavior is concerned. We stress again that this distinction is present only for $\mathcal{A} = \{1, -1\}$ and becomes irrelevant if $\mathcal{A} = \mathbf{C}^\times$ is considered, which is the relevant point of view for quantum mechanical states. However, in the latter situation one must stick to one chosen possibility and is not allowed to make superpositions or direct sums of spinors belonging to different phase conventions. The significance of the covering groups appears in geometric spinor theory.

We refrain here from giving the analogous discussion for the full Lorentz group $\mathcal{L}$, in particular in view of the fact that the representation of time reversal in quantum mechanics brings in a further complication (sect. 9.2) which will be dealt with for the full Poincaré group in sect. 9.6. Suffice it to mention that there are eight nonisomorphic covering groups to $\mathcal{L}$ that restrict to the universal covering of $\mathcal{L}_+^\uparrow$; four of them can be faithfully represented in bispinors—cf. eq. (9.1.27). Again, any of them may be chosen for the purposes of quantum mechanics as a phase convention (cf., e.g., Cornwell 1985).

In closing this section, it should be stressed that it is an *experimental* question whether the laws of nature possess space and/or time reversal as a symmetry. This does not follow from the invariance under $\mathcal{L}_+^\uparrow$ alone but requires a separate check. While this was basically clear, physicists were, perhaps under the impression of electrodynamics, not always conscious about it. So it came to be a great (Nobel prize decorated) achievement of C. N. Yang and T. D. Lee to envisage a violation of symmetry under P to solve a certain paradox in elementary particle physics (the so-called 'τ-θ-puzzle') and to suggest experiments which (1957) indeed demonstrated P-violation in the domain of weak interactions. In this, the combination of space reversal and charge conjugation still remained a symmetry—i.e., it was still impossible to tell the 'man behind the moon' a *local* experiment to know what we mean by right and left without telling him at the same time which particles we call electrons and which we call positrons. In 1964 a violation of this combined symmetry was discovered by Fitch and Cronin (see, e.g., Kabir 1968; Ho-Kim and Pham 1998).

What could then be the purpose of a formalism using representations of the Lorentz group including reversals when those symmetries are violated in nature? The answer is, first of all, that there are wide areas of physics where we do have those symmetries, and second, that it is sometimes easier in a covariant formalism to make the violation of part of the symmetry explicit than to make a bigger symmetry explicit in a formalism adapted to a lower one.

Exercises

1. Let $\mathcal{G}$ be a group with subgroup $\mathcal{G}_1$ of index 2, and consider a finite-dimensional reducible representation of the former whose restriction to the latter is decomposable. Show that the representation is decomposable for the whole group. It follows that if all finite-dimensional representations of $\mathcal{G}_1$ are completely reducible, this then holds also for the whole group $\mathcal{G}$.

   Corrolary: The finite-dimensional representations of $\mathcal{L}^{\uparrow}$, $\mathcal{L}_o$, $\mathcal{L}$ are completely reducible.

   Hints: Let $g \mapsto D(g) = \begin{pmatrix} D_1(g) & K(g) \\ 0 & D_2(g) \end{pmatrix}$ be the reducible representation of $\mathcal{G}$ considered, and assume that $K(g) = 0$ for $g \in \mathcal{G}_1$ has been achieved already. Fix some $g_2 \in \mathcal{G}_2$ and let $g_1 \in \mathcal{G}_1$: then $g_2^{-1} g_1 g_2 \in \mathcal{G}_1$, $g_2^2 \in \mathcal{G}_1$, and from the representation property of $D(g)$ one deduces the relations $D_1(g_1)\,K(g_2) = K(g_2)\,D_2^{-1}(g_2)\,D_2(g_1)\,D_2(g_2)$ and $D_1(g_2)\,K(g_2) + K(g_2)\,D_2(g_2) = 0$. They suffice to verify that $S\,D(g)\,S^{-1} = D_1(g) \oplus D_2(g)$ may be achieved on choosing $S = \begin{pmatrix} \mathrm{E} & \mathrm{X} \\ 0 & \mathrm{E} \end{pmatrix}$, where $\mathrm{X} := -\frac{1}{2} K(g_2)\,D_2^{-1}(g_2)$.

   Remark: For the 'cohomological' aspect of this problem, which here, however, would bring only a minor simplification in writing, see, e.g., Kirillov (1976).

2. Exercise 2 of sect. 1.5 showed that an element $L \in \mathcal{L}_+^{\uparrow}$ remains fixed under the 'involutive automorphism' $L \mapsto P\,L\,P^{-1}$ iff it is a rotation. Near the identity, and in particular infinitesimally, there is a similar characterization of boosts: they are carried into their inverses by the automorphism. Compare to the discussion at the end of sect. 1.5! What is the corresponding automorphism for $\mathrm{SL}(2,\mathbf{C})$, related to the decomposition (8.2.9), and why does the ensuing boost criterion work here 'globally' as well?

3. Prove Theorem 2 of sect. 7.9: Let g_1, g_2 be as in exercise 1, let $(\mathbf{V}, T)$ be an irreducible representation of $\mathcal{G}$ that becomes reducible upon restriction to $\mathcal{G}_1$. Let $\mathbf{V}' \subset \mathbf{V}$ be a subspace invariant and irreducible under the restriction to $\mathcal{G}_1$; call the arising subrepresentation T'. Put $\mathbf{V}'' := T_{g_2} \mathbf{V}' \subset \mathbf{V}$ and use $g_2^{-1} g_1 g_2 \in \mathcal{G}_1$ to show that this subspace is also invariant under $\mathcal{G}_1$, giving rise to a subrepresentation T'' which is equivalent to the representation in $\mathbf{V}'$ given by $g_1 \mapsto T'_{g_2^{-1} g_1 g_2}$ (called conjugate to T') and thus is irreducible and of same dimension. From the invariance of the span $\prec \mathbf{V}', \mathbf{V}'' \succ$ under $\mathcal{G}$ and of the intersection $\mathbf{V}' \cap \mathbf{V}''$ under $\mathcal{G}_1$ conclude $\mathbf{V} = \mathbf{V}' \oplus \mathbf{V}''$. Now show: (1) The representations T', T'' of $\mathcal{G}_1$ are inequivalent; (2) any other representation D of $\mathcal{G}$ on $\mathbf{V}$ that restricts to the same representations T', T'' of $\mathcal{G}_1$ on $\mathbf{V}', \mathbf{V}''$ is equivalent to T.

   Hints: Ad 1. Assume there were an equivalence map $A: \mathbf{V}' \to \mathbf{V}''$, i.e., $A\,T'_{g_1} = T''_{g_1} A$ for $g_1 \in \mathcal{G}_1$, then also (compare eq. (6.6.20)) the subspaces $\mathbf{V}(a) := \{v = v' + a\,A\,v' \,|\, v' \in \mathbf{V}'\} \subset \mathbf{V}$ would be invariant under $\mathcal{G}_1$ for all $a \in \mathbf{C}$. One can then find two values for a for which $\mathbf{V}(a)$ becomes invariant also under $\mathcal{G}$, contrary to the assumption. To see this, study the action of T_{g_2} on vectors from

$\mathbf{V}(a)$. Since $\mathbf{V}''$ does not depend on the special choice of $g_2 \in \mathcal{G}_2$, and since also $g_2^{-1} \in \mathcal{G}_2$, we have $T_{g_2}\mathbf{V}'' = \mathbf{V}'$; thus T_{g_2} defines two maps $U: \mathbf{V}' \to \mathbf{V}''$, $W: \mathbf{V}'' \to \mathbf{V}'$, and $T_{g_2}^{-1}$ defines $W^{-1}: \mathbf{V}' \to \mathbf{V}''$, $U^{-1}: \mathbf{V}'' \to \mathbf{V}'$, so that $T_{g_2}(v' + a\,A\,v') = Uv' + a\,WA\,v'$ for $v' \in \mathbf{V}'$. In $A\,T'_{g_1} = T''_{g_1}A$ now replace $g_1 \in \mathcal{G}_1$ by $g_2^{-1}\,g_1\,g_2 \in \mathcal{G}_1$ and use the definition of T', T'' to see that $WAU^{-1}A$ commutes with the T'_{g_1}. Schur's lemma implies $U = \lambda AWA$, and for the choice $a = \pm\sqrt{\lambda}$ we indeed have $T_{g_2}\mathbf{V}(a) = \mathbf{V}(a)$.

Ad 2. D_{g_2} likewise defines maps $R: \mathbf{V}' \to \mathbf{V}''$, $S: \mathbf{V}'' \to \mathbf{V}'$. Replacing g_1 by $g_2^{-1}\,g_1\,g_2 \in \mathcal{G}_1$ we get $R^{-1}\,T''_{g_1}\,R = U^{-1}\,T''_{g_1}\,U$ and thus by Schur $R = rU$, and similarly $S = sW$. The numbers r, s depend only on D but not on the special $g_2 \in \mathcal{G}_2$ chosen, as follows from replacing g_2 by $h_2 = h_1\,g_2$ with $h_1 \in \mathcal{G}_1$. Replacing g_2 by $g_2^{-1} \in \mathcal{G}_2$ then implies $s = 1/r$, and therefore $A := r\,\mathrm{id}_{\mathbf{V}'} \oplus \mathrm{id}_{\mathbf{V}''}$ is an equivalence map: $T_g\,A = A\,D_g$ for all $g \in \mathcal{G}_1$ and $g \in \mathcal{G}_2$.

Remark: Theorems 1, 2 of sect. 7.9 and supplements a, b in exercise 7 of that section together with the result of exercise 1 above completely clear up the situation for finite-dimensional representations of $\mathcal{G} = \mathcal{G}_1 \cup \mathcal{G}_2$ when the finite-dimensional representations of $\mathcal{G}_1$ are known and are fully reducible.

9 Representation Theory of the Poincaré Group

In this chapter we develop the relation between the relativistic wave equations for free fields and the representation theory of the Poincaré group $\mathcal{P}$. We then give a brief discussion of the description of invariance or covariance in the formalism of *quantum theory*, thereafter turning to a systematic theory of unitary irreducible representations of $\mathcal{P}$.

$\mathcal{P}$ is the biggest invariance group of the line element $ds^2 = \eta_{ik}\, dx^i\, dx^k$; it is generated by the Lorentz group $\mathcal{L}$ and the group $\mathcal{T}$ of space-time translations (semidirect product). Just as $\mathcal{L}$, it is composed of four connected pieces $\mathcal{P}_+^\uparrow, \ldots$, and most of the time we will limit ourselves to the doubly connected component of the unit element, $\mathcal{P}_+^\uparrow$. Consideration of the reflections, which are not contained in $\mathcal{P}_+^\uparrow$, is of particular importance in the quantum context, but their full significance appears there only when a further discrete operation (charge conjugation) is added to the game; however, we cannot enter into this here.[1]

New mathematical techniques are necessary to deal with $\mathcal{P}_+^\uparrow$: this group is neither compact nor semisimple, so that the general theorems used so far do not suffice. Also, when symmetry in the quantum domain is considered, the representation concept itself must be extended in the way described already in sect. 7.10, and still in another new way when time reversals are to be included.

9.1 Fields and Field Equations. Dirac Equation

Up to now, the space-time *translations* $x \mapsto x + a$ have been disregarded almost entirely. They belong to the invariance operations of Minkowski's line element but do not appear in the transformation behavior (3.2.2) of the coordinate differentials, on which latter the formalism of 4-vectors, tensors and spinors was based. All representations of $\mathcal{L}$ considered so far may thus be also considered as representations of $\mathcal{P}$ in which the translations are represented trivially, i.e., by unit operators of the representation spaces in question.

As mentioned before, this is reflected in a group theoretical property of $\mathcal{P}$: the translation subgroup $\mathcal{T} \subset \mathcal{P}$ forms an invariant subgroup in $\mathcal{P}$ which in those representations is mapped to the unit operator. $\mathcal{T}$ being a connected Abelian invariant subgroup of $\mathcal{P}_+^\uparrow$, the latter is not semisimple any more (cf. sect. 7.4).

Our work on finite-dimensional representations of $\mathcal{L}$ is, nevertheless, not useless as regards $\mathcal{P}$, since we already know physical objects where these representations occur but which have a nontrivial translation behavior: vector and tensor *fields* (cf. eqs. (3.4.10), (5.6.1,2)). Spinor fields are to be defined analogously. In the linear space of fields of a given type (as specified by the finite-dimensional representation D of $\mathcal{L}_+^\uparrow$

---

[1] See any text on elementary particle physics, but in particular, Streater and Wightman (1964) and Weinberg (1995)

or $\mathcal{L}$ under consideration) we obtain, just as in the case of the rotation group (sects. 7.7 and 7.8), an infinite-dimensional representation of $\mathcal{L}_+^\uparrow$ or $\mathcal{L}$ which can be easily extended to become a representation of $\mathcal{P}_+^\uparrow$ or $\mathcal{P}$. Namely, if we write the elements of $\mathcal{P}$ as pairs (a, L) as done in sect. 3.1, then to each pair we can assign the linear operator $T_{(a,L)}$ which sends the field Φ to the field $\Phi' = T_{(a,L)}\Phi$, where

$$\Phi'(x) = \mathrm{D}(L)\,\Phi(L^{-1}(x - a)). \tag{9.1.1}$$

Such representations are reducible; but this cannot be concluded from a general theorem as in the case of compact groups where the irreducible representations had to be finite-dimensional. Rather, reducibility follows from the existence of $\mathcal{P}$-covariant systems of linear homogeneous differential equations for such fields, whose solutions therefore form invariant subspaces.

To illustrate this point, consider, inside the space of fields of a certain type D, the subspace formed by solutions of the free wave equation (*d'Alembert's equation*)

$$\Box\Phi = 0 \tag{9.1.2}$$

or of the free *Klein-Gordon equation*[1]

$$\left(\Box + \kappa^2\right)\Phi = 0. \tag{9.1.3}$$

Similarly, the free Maxwell equations

$$F^{ik}{}_{,k} = 0 = {*F}^{ik}{}_{,k} \tag{9.1.4}$$

and the equations for the 4-potential in Lorenz' gauge

$$\Box A^i = 0, \qquad\qquad \partial_i A^i = 0, \tag{9.1.5}$$

select an invariant subspace in the space of sixtor and 4-vector fields, respectively.[2]

From a systematic point of view, then, there arise the following questions. How do all $\mathcal{P}_+^\uparrow$- or $\mathcal{P}$-covariant field equations look like? What is their detailed group theoretic role? How does one get irreducible representations? We shall not answer these questions completely but shall be content with important special cases. The linear representation theoretic point of view on wave equations should also not be overly stressed in field theory, since in reality fields are interacting, i.e., are coupled by nonlinear terms; free fields serving to describe in- and outgoing waves (particles) in scattering processes.

To be able to write down covariant wave equations also for spinor fields, from the 4-gradient operator ∂_i we form the operator

$$\partial_{A\dot{X}} := \sigma^i{}_{A\dot{X}}\,\partial_i, \tag{9.1.6}$$

---

[1]For its significance in particle physics see any textbook on this subject; $\kappa = mc/h$ is the reciprocal Compton wavelength of the particles described by eq. (9.1.3); cf. sect. 4.3.

[2]In the latter case, the field A_i and the *gauge-transformed* field $A_i + \partial_i\Lambda$ with $\Box\Lambda = 0$ (cf. sect. 5.2) carry the same physical information; but there is no covariant linear homogeneous differential equation which could serve to fix the gauge completely. The representation space must therefore be taken as the space of gauge equivalence classes; see sect. 9.5

for which because of eq. (8.4.5) we have

$$\partial_{A\dot{X}}\,\partial^{A\dot{Y}} = \frac{1}{2}\,\delta_{\dot{X}}{}^{\dot{Y}}\,\Box \qquad\qquad \partial_{A\dot{X}}\,\partial^{B\dot{X}} = \frac{1}{2}\,\delta_{A}{}^{B}\,\Box. \tag{9.1.7}$$

The simplest cases of spinorial wave equations, then, are the *Weyl equations*

$$\partial_{A\dot{X}}\,\Phi^{A} = 0 \qquad\qquad \text{and} \qquad\qquad \partial_{A\dot{X}}\,\Psi^{\dot{X}} = 0 \tag{9.1.8a, b}$$

for a $D^{(1/2,0)}$ and $D^{(0,1/2)}$ spinor field, respectively. They are obviously $\mathcal{P}_{+}^{\uparrow}$- but not $\mathcal{P}$-covariant, this being the reason for their initial dismissal and later (following the discovery of parity violation in the realm of weak interactions) use in the description of free massless neutrinos and antineutrinos. It follows from eq. (9.1.7) that each component of the Weyl field satisfies d'Alembert's equation (9.1.2).

If the Weyl equation is written in the equivalent form $\partial_{[A}{}^{\dot{X}}\,\Phi_{B]} = 0$ (using eq. (8.3.7)), one could think of considering also the covariant equation $\partial_{(A}{}^{\dot{X}}\,\Phi_{B)} = 0$. However, just as the Killing equation (5.9.29), this equation is *overdetermined* in that it has rather restrictive integrability conditions that admit for solutions only $\Phi_{B}(x) = a_{B} + x^{i}\sigma_{iB\dot{X}}b^{\dot{X}}$ with constant spinors a_{B}, $b^{\dot{X}}$—and these solutions do not satisfy the usual ($\mathcal{P}_{+}^{\uparrow}$-invariant!) *boundary conditions* for physical fields at infinity. Nevertheless this 'twistor equation' and its solutions, just as the Killing equation and its solutions, have geometrical significance: cf. Penrose and Rindler (1986). But it is thus evident that covariance alone cannot be a criterion for meaningful wave equations describing propagation processes consistent with the principle of relativity: one must add here the condition that the equation is *hyperbolic* in the sense of the theory of partial differential equations (cf. Courant and Hilbert 1962).

If instead of the Weyl equation we were looking for a spinor equation containing a term without derivative, just as eq. (9.1.3), then that term must be a dotted spinor. For this, $\Phi^{*}_{\dot{X}}$ cannot be used, since complex conjugation is no (complex-)linear operation. We are then forced to introduce a second independent spinor $\Psi_{\dot{X}}$, for which another field equation must be written. The simplest closed system of this kind is

$$\begin{aligned} \partial_{A\dot{X}}\,\Phi^{A} &= \frac{\kappa}{i\sqrt{2}}\,\Psi_{\dot{X}}, \\[2mm] \partial^{A\dot{X}}\,\Psi_{\dot{X}} &= \frac{\kappa}{i\sqrt{2}}\,\Phi^{A}, \end{aligned} \tag{9.1.9}$$

where the equality of the constant factors on the right was achieved by suitable normalization, and where $\sqrt{2}$ is for later convenience. When we substitute the right-hand side of one of these equations into the left-hand side of the other, we get from eq. (9.1.7) the compatibility conditions

$$\left(\Box + \kappa^{2}\right)\Phi^{A} = 0 = \left(\Box + \kappa^{2}\right)\Psi_{\dot{X}} : \tag{9.1.10}$$

each component of the *bispinor (Dirac spinor)*

$$\psi = \begin{pmatrix} \Phi^{A} \\ \Psi_{\dot{X}} \end{pmatrix} \tag{9.1.11}$$

must thus satisfy the Klein-Gordon equation (9.1.3).

Equation (9.1.9) is nothing but the free *Dirac equation*, which is usually written, using the four-component field quantity $\psi(x)$, as

$$\left(i\gamma^k \partial_k - \kappa\right)\psi = 0. \tag{9.1.12}$$

Here the γ^k are the 4×4 *Dirac matrices*, which from eqs. (9.1.9) and (8.4.2) result as

$$\gamma^k = \begin{pmatrix} 0 & \sigma^k \\ \tilde{\sigma}^k & 0 \end{pmatrix}; \tag{9.1.13}$$

from eq. (8.2.19) it follows that they satify the *anticommutation relations*[1]

$$\{\gamma^i, \gamma^k\} := \gamma^i\gamma^k + \gamma^k\gamma^i = 2\,\eta^{ik}\,E. \tag{9.1.14}$$

From these relations we get

$$\left(i\gamma^k\partial_k + \kappa\right)\left(-i\gamma^k\partial_k + \kappa\right) = \gamma^k\,\gamma^l\,\partial_k\,\partial_l + \kappa^2 = \frac{1}{2}\left(\gamma^k\,\gamma^l + \gamma^l\,\gamma^k\right)\partial_k\,\partial_l + \kappa^2 = \Box + \kappa^2, \tag{9.1.15}$$

which gives eq. (9.1.10) again.

It is well-known that Dirac arrived at his wave equation from the physically motivated attempt of splitting the Klein-Gordon operator $\Box + \kappa^2$ into factors of first order, corresponding to reading eq. (9.1.15) from the right to the left, in order to arrive at a relativistically covariant wave equation for electrons that would give the correct fine structure splitting of energy levels in the hydrogen atom. But it turned out that the decisive physical reason why Dirac's equation works for the electron, while the Klein-Gordon equation does not, is not the reason which put Dirac on its track, rather than the fact that electrons have spin $\hbar/2$, which is impossible to describe by a scalar field. Since de Broglie's considerations on matter waves had been relativistic, Schrödinger had initially also worked with the Klein-Gordon equation, but had discarded it as it gave a wrong fine structure formula; so he published only its N.R. limit now bearing his name. Pauli regretted his earlier lack of knowledge about the spinor representations of the Lorentz group: a knowledge that would have allowed him to arrive at the relativistic version of the Pauli equation—i.e., Dirac's equation—from the 'correct' side, and earlier than Dirac. It turned out then that Dirac's equation still shares some difficulties with the Klein-Gordon equation (the question of the interpretation of the negative energy states), and when these were overcome for the former by the introduction of antiparticles, Pauli and Weisskopf pointed out in what they called their 'anti-Dirac' paper (Helv. Phys. Acta 7, 709 (1934)) that if this is done using the formalism of field quantization then also the initial difficulty of the Klein-Gordon is resolved.

Dirac's equation is—in contrast to Weyl's equation—also covariant under space reversals, since the bispinor ψ transforms under $\mathcal{L}_+^\uparrow$ according to $\mathrm{D}^{(1/2,0)} \oplus \mathrm{D}^{(0,1/2)}$, and this reducible representation can be extended—cf. eqs. (8.5.8,9)—to an irreducible representation of $\mathcal{L}$ under which eq. (9.1.9) is covariant. (Below, we shall show this also in the bispinor formalism.)

Using the operators $\partial_{A\dot{X}}$ and $\gamma^k\,\partial_k$ we can construct covariant wave equations for spinor fields of higher degrees, but we do not follow this here systematically, since the pertinent fields are to serve for a quantum mechanical description of particles and since the probability interpretation of quantum mechanics requires an additional

---

[1] E is the 4×4 unit matrix, whose multiples like κE in the following will be written simply κ.

condition on our representations: *unitarity*. This will be explained in quite general terms in the next section. Thereafter, the infinitesimal structure of the Poincaré group will be analyzed and its relevant unitary irreducible representations found. Only after this we return to wave equations.

Appendix: Dirac Spinors and Clifford-Dirac Algebra

Usually the bispinor representation of $\mathcal{L}$ is constructed directly from the algebra of the γ^k matrices without using the representations $D^{(1/2,0)}$, $D^{(0,1/2)}$. One then usually speaks of (Dirac) spinors rather than bispinors, calling 2-component spinors semi-, or half, or Weyl, or chiral, or reduced, spinors. In this approach, one introduces an associative algebra with unit element e whose elements are linear combinations of power products

$$(\alpha_1)^{e_1} (\alpha_2)^{e_2} \ldots (\alpha_n)^{e_n} \tag{9.1.16}$$

of n 'generating' elements $\alpha_1, \ldots, \alpha_n$ which are subject to the the anticommutation relations

$$\{\alpha_i, \alpha_k\} := \alpha_i \alpha_k + \alpha_k \alpha_i = 2 Q_{ik} e \tag{9.1.17}$$

(and no other relations except the consequences of them), where Q_{ik} is the (symmetric) matrix of a quadratic form that characterizes the algebra, which is then called the *Clifford algebra* for Q. We will assume that Q_{ik} is nonsingular and has been diagonalized. Because of the anticommutation relations the exponents e_i may be restricted to take only the values 0 and 1, and any product of monomials (9.1.16) may be reordered to become such a monomial again, perhaps up to sign. (Of course, as usual, exponentiation by 0 is meant to give the unit element e.) It is natural to take the coefficients of the linear combinations from the ground field that contains the Q_{ik}, or an extension thereof if necessary. The Dirac matrices γ^k supply a 4-dimensional irreducible complex representation of the generators of the real Clifford algebra determined by the Minkowski metric η_{ik}.

The monomials (9.1.16) provide 2^n (=16 in our case) linearly independent elements, among which besides e and the generators $\alpha_1, \ldots, \alpha_n$ the element

$$\alpha := \alpha_1 \ldots \alpha_n = \frac{1}{n!} \epsilon(i_1 \ldots i_n) \alpha_{i_1} \ldots \alpha_{i_n} \tag{9.1.18}$$

is of particular importance. It satisfies

$$\alpha \alpha_i = (-1)^{n-1} \alpha_i \alpha, \tag{9.1.19}$$

thus commuting with all elements of the algebra when n is odd, while it is anticommuting with all odd ($\sum e_i$ odd) elements and commuting with the even ones when n is even.

The significance of the Clifford algebra for the theory of representations of (pseudo)orthogonal groups derives from the fact that *for even $n = 2m$ there is, up to equivalence, only one complex irreducible representation of the Clifford algebra; it is faithful, its dimension is 2^m; the representing matrices of the algbra elements range over the set of all $2^m \times 2^m$ matrices.* (Note that the number of independent matrices of this kind is 2^n, which is the same as the number of linearly independent algebra elements. In the exercises, the reader is guided to an informative proof of this theorem.)

We illustrate the use of this theorem in our case $n = 4$, $Q_{ik} = \eta_{ik}$. Let γ^i ($i = 0, 1, 2, 3$) be four arbitrary 4×4 matrices satisfying eq. (9.1.14). Then for any other quadruple also satisfying eq. (9.1.14) we must have a relation

$$\gamma'^i = S^{-1} \gamma^i S, \tag{9.1.20}$$

where the nonsingular matrix S is unique up to a complex factor (exercise 4 of sect. 6.6). When $L^i{}_k$ is a Lorentz transformation from $\mathcal{L}$, then the matrices $\gamma'^i = L^i{}_m \gamma^m$ also satisfy eq. (9.1.14), as is easily checked. There must then exist an $S(L)$, unique up to a factor $\lambda(L) \in \mathbf{C}$, such that

$$L^i{}_m \gamma^m = S^{-1}(L) \gamma^i S(L). \tag{9.1.21}$$

266   9 Representation Theory of the Poincaré Group

Due to this open factor, the assignment $L \mapsto S(L)$ is only a *multiplier representation* in the sense discussed in sect. 7.10: from the last equation we can only conclude that

$$S(L')\, S(L) = \omega(L', L)\, S(L'\, L), \quad 0 \neq \omega(L', L) \in \mathbf{C}. \tag{9.1.22}$$

We shall, however, show that one can choose the $\lambda(L)$ such as to get a double-valued representation only. This is done by observing that the contragredient assignment $L \mapsto \widetilde{S}(L)$ possesses the cocycle $1/\omega(L', L)$ in place of $\omega(L', L)$ (exercise 7 of sect. 7.10), and by showing that the $\lambda(L)$ may be chosen so that these two representations become equivalent, implying $\omega = 1/\omega$, so $\omega = \pm 1$.

To see this, one notes that the matrices $-\gamma^{i\,\top}$ also satisfy $\{-\gamma^{i\,\top}, -\gamma^{k\,\top}\} = 2\,\eta^{ik} E$, so that there exists a matrix B with[1]

$$\gamma^{i\,\top} = -B\,\gamma^i\, B^{-1}. \tag{9.1.23}$$

Transposing eq. (9.1.21) and using eq. (9.1.23) we get

$$L^i{}_m\, \gamma^m = B^{-1}\, S^{\top}(L)\, B\, \gamma^i\, B^{-1}\, \widetilde{S}(L)\, B. \tag{9.1.24}$$

$B^{-1}\, \widetilde{S}(L)\, B$ thus does the same job as $S(L)$ and is therefore proportional to $S(L)$. When now $S(L)$ is changed by a suitable factor $\lambda(L)$ we can indeed achieve the equivalence

$$B^{-1}\, \widetilde{S}(L)\, B = S(L), \tag{9.1.25}$$

the factor becoming determined up to sign.

Using eq. (9.1.21) it is easy to demonstrate the $\mathcal{L}$-covariance of the Dirac equation in the present formalism: if ψ satisfies $i\gamma^k \partial_k \psi = \kappa \psi$, then $\psi' = S(L)\,\psi$ satisfies

$$i\gamma^k \partial_k'\, \psi' = i\gamma^k L_k{}^j \partial_j\, S\,\psi = i\,(L^{-1})^j{}_k\, \gamma^k \partial_j\, S\,\psi = i\, S\, \gamma^j\, S^{-1}\, \partial_j\, S\,\psi = \kappa\, S\,\psi = \kappa\,\psi'. \tag{9.1.26}$$

We emphasize, however, that time reversal requires, in the context of quantum mechanics, an essential modification (cf. sect. 9.6 and Appendix C.2).

Instead of eq. (9.1.25) one may achieve, by other choices for the $\lambda(L)$, the modified condition

$$B^{-1}\, \widetilde{S}(L)\, B = d(L)\, S(L), \tag{9.1.27}$$

where $L \mapsto d(L)$ is one of the three nontrivial 1-dimensional representations (8.5.7) of $\mathcal{L}$. Also from eqs. (9.1.27,21) there result double-valued representations, and from the point of view of quantum mechanics it is a matter of convention which transformation law for spinors under $\mathcal{L}$ is adopted. The $S(L)$ thus defined form three further mutually non-isomorphic covering groups of $\mathcal{L}$. One frequently chosen convention is $d(L) = \operatorname{sign}\det L$ (cf. Bjørken and Drell 1964, Pietschmann 1974), while $d(L) = \operatorname{sign}\det L\, \operatorname{sign} L^0{}_0$ allows for an interesting alternative description of the respective covering group, closely related to the use of the abstract real Clifford algebra: these $S(L)$ are just all *real* linear combinations of the matrices corresponding to the elements in eq. (9.1.16) satisfying two conditions: (i) $\det S = 1$, (ii) $S^{-1}\, \gamma^i\, S$ is a *real* linear combination of the γ^i. Indeed, one verifies that for infinitesimal $L^i{}_k \approx \delta^i{}_k + \ell^i{}_k$ eqs. (9.1.21,27) are satisfied by

$$S \approx E + \frac{1}{8}\ell_{ik}[\gamma^i, \gamma^k]. \tag{9.1.28}$$

The $S(L)$ for $L \in \mathcal{L}_+^{\uparrow}$ therefore are of the form $\exp\left(\frac{1}{8}\ell_{ik}[\gamma^i, \gamma^k]\right)$, which involves only real coefficients. For space reversal, eq. (9.1.21) is solved by multiples of γ^0 for S; eq. (9.1.27) with

---

[1] In particle physics, there is a tradition to use the letter C and to call this matrix 'charge conjugation matrix', although charge conjugation involves one more matrix (A below) for its definition and is an *antilinear* operation on the level of first quantization (cf. Appendix C.2). We prefer the notation used by Jauch and Rohrlich (1976), Budinich and Trautman (1988) with the mnemonic 'B for *bilinear*' (explained below). For the purpose here the minus sign in eq. (9.1.23) is not necessary; indeed the generalization to arbitrary even *and* odd values $n = 2m$ and $n = 2m + 1$ requires the sign $(-1)^m$.

$d(L) = \text{sign} \det L \, \text{sign} \, L^0{}_0$ then restricts S to be $\pm\gamma^0$. Similarly it is seen that space-time reversal is represented by the matrix $\pm\gamma$,

$$\gamma := \gamma^0\gamma^1\gamma^2\gamma^3 = \frac{1}{4!}\,\epsilon_{ikmn}\,\gamma^i\gamma^k\gamma^m\gamma^n. \tag{9.1.29}$$

(For the convention $d(L) = \text{sign}\det L$ one would have to use the matrix

$$\gamma^5 := i\gamma \tag{9.1.30}$$

instead, while $\pm\gamma^0$ remains for space reversal.) So property ii has been checked; as for i, this follows (for all conventions) from $\det\exp\left(\frac{1}{8}\ell_{ik}[\gamma^i,\gamma^k]\right) = \exp\left(\frac{1}{8}\ell_{ik}\,\text{Tr}[\gamma^i,\gamma^k]\right) = 1$ (since the trace of any commutator vanishes) and from $\det\gamma^i = 1$. The latter obtains directly from the matrix realization (9.1.13) or, without using any special matrix realization of the anticommutation rules (9.1.14) as follows: eqs. (9.1.14) imply $(\gamma^i)^2 = +E$ or $-E$, the eigenvalues therefore are ± 1 or $\pm i$, the positive and negative signs occurring in equal number as we have, from eq. (9.1.18),

$$\text{Tr}\,\gamma^i = \text{Tr}\,\gamma^i\gamma\gamma^{-1} = \text{Tr}\,\gamma^{-1}\gamma^i\gamma = -\text{Tr}\,\gamma^{-1}\gamma\gamma^i = -\text{Tr}\,\gamma^i \Rightarrow \text{Tr}\,\gamma^i = 0, \tag{9.1.31}$$

so that in both cases the product of all eigenvalues is 1. Conversely, every S satisfying properties i and ii defines a Lorentz transformation—read eq. (9.1.21) from right to left—pseudo-orthogonality being implied by eq. (9.1.14). The covering group of $\mathcal{L} = O(1,3)$ thus described is called Pin(1,3); by definition, the subgroups covering $\mathcal{L}_+$, $\mathcal{L}^\uparrow$, $\mathcal{L}_+^\uparrow$ are called Spin(1,3), $\text{Pin}^\uparrow(1,3)$, $\text{Spin}^\uparrow(1,3) = \text{Pin}_e(1,3)$ (component of the unit element), respectively.

The characterization of the group Pin(1,3) just given stresses its relation to the *real* Clifford algebra, spanned by the real linear combinations of the products (9.1.16), and in particular stresses the role of the signature of Q_{ik} ($=\eta_{ik}$ in our case). Its applications are mainly in the field of geometry and topology. It is remarkable that the group Pin(3,1) belonging to the opposite signature, i.e., belonging to $Q_{ik} = -\eta_{ik} = \text{diag}(-+++)$, is *not* isomorphic to Pin(1,3) but isomorphic to that covering group for $\eta_{ik} = \text{diag}(+---)$ which has $d(L) = \text{sign}\, L^0{}_0$ in eq. (9.1.27): and this is in spite of the fact that the pseudo-orthogonal groups O(1,3) and O(3,1) are of course isomorphic, as are the groups Spin(1,3) and Spin(3,1). As we remarked in sect. 1.5, there are attempts to derive physical consequences from this phenomenon. (*Warning*: Some authors include an additional minus sign on the right of the defining equation (9.1.17)! It is also customary in mathematics to include an additional sign factor on the right of eq. (9.1.21), since then the assignment $S \mapsto L$ already works in the abstract Clifford algebra, yielding a covering group of the *full* pseudo-orthogonal group also in the case of an odd-dimensional space; unfortunately this conflicts with the covariance of the Dirac equation.)

To return, in the present framework, to the 2-component formalism when the restriction to $\mathcal{L}_+$ is made, one remarks that the relevant S commute with γ—in general, we have, from eqs. (9.1.21,29),

$$S^{-1}\gamma S = \det L\,\gamma. \tag{9.1.32}$$

Because of $(\gamma)^2 = -E$ the eigenvalues of γ are $\pm i$, and the projectors upon the two eigenspaces that effect the decomposition into irreducible parts—the *chiral projectors*—are thus given by $(1 \pm i\gamma)/2$. In the matrix realization (9.1.13)—a *'chiral representation'*—they are diagonal, as is γ.

Invariant bilinear and sesquilinear forms. It follows from eq. (9.1.27) that the bilinear form $\varphi^\top B\psi$ under $\psi \mapsto S(L)\psi$, $\varphi \mapsto S(L)\varphi$ behaves as

$$\varphi^\top B\psi \overset{L}{\mapsto} d(L)\,\varphi^\top B\psi, \tag{9.1.33}$$

and that, by eq. (9.1.32), we further have

$$\varphi^\top B\gamma\psi \overset{L}{\mapsto} d(L)\det L\,\varphi^\top B\gamma\psi. \tag{9.1.34}$$

Both bilinear forms are antisymmetric (thus defining a symplectic geometry in spin space) in the 2-component framework, writing $\varphi^\top = (\alpha^A, \beta_{\dot X})$, $\psi^\top = (\kappa^A, \mu_{\dot X})$ as in eq. (9.1.11), they are given by the expressions $\alpha^A \kappa_A \pm \beta_{\dot X}\, \mu^{\dot X}$, whose antisymmetry is clear from eq. (8.3.4). (A basis-free argument staying in the bispinor context and generalizing to higher dimensions, where B is sometimes symmetric and sometimes antisymmetric, is sketched in the exercises.)

While B makes sense even for the complex Lorentz group (where $d(L) = 1$ or $=\det(L)$ only), for the real Lorentz group there are also two invariant *Hermitian* sesquilinear forms. These are given, in the 2-component formalism, by $\alpha^{*\dot X}\mu_{\dot X} \pm \beta^*_A\,\kappa^A$. In the present formalism, generalizing to arbitrary dimensions, they are described as follows. Since the matrices $\gamma^{i\dagger}$ also satisfy eq. (9.1.14), there must exist a matrix A, unique up to a complex factor, such that

$$\gamma^{i\dagger} = A\,\gamma^i\,A^{-1}. \tag{9.1.35}$$

By an argument entirely analogous to the one sketched in sect. 7.5, exercise 5, one can choose A to be Hermitian, leaving it unique up to a real factor. It will be important later to know that then $A\gamma_i$ is Hermitian as well—which follows directly from eq. (9.1.35)—and that, furthermore, $A\gamma_0$ is Hermitian definite, so that the free real factor can be narrowed down to be positive by requiring $A\gamma_0$ to be *positive-definite*. To see this without the use of a special matrix representation, one derives from eq. (9.1.35) that $\gamma_i^{\dagger}(A\gamma_0)\gamma_i = A\gamma_0$ (no sum), so that the associated Hermitian form is invariant under the finite group generated multiplicatively by the γ_i—see exercise 4 for that group. Since this group acts irreducibly, a Hermitian form invariant under it is unique up to a scalar factor by the same argument as for A; since the group is finite, there exists a positive-definite invariant Hermitian form by the argument presented around eq. (7.5.9); so $A\gamma_0$ is definite.

When eq. (9.1.35) is now substituted into the Hermitian conjugate of eq. (9.1.21), one concludes that (similar to the procedure for B) $A^{-1}\,S^{\dagger\,-1}\,A \propto S$ or that $S^{\dagger}A S = fA$ for some complex number f dependent on S. It is seen immediately that these factors furnish a 1-dimensional representation of the covering group chosen; but since there is the same factor for S and $-S$, one actually gets a 1-dimensional representation of $\mathcal{L}$ which, due to the appearance of $S^{\dagger}$, is in fact the same for all conventions. From the representations for the reversals given above one finds

$$S^{\dagger}(L)\,A\,S(L) = \operatorname{sign} L^0{}_0\,A. \tag{9.1.36}$$

Defining the *Dirac adjoint* spinor $\bar\psi$ as

$$\bar\psi := \psi^{\dagger}\,A \tag{9.1.37}$$

we obtain from eqs. (9.1.36,32) two sesquilinear forms with transformation behavior

$$\bar\varphi\,\psi \overset{L}{\mapsto} \operatorname{sign} L^0{}_0\,\bar\varphi\,\psi \qquad\qquad \text{(time-pseudoscalar)} \tag{9.1.38}$$

$$\bar\varphi\,\gamma\,\psi \overset{L}{\mapsto} \operatorname{sign} L^0{}_0\,\det L\,\bar\varphi\,\gamma\,\psi \qquad\qquad \text{(space-pseudoscalar)}. \tag{9.1.39}$$

In particle physics, where time reversal has to be treated differently—see Appendix C—the expressions (9.1.38) and (9.1.39) are simply called scalar and pseudoscalar, respectively. It is also customary there to consider only a subclass of all possible matrix representations of the basic anticommutation relations (9.1.17) in which one can take $A = \gamma^0$ to satisfy eq. (9.1.35). However, this is very nongeometrical: the γ^i represent linear *operators* $\Sigma \to \Sigma$ in bispinor space Σ, while A represents a sesquilinear *form* on that space, which certainly is something conceptually different. The necessity of a clean separation shows up when one tries to develop the formalism for other dimensions and signatures—the Euclidean signature being particularly relevant in Quantum Field Theory—or in the curved spacetime of General Relativity, as first pointed out by V. Bargmann (Sitzungsber. Preuss. Akad. Wiss. Math. Naturwiss. Kl. 1932, p. 346). Again, our notation follows Jauch and Rohrlich (1976), Budinich and Trautman (1988) with the mnemonic 'A for (Dirac) adjoint'.)

With these remarks (and the exercises below) we close our discussion of the formal properties of the bispinor representation. For a further development of calculational techniques as well as a discussion of the physical properties of the Dirac equation we refer the reader to textbooks on particle physics. For time reversal, charge conjugation and *Majorana spinors* see also Appendix C.

Exercises

1. Show, for two complex-valued scalar solutions Φ, Ψ of the Klein-Gordon equation, the vanishing of the 4-divergence of

$$\Phi^* \overset{\leftrightarrow}{\partial_k} \Psi := \Phi^* \partial_k \Psi - (\partial_k \Phi^*) \Psi. \tag{9.1.40}$$

2. Show that, for a free Dirac field ψ, the Dirac-adjoint spinor (9.1.37) satisfies

$$i \partial_k \bar\psi \gamma^k + \kappa \bar\psi = 0. \tag{9.1.41}$$

Deduce from this that for two solutions φ, ψ of the Dirac equation one has

$$\partial_k (\bar\varphi \gamma^k \psi) \equiv 0. \tag{9.1.42}$$

3. In the special representation of the γ_i provided by eq. (9.1.13), calculate γ^5, $(1 \pm \gamma^5)/2$ and find matrices for A and B!

   The theorem on complex irreducible representations of the Clifford algebra over an even-dimensional vector space quoted above follows from general theorems on representations of so-called semisimple algebras. However, it is worthwhile to provide a special proof, due to W. Pauli (Zeeman Verhandelingen 1935, p. 31; Ann. Inst. Henri Poincaré 6, 109 (1936)), since it gives rise to a certain identity ('completeness relation') which is useful in applications. We break up the proof into a number of steps and give a few of the applications. At first we let the number n of generating elements of the Clifford algebra be arbitrary but assume that the matrix Q_{ik} is nondegenerate and has been diagonalized as diag$(1,...1,-1,...-1)$ with p pluses and q minuses. Apart from specifying the sequences $(e_1, ..., e_n)$ of exponents associated with the basis elements (9.1.16), there is another, related, way to index these basis elements. One defines $\alpha_{h_1...h_r} := \alpha_{[h_1}...\alpha_{h_r]}$, and out of these one takes as the independent ones $\alpha_H := \alpha_{h_1...h_r}$ whenever $H = \{h_1, ..., h_r\}$ is a subset of cardinality $|H| = r$ of $N := \{1, ..., n\}$, ordered by the condition $h_1 < ... < h_n$; to the empty subset we associate the unit element of the algebra: $\alpha_\emptyset = e$. So a basis is given by the α_H where H runs through the set of all subsets of N (the 'power set' of N), which are 2^n in number.

4. Convince yourself that $(\alpha_H)^{-1} = \alpha^H := \alpha^{h_r...h_1}$ and that the $\pm\alpha_H$ form a multiplicative group of 2^{n+1} elements. Thus for any $K \subset N$ the product $\pm\alpha_H\alpha_K$ with a suitable choice of sign runs through the basis if α_H does.

5. Show that if $H \neq \emptyset, H \neq N$ there is an $h \in N$ such that $\{\alpha_h, \alpha_H\} = 0$. When n is even, this works also for $H = N$; what about α_N when n is odd?

6. Consider a (nontrivial) representation $\alpha_H \mapsto \gamma_H$ of the algebra by linear operators in a vector space Σ. Use the previous result to conclude, in a way similar to eq. (9.1.31), that $\operatorname{Tr}\gamma_H = 0$ under the previous conditions on H.

7. From now on, assume $n = 2m$ even: show that the γ_H are 2^n linearly independent operators, entailing the representation to be faithful.
   Hint: To see that in a possible relation $\sum_H c_H\gamma_H = 0$ all coefficients are zero, multiply by γ_K and take the trace.

8. Let $\alpha_h \mapsto \gamma_h$, $\mapsto \gamma'_h$ be representations of the Clifford algebra in vector spaces Σ, Σ', and let $F \in L(\Sigma, \Sigma')$ (the space of linear maps $\Sigma \to \Sigma'$) be arbitrary. Form

$$\hat{F} := \sum_H \gamma'^H F \gamma_H \in L(\Sigma, \Sigma') \tag{9.1.43}$$

and show that $(\gamma'_h)^{-1} \hat{F} \gamma_h = \hat{F}$ for all h (no sum), so that $\hat{F}$ intertwines the representations. Now let Σ, Σ' both be irreducible and nontrivial. Apply Schur I and exclude, under the standing assumption $n = 2m$ even, the possibility that $\hat{F} = 0$ for all F: it follows that there is only one nontrivial equivalence class of irreducible representations.
Hint: Employing indices, $\hat{F} = 0$ for all F means

$$\sum_H \gamma'^{H\,\alpha'}{}_{\mu'} \gamma_H{}^{\nu}{}_\beta = 0. \tag{9.1.44}$$

Now use the result of exercise 6.
Accepting the equivalence of the irreducible representations in Σ, Σ', let $G \in L(\Sigma, \Sigma')$ be any fixed equivalence map. Assume further now that the representations are complex, as usual: then, from Schur II, G is unique up to a complex factor, whence $\hat{F} = fG$, where f depends linearly on F, $f = f^\nu_{\mu'} F^{\mu'}_\nu$. The $f^\nu_{\mu'}$ is determined in the same way as eq. (9.1.44) was excluded. This results in the first of the equalities

$$\sum_H \gamma'^{H\,\alpha'}{}_{\mu'} \gamma_H{}^{\nu}{}_\beta = (\dim\Sigma) G^{-1\,\nu}_{\mu'} G^{\alpha'}_\beta = 2^m G^{-1\,\nu}_{\mu'} G^{\alpha'}_\beta. \tag{9.1.45}$$

Now use $G\gamma_H G^{-1} = \gamma'_H$ or simply specialize to the case $\Sigma = \Sigma'$, $\gamma_H = \gamma'_H$ to get the first of the equalities

$$\sum_H \gamma^{H\,\alpha}{}_\mu \gamma_H{}^{\nu}{}_\beta = (\dim\Sigma)\delta^\nu_\mu \delta^\alpha_\beta = 2^m \delta^\nu_\mu \delta^\alpha_\beta. \tag{9.1.46}$$

Sum over $\nu = \beta$ to get $(\dim\Sigma)^2 = 2^n$ and thus the second equalities in eqs. (9.1.45,46). The latter is called the Pauli *completeness relation* for the γ_H. At the same time, the theorem in the text is now proved. Note that eq. (9.1.45) can be obtained from eq. (9.1.46) as well and constructs the equivalence map G for a given pair of complex irreducible representations of the algebra (up to a factor, of course; just specialize the indices ν, μ' in eq. (9.1.45) in some way).
Finally multiply eq. (9.1.46) by F^μ_α to get, for all $F \in L(\Sigma, \Sigma')$,

$$F = 2^{-m} \sum_H \mathrm{Tr}(F\gamma^H)\gamma_H, \tag{9.1.47}$$

which, together with the linear independence result of the previous exercise, tells us that the γ_H form a basis for $L(\Sigma, \Sigma)$. Show this to entail the formula

$$2^{-m}\mathrm{Tr}\gamma^H \gamma_K = \delta^H_K \tag{9.1.48}$$

with the obvious meaning of the right-hand side.

9. Show that the matrix B introduced in eq. (9.1.23) is antisymmetric for $n = 4$ and has

$$B^\top = (-1)^{m(m+1)/2} B \tag{9.1.49}$$

for general even $n = 2m$.
Hint: To avoid a special matrix representation, proceed as in exercise 4 of sect. 7.5 to see that $B^\top = bB$ with $b^2 = 1$. Now write eq. (9.1.45) for B and perform a suitable contraction. Finally apply the trick $(-1)^{r(r+1)/2} \equiv \mathrm{Re}(1+i)i^r$, $(1+i)^2 \equiv 2i$ to do the sum via the binomial theorem.

10. Reduce the Kronecker square $[\mathrm{D}^{(1/2,0)} \oplus \mathrm{D}^{(0,1/2)}] \otimes [\mathrm{D}^{(1/2,0)} \oplus \mathrm{D}^{(0,1/2)}]$ of the bispinor representation with respect to $\mathcal{L}$ and interpret the *bilinear* and *sequilinear concomitants* of two bispinors φ, ψ,

$$\varphi^{\mathsf{T}} B \psi, \quad \varphi^{\mathsf{T}} B \gamma_h \psi, \quad \varphi^{\mathsf{T}} B \gamma_{hk} \psi, \dots \tag{9.1.50}$$

$$\bar{\varphi}\psi, \quad \bar{\varphi}\gamma h\,\psi, \quad \bar{\varphi}\gamma_{hk}\,\psi, \dots \tag{9.1.51}$$

correspondingly.

11. The concomitants just introduced are usually rewritten in part using eq. (9.1.29) and the relation

$$\gamma_{ijk} = \epsilon_{ijkh}\gamma^h\gamma. \tag{9.1.52}$$

Verify this relation and also

$$\gamma\,\gamma_{ik} = -\frac{1}{2}\,\epsilon_{ikmn}\gamma^{mn}. \tag{9.1.53}$$

Note that eqs. (9.1.14,43) comprise eq. (8.4.8) and its dotted version! Using eq. (9.1.46) you can make the decomposition of $\psi \otimes \varphi$ into irreducible parts and thus the projection operators onto these quite explicit within the bispinor formalism. From their role as intertwiners, the Lorentz invariance of the $\gamma_{i\dots k}$ under the combined action on the spacetime and spinor indices is now clear. (Of course, we could have read eq. (9.1.21) this way.)

12. There are algebraic identities between the concomitants just considered. Deduce some of them for the special case $\varphi = \psi$ of the expressions (9.1.51)! (Which of these expressions are real and which of them are purely imaginary?) In particular, show that the vector $\bar{\psi}\,\gamma^h\,\psi$ and the axial vector $\bar{\psi}\,\gamma^h\,\gamma^5\,\psi$ (cf. eq. (9.1.30)) are orthogonal with their 4-squares differing in sign only, the 4-square of the vector being nonnegative, vanishing iff ψ is chiral. With our sign convention on $A\gamma_0$, the vectors $\bar{\psi}\,\gamma^h\,\psi$ are *future*-directed.

Hint: To get along without a special realization of the γ^i you can use the *Fierz rearrangement*

$$(\bar{\varphi}\, M\,\psi)\,(\bar{\varphi}'\, M'\,\psi') \equiv \frac{1}{4}\sum_H (\bar{\varphi}\,\gamma^H\,\psi')\,(\bar{\varphi}'\, M'\,\gamma_H\, M\,\psi), \tag{9.1.54}$$

to be proved using eq. (9.1.46).

9.2 Relativistic Covariance in Quantum Mechanics

Up to now, the following vague description of covariance under some group $\mathcal{G}$ in physics has emerged. The physical laws are written, and the physical phenomena described, in terms of certain mathematical objects which are thus restricted by those laws; and these objects have a well-defined active or passive transformation behavior under the group, such that the restricting laws are invariant under these transformations. An ambitious goal would be to classify 'all' possibilities for such situations for a given group. We were mainly considering objects with a linear transformation law, because of mathematical simplicity and a multitude of examples and applications. But we have also seen nonlinear realizations, either in a form which is not manifestly covariant—cf. the behavior (2.9.2) of 3-velocity components under Lorentz transformations—or in a manifestly covariant form, where linear objects are used that are restricted by a nonlinear but manifestly covariant constraint, as happens when 4-velocities are used instead of the 3-velocities. (We do not investigate here whether for a given group all nonlinear realizations can be obtained by imposing an invariant nonlinear constraint upon a suitable linear representation, which is indeed sometimes the case.) In the last section, we wrote down some linear Poincaré covariant field equations, thus defining linear representations (subrepresentations of the representation furnished by all fields of given types—cf. eq. (9.1.1)). For interacting fields, one usually writes down some manifestly Poincaré covariant but nonlinear equations. The set of solutions of these do form a space on which the group acts, transforming allowed fields into allowed ones, but this space is then not a vector space (there is no superposition principle). Basically, in the nonlinear case one might imagine all constraining equations solved in terms of some free data on which one has a nonlinear realization; but this is not always convenient, as the example of the 3-velocities shows.

Let us indicate the 'abstract nonsense' argument leading from the Principle of Relativity to 'covariance under the Poincaré group' as described above; it closely parallels the argument in the smallprint paragraph of the introduction to chap. 3, which the reader is now urged to (re)read. Physical objects ϕ, ψ, etc., of some specific kind are described in the inertial frames I_1 and $I_2 \in \mathcal{I}$ by the mathematical descriptors $\phi_{(1)}$, $\psi_{(1)}$, etc., and $\phi_{(2)}$, $\psi_{(2)}$, etc. All I_i being on equal footing, the corresponding observers must be able to choose their descriptors from the *same* total set $\mathbf{M}_1 = \mathbf{M}_2 = \ldots =: \mathbf{M}$. Of course, the single descriptor $\phi_{(1)}$ will not be the same as $\phi_{(2)}$, etc., so that to the transition $I_i \mapsto I_j = I_i \circ f_{ij}$ there will correspond a bijection F_{ij} of $\mathbf{M}$ to itself sending $\phi_{(i)}$ to $\phi_{(j)}$ etc. The point is now that—by a reasoning completely analogous to the one that showed the transition maps f_{ij} to form a group $(=\mathcal{P})$—the F_{ij} form a group of bijections of $\mathbf{M}$ homomorphic to that formed by the f_{ij} and thus homomorphic to $\mathcal{P}$. (In particular, F_{ij} depends only on f_{ij} but not on the special pair I_i, I_j related by f_{ij}.) In other words, $\mathcal{P}$ acts on the descriptor set $\mathbf{M}$ as a transformation group, or, still in other words, $\mathbf{M}$ is a $\mathcal{P}$-space. (Following the latter terminology, $\mathcal{I}$ is called a *principal* $\mathcal{P}$-space: the action of the group here is transitive and *free* in that only the unit element leaves 'points' (=frames) fixed.)

Clearly, one also has an active interpretation. Formally, it is given as follows: calling the pairs (I, ψ) and $(I \circ f^{-1}, F\psi)$ equivalent when $f \in \mathcal{P}$ and F is the corresponding transformation of $\mathbf{M}$, we can form the quotient space $(\mathcal{I} \times \mathbf{M})/\mathcal{P} =: \mathcal{M}$; it is called the *associated* $\mathcal{P}$-space modelled after $\mathbf{M}$. Every I defines a bijective map $\mathbf{M} \to \mathcal{M}$ by assigning to $\psi \in \mathbf{M}$ the equivalence class of (I, ψ); denoting this map by the same letter I, the active transformations of $\mathcal{M}$ are given by $I \circ F \circ I^{-1}$. Structures on $\mathbf{M}$ preserved by the F correspond to structures on $\mathcal{M}$ preserved by the active transformations.

A special case of nonlinear realization—namely *projective representations*—arises when the objects are to be *quantum mechanical states*. Covariance of the quantum mechanical formalism under a group requires the group to be realized in a way close to, but in general not identical with, linear realization: it turns out that the field equations written in the last section—which in part have only significance in the microscopic, i.e., quantum, domain— are related to this. Before we discuss the realizations of the Poincaré group in the quantum domain, let us sketch the kind of realization in general. (There is a classical analog helping to see what the issue is: covariance of the classical Hamiltonian formalism under a group would just require the group to be realized by canonical transformations. Note that we are not talking here about invariance of a specific Hamiltonian under such transformations! For the reasons indicated at the end of sect. 5.1 we do not, however, discuss relativistic symmetry in the framework of classical Hamiltonian particle dynamics.)

In quantum mechanics, the (pure) *states* of a physical system are described by wave functions, or, more generally but also more abstractly, by the nonzero vectors $|\psi\rangle$ of a complex Hilbert space $\mathbf{H}$, where, however, every nonzero complex multiple of a vector describes the *same* state. The pure states thus correspond to the *rays* (=1-dimensional subspaces) $\mathbf{C}|\psi\rangle$ of the space, i.e., to the points of the pertinent projective space $\mathbf{PH}$. Usually one normalizes the state vectors, so that states correspond to normalized vectors up to phase factors (unitary rays, topologically circles). *Observables* are described by Hermitian operators O; they also map rays to rays. The possible values for these observables measurable in experiments are eigenvalues o of the operators: $O|o\rangle = o|o\rangle$. The probability to measure, in the state $|\psi\rangle$, the value o of the observable O is given by $w(o) = \langle\psi|o\rangle\langle o|\psi\rangle/\langle\psi|\psi\rangle\langle o|o\rangle$ (assuming the general case where the eigenvalue is not degenerate; Schwarz' inequality ensures $0 \le w(o) \le 1$).

Covariance of the quantum mechanical formalism with respect to the Poincaré group $\mathcal{P}$ thus requires the group to be realized as a transformation group on the state space $\mathbf{PH}$, the transformations being further restricted by the probability interpretation: the 'transition probabilities' $\langle\phi_{(1)}|\psi_{(1)}\rangle\langle\psi_{(1)}|\phi_{(1)}\rangle/\langle\phi_{(1)}|\phi_{(1)}\rangle\langle\psi_{(1)}|\psi_{(1)}\rangle$ between states as measured in I_1 have to remain invariant under changing from I_1 to I_2, since they are obtained experimentally by pure counting. (Note that this expression depends only on the rays $\mathbf{C}|\phi_{(i)}\rangle$ and $\mathbf{C}|\psi_{(i)}\rangle$ involved!)

Also note that this requirement does not conflict with the fact that scattering cross sections— which are an example of probabilities—do depend on whether they are measured in the CM or the lab frame. What is at issue here is, of course, that the result must not depend on the frame containing the *whole* experimental setup, not just target and counters.

Now a fundamental theorem of E. Wigner states that every bijective ray map $\mathbf{C}|\psi\rangle \mapsto \mathbf{C}|\psi'\rangle$ of $\mathbf{PH}$ satisfying the above requirement may be extended to a map of vectors $|\psi\rangle \mapsto |\psi'\rangle = U|\psi\rangle$ furnished by a *semiunitary*, i.e., *unitary* or *antiunitary*, operator $U : \mathbf{H} \to \mathbf{H}$ which is *unique up to a phase factor*. (An antiunitary operator is antilinear (App. B.1) and satisfies $\langle Ux, Uy\rangle = \langle y, x\rangle$.)

A complete proof of this theorem is given by V. Bargmann (J. Math. Phys. *5*, 862 (1964)); we also point out the proof given by U. Uhlhorn (Ark. Fys. *23*, 307 (1962)) quoted by Bargmann, which relates Wigner's theorem to the so-called second fundamental theorem of projective geometry. The fact that the dimension of the Hilbert spaces involved is infinite in general is welcome here: Wigner's theorem holds when $\dim \mathbf{H} \ge 3$. A modern treatment of the whole setup is given by Varadarajan (1985).

It should be added that relevant quantum systems possessing relativistic symmetry have infinitely many degrees of freedom, for reasons sketched in sect. 9.5. For these one should use, at least for general considerations including symmetry and covariance, the formalism of C^* algebras, which is, however, beyond our scope here. We refer the reader to Bogolubov et al. (1990), where a formulation and a proof of Wigner's theorem is given within that framework.

It now follows from the theorem of Wigner that covariance of some quantum mechanical system with Hilbert space $\mathbf{H}$—we now think of an active interpretation but do not distinguish this notationally—under a group $\mathcal{G}$ can be described by assigning to each group element $g \in \mathcal{G}$ a unitary or antiunitary operator, unique up to a phase factor, i.e., an *operator ray*: $g \mapsto \exp(i\alpha)\,U(g)$, where $\alpha \in \mathbf{R}$ and $U(g)$ is a special choice of an operator in the ray, made for every g. Since the phase factors are arbitrary, in composing the operators one may conclude only that

$$U(g_1)\,U(g_2) = \omega(g_1, g_2)\,U(g_1\,g_2),\tag{9.2.1}$$

where $|\omega(g_1, g_2)| = 1$. Such a situation is called *(semi)unitary ray representation.* Every quantum physical system admitting $\mathcal{G}$ as a symmetry group associates to $\mathcal{G}$ such a semiunitary ray representation, and one may thus classify the systems by classifying the ray representations.

The remarks made about multivalued representations in sect. 7.10 do not suffice for the discussion of the ray representations of $\mathcal{P}$ for two reasons: first, we considered only linear, but no antilinear operators there; and second, $\mathcal{P}$ is not compact.

For connected groups $\mathcal{G} = \mathcal{G}_e$ like $\mathcal{P}_+^\uparrow$, the first of these additional difficulties is absent: as mentioned in sect. 7.4, every group element may be written as a product of finitely many elements that lie in 1-parameter subgroups $g(\tau)$, and with $g(0) = e$, $g(1) = g$, $g(\tau + \tau') = g(\tau)\,g(\tau')$ we have $g = g(1/2)\,g(1/2)$; but the square of an antilinear operator is linear.

For the full group $\mathcal{P}$ this argument cannot be applied, and for the reversals both possibilities, linear or antilinear, are open a priori. As will be explained in sect. 9.6, in physics space and time reversal have to be represented linearly and antilinearly, respectively, on grounds of the postulate of positivity of energy (cf. Wigner 1959 for the N.R. case).

Let us first restrict to $\mathcal{P}_+^\uparrow$ or to connected Lie groups $\mathcal{G}$ in general. V. Bargmann (Ann. Math. **59**, 1 (1954)) has shown that the factors λ_g in eq. (7.10.4)—which now because of unitarity have to be phase factors—may be chosen near the unit element in such a way that the $U(g)$ depend differentiably on g and (anti-Hermitian) generators become definable as in sect. 7.4; one can then proceed in a way similar to that section to derive, for a set $\{t_A\}$ of generators associated to a basis of the Lie algebra of $\mathcal{G}$, commutation relations of the form

$$[t_A, t_B] = C_{AB}^D\, t_D + i\, C_{AB}\, \mathrm{id}_{\mathbf{H}},\tag{9.2.2}$$

where the C_{AB}^D are the structure constants of the group, while the real constants $C_{AB} = -C_{BA}$ stem from the additional factor ω appearing in eq. (9.2.1). Because of the identity $[t_{[E}, [t_A, t_{B]}]] \equiv 0$ and eq. (7.2.17) they have to satisfy the so-called ('infinitesimal') 2-*cocycle condition*

$$C_{[AB}^D\, C_{E]D} = 0.\tag{9.2.3}$$

To another choice of phase factors there corresponds, infinitesimally, the replacement

$$t_A \mapsto t_A' = t_A + i\, C_A\, \mathrm{id}_{\mathbf{H}}\tag{9.2.4}$$

with real constants C_A, and for the t_A' there is a 'primed' version of eq. (9.2.2) with

$$C_{AB}' = C_{AB} - C_{AB}^D\, C_D.\tag{9.2.5}$$

Therefore, the solvability of the system of linear equations

$$C_{AB} = C_{AB}^{D} \, C_{D} \qquad (9.2.6)$$

for the C_A is a necessary condition to be able to lift the ray representation to an ordinary representation. It then follows by exponentiation that one may lift to an ordinary representation of the universal covering group $\widetilde{\mathcal{G}}$ of the connected group $\mathcal{G}$. When eq. (9.2.6) can be solved for *all* 2-cocycles C_{AB} one obtains all ray representations of $\mathcal{G} = \widetilde{\mathcal{G}}/\mathcal{Z}$ (where $\mathcal{Z} = $ discrete central subgroup) from those ordinary representations of $\widetilde{\mathcal{G}}$ in which $\mathcal{Z}$ gets represented by multiples of the identity operator. The latter is the case, e.g., in irreducible representations. (A slightly less abridged account of Bargmann's work is given in Hamermesh (1962).)

Generally, a ray representation of $\mathcal{G}$ gives, by composition with the homomorphism $\widetilde{\mathcal{G}} \to \mathcal{G}$, a ray representation of $\widetilde{\mathcal{G}}$ in which $\mathcal{Z}$ acts trivially—so the last proviso is necessary; but it is easily seen to be sufficient as well for a ray representation of $\widetilde{\mathcal{G}}$ to yield one for $\mathcal{G}$. To illustrate the proviso, consider the group SU(2), which is the universal covering group of SO(3) $\cong$ SU(2)$/\mathcal{Z}_2$, and the following of its representations. (i) $D^{(1/2)}\ldots$ this irreducible representation yields a ray representation of SO(3). (ii) $D^{(1/2)} \oplus D^{(3/2)} \ldots$ this also gives a ray representation of SO(3), since $-\mathbf{1}_2 \in \mathcal{Z}_2 \subset$ SU(2) is represented by $-\mathbf{1}_2 \oplus (-\mathbf{1}_4) = -(\mathbf{1}_2 \oplus \mathbf{1}_4)$. (iii) $D^{(1/2)} \oplus D^{(1)} \ldots$ this does *not* give a ray representation of SO(3), since $-\mathbf{1}_2$ is represented by $(-\mathbf{1}_2) \oplus \mathbf{1}_3$, which is not a multiple of $\mathbf{1}_2 \oplus \mathbf{1}_3$. This state of affairs is related to the occurrence of *'superselection rules'* (cf. G. C. Wick, A. S. Wightman, E. P. Wigner, Phys. Rev. *88*, 101 (1952); P. P. Divakaran, Rev. Math. Phys. *6*, 167 (1994); Weinberg 1995).

Under the assumed validity of eq. (9.2.3), eq. (9.2.6) turns out to be always solvable for all semisimple Lie groups. Even though $\mathcal{P}_+^\uparrow = \widetilde{\mathcal{P}}_+^\uparrow/\mathcal{Z}_2$ is not semisimple, we shall—following V. Bargmann, loc. cit.—verify in the next section that these equations can always be solved for this group; here the universal covering group $\widetilde{\mathcal{P}}_+^\uparrow$ is the semidirect product (see Appendix A for definition) of $\widetilde{\mathcal{L}}_+^\uparrow = $ SL(2,C) with the translation group $\mathcal{T}$, on which it acts via the 4-vector representation (8.2.8); the double connectivity coming from the rotation subgroup.

From these considerations[1] it follows, writing the elements g of the Poincaré group as in eq. (3.1.9), that by a suitable choice of phases in eq. (9.2.1) we can reach for $\mathcal{P}_+^\uparrow$

$$\omega((a_1, E), (a_2, E)) = 1 = \omega((0, L), (a, E)), \qquad \omega((0, L_1), (0, L_2)) = \pm 1. \qquad (9.2.7)$$

Further, all the operators $U(a, L)$ have to be unitary. The physically relevant irreducible unitary ray representations shall be analyzed in sect. 9.4.

We now add some general remarks concerning semilinear ray representations of nonconnected groups $\mathcal{G}$ such as $\mathcal{P}^\uparrow$, $\mathcal{P}_+$, $\mathcal{P}_0$ and $\mathcal{P}$. Those elements of $\mathcal{G}$ that get represented linearly form a subgroup $\mathcal{G}_1$, the elements represented antilinearly form the only coset $\mathcal{G}_2$ of it (exercise 2). Given $\mathcal{G}_1$, the problem of determining the *irreducible* semilinear representations of $\mathcal{G}$ may be reduced to the one of finding the linear irreducible representations of $\mathcal{G}_1$. This is done with the help of two theorems and their supplements, which are are similar to the theorems 1, 2 of sect. 7.9 and their supplements in exercise 7 of that section. (Also the proofs are similar; but antilinearity introduces some characteristic differences.) When $\omega \equiv 1$ (ordinary semilinear representations) they are as follows:

---

[1] For a complete, modern mathematical presentation of these matters, see Varadarajan (1985).

1. *If an irreducible semilinear representation of $\mathcal{G}$ restricts to an irreducible (linear) representation of $\mathcal{G}_1$, then the former is uniquely determined by the latter. (Type I.)*

2. *If a semilinear representation of $\mathcal{G}$ restricts to a reducible (linear) representation of $\mathcal{G}_1$, then the latter decomposes into two irreducible representations of the same dimension which determine the former representation uniquely up to equivalence; these two representations of $\mathcal{G}_1$ may be equivalent (type II) or inequivalent (type III).*

To be able to formulate the supplements in the case of semilinearity we need a preliminary consideration. Let $(\mathbf{V}, T)$ be a complex irreducible representation of $\mathcal{G}_1$; choose some fixed $g_2 \in \mathcal{G}_2$ and put $g_0 := g_2^2 \in \mathcal{G}_1$. We consider the *conjugate* representation of $\mathcal{G}_1$ on $\mathbf{V}$ given by $g_1 \mapsto T'_{g_1} := T_{g_2^{-1} g_1 g_2}$ (whose equivalence class is not changed by changing the choice of $g_2 \in \mathcal{G}_2$: taking $g_2' \in \mathcal{G}_2$ instead of g_2, we have that $g_2^{-1} g_2' \in \mathcal{G}_1$, and $T_{g_2^{-1} g_2'}$ furnishes an equivalence map.) If, as in type I, T comes from restricting an irreducible semilinear representation (also called T) of $\mathcal{G}$ in $\mathbf{V}$, then the representations T and T'^* of $\mathcal{G}_1$ are equivalent, since then there is the (antilinear) operator T_{g_2} operating on the same space, and on using the (antilinear!) complex conjugation $\mathcal{K} : \mathbf{V} \to \mathbf{V}^*$ (see Appendix B.3,4) we have $T'^*_{g_1} := \mathcal{K} T'_{g_1} \mathcal{K}^{-1} = \mathcal{K} T_{g_2}^{-1} T_{g_1} T_{g_2} \mathcal{K}^{-1}$; the (linear) operator $S := T_{g_2} \mathcal{K}^{-1}$ thus furnishing an equivalence map.

Conversely now, assume the representations T_{g_1} and $T'^*_{g_1}$ of $\mathcal{G}_1$ to be equivalent, $T'^*_{g_1} = S^{-1} T_{g_1} S$. As opposed to the situation encountered in exercise 7a of sect. 7.9, a further distinction has to be made. From the assumed equivalence it follows

$$(SS^*)^{-1} T_{g_1} SS^* = S^{*-1} T'^*_{g_1} S^* = \left(S^{-1} T_{g_2^{-1} g_1 g_2} S\right)^* = T'_{g_2^{-1} g_1 g_2} = T_{g_0^{-1} g_1 g_0} = T_{g_0}^{-1} T_{g_1} T_{g_0},$$

and thus by Schur II $SS^* = s T_{g_0}$, $0 \neq s \in \mathbf{C}$. We can take the complex conjugate of this relation: $S^* S = s^* T_{g_0}^*$. On the other hand, we can get from it

$$S^* S = s\, S^{-1} T_{g_0} S = s\, T_{g_2^{-1} g_0 g_2}^* = s\, T_{g_0}^*;$$

therefore s has to be *real*. Changing S by a complex factor will change only the absolute value of s, so that we may assume that in the case of equivalent T and T'^* we have achieved that $SS^* = +T_{g_0}$ or that $SS^* = -T_{g_0}$. We are now in a position to formulate the *supplements* in question:

a. If T_{g_1} and $T'^*_{g_1} = S^{-1} T_{g_1} S$ are equivalent and if

a_{I}. $SS^* = +T_{g_0}$, then the assignment $g_1 \mapsto T_{g_1}$, $g_2 \mapsto S\mathcal{K}$ may be extended to yield a semilinear representation (of type I) of $\mathcal{G}$ on $\mathbf{V}$;

a_{II}. $SS^* = -T_{g_0}$, then the assignment

$$g_1 \mapsto T_{g_1} \oplus T'_{g_1}, \qquad g_2 \mapsto \begin{pmatrix} 0 & -S\mathcal{K} \\ S\mathcal{K} & 0 \end{pmatrix}$$

may be extended to yield an irreducible semilinear representation (of type II) of $\mathcal{G}$ on $\mathbf{V} \oplus \mathbf{V}$.

b. If T_{g_1} and $T_{g_1}'^*$ are inequivalent, then the assignment

$$g_1 \mapsto T_{g_1} \oplus \mathcal{E}T_{g_1}'\mathcal{E}^{-1}, \qquad g_2 \mapsto \begin{pmatrix} 0 & T_{g_0}\mathcal{E}^{-1} \\ \mathcal{E} & 0 \end{pmatrix}$$

may, for every choice of an invertible antilinear map $\mathcal{E}\colon \mathbf{V} \mapsto \mathbf{V}$, be extended to yield an irreducible semilinear representation (of type III) of $\mathcal{G}$ on $\mathbf{V} \oplus \mathbf{V}$. (Changing $\mathcal{E}$ does not change the equivalence class of the representation obtained.)

Thus far the theorems and supplements were formulated for ordinary semilinear representations; they may, however, be modified to hold for ray representations as well, such as may be done for the theorems 1,2 and their supplements from sect. 7.9—it is only necessary to apply them to the pertinent extension groups.

In this modification, the definition of T_{g_1}' receives an additional factor

$$\frac{\omega^*(g_1, g_2)}{\omega^*(g_2, g_2^{-1}g_1g_2)}, \tag{9.2.8}$$

and a factor $\omega(g_2, g_2)$ has to be inserted in front of T_{g_0} in the relation to be satisfied by S as well as in the matrix assigned to g_2. The modifications for the analogous supplements in sect. 7.9 are the same, apart from the absence of the complex conjugations in the factor (9.2.8). For details and proofs we refer the reader to the particularly clear article by R. Shaw and J. Lever, Commun. Math. Phys. *38*, 257 (1974).

It should be pointed out that, due to the modification of the cocycle relation necessitated by semilinearity (exercise), it follows for an *involutory* element $g_2 \in \mathcal{G}_2$, i.e., one having $g_2^2 = e$ (such as T or PT in $\mathcal{P}$)—by putting $g_1 = g_2 = g_3$ in the cocycle relation—that

$$\omega(g_2, g_2) = (\omega(g_2, g_2))^* \qquad \text{for} \quad g_2^2 = e \tag{9.2.9}$$

is *real*, and that the modified concept of equivalence of ray representations (exercise) allows to change, by the complex rescaling $T_{g_2} \mapsto \lambda_{g_2} T_{g_2}$, the cocycle value (9.2.9) only by the *positive* factor $|\lambda_{g_2}|^2$. We thus can achieve *only* $\omega(g_2, g_2) = \pm 1$, the phase of λ_{g_2} remaining undetermined. (Confront this with the state of affairs in sect. 7.10 where it was possible to achieve $\omega(P, P) = 1$, λ_P becoming determined up to sign.)

Also the determination of the inequivalent 2-cocycles ω for $\mathcal{G} = \mathcal{G}_1 \cup \mathcal{G}_2$ may, in many cases, be reduced to the determination of those for $\mathcal{G}_1$. For a discussion of the general mathematical methods at disposal we refer to the article of L. Michel in Gürzey (1964). We shall carry this out for $\mathcal{P}$ in sect. 9.6.

Exercises

1. To a semilinear ray representation (9.2.1) there belongs

   a. a 2-cocycle relation modifying eq. (7.10.2)

   b. a definition of extension group modifying eq. (7.10.3)

    c. an equivalence concept modifying eq. (7.10.5).

Establish these!
Hint: To formulate your answer in a concise way, let σ_g be the identity or complex conjugation on $\mathbf{C}$, depending on whether U_g is linear or antilinear.

2. Show that in a semilinear representation of a group $\mathcal{G}$ the elements that are represented linearly form a subgroup $\mathcal{G}_1$ with one coset $\mathcal{G}_2$ only.

3. Show that for the group SO(3) the condition (9.2.3) is always satisfied and that eq. (9.2.6) can always be solved.

4. For the subgroup $\mathcal{V}_4 = \{E, P, T, PT\} \subset \mathcal{P}$ generated by the reversals P, T, determine, up to equivalence (see exercise 1c!), all 2-cocycles $\omega(\,.\,,\,.\,)$ with values in U(1)= {phase factors}, where $\{E, P\}$ and $\{T, PT\}$ are to be represented linearly and antilinearly, respectively.
Hints: Assume $\omega(E, E) = \omega(E, P) = \omega(P, E) = \omega(P, P) = 1$ has been reached as in sect. 7.10; put $\omega(T, T) = \alpha\ (=\pm 1)$, $\omega(PT, PT) = \beta\ (=\pm 1)$ and assume $\omega(P, T) = 1$ by choice of λ_{PT}, where the phase factor λ_T still remains free. Now verify that the remaining cocycle conditions determine $\omega(\,.\,,\,.\,)$ completely: $\omega(P, T) = \omega(P, PT) = 1$, $\omega(PT, T) = \alpha$, $\omega(T, PT) = \beta$, $\omega(T, P) = \omega(PT, P) = \alpha\beta$. Here α, β may independently take their allowed values ± 1, so that four different extension groups of $\mathcal{V}_4$ result.

5. Verify the supplements to theorems 1, 2 of the present section.

6. Try to prove theorems 1, 2 themselves along the pattern of the proofs for theorems 1, 2 in sect. 7.9 as indicated in the exercises to sects. 7.9 and 8.5, strictly keeping track of antilinearity.

9.3   Lie Algebra and Invariants of the Poincaré Group

The Lie algebra of the Poincaré group $\mathcal{P}$ is given by the commutation relations (8.1.3) of the homogeneous Lorentz group, by the trivial commutation relations of the translation subgroup, and by the commutators between translation and Lorentz generators yet to be determined. For this we consider the adjoint action of $\mathcal{P}$. Let $(a, L) \in \mathcal{P}_+^\uparrow$ be an infinitesimal transformation in which we take as parameters the a^i and the six independent elements of the matrix $\omega^i{}_k := L^i{}_k - \delta^i{}_k$. From $L^\mathsf{T} \eta L = \eta$ follows $\omega^\mathsf{T}\eta + \eta\omega = 0$, i.e.,

$$\eta_{ij}\,\omega^j{}_k =: \omega_{ik} = -\omega_{ki}; \tag{9.3.1}$$

cf. eq. (6.1.3). In the active interpretation of (a, L), a^k and ω_{ik} are the components of an infinitesimal 4-vector and a sixtor, respectively. (The latter relates to $\boldsymbol{\alpha}$, $\mathbf{v}$ in eq. (6.1.5) as does the field tensor F_{ik} to the field strengths $\mathbf{B}$, $\mathbf{E}$ of the electromagnetic field.)

Now let $(a, L) \mapsto U(a, L)$ be a faithful representation in the space $\mathbf{H}$, where infinitesimally

$$U(a, L) \approx \mathrm{id}_{\mathbf{H}} - \frac{i}{2}\,\omega_{ab}\, M^{ab} + ia^c P_c. \qquad (9.3.2)$$

Here $M^{ab} = -M^{ba}$ and P_c are the generators of Lorentz transformations and translations in that representation; a factor i was taken outside to get Hermitian generators when the representation is unitary.

We obtain the adjoint representation, according to sect. 7.4, when in the relations[1]

$$U^{-1}(L)\, U(a', L')\, U(L) = U(L^{-1}a',\, L^{-1} L' L), \qquad (9.3.3a)$$

$$U^{-1}(a)\, U(a', L')\, U(a) = U(L'a + a' - a,\, L') \qquad (9.3.3b)$$

the element (a', L') is made infinitesimal and eq. (9.3.2) is used (the adjoint representation of the general element $(a, L) = (a, E)\,(0, L)$ is obtained by composition). For the right hand sides of eqs. (9.3.3), eq. (9.3.2) becomes

$$U(L^{-1}a',\, L^{-1}L'L) \;\approx\; \mathrm{id}_{\mathbf{H}} - \frac{i}{2}(L^{-1}\omega'L)_{mn}\, M^{mn} + i(L^{-1}a')^d\, P_d =$$

$$= \mathrm{id}_{\mathbf{H}} - \frac{i}{2}\omega'_{ab}\, L^a{}_m\, L^b{}_n\, M^{mn} + i\,a'^c\, L_c{}^d\, P_d,$$

$$U(L'a + a' - a,\, L') \approx \mathrm{id}_{\mathbf{H}} - \frac{i}{2}\omega'_{ab}\, M^{ab} + i(\omega'^c{}_d\, a^d + a'^c)P_c,$$

and we can read off from the coefficients ω'_{ik}, a'^c the adjoint representation

$$U^{-1}(L)\, M^{ik}\, U(L) = L^i{}_m\, L^k{}_n\, M^{mn} \qquad (9.3.4a)$$

$$U^{-1}(L)\, P_c\, U(L) = L_c{}^d\, P_d \qquad (9.3.4b)$$

$$U^{-1}(a)\, P_c\, U(a) = P_c \qquad (9.3.4c)$$

$$U^{-1}(a)\, M^{ik}\, U(a) = M^{ik} + 2\, a^{[i}\, P^{k]} \qquad (9.3.4d)$$

(observe the antisymmetry of the ω'_{ik} in order to arrive at eq. (9.3.4d)!). The first two equations mean that M^{ik} is a sixtor operator and P_c is a 4-vector operator under $\mathcal{L}^{\uparrow}_{+}$; the third one expresses the commutativity of translations; we shall encounter the last one again in chap. 10: it describes, among other things, the dependence of angular momentum on the reference point.

We now get the commutation relations for the generators of the Poincaré group by making L, a infinitesimal in eq. (9.3.4): $L = E + \omega$, $U(L) = \mathrm{id}_{\mathbf{H}} - \frac{i}{2}\omega_{ab}\, M^{ab}$, $U(a) = \mathrm{id}_{\mathbf{H}} + ia^k P_k$. Comparison of factors of ω_{ab}, a^c on the right and left gives, observing the antisymmetry of the ω_{ab}:

---

[1]From here on we will write $U(a, E) =: U(a)$ in case of pure translations and $U(0, L) =: U(L)$ for homogeneous transformations; for infinitesimal (a', L') there will be a $+$ sign on the right even in double-valued representations. One may regret that what is called action by conjugation on the group level, eq. (9.3.3), is called adjoint action on the Lie algebra level.

$$i\,[M^{ab}, M^{ik}] = \eta^{ai}\,M^{bk} - \eta^{bi}\,M^{ak} + \eta^{ak}\,M^{ib} - \eta^{bk}\,M^{ia} \qquad (9.3.5a)$$

$$i\,[M^{ab}, P^{c}] = \eta^{ca}\,P^{b} - \eta^{cb}\,P^{a} \qquad (9.3.5b)$$

$$i\,[P_a, P_b] = 0. \qquad (9.3.5c)$$

(Eq. (9.3.4d) also leads to eq. (9.3.5b); relations (9.3.5a) are, of course, nothing but relations (8.2.3) in a 4-dimensional notation).

One may verify the commutation relations (9.3.5) also directly, using a concrete representation, e.g., using the following 5×5 matrix representation

$$(a, L) \mapsto \begin{pmatrix} L & a^{\top} \\ 0 & 1 \end{pmatrix}, \qquad (9.3.6)$$

which incidentally is reducible but not decomposable, like the example (9.3.4), as may happen in non-semisimple groups like $\mathcal{P}$.

We now come to the invariant, or Casimir, operators for $\mathcal{P}_+^{\uparrow}$. Since the group is not semisimple, we cannot apply the recipe given in sect. 7.4. However, we can make use of the tensor and the vector operator nature of M^{ik} and P^{c} with respect to $\mathcal{L}_+^{\uparrow}$ in forming expressions that at least are $\mathcal{L}_+^{\uparrow}$-invariant. We then have only to take care of translation invariance.

The first tensor operator which is translation invariant is P^{c} itself; so its 4-square commutes with all $U(a, L)$:

$$P_c\,P^{c} =: M^2, \qquad\qquad [M^2, U(a, L)] = 0. \qquad (9.3.7)$$

In an irreducible representation, M^2 must be a multiple of the unit operator, $M^2 = m^2\,\mathrm{id}_{\mathbf{H}}$.

The next $\mathcal{L}_+^{\uparrow}$-invariant operators $M_{ik}\,M^{ik}$, $*M_{ik}\,M^{ik}$ that offer themselves (essentially agreeing with eq. (8.1.6)) are not—because of eq. (9.3.4d)—translationally invariant. However, we can find from M^{ik} by antisymmetric multiplication with P^{j} a second translationally invariant (pseudo-)vector operator, since the disturbing term in eq. (9.3.4d) vanishes in this combination: we thus define the $Pauli\text{-}Lubanski\ vector$

$$W_d := -\frac{1}{2}\,\epsilon_{abcd}\,M^{ab}\,P^{c}, \qquad (9.3.8)$$

which is orthogonal to P^{d}:

$$W_d\,P^{d} = 0, \qquad\qquad [P_c, W_d] = 0. \qquad (9.3.9)$$

Its 4-square

$$W^2 := W^d\,W_d, \qquad\qquad [W^2, U(a, L)] = 0 \qquad (9.3.10)$$

is therefore a further $\mathcal{P}_+^{\uparrow}$-invariant operator, whose eigenvalues w^2 may be used to classify irreducible representations.

Although no further independent invariants exist, the eigenvalues of M^2, W^2 turn out not to be sufficient to completely classify the irreducible representations of $\mathcal{P}_+^{\uparrow}$, as may happen in non-semisimple groups. It will, e.g., be found that—while for $\mathcal{L}_+^\uparrow$ the (possible) eigenvalues 0, 0 of the invariants $(\mathbf{M}^+)^2$, $(\mathbf{M}^-)^2$ characterize the trivial representation—there exists a whole series of nontrivial irreducible representations in which M^2, W^2 have the eigenvalues 0, 0. For these, P_c, W_c are lightlike orthogonal vector operators, Hermitian in unitary representations and thus (cf. exercise 2 of sect. 3.2) having to be proportional:

$$W_c = \lambda\, P_c. \tag{9.3.11}$$

Here, the factor of proportionality, λ, is a further (pseudoscalar) invariant, an 'ametric' quantity in the sense that it cannot be computed by forming scalar products of the vector operators involved. As we shall see, the reason for its occurrence is that these representations 'live' on the light cone, which has an invariance group bigger than $\mathcal{L}$. λ serves as a further parameter to classify these representations.

Let us now consider all these operators for representations given in spaces of tensor and spinor fields defined on Minkowski space. An infinitesimal transformation $x \mapsto x + a$ effects $\Phi \mapsto \Phi'$, where

$$\Phi'(x) = \Phi(x - a) \approx \Phi(x) - a^k\, \partial_k\, \Phi(x) = (\mathrm{id} - a^k\, \partial_k)\, \Phi,$$

thus

$$P_k = i\partial_k\ ; \tag{9.3.12}$$

here Φ may carry arbitrary spinor indices. The operators $\hbar P_k$ form the 4-dimensional analog of the wave mechanical momentum operator when the $\Phi(x)$ is regarded as the wave function of some particle. (This is not to be confused with a quantum field! Also, there are some difficulties with the interpretation of x as a particle position.) In what follows we shall employ units such that $\hbar = 1 = c$: then eq. (9.3.12) gives the *operator of 4-momentum*. Correspondingly (cf. (4.1.7)),

$$M^2 = P^k P_k = -\partial^k \partial_k = -\Box \tag{9.3.13}$$

is the operator of *mass square*. For wave functions that belong to an *irreducible* representation of $\mathcal{P}_+^\uparrow$ we must necessarily have $M^2\Phi = m^2\Phi$ or

$$(\Box + m^2)\Phi = 0. \tag{9.3.14}$$

This is identical with the Klein-Gordon equation (9.1.3). For physical reasons, one restricts to eigenvalues $m^2 \geq 0$, although negative values would be possible mathematically.[1]

An infinitesimal Lorentz transformation $x \mapsto Lx$ effects, in the space of *scalar* fields, a transformation $\Phi \mapsto \Phi'$, where

$$\Phi'(x) = \Phi(L^{-1}x) \approx \Phi(x) - \omega^i{}_k\, x^k\, \partial_i\, \Phi(x) = \mathrm{id} - \frac{1}{2}\, \omega_{ik}\, (x^k \partial^i - x^i \partial^k)\, \Phi,$$

---

[1] For *tachyons*, i.e., particles having $m^2 < 0$, see G. Ecker, Ann. Phys. (N.Y.) *58*, 303 (1970), and references quoted there.

so that we here have $M^{ik} = L^{ik}$ with

$$L^{ik} := \frac{1}{i}\left(x^k \partial^i - x^i \partial^k\right). \tag{9.3.15}$$

The L^{ik} are obviously the relativistic counterpart of the wave mechanical *angular momentum operators*—more precisely, of the *orbital angular momentum* operators. Had we taken a tensor or spinor field instead of a scalar one,

$$M^{ik} = L^{ik} + S^{ik} \tag{9.3.16}$$

would be the sum of an orbital part and a spin part, i.e., the sum of L^{ik} and the generator $S^{ik} = -S^{ki}$ of the tensor or spinor representation involved. (Here we are employing the physicist's abbreviated notation explained after eq. (7.8.6).) For instance, for 4-vector fields, the S^{jk} are 4×4 matrices with elements

$$\left(S^{jk}\right)^m{}_n = i\left(\eta^{jm}\,\delta^k{}_n - \eta^{km}\,\delta^j{}_n\right), \tag{9.3.17}$$

since then $-\frac{i}{2}\,\omega_{ik}(S^{ik})^m{}_n\, x^n = \omega^m{}_n\, x^n$. For Dirac spinors we have, according to eq. (9.1.28),

$$S^{jk} = \frac{i}{2}\,\gamma^{[j}\gamma^{k]}, \tag{9.3.18}$$

while with 2-component spinors we have (see exercise)

$$S^{jk} = \frac{i}{2}\,\sigma^{[j}\,\tilde{\sigma}^{k]} \qquad \text{for} \quad \mathrm{D}^{(1/2,0)} \tag{9.3.19a}$$

$$= \frac{i}{2}\,\tilde{\sigma}^{[j}\,\sigma^{k]} \qquad \text{for} \quad \mathrm{D}^{(0,1/2)}. \tag{9.3.19b}$$

When we form the operators W_d we see that L^{ik} does not contribute: in the space of scalar fields we have $W_d \equiv 0$, $W^2 \equiv 0$. The S^{jk}, the relativistic generalization of the spin matrices, determine the form of W_d and W^2. The second invariant is therefore related to the *spin* of quantum particles, as will be explained in more detail later.

The condition that the operators M^2, W^2 are proportional to the unit operator is necessary for an irreducible representation, but is not sufficient, due to possible multiplicities (isotypic representations). Let us illustrate this point by considering the space of Dirac spinor fields $\psi(x)$. In it we have identically $W^2 = -\frac{1}{2}\left(\frac{1}{2}+1\right)M^2$ (see exercise), implying that if we impose the Klein-Gordon equation (9.3.14), W^2 automatically becomes a multiple of the unit operator. But the space of solutions of the Dirac equation (9.1.12) forms a genuine subspace, since the Dirac equation is *not* implied by the Klein-Gordon equation. (The converse is true, cf. eq. (9.1.10).) In this subspace—which turns out to be irreducible under $\mathcal{P}$, decomposing into two irreducible subspaces under $\mathcal{P}^\uparrow_+$, as we shall see—we get a *unitary* representation, in conformity with the quantum mechanical significance of the Dirac equation. The invariant scalar product is given by the integral

$$\int_\sigma d\sigma^k\, \bar{\varphi}\,\gamma_k\,\psi, \tag{9.3.20}$$

which, because of $\partial^k(\bar\varphi\,\gamma_k\,\psi) = 0$ for any two solutions of the Dirac equation φ, ψ (cf. exercise 2 of sect. 9.1), is independent of the special spacelike hypersurface chosen (cf. sect. 5.7). To see the definiteness, choose $d\sigma^k = (d^3x, \mathbf{0})$ and remember the definition (9.1.37) as well as the definiteness of $A\gamma_0$ proved there.

As we are interested in unitary *ray* representations, let us here write out the analysis sketched in sect. 9.2 for lifting a ray representation. The modification (9.2.2) of the commutation relations means in our present case that we admit additive terms

$$C^{ab,ik}\,\mathrm{id}_{\mathbf{H}}, \qquad\qquad C^{ab,c}\,\mathrm{id}_{\mathbf{H}}, \qquad\qquad C_{a,b}\,\mathrm{id}_{\mathbf{H}} \qquad\qquad (9.3.21a,b,c)$$

on the right-hand sides of eqs. (9.3.5a, b, c), where

$$C^{ab,ik} = -C^{ba,ik} = -C^{ab,ki} = -C^{ik,ab} \qquad\qquad (9.3.22a)$$

$$C^{ab,c} = -C^{ba,c} \qquad\qquad C_{a,b} = -C_{b,a}. \qquad\qquad (9.3.22b,c)$$

The cocycle condition (9.2.3) for admissible terms consists of

$$\begin{aligned}
&\eta^{ai}C^{rs,bk} - \eta^{bi}C^{rs,ak} + \eta^{ak}C^{rs,ib} - \eta^{bk}C^{rs,ia} + \\
&+\eta^{ra}C^{ik,sb} - \eta^{sa}C^{ik,rb} + \eta^{rb}C^{ik,as} - \eta^{sb}C^{ik,ar} + \\
&+\eta^{ir}C^{ab,ks} - \eta^{kr}C^{ab,is} + \eta^{is}C^{ab,rk} - \eta^{ks}C^{ab,ri} = 0
\end{aligned} \qquad\qquad (9.3.23a)$$

$$\begin{aligned}
&\eta^{ca}C^{rs,b} - \eta^{cb}C^{rs,a} - \eta^{cr}C^{ab,s} + \eta^{cs}C^{ab,r} - \\
&-\eta^{ra}C^{sb,c} + \eta^{sa}C^{rb,c} + \eta^{rb}C^{sa,c} - \eta^{sb}C^{ra,c} = 0
\end{aligned} \qquad\qquad (9.3.23b)$$

$$\eta^{ca}C^{b,d} - \eta^{cb}C^{a,d} - \eta^{da}C^{b,c} + \eta^{db}C^{a,c} = 0. \qquad\qquad (9.3.23c)$$

The infinitesimal change of phase (9.2.4) here means

$$M^{ab} \mapsto M^{ab} + C^{ab}\mathrm{id}_{\mathbf{H}}, \qquad\qquad P^a \mapsto P^a + C^a\mathrm{id}_{\mathbf{H}}. \qquad\qquad (9.3.24a,b)$$

By it, we want to achieve the validity of eq. (9.2.6), i.e.,

$$C^{ab,ik} = \eta^{ai}C^{bk} - \eta^{bi}C^{ak} - \eta^{ak}C^{bi} + \eta^{bk}C^{ai} \qquad\qquad (9.3.25a)$$

$$C^{ab,c} = \eta^{ca}C^b - \eta^{cb}C^a \qquad\qquad (9.3.25b)$$

$$C_{a,b} = 0 \qquad\qquad (9.3.25c)$$

for some given system of constants $C^{ab,ik}$, $C^{ab,c}$, $C^{a,b}$ satisfying eq. (9.3.23). Luckily, one obtains from eq. (9.3.23c) by contraction with η_{ca}, because of eq. (9.3.22c),

$$(4-1-1)C^{b,d} = 0, \qquad\qquad (9.3.26c)$$

so that eq. (9.3.25c) is satisfied. (At this point it is essential to have a spacetime dimension greater than 2!) Contracting eq. (9.3.23b) with η_{cs}, we obtain, due to eq. (9.3.22b),

$$(4-1)C^{ab,r} - \eta^{ra}C_s^{b,s} + \eta^{rb}C_s^{a,s} = 0, \qquad\qquad (9.3.26b)$$

so that (9.3.25b) can be satisfied on choosing $C^b := \frac{1}{3}C_s^{\,b,s}$. Finally, contracting eq. (9.3.23a) with η_{is}, we get, due to eq. (9.3.22a),

$$(4-1-1)C^{ab,rk} - \eta^{ar}C^{ks,\,b}_{\quad\ \ s} + \eta^{br}C^{ks,\,a}_{\quad\ \ s} + \eta^{ak}C^{rs,\,b}_{\quad\ \ s} - \eta^{bk}C^{rs,\,a}_{\quad\ \ s} = 0, \qquad (9.3.26a)$$

so that eq. (9.3.25a) can be satisfied on choosing $C^{ai} := \frac{1}{2}C^{is,\,a}_{\quad\ \ s} = -C^{ia}$. (Note again that we needed a spacetime dimension greater than 2.) This analysis shows that near the identity element of the group $\mathcal{P}_+^\uparrow$ every ray representation is equivalent to an ordinary one, implying that for the global group the ray representations are given by ordinary representations of the universal covering group $\widetilde{\mathcal{P}_+^\uparrow}$. This is the *physical* justification for considering the spinor representations, announced several times.

In the next section we shall classify the unitary irreducible representations of $\widetilde{\mathcal{P}_+^\uparrow}$ systematically.

Exercises

1. Verify eq. (9.3.19) and show that eq. (9.3.18) is the direct sum of formulae (9.3.19). Observe that, in eq. (9.3.19), $\omega_{ik}S^{ik}$ is the $D^{(1,0)}$ or $D^{(0,1)}$ part of the sixtor ω_{ik}—this being the only possibility to remain consistent with $D^{(1/2,1/2)} = D^{(1/2,0)} \otimes D^{(0,1/2)}$.

2. Calculate the operator W^2 in the space of
   a. 2- and 4-component spinor fields:

   $$W^2 = \frac{3}{4}\Box = -\frac{1}{2}\left(\frac{1}{2}+1\right)M^2 \qquad (9.3.27a)$$

   b. sixtor fields:

   $$W^2 = 2\,\Box = -1(1+1)M^2 \qquad (9.3.27b)$$

   c. 4-vector fields:

   $$\left(W^2\right)^i_{\ k} = 2\left(\delta^i_{\ k}\Box - \partial^i\,\partial_k\right). \qquad (9.3.27c)$$

   Observe in case c that in the subspace of divergence-free vector fields eq. (9.3.27b) also holds!

3. Evaluate the condition $W_c = \lambda P_c$, which in the case $m^2 = w^2 = 0$ is necessary for unitary irreducibility, for the space of solutions of the (massless) Dirac equation.
   Hint: Use eq. (9.1.53) to convert $W_c = \lambda P_c$ into $\lambda\,\partial_c\,\psi = \frac{1}{2}\gamma^5\,\partial_c\,\psi$. This then means that $\partial_c\,\psi$ is an eigenvector of γ^5. Because of $(\gamma^5)^2 = 1$ it follows that $\lambda = \pm 1/2$, and the matrices $(1\pm\gamma^5)/2$ project to the eigenspaces, in which the Weyl equations hold.

4. Show that in the space of all solutions of the Maxwell equations in vacuum, $\partial_k F^{ik} = 0 = \partial_k {}^*F^{ik}$, one has $M^2 = 0 = W^2$. Work out the conditions to which $W_c = \lambda P_c$ leads in this space, and determine the possible values of λ, imposing nonconstant F^{ik}.
   Solution: $\lambda = \pm 1$; the field tensors belonging to these values are selfdual or anti-selfdual.

5. Investigate the conditions $M^2 = m^2 \,\mathrm{id}$, $W^2 = w^2 \,\mathrm{id}$, $W_c = \lambda\, P_c$ in the space of vector fields $A^i(x)$!
   Solution: For $w \neq 0$ it follows that $\partial_i\, A^i = 0$, and for $A^i \neq 0$: $w^2 = -1(1+1)m^2$. A^i in this case satisfies the *Proca equations*

$$(\Box + m^2)A^i = 0, \qquad\qquad \partial_i\, A^i = 0. \qquad (9.3.28)$$

The representation defined by these equations can be made unitary for $m^2 > 0$; it then decomposes under $\mathcal{P}_+^\uparrow$ into two irreducible parts, as will turn out later.

For $w^2 = 0$, $m^2 \neq 0$ it follows that $\partial_j\, A_i = \partial_i\, A_j$, i.e., A_i is a 4-gradient field: $A_i = \partial_i\, \Lambda$, $\Lambda = const. + \Phi$, where Φ satisfies the Klein-Gordon equation.

For $w^2 = 0$, $m^2 = 0$ every solution has the form $A^i = \bar{A}^i + c\,x^i$, where $\Box \bar{A}^i = 0$, $\partial_i\, \bar{A}^i = 0$, i.e., $\bar{A}^i$ is a 4-potential of a vacuum Maxwell field in Lorenz gauge. The fields $\bar{A}^i$ form an invariant subspace; on the other hand, the space of fields $c\,x^i$ is not translationally invariant (reducibility without decomposability), but these fields do not satisfy the usual conditions at infinity.

Finally, in the space of solutions of $\Box A^i = 0$, $\partial_i\, A^i = 0$, the condition $W_c = \lambda\, P_c$ leads to $\epsilon_{abcd}\, \partial^c A^b = -i\,\lambda\, \partial_d\, A_a$. For $\lambda \neq 0$ it follows that $\partial_d\, A_a + \partial_a\, A_d = 0$, from which—cf. eq. (5.9.29)—$A_d = a_d + a_{dc}\, x^c$ with constants a_d, $a_{dc} = -a_{cd}$. Going back yields selfduality or anti-selfduality of a_{cd} and $\lambda = \pm 2$, but this solution does not satisfy the usual conditions at infinity (it would be the 4-potential for constant electromagnetic fields). The case $\lambda = 0$ gives $\partial^c A^b = \partial^b A^c$, i.e., $A^b = \partial^b \Lambda$ is the 4-gradient of a solution of the scalar wave equation. Observe that the subspaces $\lambda = \pm 2$ and $\lambda = 0$ have the solution $A_d = const.$ in common, so again there is no direct sum.

These naive considerations illustrate the complications that may occur in nonunitary representations. It also emerges that the value $\lambda = \pm 1$, expected for the electromagnetic radiation field, does not appear. However, if we distinguish some auxiliary vector n^c and pass from the condition $W_c = \lambda\, P_c$ to $n^c W_c = \lambda\, n^c P_c$, from which λ may also be calculated, then we get for $\lambda \neq 0$ by transvection $n^c\partial_c\,(n^a A_a) = 0$ and by iteration $(\lambda^2 - 1)\,(n^c\partial_c)^2 A_a = 0$, and thus also the eigenvalues $\lambda = \pm 1$. The gauge transformation $A_i \mapsto \bar{A}_i + \partial_i\, \Lambda$, $\Lambda = -(m_a x^a)(n_b A^b)$ with $m_a n^a = 1$ leads to the noncovariant gauge condition $n^a A_a = 0$ (n has been specified at will! When $n^2 > 0$ it is called radiation gauge, when $n^2 < 0$ it is called axial gauge, and when $n^2 = 0$ it is called lightlike gauge).

6. Verify eqs. (9.3.23,25,26) in detail!

9.4 Irreducible Unitary Representations of the Poincaré Group

To classify the unitary irreducible representations of $\widetilde{\mathcal{P}_+^\uparrow}$ we imagine any of them as being given and analyze it; the vectors of the representation space $\mathbf{H}$ we denote as $|\ldots\rangle$, using Dirac's notation.

The unitary operators $U(a)$ representing translations all commute among each other, so there is a complete set of eigenvectors common to all $U(a)$, which we can use as a basis for **H**. If $|\ \rangle$ is one of these, we must have $U(a)|\ \rangle = \chi(a)|\ \rangle$ for all translations a, where the dependence of the eigenvalue χ on a has been indicated; i.e., we can consider χ as a map from the translation subgroup $\mathcal{T}$ to the group U(1) of phase factors, also known as a *character* of $\mathcal{T}$. (Remember that the eigenvalues of unitary operators are phase factors.) From $U(a)U(a') = U(a+a')$ we get $\chi(a)\chi(a') = \chi(a+a')$, and the continuous solutions of this functional equation are given by $\chi(a) = \exp(ip(a))$, where $p(a) = p_i a^i$ is linear in a, thus defining a 4-vector p characterizing the character χ and the eigenvector $|\ \rangle = |p, \alpha\rangle$—in the latter case we have added a further parameter α that may (and will) be necessary for a *unique* characterization in case of degenerate eigenvalues:[1]

$$U(a)|p, \alpha\rangle = \exp(ip_k a^k)|p, \alpha\rangle$$
$$P_k|p, \alpha\rangle = p_k|p, \alpha\rangle. \tag{9.4.1}$$

Here the second equation is the infinitesimal version of the first. The eigenspace spanned by the $|p, \alpha\rangle$ will be denoted $\mathbf{H}_p$. We see that the characters form a group, which appears additive when written in terms of the variables p; we will see immediately that it is natural to identify this 'dual' group of $\mathcal{T}$ with a Minkowski vector space.

Basically, all values of p are admitted, but for irreducibility we try to get along with as few as possible. To see how many are necessary, we now investigate the effect of the operators[2] $U(L)$ when applied to the vectors $|p, \alpha\rangle$, observing that the P_k are vector operators. It follows that the vector $|\ \rangle = U(L)|p, \alpha\rangle$ has

$$P_k|\ \rangle = P_k\, U(L)|p, \alpha\rangle = U(L)\, L_k{}^j\, P_j|p, \alpha\rangle = U(L)\, L_k{}^j\, p_j|p, \alpha\rangle =$$
$$= L_k{}^j\, p_j\, U(L)|p, \alpha\rangle = L_k{}^j\, p_j|\ \rangle, \tag{9.4.2}$$

i.e., $|\ \rangle$ is an eigenvector of P_k for the eigenvalue $L_k{}^j\, p_j$ and thus is contained in the eigenspace $\mathbf{H}_{Lp}$ of P_k:

$$U(L)|p, \alpha\rangle = \sum_\beta Q^\beta{}_\alpha(L, p)|Lp, \beta\rangle. \tag{9.4.3}$$

Here we have indicated that the matrix Q that gives the expansion with respect to a basis of the eigenspace may depend not only on L but also on p. Also, we must require Q to be unitary if the indices refer to orthonormal bases. In geometrical terms, we have shown that $U(L)\mathbf{H}_p = \mathbf{H}_{Lp}$.

To illustrate what we did so far in a completely abstract setting, consider the representation obtained in the space of fields Φ of some definite type (spinor field, tensor field). Then P_k has the form $+i\partial_k$, the 'eigenfields' are of the form $\tilde\Phi\exp(-ip_k x^k)$, where $\tilde\Phi = const.$ (spinor, tensor). The

---

[1] As is usual in the physics literature, a continuous spectrum will be treated in formal analogy to a discrete one; see Naimark (1960) or Reed and Simon (1978) for exact formulations, in particular for direct integrals.

[2] We should actually be writing $A \in \mathrm{SL}(2, \mathbf{C})$ instead of $L \in \mathcal{L}_+^\uparrow$ or else take account of eq. (9.2.7). However, we shall only return to SL(2,**C**) when it becomes necessary.

decompsition of an arbitrary field of the type under consideration with respect to these eigenfunctions is

$$\Phi(x) = \int \frac{d^4 p}{(2\pi)^4}\, e^{-ipx}\, \tilde{\Phi}(p), \tag{9.4.4}$$

and so is a Fourier expansion. The extra indices $\alpha, \ldots$ occurring in the $|\,p, \alpha\,\rangle$ may thus be chosen here as the spinor or tensor indices of $\tilde{\Phi}$ (spinor basis, tensor basis); in doing so, however, we apparently give up orthonormality in the sense of a positive-definite scalar product. If the type of the field is given by the representation D of $\mathcal{L}_+^\uparrow$, the effect of a Lorentz transformation L on the basis functions is given, according to eq. (9.1.1), by

$$\tilde{\Phi}\, e^{-ipx} \to \mathrm{D}(L)\, \tilde{\Phi}\, e^{-ip(L^{-1}x)} = \mathrm{D}(L)\, \tilde{\Phi}\, e^{-i(Lp)x}, \tag{9.4.5}$$

i.e., the matrix Q in eq. (9.4.3) is given by $D(L)$ and is independent of p but non-unitary except in the case of a scalar field. We shall see soon that for the classification problem another basis is more suitable, where Q becomes p-dependent but unitary. The transformation from the spinor or tensor basis to this one will involve p-dependent coefficients.

We can now already give a partial classification of the representations. Namely, from eq. (9.4.2) it follows that the spectrum of the vector operator P_k in $\mathbf{H}$ is a $\mathcal{L}_+^\uparrow$-invariant subset of the 4-vector space of all characters (candidates for the eigenvalues). Such an invariant subset is a union of 'minimal' invariant subsets; each of the latter is obtained by applying all the L to some element of it and is also called *orbit* of that element under $\mathcal{L}_+^\uparrow$. (Clearly, $\mathcal{L}_+^\uparrow$ acts transitively on such an orbit which thus is a homogeneous space of $\mathcal{L}_+^\uparrow$ in the sense discussed in sects. 7.7 and 7.8.) It should be clear that the linear span of the $|\,p, \alpha\,\rangle$, where the p are just taken from such an orbit $\mathbf{O}$, is a $\mathcal{P}_+^\uparrow$-invariant subspace of the representation space $\mathbf{H}$ and must thus coincide with it in case of irreducibility. The announced partial classification now comes simply from a classification of the orbits $\mathbf{O}$ of $\mathcal{L}_+^\uparrow$ in the space of 4-vectors p.

Before we write this down in detail, let us first remark that a Lorentz invariant quantity is constant along an orbit. Therefore, along an orbit, the 4-square p^2 takes a constant value which we write m^2 (without implying this to be nonnegative a priori). It follows that the Casimir operator $M^2 = P_k P^k$ becomes $M^2 = m^2\, \mathrm{id}_{\mathbf{H}}$, as we found earlier.

We come to the classification of orbits, which for non-spacelike vectors also needs consideration of the (discontinuous) invariant $\operatorname{sign} p_0$ (Fig. 9.1):

$$
\begin{aligned}
&\text{a}_+)\ \ p^2 = m^2 > 0, && \operatorname{sign} p_0 = +1 && \text{(mass shell)}\\
&\text{a}_-.\ \ p^2 = m^2 > 0, && \operatorname{sign} p_0 = -1 &&\\
&\text{b}_+.\ \ p^2 = 0, && \operatorname{sign} p_0 = +1 && \text{(future light cone)}\\
&\text{b}_-.\ \ p^2 = 0, && \operatorname{sign} p_0 = -1 && \text{(past light cone)}\\
&\text{c.}\ \ \ p = 0 && && \text{(zero vector)}\\
&\text{d.}\ \ \ p^2 < 0 && && \text{(timelike hyperboloid)}
\end{aligned}
$$

A unitary irreducible representation of $\widetilde{\mathcal{P}_+^\uparrow}$ has to fall into one of these six classes. Note that except for case c the spectrum of P_k in an irreducible representation is purely continuous. For physical reasons, in what follows we shall not consider the

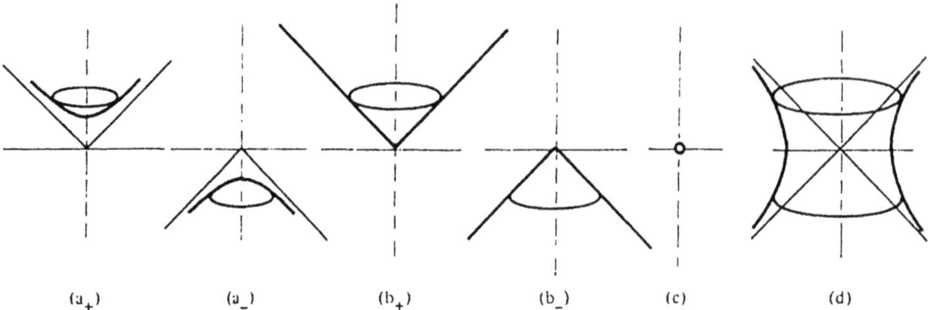

Fig. 9.1. Orbits of $\mathcal{L}_+^\uparrow$ in p-space

cases c, d any further: particles having $p = 0$ or $p^2 < 0$ have not played a role in physics.

Let us illustrate this partial classification by the example of fields Φ. The condition $p^2 = m^2$ requires the Fourier components $\tilde{\Phi}(p)$ in eq. (9.4.4) to be nonvanishing only if $p^2 = m^2$, which is the same as requiring $(\Box + m^2)\,\Phi(x) = 0$. The further condition sign $p_0 = +1$ or $= -1$ that appears in cases a, b requires the Fourier components to be nonvanishing only for positive or only for negative frequencies (energies). This causes some troubles for the idea of a relativistic wave mechanics and in the end requires the transition to quantum field theory. Although that theory is not the subject of this book, we will come back to this in the next section.

Our next step will be the classification of the possible Q in eq. (9.4.3.). They are subject to the condition

$$Q^\gamma{}_\alpha(L'L,p) = \sum_\beta Q^\gamma{}_\beta(L',Lp)\,Q^\beta{}_\alpha(L,p), \qquad (9.4.6)$$

which arises from further application of a Lorentz transformation $U(L')$, using the representation property. Condition (9.4.6) looks almost like a representation property—in particular, it implies $Q(E,p) = \mathrm{id}_\mathbf{H}$—and goes over into that property in two cases. One is the case mentioned above with fields on spacetime, where Q does not depend on p; however, this case will be less important in what follows, since it is not immediate how the representation should become unitary.

The other case where eq. (9.4.6) becomes a representation property is when the representation is restricted to the subgroup of those elements $K \in \mathcal{L}_+^\uparrow$ that leave fixed some arbitrarily selected 4-vector $\bar{p}$ in the orbit, $K\bar{p} = \bar{p}$, at the same time restricting the representing operators $U(K)$ to the eigenspace $\mathbf{H}_{\bar{p}}$. This subgroup will be written $\mathcal{K}_{\bar{p}} \subset \mathcal{L}_+^\uparrow$ and will be called the *little group* for the *standard vector* $\bar{p}$.

In the mathematical literature one finds the names isotropy subgroup, stable subgroup, or stabilizer, of the selected *origin* $\bar{p}$ in the orbit $\mathbf{O}$; cf. also remarks at the end of sect. 7.7.

We thus get the condition that the $Q(K,\bar{p})$ furnish a unitary representation of the little group $\mathcal{K}_{\bar{p}}$ in the eigenspace $\mathbf{H}_{\bar{p}}$, to be referred to as the *little* vector space. We now want to show, following E. Wigner, that this representation of the little group already determines the representation of the whole group (up to equivalence), so that our classification problem is reduced to the one of classifying the unitary irreducible representations of the little group. What we have to show is that we can construct the general $Q(L, p)$ from the special $Q(K, \bar{p})$.

The proof rests upon a factorization of L which depends on the vector $|\, p, \alpha \,\rangle$ to which the operator $U(L)$ is going to be applied. For this purpose we choose, for each $p \in \mathbf{O}$, a transformation $\Lambda_p \in \mathcal{L}_+^\uparrow$ sending $\bar{p}$ to p, depending continuously on p and satisfying $\Lambda_{\bar{p}} = E$:

$$\Lambda_p \in \mathcal{L}_+^\uparrow \ : \ \Lambda_p \bar{p} = p, \qquad \Lambda_{\bar{p}} = E. \tag{9.4.7}$$

This is clearly possible by the assumption that all p belong to the same orbit.

When $p^2 > 0$, $p^0 > 0$, the transformation given by eq. (6.3.6) does what we want and is uniquely determined by the condition that it be a boost for an observer whose 4-velocity $\underset{\sim}{u}$ is collinear with the standard vector $\bar{p}$. Since we are actually interested in representing the group $\widetilde{\mathcal{P}}_+^\uparrow$, and therefore, at this moment, in $\widetilde{\mathcal{L}_+^\uparrow} \cong \mathrm{SL}(2, \mathbf{C})$, what we need is some $A_p \in \mathrm{SL}(2, \mathbf{C})$ yielding Λ_p via eq. (8.2.8). Accordingly, we might take eq. (8.4.30). The continuous dependence is explicit.

When $p^2 = 0$, $p^0 > 0$, however, the choice of Λ_p, or A_p, as a boost for some fixed observer u, combining Doppler effect and aberration, does not work for the special $p \neq \bar{p}$ that is coplanar with u and $\bar{p}$. Of course, one might simply change u or admit an additional rotation; but, given $\bar{p}$, it turns out to be impossible to make the choice depend on p *continuously everywhere* on $\mathbf{O}$ on topological grounds, whatever the choice. (Namely, if a continuous choice is possible it follows that the full group $\mathrm{SL}(2,\mathbf{C})$ is topologically the product of the orbit $\mathbf{O}$ and the little group. While this works in the case of the mass shell, it does not in the case of the light cone whose topology is $\mathbf{S}_2 \times \mathbf{R}$, as we shall see later that in the present case the topology of $\mathcal{K}_{\bar{p}}$ is $\mathbf{R}^2 \times \mathbf{S}_1$, so that the product space would be infinitely connected, while $\mathrm{SL}(2,\mathbf{C})$ is simply connected.) However, if one does the functional analysis needed in infinite-dimensional representations—but omitted by us—correctly, it turns out that violation of continuity is allowed on a set of measure zero on the orbit, in the sense of the measure d^3p/p^0 to be considered later.

The factorization of L in question is now given by

$$L = \Lambda_{Lp}\, K(L, p)\, \Lambda_p^{-1}, \tag{9.4.8}$$

where $K(L, p)$ is defined by this equation, i.e.,

$$K(L, p) := \Lambda_{Lp}^{-1} L \Lambda_p. \tag{9.4.9}$$

The point here is that $K(L, p)$ belongs to the little group, since by construction of the Λ_p

$$\Lambda_p \bar{p} = p, \qquad \Lambda_{Lp}^{-1} Lp = \bar{p} \Rightarrow K(L, p)\, \bar{p} = \bar{p}.$$

It is useful to imagine the total space $\mathbf{H}$, which is the direct sum (or rather, direct integral) of the eigenspaces $\mathbf{H}_p$, as a *vector bundle* over the orbit $\mathbf{O}$ (cf. sect.7.8) with the eigenspaces $\mathbf{H}_p$ as fibers and the elements $|\,\psi\,\rangle$ of $\mathbf{H}$ as *cross sections*, associating with each $p \in \mathbf{O}$ the projection of $|\,\psi\,\rangle$ into the eigenspace $\mathbf{H}_p$. When a basis $|\,p, \alpha\,\rangle$ is selected in each fiber, a cross section is specified by the component functions $\psi_\alpha(p) = \langle\, p, \alpha\,|\,\psi\,\rangle$. This is not only a useful picture but also helps the mathematics if one wants to avoid the use of 'improper' eigenvectors $|\,p, \alpha\,\rangle$ associated with a continuous spectrum: one simply takes the $\mathbf{H}_p$ as isomorphic Hilbert spaces without thinking of them as subspaces of $\mathbf{H}$, but *defines* the latter to be the space of cross sections of the vector bundle formed by the $\mathbf{H}_p$, square-integrable in the sense of some group-invariant measure (see eqs.

(9.4.26,27)) on **O**. Then eqs. (9.4.2,3) are to be interpreted as saying that L *acts* on the bundle—which thus becomes a homogeneous vector bundle in the sense of sect. 7.8—transforming each fiber $\mathbf{H}_p$ to $\mathbf{H}_{Lp}$ by some unitary map $Q(L,p)$ whose matrix description when referred to bases is given in eq. (9.4.3). From this derives an action on the cross sections and thus on **H**.

It will be important to notice that we have not yet committed ourselves as to what the basis in each fiber is, except perhaps in the little vector space—the fiber—over the standard vector $\bar{p}$, where it may refer to some standard form of matrix representations of the little group. In what follows, we are going to use the freedom of choice for bases in the other fibers to facilitate the classification without changing the equivalence class of the representation. It should be noticed that the definition (9.4.10) actually also specifies the (local) *topology* of the vector bundle: so far, the bundle was only defined as a set, namely as the disjoint union of the fibers $\mathbf{H}_p$; it is here that the continuous choice of the Λ_p comes in as well as the assumption that our representation is continuous. The difficulty in choosing the Λ_p in a continuous fashion globally all over the orbit, noted before in one case, necessitates to work with two different choices for $\bar{p}$ and makes the vector bundle globally 'nontrivial', i.e., twisted in a way similar to a Moebius band.

We now *define* the basis vectors in each $\mathbf{H}_p$ by setting

$$|\,p,\alpha\,\rangle = U(\Lambda_p)\,|\,\bar{p},\alpha\,\rangle \qquad\qquad (Wigner\,basis). \qquad\qquad (9.4.10)$$

In this way every given irreducible unitary representation defines, after choosing $\bar{p}$, $|\,\bar{p},\alpha\,\rangle$ and the Λ_p, a special basis in **H**, with respect to which we have

$$U(L)\,|\,p,\alpha\,\rangle = \sum_\beta Q^\beta{}_\alpha(K(L,p),\bar{p})\,|\,Lp,\beta\,\rangle, \qquad\qquad (9.4.11)$$

as follows from eqs. (8.4.3,8,10), where, as we noted before, the $Q(K,\bar{p})$ are a unitary representation of the little group. Unitarity of the map (9.4.10) requires the usual 'continuum normalization', however using a δ-function which is defined on the orbit **O** in a Lorentz invariant fashion. We shall do this later explicitly.

It should be evident at this point that irreducibility of the representation of the little group is necessary and sufficient for the irreducibility of the total representation. Note that we also found

$$Q^\beta{}_\alpha(L,p) = Q^\beta{}_\alpha(K(L,p),\bar{p})$$

if referred to Wigner bases; the validity of eq. (9.4.6) can be checked from this.

One says that the representation $L \mapsto U(L)$ of the Lorentz group $\mathcal{L}_+^\uparrow$ is *induced* by the representation $K \mapsto Q(K)$ of the little group $\mathcal{K}_{\bar{p}} \subset \mathcal{L}_+^\uparrow$. This inducing construction is of great generality, since the orbit **O** may be interpreted abstractly as the coset space $\mathcal{L}_+^\uparrow/\mathcal{K}_{\bar{p}}$ (cf. Hermann 1966, Mackey 1968). Note, however, that the representations obtained by the inducing construction are reducible for $\mathcal{L}_+^\uparrow$ (cf. the remarks in sect. 7.8); it is only for $\mathcal{P}_+^\uparrow$ that we have irreducibility! We also point out that the reformulation, indicated in that section, in terms of functions on the group $(\mathcal{L}_+^\uparrow)$ with values in the representation space of the little group automatically shows that the equivalence class of the representation does not depend on the choice of $\bar{p}$ and the Λ_p (representatives of the cosets). We therefore refrain from checking this explicitly in the formulation given above.

Before we determine the little groups and their irreducible representations in detail, at least in the physically interesting cases a, b, we characterize them infinitesimally, obtaining an interpretation of the Pauli-Lubanski vector operator. The infinitesimal Lorentz transformation $K^i{}_k = \delta^i{}_k + \omega^i{}_k$ belongs to $\mathcal{K}_{\bar{p}}$ iff $\omega^i{}_k\,\bar{p}^k = 0$; for

$\bar{p} \neq 0$ the general solution is $\omega_{ik} = \epsilon_{ikjm}\, k^j\, \bar{p}^m$, where k^j is an arbitrary infinitesimal 4-vector. Now in the representation $L \mapsto U(L)$ we have

$$U(K) = \mathrm{id} - \frac{i}{2}\, \omega_{ab}\, M^{ab} = \mathrm{id} - \frac{i}{2}\, \epsilon_{abcd}\, k^c\, \bar{p}^d\, M^{ab}. \qquad (9.4.12)$$

We thus see that $U(K)$ acts on the vectors $|\,\bar{p}, \alpha\,\rangle$ in the same way as does the operator

$$\mathrm{id} - i k^c W_c, \qquad (9.4.13)$$

where W_c is given by eq. (9.3.8). Thus the Pauli-Lubanski vector just generates the transformations of the little group in the little vector space $\mathbf{H}_{\bar{p}}$. The number of parameters in the little group is only 3, however, since one of the components of k^j may be eliminated due to the relation $P^c W_c = 0$ ('inessential' parameter). The structure of the little groups results from the commutation relations—to be verified as an exercise—

$$[W_d, W_e] = -\frac{1}{2}\epsilon_{abcd}\,[M^{ab}, W_e]\, P^c = -i\,\epsilon_{debc}\, W^b\, P^c \qquad (9.4.14)$$

by applying them to the vectors $|\,\bar{p}, \alpha\,\rangle$.

We now treat the classes a, b separately and recommend the determination of the little groups for the remaining classes c, d as an exercise.

a. Case $p^2 = m^2 > 0$.
We assume $\mathrm{sign}\, p_0 > 0$, the other sign being completely analogous. For the standard vector $\bar{p}$ we take the normal form (3.2.7), i.e., $\bar{p}^i = (m, \mathbf{0})^\top$. The little group $\mathcal{K}_{\bar{p}}$ has then obviously the structure of the rotation group $SO(3)$, or rather its covering group $SU(2)$. When we interpret the vectors $|\,p, \alpha\,\rangle$ quantum mechanically as momentum eigenstates of a particle with rest mass m, then $|\,\bar{p}, \alpha\,\rangle$ will describe possible states of the particle at rest ($\mathbf{p} = \mathbf{0}$), the index α therefore referring to some 'inner' rotational degree of freedom—its *spin*. The irreducible unitary representations of $SO(3)$ and $SU(2)$ were obtained in chap. 7; they are characterized uniquely up to equivalence by their highest weight, which will be denoted here by s and which may take the values $s = 0,\ 1/2,\ 1,\ 3/2,\ \ldots$. (Note that the half-integer values have to be admitted here!) To every pair of values $m^2 > 0$ and s there belong two equivalence classes of unitary irreducible representations of $\mathcal{P}_+^\uparrow$, one having $\mathrm{sign}\, p_0 = +1$, the other one $\mathrm{sign}\, p_0 = -1$.

Let us determine the value of the invariant W^2 for these representations! Because of $\bar{p} = (m, \mathbf{0})^\top$, we get from the relation $P^c W_c = 0$ as applied to the subspace $\mathbf{H}_{\bar{p}}$:

$$W_0\,|\,\bar{p}, \alpha\,\rangle = 0,$$

while evaluating eq. (9.4.14) on that subspace gives

$$[W_0, W_\mu]\,|\,\bar{p}, \alpha\,\rangle = 0, \qquad\qquad [W^\mu, W^\nu]\,|\,\bar{p}, \alpha\,\rangle = i\, m\, \epsilon_{\mu\nu\lambda}\, W^\lambda\,|\,\bar{p}, \alpha\,\rangle.$$

The operators

$$S^\mu := W^\mu/m \qquad (9.4.15)$$

thus satisfy on this subspace the commutation relations of ordinary angular momentum, so that $S^\mu S^\mu$ takes, in an irreducible representation of highest weight s, the value $s(s+1)$. Therefore we have

$$W^2 \, | \bar{p}, \alpha \rangle = -W^\mu W^\mu \, | \bar{p}, \alpha \rangle = -m^2 \, s(s+1) \, | \bar{p}, \alpha \rangle,$$

and since generally in irreducible representations $W^2 = w^2 \, \mathrm{id}$, it follows that

$$W^2 = -m^2 \, s(s+1) \, \mathrm{id} \tag{9.4.16}$$

in all representations of the class a.

In addition to the eigenvalues m^2 and w^2 we needed the value sign p_0 to nail down the equivalence class completely. The latter quantity does not correspond to the eigenvalue of a Casimir operator in the algebraic sense.

From our results on SO(3) and SU(2) in chap. 7 it follows that the $| \bar{p}, \alpha \rangle$ span a $(2s+1)$-dimensional space $\mathbf{H}_{\bar{p}}$ in which we can construct the canonical basis (7.5.35): $| \bar{p}, \alpha \rangle = | \bar{p}, \sigma \rangle$, $\sigma = -s, -s+1, \ldots, s$, where

$$S^\mu S^\mu \, | \bar{p}, \sigma \rangle = s(s+1) \, | \bar{p}, \sigma \rangle, \tag{9.4.17}$$

$$S^3 \, | \bar{p}, \sigma \rangle = \sigma \, | \bar{p}, \sigma \rangle, \qquad (S^1 \pm iS^2) \, | \bar{p}, \sigma \rangle = \sqrt{s(s+1) \mp \sigma - \sigma^2} \, | \bar{p}, \sigma \pm 1 \rangle.$$

When $| \bar{p}, \sigma \rangle$ is interpreted as a state vector of a massive particle at rest, we get the interpretation of $\mathbf{S}$ as the operator of spin and of s as the spin of the particle.

Let us now write the vectors $| \bar{p}, \sigma \rangle$ in a representation with $M^2 = m^2 \, \mathrm{id}$, $W^2 = -s(s+1) \, m^2 \, \mathrm{id}$ somewhat more completely as $| m, s; p, \sigma \rangle$. Then we can collect our results for the case $m^2 > 0$, sign $p_0 > 0$ as follows:

$$M^2 \, | m, s; p, \sigma \rangle = m^2 \, | m, s; p, \sigma \rangle \tag{9.4.18a}$$

$$\mathcal{W}^2 \, | m, s; p, \sigma \rangle = -s(s+1) \, m^2 \, | m, s; p, \sigma \rangle \tag{9.4.18b}$$

$$P_k \, | m, s; p, \sigma \rangle = p_k \, | m, s; p, \sigma \rangle, \quad \text{where} \quad p_k \, p^k = m^2, \quad p_0 > 0 \tag{9.4.18c}$$

$$U(a) \, | m, s; p, \sigma \rangle = \exp(ipa) \, | m, s; p, \sigma \rangle \tag{9.4.18d}$$

$$U(L) \, | m, s; p, \sigma \rangle = \sum_{\sigma'=-s}^{s} \mathrm{D}^{(s)}_{\sigma'\sigma}(K(L,p)) \, | m, s; Lp, \sigma' \rangle, \tag{9.4.18e}$$

where

$$K(L,p) = \Lambda_{Lp}^{-1} L \Lambda_p \tag{9.4.9bis}$$

is the *Wigner rotation* for L, p, and where Λ_p may be chosen as the boost (6.3.6), with respect to an observer $u \propto \bar{p}$, that takes $\bar{p} = (m, \mathbf{0})^\top$ to p; $\mathrm{D}^{(s)}(K)$ is the irreducible representation for the highest weight s of the Wigner rotation.

Having specified the basis vectors of our representation, we still want to see how the invariant scalar product looks like, with respect to which they are orthonormal. Since we are in a continuous spectrum, we will have a δ-like continuum normalization on the orbit $\mathbf{O}$ which, however, is not simply Euclidean space but the curved mass shell, given in p-space by the implicit equation $p^2 = m^2$. Let us take the space components $\mathbf{p}$ as independent variables and write, as in eq. (4.5.5),

$$|p^0| = \sqrt{m^2 + \mathbf{p}^2} =: E(\mathbf{p}). \tag{9.4.19}$$

We further write $L\mathbf{p}$ for the space components of the 4-vector Lp, where $p = (E(\mathbf{p}), \mathbf{p})$. The orthogonality relations for the basis vectors will then certainly be of the form

$$\langle \mathbf{p}', \sigma' \,|\, \mathbf{p}, \sigma \rangle = A(\mathbf{p})\, \delta^3(\mathbf{p} - \mathbf{p}')\, \delta_{\sigma\sigma'}, \tag{9.4.20}$$

where, however, the normalizing factor $A(\mathbf{p})$ cannot be chosen to be $\equiv 1$ but must guarantee the unitarity condition

$$\langle \mathbf{p}', \sigma' \,|\, U^\dagger(L)\, U(L) \,|\, \mathbf{p}, \sigma \rangle = \langle \mathbf{p}', \sigma' \,|\, \mathbf{p}, \sigma \rangle; \tag{9.4.21}$$

translational invariance being already guaranteed by eq. (9.4.20). Now from eqs. (9.4.18) and the unitarity of the $D^{(s)}_{\sigma\sigma'}$ we first obtain

$$\langle L\mathbf{p}', \sigma' \,|\, L\mathbf{p}, \sigma \rangle = \langle \mathbf{p}', \sigma' \,|\, \mathbf{p}, \sigma \rangle, \tag{9.4.22}$$

which is compatible with eq. (9.4.20) if

$$A(L\mathbf{p})\, \delta^3(L\mathbf{p} - L\mathbf{p}') = A(\mathbf{p})\, \delta^3(\mathbf{p} - \mathbf{p}'). \tag{9.4.23}$$

It follows that the ratio $A(L\mathbf{p}) : A(\mathbf{p})$ is given by the Jacobian of $\mathbf{p} \mapsto L\mathbf{p}$ and that $A(\mathbf{p})$ is unique up to a constant factor (the choice of $A(\bar{\mathbf{p}})$) if it exists. To evaluate the condition further, an elegant way to calculate the Jacobian is as follows. In the invariant δ-function $\delta^4(p - p') = \delta(p^0 - p'^0)\, \delta^3(\mathbf{p} - \mathbf{p}')$—whose invariance we know from eq. (4.5.25)—we introduce instead of p^0, p'^0 the invariants $m^2 = (p^0)^2 - \mathbf{p}^2$, $m'^2 = (p'^0)^2 - \mathbf{p}'^2$ as new variables: using well-known rules for the δ-function we get

$$\delta^4(p - p') = \delta(m^2 - m'^2) \cdot 2\, E(\mathbf{p})\, \delta^3(\mathbf{p} - \mathbf{p}'), \tag{9.4.24}$$

which shows that the expression $2\, E(\mathbf{p})\, \delta^3(\mathbf{p} - \mathbf{p}')$ is invariant. Therefore, a possible candidate for the normalization is $A(\mathbf{p}) := 2\, E(\mathbf{p})$. Adopting it—remember that only a numerical factor remains conventional—the orthonormalization condition becomes

$$\langle \mathbf{p}', \sigma' \,|\, \mathbf{p}, \sigma \rangle = 2\, E(\mathbf{p})\, \delta^3(\mathbf{p} - \mathbf{p}')\, \delta_{\sigma\sigma'}. \tag{9.4.25}$$

The completeness relation is then

$$\sum_{\sigma=-s}^{s} \int \frac{d^3p}{2E(\mathbf{p})}\, |\, \mathbf{p}, \sigma \rangle \langle \mathbf{p}, \sigma \,| = \mathrm{id}, \tag{9.4.26}$$

as is immediate when applied to some basis vector $|\, \mathbf{p}', \sigma' \rangle$. (The invariant $d^3p/E(\mathbf{p})$ was encountered already in relativistic phase space; cf. also exercise 1 of sect. 5.6!)

The expression for the scalar product between two arbitrary vectors $|\,\varphi\,\rangle$, $|\,\psi\,\rangle$ in the representation space—i.e., between two arbitrary cross sections of our vector bundle, given by the component functions ('wave functions in p-representation', quantum mechanically speaking) $\langle \mathbf{p}, \sigma \,|\, \varphi \rangle = \varphi_\sigma(\mathbf{p})$, $\langle \mathbf{p}, \sigma \,|\, \psi \rangle = \psi_\sigma(\mathbf{p})$—becomes, from eq. (9.4.26),

$$\langle \varphi \,|\, \psi \rangle = \sum_\sigma \int \frac{d^3 p}{2E(\mathbf{p})}\, \varphi_\sigma^*(\mathbf{p})\, \psi_\sigma(\mathbf{p}). \tag{9.4.27}$$

The action of U_a is given by multiplication with $\exp(ipa)$, while U_L sends $|\psi\rangle$ to $|\psi'\rangle$, where

$$\psi'_\sigma(\mathbf{p}) = \sum_{\sigma'=-s}^{s} \mathrm{D}^{(s)}_{\sigma\sigma'}\big(K(L, L^{-1}p)\big)\psi_{\sigma'}(L^{-1}\mathbf{p}). \tag{9.4.28}$$

This lends itself to a rigorous *construction* of the representation in the Hilbert space of cross sections of the bundle, square-integrable in the sense of the measure on $\mathbf{O}$ that appears in eq. (9.4.27).

b. Case $p^2 = 0$, $p \neq 0$

Let us again choose sign $p_0 > 0$, and take $\bar{p}^i = (1,0,0,1)^\top$ as the standard vector. In this situation, the spatial rotations in the 1,2-plane obviously form a subgroup of the little group $\mathcal{K}_{\bar{p}}$. To determine all elements of $\mathcal{K}_{\bar{p}} \subset \mathrm{SL}(2,\mathbf{C})$ we remember that to the lightlike 4-vector $\bar{p}$ there corresponds, according to eq. (8.4.23), the spinor $\bar{\pi}^A = 2^{1/4}(1,0)^\top$; it is unique up to a phase factor. $\mathcal{K}_{\bar{p}}$ therefore consists of complex unimodular matrices A such that $\mathrm{A}\bar{\pi} = e^{i\alpha/2}\bar{\pi}$, where $e^{i\alpha/2}$ is the undetermined phase factor. It follows that the required A can be written

$$\begin{pmatrix} e^{i\alpha/2} & b\,e^{-i\alpha/2} \\ 0 & e^{-i\alpha/2} \end{pmatrix} =: \mathrm{A}(b, \alpha), \tag{9.4.29}$$

where $0 \leq \alpha < 4\pi$, $b \in \mathbf{C}$. The group of matrices so obtained, which we denote by $\widetilde{\mathrm{E}(2)}$, has the multiplication rule

$$\mathrm{A}(b', \alpha')\,\mathrm{A}(b, \alpha) = A(b' + e^{i\alpha'}b, \alpha' + \alpha), \tag{9.4.30}$$

which is the same as the multiplication rule in the group $\mathrm{E}(2)$ of translations and rotations in a Euclidean 2-plane when written with the help of complex numbers; indeed our group is a double covering of the Euclidean group, due to the range of the angle α. Topologically, it is homeomorphic to $\mathbf{R}^2 \times \mathbf{S}_1$, a fact that was already anticipated in discussing the choice of the transformations Λ_p, eq. (9.4.7).

The 2-dimensional Euclidean plane on which our group operates in this interpretation is the Argand plane for the ratio of the spinor components on which the $\mathrm{A}(b,\alpha)$ act—see Penrose and Rindler (1984) for geometrical details. More significant for our purposes here is the behavior of the null flag associated with the spinor $\bar{\pi}$, defined after eq. (8.4.24): it gets rotated by the angle α. A still more physical interpretation of this is as follows: form the selfdual sixtor $\bar{f}^{ab}_+ := \sigma^a_{\ A\dot{X}}\,\sigma^b_{\ B\dot{Y}}\,\bar{\pi}^A\,\bar{\pi}^B\,\epsilon^{\dot{X}\dot{Y}}$ and from it the right circularly polarized plane electromagnetic wave (cf. exercise 8 of sect. 5.5)

$$F^{ab}(x) = \mathrm{Re}\,[\bar{f}^{ab}_+ e^{-i\bar{p}x}]. \tag{9.4.31}$$

Under A it only changes its phase by α, since also $\bar{p}x$ stays invariant. Similarly one can form a left circularly polarized wave from the antiselfdual sixtor $\bar{f}^{ab}_- := \sigma^a_{\ A\dot{X}}\,\sigma^b_{\ B\dot{Y}}\,\bar{\pi}^{*\dot{X}}\,\bar{\pi}^{*\dot{Y}}\,\epsilon^{AB}$, whose phase changes by $-\alpha$. When these circularly polarized waves are superposed to give a linearly polarized wave, the effect of $\mathrm{A}(b,\alpha)$ is to rotate the plane of polarization through the angle α.

In particular, the transformations $\mathrm{A}(b,0)$ have no effect on these waves, thus belonging, together with the translations in the 1- and 2-direction and in the lightlike direction $\bar{p}$, to the *symmetries* of the plane wave. These Lorentz transformations were called lightlike, or null, rotations in the intrinsic classification of Lorentz transformations given in appendix 2 to sect. 8.4. In the isomorphism $\mathcal{L}_+^\uparrow \cong \mathrm{SO}(3,\mathbf{C})$ they correspond to complex rotations around axes $\boldsymbol{\alpha}$ having $\boldsymbol{\alpha}^2 = 0$ ('isotropic axes').

Let us mention here a related but slightly larger subgroup than $\widetilde{\mathrm{E}(2)}$, consisting of all A that leave invariant the *direction* of $\bar{p}$ only, thus leaving $\bar{\pi}$ invariant up to an arbitrary nonzero complex factor. These A are just upper triangular as in eq. (9.4.29) but with the diagonal elements a, a^{-1} not restricted to be phase factors. What is added here are just boosts preserving the direction of $\bar{p}$ but changing its extension by the Doppler effect. As a consequence, this subgroup is a *complex* 2-parameter subgroup with compact quotient (homeomorphic to $\mathbf{S}_2$)—in fact the smallest subgroup with that property (a so-called *Borel subgroup*). It is used extensively in certain approaches towards exact solutions of the field equations of General Relativity; cf. Kramer et al. (1979).

Since $\widetilde{\mathrm{E}(2)}$ is again a semidirect product of a rotation subgroup and an Abelian subgroup of 'translations' (the null rotations), the required classification of its unitary irreducible representations can proceed as in the initial stages of the Poincaré group. We leave the detailed execution of this program as an exercise and just make the following remarks. As before, the rotation subgroup acts on the set of characters of the 'translation' subgroup, but this time the geometry there is Euclidean. This leaves us with only two types of orbits to be discussed: circles, corresponding to a continuous spectrum of the 'translation' generators (analogous to the situation encountered above in all cases except case c of Fig. 9.1), and, as the second type of orbit, the zero point. For the first type we refer to the exercises; until now, it has not played any role in physics ('representations with continuous spin'); to the contrary, admitting them would create difficulties in thermodynamical considerations.

In the second case, the 'translations' (null rotations) are represented trivially, and the stabilizer of the only point of the orbit (the 'little group of the little group') is the rotation subgroup of $\widetilde{\mathrm{E}(2)}$, consisting of all $\mathrm{A}(0,\alpha)$, $0 \le \alpha < 4\pi$. We need its unitary irreducible representations, which are 1-dimensional, since the group is Abelian. Locally, they are given by $\alpha \mapsto \exp(i\lambda\alpha)$, where λ is real. Globally, we have to observe that $\alpha = 0$ and $\alpha = 4\pi$ give the same element, so continuity on the global level obtains iff λ is restricted to integer or half integer values. The representations of $\mathcal{K}_{\bar{p}}$ we are looking for are therefore finally given by

$$\mathrm{A}(b, \alpha) \mapsto \exp(i\lambda\alpha) \qquad\qquad \lambda = 0, \pm 1/2, \pm 1, \dots . \qquad (9.4.32)$$

The classification is thus by the value of λ, which will be interpreted below.

Note that the restriction to the values of λ just written comes from the requirement that we are heading here for a genuine, i.e., single-valued, representation of $\mathrm{SL}(2,\mathbf{C})$, i.e., a double-valued representation of $\mathcal{L}_+^\uparrow$. The subgroup of rotations $\mathrm{A}(0,\alpha)$, isomorphic to $\mathrm{U}(1)$, per se has multivalued representations that result when λ is real but not restricted as above. These are single-valued representations of its universal covering group, which is topologically $\mathbf{R}$, the real line. Note that the construction of this covering group is encountered in everyday life where time is measured by periodic processes, and in the need to introduce an international date line; but it is of no relevance in the problem considered here!

The vector bundles that arise from these representations of $\mathcal{K}_{\bar{p}}$ all have complex 1-dimensional fibers and are called *complex line bundles*. In the language of differential geometry, they are associated to the 'principal fiber bundle' $\mathrm{SL}(2,\mathbf{C}){\to}\mathrm{SL}(2,\mathbf{C})/\widetilde{\mathrm{E}(2)}$, which is topologically nontrivial (not a product), as we remarked earlier. As a consequence, the line bundles in question are also nontrivial ('twisted'), except for $\lambda = 0$. In a definite sense, λ describes the topological twist in these line bundles, and thus is geometrically related to the quantum number associated with magnetic monopoles.

Let us determine the value of the invariant W^2 in the representations of $\mathcal{P}_+^\uparrow$ induced by the representations (9.4.32) of $\mathcal{K}_{\bar{p}}$! It is easy to see (exercise) that infinitesimally the parameters b, α are related to the parameters k^c introduced in eq. (9.4.12) by

$$\alpha = k^3 - k^0, \qquad\qquad b = k^2 - i\,k^1, \qquad\qquad (9.4.33)$$

so that W_1, W_2 generate the null rotations, while W_3 generates rotations about the 3-axis. Generally we have, from eq. (9.3.9),

$$0 = W_c\,P^c\,|\,\bar{p}\,\rangle = (W_0 + W_3)\,|\,\bar{p}\,\rangle. \qquad\qquad (9.4.34)$$

In addition—since we are considering only representations in which the null rotations act trivially—we have

$$W_1\,|\,\bar{p}\,\rangle = W_2\,|\,\bar{p}\,\rangle = 0. \qquad\qquad (9.4.35)$$

Also, eq. (9.4.14) tells us that W_0, W_3 commute on $|\,\bar{p}\,\rangle$. We thus calculate $W^2\,|\,\bar{p}\,\rangle = (W_0)^2\,|\,\bar{p}\,\rangle - (W_3)^2\,|\,\bar{p}\,\rangle = 0$, and therefore

$$W^2 = 0. \qquad\qquad (9.4.36)$$

Furthermore, comparing eqs. (9.4.34,35) with $P_c\,|\,\bar{p}\,\rangle = \bar{p}_c\,|\,\bar{p}\,\rangle$ we see immediately that $W_c\,|\,\bar{p}\,\rangle \propto P_c\,|\,\bar{p}\,\rangle$, as must hold generally for two commuting Hermitian lightlike orthogonal vector operators. The factor of proportionality obtains from a calculation of the eigenvalue of W_3: on the one hand we have

$$U(K(b,\alpha))\,|\,\bar{p}\,\rangle \approx (1 - i\,k^c\,W_c)\,|\,\bar{p}\,\rangle = (1 - i\,\alpha\,W_3)\,|\,\bar{p}\,\rangle,$$

and on the other, from eq. (9.4.32),

$$U(K(b,\alpha))\,|\,\bar{p}\,\rangle = \exp(i\lambda\alpha)|\,\bar{p}\,\rangle \approx (1 + i\lambda\alpha)\,|\,\bar{p}\,\rangle,$$

so that $W_3\,|\,\bar{p}\,\rangle = -\lambda\,|\,\bar{p}\,\rangle = \lambda P_3\,|\,\bar{p}\,\rangle$. Consequently, we have

$$W_c = \lambda P_c, \qquad\qquad (9.4.37)$$

at first in the subspace $\mathbf{H}_{\bar{p}}$, but then also on the whole representation space $\mathbf{H}$ as an equation between vector operators. Thus eq. (9.3.11) holds, with the same meaning of λ.

The physical meaning of the invariant λ can be obtained by going to a specific inertial system. Consider there the time component of eq. (9.4.37), inserting the definition (9.3.8) of W_c:

$$\lambda\,P_0 = \frac{1}{2}\,\epsilon_{0\mu\nu\rho}\,M^{\mu\nu}\,P^\rho = \mathbf{MP},$$

where we introduced the angular momentum vector operator $\mathbf{M}$ with respect to that system, given by the components

$$M^\rho := \frac{1}{2}\,\epsilon_{\mu\nu\rho}\,M^{\mu\nu}. \tag{9.4.38}$$

Application to $|\,p\,\rangle$—observe $(p^0)^2 = \mathbf{p}^2$!— then gives

$$\lambda = \frac{\mathbf{M}\mathbf{p}}{|\mathbf{p}|}. \tag{9.4.39}$$

In a particle interpretation, by eqs. (9.3.12,15,16) $\mathbf{M}$ is the operator of total angular momentum, so that this equation says that λ is the projection of total angular momentum onto the direction of motion (momentum). This quantity is called *helicity*. Its (space reversal invariant) absolute value $|\lambda|$ serves, for massless particles, as a substitute for the concept of spin, which, in the form 'angular momentum in the rest system' makes sense only for massive particles: massless particles have no rest system. Kinematically, eq. (9.4.37) when contracted with a spacelike vector says that a translational shift in its direction has the same effect as a rotation about it as an axis through an angle proportional to the shift—which is just the characteristic symmetry of a helix, or screw, with pitch given by $1/\lambda$.

When helicity is defined by eq. (9.4.39) it makes sense also in the 'massive' representations having $W^2 \neq 0$, but it is not a Poincaré invariant quantity any more. It is intuitively clear that for a massive particle a state of positive helicity may be boosted into one with negative helicity. One can therefore speak, in the massive case, only of the helicity of some *state* of the particle, but not of the helicity of the particle itself. It is, however, invariant under spatial rotations also in the massive case and has spectrum $\lambda = -s,\,\ldots,+s$. It is useful sometimes to work with helicity eigenstates in practical problems, as pointed out in M. Jacob, G.-C. Wick, Ann. Phys. (N.Y.) 7, 404 (1959), or Halpern (1968), Gasiorowicz (1966).

For the representations with $M^2 = 0 = W^2$, to stress again, helicity is a $\mathcal{P}_+^\uparrow$-invariant quantity whose values $\lambda = 0,\,\pm 1/2,\,\pm 1,\,\ldots$ classify these representations. In particular, its Lorentz invariance says that, e.g., a right circularly polarized light wave ($\lambda = 1$, cf. eq. (9.4.31)) will remain so under all boosts, massless neutrinos ($\lambda = -1/2$) are 'left-handed' in all frames.

The invariant scalar product for the representations just under consideration is obtained from eq. (9.4.38) by putting $m = 0$; the sum over σ is absent since all subspaces $\mathbf{H}_p$ are 1-dimensional. If we write the basis vectors $|\,p\,\rangle$ of the representation characterized by $m^2 = 0$, $\operatorname{sign} p_0 = 0$ and helicity λ more completely as $|\,\lambda,p\,\rangle$, we can put together our findings as follows:

$$M^2\,|\,\lambda,p\,\rangle = 0 = W^2\,|\,\lambda,p\,\rangle \tag{9.4.40a}$$

$$W_k\,|\,\lambda,p\,\rangle = \lambda P_k\,|\,\lambda,p\,\rangle = \lambda p_k\,|\,\lambda,p\,\rangle \qquad p^2 = 0, \quad p_0 > 0 \tag{9.4.40b}$$

$$U(a)\,|\,\lambda,p\,\rangle = \exp(ipa)\,|\,\lambda,p\,\rangle \tag{9.4.40c}$$

$$U(L)\,|\,\lambda,p\,\rangle = \exp(i\lambda\alpha(L,p))\,|\,\lambda,Lp\,\rangle, \tag{9.4.40d}$$

where $\alpha(L,p)$ is the rotation angle about the 3-axis contained in the little group element $K(L,p)$ according to eq. (9.4.29).

Here we end the abstract theory of representations of $\mathcal{P}_+^\uparrow$. We just add that for the transition to $\mathcal{P}^\uparrow$ the space reversal operation can be hosted within the massive representations, while—due to the pseudoscalar nature of helicity—the massless representations with $\lambda \neq 0$ have to be combined (direct sum) with the ones with helicity $-\lambda$ to host the space reversal. Thus, in a space reversal invariant theory of massless spin $|\lambda| = 1$ particles (photons) one has right circularly polarized and left circularly polarized states and their *complex* superpositions (elliptic polarization, linear polarization). To pass to all of $\mathcal{P}$ one would have to add, in strictly linear (ray) representations, the representations above to their sign $p_0 = -1$ counterparts (direct sum); however—as was mentioned before—for physical reasons one has to proceed differently, as will be sketched in sect. 9.6.

Exercises

1. a. Prove eqs. (9.4.27,28)!

   b. Define, according to what was sketched in sect. 7.8, from $\psi(p)$ a $\mathbf{H}_{\bar{p}}$-valued function on $\mathcal{L}_+^\uparrow$, or SL(2,$\mathbf{C}$), by $\hat{\psi}(L') := (Q(L',\bar{p}))^{-1}\psi(L'\bar{p})$. Show that $\hat{\psi}$ satisfies, for all $K \in \mathcal{K}_{\bar{p}}$, the 'equivariance condition' $\hat{\psi}(L'K) = (D(K))^{-1}\hat{\psi}(L')$, where D is the representation of the little group involved. Show then that to eq. (9.4.28) there simply corresponds $\widehat{\psi'}(L') = \hat{\psi}(L^{-1}L')$, again satisfying the equivariance condition. Note that this alternative description of the inducing construction removes the arbitrariness involved in the choice of the Λ_p. How are the translation operators described here?

2. Verify eq. (9.4.14)!

3. Determine the little groups in the cases c, d!

4. Verify the representation property of eq. (9.4.11) directly! Also verify directly that the ensuing solution for Q satisfies eq. (9.4.6).

5. Determine the irreducible unitary representations of the Euclidean group of motions (rotations and translations) in a 2- and a 3-dimensional (real) Euclidean space, using the method of induced representations.

6. Verify the parameter assignment given by eq. (9.4.33).

7. Analyze the representation of $\mathcal{P}_+^\uparrow$ that is realized in the space of (complex-valued) vector fields $A^i(x)$ satisfying the Proca equations (9.3.22) for its content of unitary irreducible representations.
   Solution: The Fourier components $\tilde{A}^i(p)$ satisfy $\tilde{A}^i(p) = 0$ for $p^2 \neq m^2$, $p_i\tilde{A}^i = 0$. Since $m^2 > 0$ is assumed, $\mathcal{K}_{\bar{p}}$ is the rotation group, for which $\tilde{A}^i(\bar{p})$ decomposes into the irreducible parts $\tilde{A}^0(\bar{p})$ and $\tilde{\mathbf{A}}(\bar{p})$, transforming as D$^{(0)}$ and D$^{(1)}$. We have $0 = \bar{p}_i\tilde{A}^i(\bar{p}) = m\,\tilde{A}^0(\bar{p})$, so that D$^{(0)}$ actually does not occur. It follows that the representation just contains the two irreducible unitary representations in which $M^2 = m^2$ id, $W^2 = -M^2 \cdot 1 \cdot (1+1)$, sign $p_0 = \pm 1$. The Lorenz condition has projected out the spin zero parts.

8. Consider the space of solutions of the generalized Weyl equation (cf. eq. (9.1.8))

$$\partial_{A\dot{X}}\,\Phi^{AB\ldots C} = 0, \qquad (9.4.41)$$

where $\Phi^{AB\ldots C}$ is a totally symmetric spinor field with r undotted indices.

a. Show that eq. (9.4.41) with $r = 2$ is equivalent to the vacuum Maxwell equations for the real sixtor field corresponding to Φ^{AB}.

b. Analyze the representation of $\mathcal{P}_+^\uparrow$ realized in the space of solutions of eq. (9.4.41) for its irreducible unitary parts.

Solution for b: For the Fourier components $\tilde{\Phi}^{AB\ldots C}(p)$ we have $p_{A\dot{X}}\,\tilde{\Phi}^{AB\ldots C} = 0$, with $p^2 = 0$. Let π be the spinor which according to eq. (9.4.23) exists such that $p_{A\dot{X}} = \operatorname{sign} p_0\,\pi_A\,\pi^*{}_{\dot{X}}$, whence $\pi_A\,\tilde{\Phi}^{AB\ldots C} = 0$. When the decomposition (8.3.20) is used, the last equation implies that all principal spinors of $\tilde{\Phi}$ are proportional to π: $\tilde{\Phi}^{AB\ldots C}(p) \propto \pi^A\pi^B\,\ldots\,\pi^C$. When $p^i = \bar{p}^i = (1,0,0,1)^\mathsf{T}$, $\bar{\pi}^A = 2^{1/4}(1,0)^\mathsf{T}$ one sees that the Fourier component $\tilde{\Phi}(\bar{p})$ under the transformations of $\mathcal{K}_{\bar{p}}$ picks up the factor $\exp(ir\alpha/2)$. Therefore, the representation of $\mathcal{P}_+^\uparrow$ contains just the irreducible unitary parts with $M^2 = 0 = W^2$, $\lambda = r/2$, $\operatorname{sign} p_0 = \pm 1$. Similarly, dotted symmetric spinors satisfying appropriate Weyl equations yield helicity $\lambda = -r/2$. The corresponding particles have—just as the circularly polarized light wave—a right ($\lambda > 0$) and a left ($\lambda < 0$) handedness.

9.5 Representation Theory of $\mathcal{P}_+^\uparrow$ and Local Field Equations

In sects. 9.1. and 9.3 the relation between relativistic linear wave equations and representations of the Poincaré group has been established. As we have seen in examples, the field equations restrict the space of fields of a given type in a way that only a few irreducible components are left.[1] Now that we have in hands all relevant irreducible unitary representations in abstract form, we may try to translate them, via the inverse Fourier transform, back into spaces of fields on Minkowski space satisfying suitable covariant field equations.

The first thing to notice here is that such a realization will by no means be unique. For instance, the representation characterized by $m^2 > 0$, $s = 0$, $\operatorname{sign} p_0 = 1$ (together with the one where $\operatorname{sign} p_0 = -1$) may be realized in the space of scalar fields Φ that satisfy $(\Box + m^2)\,\Phi = 0$, but also in the space of vector fields A_i which besides $(\Box + m^2)A_i = 0$ satisfy the constraint

$$\partial_i A_k - \partial_k A_i = 0. \qquad (9.5.1)$$

This is because eq. (9.5.1) means for the Fourier coefficients $\tilde{A}_i(p)$ that $p_i\tilde{A}_k = p_k\tilde{A}_i$ or $\tilde{A}_i(p) \propto p_i$, so that $\tilde{A}_i(\bar{p}) \propto \bar{p}_i$ transforms according to the trivial representation of the little group, and we have $s = 0$. The constraint (9.5.1) has projected away the

---

[1] Irreducible parts of free fields propagate independently of each other and may be coupled independently. It thus suggests itself to assign 'particles' to them which are 'elementary' in some sense. However, elementary particle physics has not finally settled the question which particles will have to be regarded as elementary in the end—there were even attempts to regard every particle as consisting of all the others.

spin 1 part. On the other hand, the Lorenz condition $\partial_i A^i = 0$ would project away the spin 0 part—see exercise 7 of the preceding section.

The occurrence of constraints in addition to a wave equation cannot be avoided in general if an irreducible representation is to be realized in a space of fields of the kind (9.1.1). This comes about as follows. The basis functions (9.4.5) transform with *p-in*dependent matrices $D(L)$; but the reduction of D with respect to the little group consists in a decomposition $D(K) = Q'(K) \oplus Q''(K) \oplus \ldots$, where Q', Q'', $\ldots$ are irreducible representations of the little group, and this gives, following eq. (9.4.11), a *p-dependent* decomposition $D(L) = Q'(K(L,p),\bar{p}) \oplus Q''(K(L,p),\bar{p}) \oplus \ldots$. Therefore one must, conversely, add some further representations to a given irreducible representation—$Q'(K)$ say—of the little group until a p-independent sum is achieved. The surplus representations have to be projected away again, using a number of p-dependent projection operators, whose translation back into spacetime by Fourier yields the extra constraints.

Given m, s one can proceed by choosing $D(L)$ such that $D^{(s)}(K)$ is safely hosted in $D(K)$. Depending on the choice of D there are various systems of constraints (and therefore various 'formalisms': Pauli-Fierz, Rarita-Schwinger, Bargmann-Wigner, …) necessary to project away the unwanted components. We do not enter into a discussion of these but refer the reader to the article by Niederer and O'Raifeartaigh in Barut (1973).

Is there a way to pick the 'right' one out of these (in principle infinitely) many possibilities? One essential remark here is that in reality the various fields are, in fact, interacting with each other—a field that couples to nothing is unobservable. It now turns out that the various possibilities of realizing the irreducible representations of $\mathcal{P}_+^\uparrow$ by fields differ quite strongly as regards their potentiality to build in interactions in a simple manner.[1] A convenient way to describe interactions is to derive coupled field equations from an action principle (cf. chap. 10). The Lagrangian resp. the related Hamiltonian formalism are also useful for the transition to the quantum mechanical treatment of the dynamics of fields (Quantum Field Theory; path integral and canonical quantization). It is thus necessary already in the case of free fields to be able to write down action principles for the field equations—and it is also here that the various possibilities mentioned differ in aptitude. It may even happen that the action principle formulation requires a choice of D where translation of the process of projecting out unwanted representations into x-space encounters difficulties.

Let us illustrate these difficulties of translating back irreducibility constraints in the simplest case, which is at the same time the most basic one. It concerns the question how the conditions $\operatorname{sign} p^0 = +1$ or $= -1$ look like in x-space. The inversion of eq. (9.4.4) is

$$\tilde{\Phi}(p) = \int d^4x \, e^{ipx} \, \Phi(x), \tag{9.5.2}$$

so that, because of $\partial_k \exp(ipx) = ip_k \exp(ipx)$, etc., irreducibility constraints which are polynomial in p like $(p^2 - m^2) \, \tilde{\Phi}(p) = 0$, $p^k A_k = 0$, $\ldots$ in x-space take the form of

---

[1] A discussion of how interactions may be introduced in the abstract formalism without realization by fields is given by S. Weinberg, Phys. Rev. *133*, 1318 (1964); *134*, 882 (1964).

differential equations, as follows from partial integration. However, it is not possible to translate the condition (which is covariant when Φ does not contain spacelike momenta)

$$\int d^4x\, e^{ipx}\, \Phi(x) = 0 \quad \text{for} \quad p^0 < 0 \tag{9.5.3}$$

into a differential equation for $\Phi(x)$; and the same for $p^0 > 0$.

It is a matter of principle whether conditions on fields must have the form of differential equations, i.e., whether propagation of effects is by local action rather than by action at a distance. Basically, one should turn to experiment to see what gives the correct description of Nature; but sometimes local interaction plus some boundary conditions may be mathematically equivalent to a suitable version of action at a distance, so that criteria of mathematical convenience, esthetical appeal and potentiality to stimulate further development also come into play. Since the time of Faraday, local action and field theory have been the winners in this competition. To illustrate the experimental side of the question, consider the propagation of sound in a gas (in some inertial system, where the gas is at rest on the average): it is described macroscopically by the wave equation $(\partial_t^2 - c_{\text{sound}}^2\Delta)\Phi = 0$, but this equation will not be relevant for problems where the atomistic structure of the gas becomes relevant. On the other hand, since according to Special Relativity electromagnetic and other fundamental fields are not regarded as excitations of some material ether, no lower bound for the size of domains where relativistic local field theory should be valid is in sight; we have agreement with experiment down to 10^{-15} cm and less.

When it is postulated that all field laws have to be local—alternative formulations are practically nonexistent—the simultaneous occurrence of representations with sign $p^0 = +1$ *and* sign $p^0 = -1$ cannot be avoided. Quantum mechanically, this means that states of *negative energy* $p^0 < 0$ will appear in the theory. As long as the field couples to nothing but stays free (and thus unobservable) this is of no harm, one could ignore them. However, when the field couples to others, there will be transitions to these states, and such processes would be a source of infinite energy. To avoid this absurdity, i.e., to maintain also a postulate of a lower bound on energy—Lorentz covariance then implies that energy is *nonnegative*, see Fig. 9.1—Dirac invented the theory of holes, subsequently replaced by the theory of *antiparticles*: the negative energy states were *reinterpreted* as positive energy states of antiparticles. These thus appear automatically in a local relativistic field theory when the wave function is interpreted in the sense of quantum mechanics: in this sense, they are 'predicted' by the theory.

With this reinterpretation, not only a difficulty was removed, but also a further proof of the utility of the idea of local action was furnished by the subsequent experimental discovery of antiparticles. However, now another mathematical difficulty appeared: the theory allowed for processes that after the reinterpretation involved two particles, while the formalism was still a one-particle formalism as in elementary N.R. wave mechanics. It became necessary—in particular in the presence of interactions without suitable restrictions—to use the quantum mechanical description of many-particle systems, adapted in a way that the particle number was not fixed, i.e., the formalism of 'second quantization'. In it, one can describe, e.g., the transition to a state of negative energy as the annihilation of a particle-antiparticle pair under the emission of radiation or other particles (where rest mass is not conserved but enters into the energy balance, as required by relativity and as made explicit in the formalism).

When this formalism is set up, it leads to a theory that looks just as if the original classical field $\Phi(x)$ had been subjected to the procedure of canonical quantization, which leads from the classical Hamiltonian description of dynamics to the corresponding quantum mechanical operator formalism (*quantum field theory*). More precisely, one obtains a specific representation—the Fock representation—of the canonical commutation relations. (The term 'second quantization' describes just this result of the treatment of the many-particle theory; by 'first quantization' one means here the transition from the classical (one-)particle description to the wave description. Conceptually, the quantization of a classical field is something different from the treatment of a quantum mechanical many-particle system with indefinite particle number using the method of second quantization.) In Appendix D we shall give a sketch of the simplest case (neutral spinless particles without interaction) and discuss Poincaré covariance within this formalism.

For bosons and fermions, second quantization is carried out using commutators and anticommutators, respectively, to take into account the symmetry properties of multiparticle states corresponding to the statistics (Pauli principle). While in N.R. quantum theory this can be done whatever the spin of the particles, in the relativistic theory it turns out that consistency of the theory requires that integer and half integer spin be *tied* to Bose and Fermi statistics, respectively, in order to cope with the basic ingredients of the theory (locality, positivity of energy). This is the famous *spin-statistics theorem* of relativistic quantum field theory. In N.R. theory, this tie has to be introduced as an extra postulate. If seen in this way, one may say that Relativity has its influence in velocity and energy regimes where none would be expected: the Pauli principle is an important ingredient in explaining the structure of matter: spectra, chemical bond, ferromagnetism, stability of matter, ...

In quantum field theory, space and time reversal, P, T, find their proper place together with a further discrete operation enabled by the introduction of antiparticles: *charge conjugation C*. Here 'charge' refers to all chargelike quantum numbers discovered in the course of time, not just to electric charge. While there are regimes in particle physics where invariance under P or even CP is violated, the basic postulates imply that in a $\mathcal{P}_+^\uparrow$-covariant quantum field theory one always has invariance under the combined operation CPT (*CPT theorem*). Thus if this were violated experimentally, one would have to give up the concept of a local relativistic quantum field theory.

Let us, after this short sketch of the concept and results of quantum field theory— for which we must refer to the literature[1]— return to the problem of translating back irreducibility constraints to x-space. Another obstacle here is to go from the efficient by not manifestly covariant transformation behavior of the Wigner basis (9.4.10) to the manifestly covariant spinor or tensor basis (9.4.5). Some of the constraints necessary to project out unwanted components do not translate to local conditions in x-space, some translate to local conditions only if manifest covariance is given up, so that additional 'gauge transformations' have to be introduced to secure Lorentz covariance, in which case the representation is carried not by fields but by gauge equivalence classes of fields (see below).

The latter phenomenon is the reason for certain technical complications in quantizing, e.g., the electromagnetic field: one either gives up manifest covariance, or one carries along unwanted 'ghost' representations that couple to nothing even in case of interaction, but then has to be careful that these do not enter physical results. It should be mentioned here that maintaining manifest Lorentz and gauge covariance historically has been a good guide in regularizing certain divergent integrals that occur in the perturbative treatment of interacting quantum fields, thus enabling a successful execution of the so-called 'renormalization program'.

---

[1]Wentzel (1949); Bogoljubov and Shirkov (1959); Roman (1960, 1969); Jost (1965); Bjørken and Drell (1964, 1965); Streater and Wightman (1964); Gasiorowicz (1966); Henley and Thirring (1962); Kastler (1961); Schweber (1961); Itzykson and Zuber (1980); Weinberg (1995).

It appears that the representation theory of the Poincaré group has its application mainly in the domain of particle physics. One may, however, try also to construct a framework for the relativistic description of the gravitational field—which is necessary in principle, since the gravitational interaction should also propagate with speed c or less. Due to its weakness, the effects of relativistic corrections to Newton's law of gravitation are, however, not found in present-day microphysics but in astronomy and astrophysics, where cumulative effects become important. (It should be mentioned here that for the GPS the corrections due to special relativistic time dilation and to gravitational time dilation in the Earth's field are of the same order!) The way how to set up such a theory on an empirical basis (light deflection, attractiveness), leading to a representation with mass zero and spin 2, and the way how, starting from this, one finally arrives at General Relativity with its curved space-time was analyzed by Thirring.[1]

The representation theoretic aspects of this analysis were made more explicit in O. Nachtmann, H. Schmidle, R. U. Sexl, Acta Phys. Aust. *29*, 289, (1969); this work gives an explicit example for the technique of decomposing a field into its spin parts. General methods are also found in Pursey, Ann. Phys. (N.Y.) *32*, 157 (1965); Moses, J. Math. Phys. *8*, 1134 (1967); *9*, 16 (1968); Langbein, Comm. Math. Phys. *5*, 73 (1967); Fonda and Ghirardi, Fortschr. Phys. *17*, 727 (1969)). The kind of coupling and the action principle require the use of a symmetric tensor field ψ_{ik}, containing besides spin 2 also parts of spin 1, 0 which have to be projected away. This leads to problems of the kind mentioned.

Pure massless spin 2 may be reached using a tensor field C_{ikmn} of degree 4 as in eq. (8.4.27), satisfying certain differential equations which are analogous to the description of mass zero and spin 1 using the Maxwell field tensor; it corresponds to the case $r = 2$ of eq. (9.4.41) when spinors are used. This tensor field describes the tidal forces of the free gravitational field, corresponding to the 'curvature tensor' of General Relativity. Using it, no local action priciple can be formulated, however, in complete analogy to the electromagnetic field tensor F_{ik}. Cf. R. H. Good Jr., Ann. Phys. (N.Y.) *62*, 590 (1971).

We now turn to the analysis of a few special fields.[2] We begin with *scalar fields* $\Phi(x)$. In their case, we have a unitary representation of $\mathcal{P}_+^\uparrow$ from the very beginning, without imposing any restrictions: the invariant scalar product is

$$\int d^4x\, \Phi^*(x)\, \Psi(x) = \int \frac{d^4p}{(2\pi)^4}\, \tilde{\Phi}^*(p)\, \tilde{\Psi}(p). \tag{9.5.4}$$

Using Parseval's theorem, we have rewritten it also in terms of the Fourier transform (9.4.4) to prepare for a decomposition into irreducible constituents. To make the reduction with respect to the mass square explicit, we further introduce instead of p^0 the variable (cf. eq. (9.4.24)!)

$$m^2 := (p^0)^2 - \mathbf{p}^2 \tag{9.5.5}$$

---

[1] W. Thirring, Fortschr. Phys. *7*, 79 (1959); Ann. Phys. (N.Y.) *16*, 96 (1961); see also Sexl and Urbantke (1995), chap. 10. O. Klein has argued that a correct incorporation of this curved spacetime into quantum field theory could have an effect on that theory comparable to the effect of special relativity, so that there could be an important influence of gravitation even in microphysics. Despite tremendous amounts of recent work in this direction, one is still far from a solution.

[2] All fields considered in the following are complex-valued, as before when they were considered as quantum mechanical wave functions. Reality conditions appear only in second quantization, imposed on the field operator, in order to describe neutral particles (cf. Appendix D).

into eqs. (9.4.4) and (9.5.4). The Jacobian of this change of variables is

$$\frac{\partial(p^0, \mathbf{p})}{\partial(m^2, \mathbf{p})} = \left(\frac{\partial(m^2, \mathbf{p})}{\partial(p^0, \mathbf{p})}\right)^{-1} = \frac{1}{2p^0} = \frac{1}{2E(\mathbf{p}, m^2)}, \tag{9.5.6}$$

where we have used the abbreviation (9.4.19), keeping the dependence on the mass parameter explicitly since it has not yet been fixed by any irreducibility constraint. To cover the whole p-space one has to observe that m^2 must vary over the full interval $-\infty < m^2 < \infty$, and, in addition, both signs of $p^0 = \pm E(\mathbf{p}, m^2)$ have to be taken into account. Writing

$$\int_{-\infty}^{\infty} dp^0 = \left(\int_0^{\infty} + \int_{-\infty}^0\right) dp^0$$

and introducing $-p^0$ instead of p^0 and also $-\mathbf{p}$ instead of $\mathbf{p}$ as variables in the second integral, eq. (9.4.4) becomes

$$\Phi(x) = \int_{-\infty}^{\infty} dm^2 \int \frac{d^3p}{2E(\mathbf{p}, m^2)} \left\{ A_+(\mathbf{p}, m^2)e^{i\mathbf{p}\mathbf{x} - iE(\mathbf{p}, m^2)x^0} + A_-(\mathbf{p}, m^2)e^{-i\mathbf{p}\mathbf{x} + iE(\mathbf{p}, m^2)x^0} \right\},$$
$$\tag{9.5.7}$$

where

$$A_\pm(\mathbf{p}, m^2) := (2\pi)^{-4} \tilde{\Phi}(\pm E(\mathbf{p}, m^2), \pm \mathbf{p}). \tag{9.5.8}$$

Equation (9.5.4) becomes

$$\int d^4x \, \Phi^*(x) \, \Psi(x) = \tag{9.5.9}$$

$$= (2\pi)^4 \int dm^2 \int \frac{d^3p}{2E(\mathbf{p}, m^2)} \left\{ A_+^*(\mathbf{p}, m^2) \, B_+(\mathbf{p}, m^2) + A_-^*(\mathbf{p}, m^2) \, B_-(\mathbf{p}, m^2) \right\},$$

when $\Psi(x)$ is decomposed similarly, with coefficient functions $B_\pm$. This already finishes the decomposition: the representation is given as a direct integral of representations with mass square m^2 and spin 0; for $m^2 > 0$, both $\mathrm{sign}\, p^0 = +1$ as well as $\mathrm{sign}\, p^0 = -1$ occur. For $m^2 < 0$, the splitting into positive and negative frequencies is not Lorentz invariant and thus unimportant; the $\mathbf{p}$-integration is only over $\mathbf{p}^2 \geq -m^2$. As for a scalar field the little group is represented trivially for all cases, it was not necessary here to know the unitary irreducible representations of $\mathcal{P}_+^\uparrow$ for $m^2 < 0$ and $m^2 = 0$, $\mathbf{p} = \mathbf{0}$, whose investigation was omitted in sect. 9.4, to perform the decomposition. However, in the following we restrict ourselves to nonnegative mass squares only and assume, in the case where $m^2 = 0$, that in $A_\pm(\mathbf{p}, m^2)$ there is no $\delta^3(\mathbf{p})$-contribution.

Equation (9.5.9) shows that our expression (9.4.27) is—up to a normalizing factor—just the scalar product induced in the irreducible subspaces. For fields Φ, Ψ satifying the Klein-Gordon equation with masses m', m'',

$$\begin{aligned} A_\pm(\mathbf{p}, m^2) &= \delta(m^2 - m'^2)A_\pm(\mathbf{p}) \\ B_\pm(\mathbf{p}, m^2) &= \delta(m^2 - m''^2)B_\pm(\mathbf{p}) \end{aligned} \tag{9.5.10}$$

we get

$$\int d^4x\, \Phi^*\, \Psi = (2\pi)^4\, \delta(m'^2 - m''^2) \int \frac{d^3p}{2E(\mathbf{p})}\{A_+^*(\mathbf{p})\, B_+(\mathbf{p}) + A_-^*(\mathbf{p})\, B_-(\mathbf{p})\}. \tag{9.5.11}$$

Thus, after splitting off the singular factor (which always occurs if continuous spectra are dealt with using eigendistributions) the expression goes over into formula (9.4.27).

The nonsingular part of the right-hand side of eq. (9.5.11) can be given, for $\operatorname{sign} p^0 = +1$ or $\operatorname{sign} p^0 = -1$, another form using x-space. For this, consider the identity

$$A \overset{\leftrightarrow}{\square} B \equiv \partial_k(A \overset{\leftrightarrow}{\partial^k} B), \tag{9.5.12}$$

where for every linear differential operator D we shall use the abbreviation

$$A \overset{\leftrightarrow}{D} B := A\, DB - (DA)\, B. \tag{9.5.13}$$

Putting $A = \Phi^*$, $B = \Psi$ and using the Klein-Gordon equation, we get from integration over some spacetime domain $\mathcal{D}$ with boundary $\partial\mathcal{D}$

$$(m'^2 - m''^2)\int_{\mathcal{D}} d^4x\, \Phi^*\, \Psi = \int_{\mathcal{D}} d^4x\, \partial_k(\Phi^* \overset{\leftrightarrow}{\partial^k} \Psi) = \int_{\partial\mathcal{D}} d\sigma_k\, \Phi^* \overset{\leftrightarrow}{\partial^k} \Psi. \tag{9.5.14}$$

If—as has been tacitly assumed up to now—the fields fall off sufficiently fast in all space-time directions, the surface integral vanishes if $\mathcal{D}$ is chosen to be all of Minkowski space, we get once again that $\int d^4x\, \Phi^*\, \Psi$ is proportional to $\delta(m'^2 - m''^2)$ (orthogonality of eigenfunctions of the Hermitian operator $\square$). Equation (9.5.14) can also be used in a slightly different way, assuming $m' = m''$ and taking for $\mathcal{D}$ a domain between two spacelike hypersurfaces. Under suitable falloff conditions at spacelike infinity we may then conclude, just as for total electric charge in sect. 5.7, that the integral

$$i\int_{\sigma} d\sigma_k\, \Phi^* \overset{\leftrightarrow}{\partial^k} \Psi \tag{9.5.15}$$

is independent of the special hypersurface σ and therefore, in particular, is Poincaré invariant. We conjecture that there should be a relation to the scalar products induced in the subspaces $m = m' = m''$, $\operatorname{sign} p^0 = \pm 1$. Indeed, substitution of the expansion (9.5.7) with the specialization (9.5.10) gives

$$\frac{i}{(2\pi)^3} \int_{\sigma} d\sigma_k\, \Phi^* \overset{\leftrightarrow}{\partial^k} \Psi = \int \frac{d^3p}{2E(\mathbf{p})}\{A_+^* B_+ - A_-^* B_-\}, \tag{9.5.16}$$

as one finds by specializing σ as $t = 0$. Therefore,

$$\pm i(2\pi)^{-3} \int d\sigma_k\, \Phi^* \overset{\leftrightarrow}{\partial}{}^k \Psi$$

agrees, for only positive, or negative, frequency content, with the scalar product written before. Observe again that here Φ, Ψ are to satisfy the Klein-Gordon equation with the same mass square, while such was not assumed in the expression (9.5.4).

For the relation between eq. (9.5.16) and the total charge of a charged quantized scalar field we refer to texts on particle physics. The possibility to define the positive and negative frequency parts (or particle and antiparticle states) without using a Fourier decomposition, just by comparing the expressions (9.5.3) and (9.5.16), was pointed out first by O. Nachtmann (Sitzungsber. Akad. Wiss. Wien Math. Naturwiss. Kl. Abt. II, *176*, 363 (1968)); an extension of this idea became recently known as 'refined algebraic quantization'. Another possibility to get along without Fourier transform comes from the remark that the vanishing of the negative or positive frequency part implies analyticity properties in certain domains of complexified Minkowski space, just as we were able to express a sharp signal form (2.3.7) in terms of analyticity properties of its Fourier transform.

For *vector fields* $A_i(x)$ a consideration of the representations of the little group becomes necessary. For $m^2 < 0$ and for $m^2 = 0$, $\mathbf{p} = \mathbf{0}$ the solution to exercise 3 of sect. 9.4 yields, as the little groups, $SO_e(2,1)$ and $\mathcal{L}_+^\uparrow$. For these noncompact semisimple groups all unitary irreducible representations are trivial or infinite-dimensional—thus in the latter case realizable only by *infinite component wave functions*. From this it follows that if $A_i(x)$ has Fourier components $\tilde{A}_i(p)$ with $p^2 < 0$ or $\propto \delta^4(p)$, the representation cannot be made unitary, since it leads to a nontrivial 4-dimensional representation of the little group—an exception being only given by fields satisfying $\partial_i A_k - \partial_k A_i = 0$, i.e., gradients of scalar fields, where $\tilde{A}_i(p) = \tilde{A}(p)p_i$ leads to the trivial representation of the little group. Also the parts having $m^2 = 0$, $\mathbf{p} \neq \mathbf{0}$ present peculiarities that will be discussed separately.

We thus first assume $p^2 > 0$. Then the reduction with respect to m^2 and sign p^0 can be effected by a decomposition analogous to eq. (9.5.7), and the reduction with respect to spin is contained in the solution to exercise 7 of sect. 9.4 and the discussion following eq. (9.5.1). Geometrically, this solution means that $\tilde{A}^i(\bar{p})$ is decomposed into parts orthogonal and tangential to the mass shells by the complementary projections

$$P_{\perp}{}^i{}_k = \frac{\bar{p}^i \bar{p}_k}{\bar{p}^2} \qquad \text{and} \qquad P_{\parallel}{}^i{}_k = \delta^i{}_k - P_{\perp}{}^i{}_k. \tag{9.5.17}$$

The orthogonal part is proportional to $\bar{p}_i$, transforming according to the trivial representation $\mathrm{D}^{(0)}$ of $\mathcal{K}_{\bar{p}}$; the tangential part transforms according to $\mathrm{D}^{(1)}$. The manifestly covariant version of eqs. (9.5.17) at the remaining points of the mass shells is obvious:

	p-space	x-space	
$\mathrm{D}^{(0)}$:	$\left(\delta^i{}_k - \dfrac{p^i p_k}{p^2}\right)\tilde{A}^k = 0$	$\left(\delta^i{}_k \Box - \partial^i \partial_k\right) A^k = 0$	
		(cf. eq. (9.3.27c)!)	(9.5.18)
	$\tilde{A}^i \propto p^i, \quad p^{[i}\tilde{A}^{k]} = 0$	$A^i = \partial^i \Lambda, \quad \partial_{[i}A_{k]} = 0$	
$\mathrm{D}^{(1)}$:	$\dfrac{p^i p_k}{p^2}\tilde{A}^k(p) = 0$	$\partial_k A^k = 0.$	(9.5.19)

The reduction with respect to spin corresponds, here, to the well-known possibility to uniquely split a vector field on $\mathbf{R}^3$ (under suitable conditions at infinity) into a divergence-free transversal part and a curl-free longitudinal part.

We discuss the form of an invariant scalar product. If we try in x-space an expression analogous to eq. (9.5.4),

$$\int d^4x\, A_i^*(x)\, B^i(x) = \int \frac{d^4p}{(2\pi)^4}\, \tilde{A}_i^*(p)\, \tilde{B}^i(p), \qquad (9.5.20)$$

this will not work in general, since the integrand is *indefinite* for $B^i = A^i$. It becomes positive-definite upon restriction to pure spin zero fields, however: with $\tilde{A}^i(p) = \tilde{A}(p)\, p^i$ we have $\tilde{A}_i^*\, \tilde{A}^i = |\tilde{A}(p)|^2 p^2 \geq 0$ because of the assumption $p^2 > 0$. On the other hand, in the subspace of spin 1 fields it follows from $p^i \tilde{A}_i = 0$ that both, $\mathrm{Re}\,\tilde{A}_i$ and $\mathrm{Im}\,\tilde{A}_i$ are orthogonal to the timelike vector p^i and thus are spacelike; therefore $\tilde{A}_i^*\, \tilde{A}^i = \mathrm{Re}\,\tilde{A}_i\, \mathrm{Re}\,\tilde{A}^i + \mathrm{Im}\,\tilde{A}_i\, \mathrm{Im}\,\tilde{A}^i$ is negative-definite, and the negative of eq. (9.5.20) yields a scalar product. In both cases the scalar product may be decomposed as in eq. (9.5.9), and, after splitting off a singular factor, for each subspace of fixed mass square and sign of p^0 be rewritten in a form analogous to eq. (9.5.15), i.e.,

$$\pm i \int_\sigma d\sigma_k\, A_i^* \overset{\leftrightarrow}{\partial^k} B^i. \qquad (9.5.21)$$

To be able to compare with expression (9.4.27), where $s = 1$ and the index σ takes 3 values only, we must compare the usual Cartesian 4-vector basis $\{e_i\}$—to which refer the indices on 4-vectors (cf. eq. (3.3.1))—with the Wigner basis vectors for the fibers over the mass shells. (If two space dimensions are neglected, the latter can be illustrated as shown in Fig. 9.2: the Wigner basis vector for $s = 0$ and the Wigner basis vector for $s = 1$ are just tangent to the curvilinear coordinate system given by the radial lines and the mass shells in the figure.) At the points $\bar{p}^i = (m, \mathbf{0})^\top$ both bases agree, or rather differ only by the fact that $\{e_1, e_2, e_3\}$ is replaced by the canonical basis $\{e_\sigma\} = \{e_{+1}, e_{-1}, e_0\}$, (7.8.14), for the representation $\mathrm{D}^{(1)}$. At the remaining points p of the mass shells the Wigner bases are obtained by 'dragging along' the basis at $\bar{p}$ using the transformations Λ_p, eq. (9.4.7), so that the timelike unit vector $(= p/m)$ remains always orthogonal to the mass shell, whereas the Wigner basis vectors for $s = 1$ always remain tangential. (The nonuniqueness in the choice of Λ_p is not borne out in Fig. 9.2 because of the absence of more space dimensions.) Since the dragged-along basis is orthogonal in the Minkowski sense, we have $\tilde{A}_i^* \tilde{A}^i = |p_i \tilde{A}^i|^2/m^2 - \tilde{A}_\sigma^* \tilde{A}_\sigma$, and the agreement with eq. (9.4.28) for pure spin 0 or 1 is now obvious.

As announced, the case $m^2 = 0$, i.e., $\square A^i = 0$, must be treated separately, with the additional assumption that in $\tilde{A}^i(p)$ there is no contribution proportional to $\delta^4(p)$. The relevant little group is then the Euclidean group (9.4.30), and we consider only representations of it where the null rotations act trivially. The solution to exercise 5 of sect. 9.4 shows that those irreducible unitary representations of this group where the null rotations act nontrivially are all infinite-dimensional with a continuous spectrum for the null rotation generators (representations with continuous spin). Now if $A^i(x)$ were subject, besides $\square A^i = 0$, to *no* further restrictions, one would obtain, in the little vector space (the fiber) over $\bar{p}$, a nontrivial action of the null rotations on the $\tilde{A}^i(\bar{p})$; since this space has finite dimension greater than 1, our representation could not be unitary.

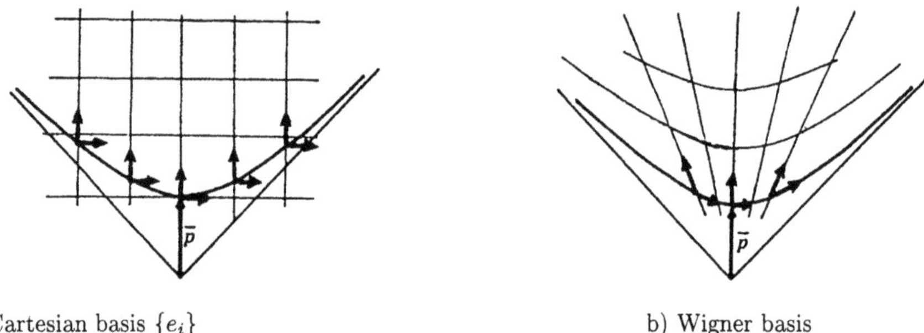

a) Cartesian basis $\{e_i\}$　　　　　　　　　　　　　　　b) Wigner basis

Fig. 9.2. Basis vectors in p-space

One possible condition leading to a trivial action of the null rotations is

$$\partial_i A_k - \partial_k A_i = 0, \qquad A_i = \partial_i \Lambda \ \text{ with } \ \Box\Lambda = 0, \tag{9.5.22}$$

already discussed before. Its effect is that the whole little group gets represented trivially, so the subspace defined by eq. (9.5.22) corresponds to helicity $\lambda = 0$; it can be split into two irreducible components (sign $p^0 = \pm 1$) as usual. An invariant scalar product may be defined with the help of the scalar field Λ, while the expression (9.5.21) simply *vanishes* under the restriction (9.5.22) (exercise).

Let us next consider the condition

$$\partial_i A^i = 0 \tag{9.5.23}$$

which for $m^2 > 0$ was complementary to $\partial_i A_k - \partial_k A_i = 0$. This is no longer the case here; the subspace defined by eq. (9.5.23) rather *contains* the fields satisfying eq. (9.2.22), as is easily seen. This is reflected in the fact that in the case $p^2 = 0$ the complementary projections (9.5.17) cannot be formed; geometrically: $\bar{p}$ is simultaneously orthogonal and tangential to the light cone $p^2 = 0$, which here takes the place of the mass shell. The representation of $\mathcal{P}_+^\uparrow$ in the space (9.5.23) is reducible but not fully reducible, as it may happen in nonunitary representations: a single invariant subspace does then not necessarily allow to define an invariant complement. The invariant sesquilinear form (9.5.21), restricted to the subspaces of positive or negative frequencies is only *semi*definite and thus degenerate—it vanishes under condition (9.5.22); and there cannot exist another invariant definite scalar product, since under the condition (9.5.23) the null rotations act nontrivially on the $\tilde{A}^i(\bar{p})$. In fact, because of $\bar{p}_i \tilde{A}^i(\bar{p}) = 0$ one easily finds the null rotation behavior

$$\tilde{A}^i(\bar{p}) \mapsto \tilde{A}^i(\bar{p}) + (\mathrm{Re}\, b\, \tilde{A}^1 - \mathrm{Im}\, b\, \tilde{A}^2)\, \bar{p}^i \tag{9.5.24}$$

(exercise), which is trivial only if $\tilde{A}^i(\bar{p}) \propto \bar{p}^i$. Therefore, unitary representations of $\mathcal{P}_+^\uparrow$ with $M^2 = 0 = W^2$, $\lambda \neq 0$ *cannot* be realized in a space of 4-vector fields.

However, one can remedy the situation by a standard construction of linear algebra and analysis, viz., the formation of *quotient* spaces (cf. exercise 7 of sect. 6.5; Naimark

1960). Let us call two vector fields satisfying $\Box A^i = 0$, $\partial_i A^i = 0$ equivalent iff they differ by a vector field satisfying eq. (9.5.22). Usually the fact that the subspace (9.5.22) is contained in the subspace (9.5.23) is expressed as follows: the equations

$$\Box A^i = 0, \qquad\qquad \partial_i A^i = 0 \qquad\qquad (9.5.25)$$

are *invariant* under the *gauge transformations*

$$A^i \mapsto A^i + \partial^i \Lambda \quad \text{with} \quad \Box \Lambda = 0, \qquad\qquad (9.5.26)$$

and vector fields equivalent in the sense just defined are said to be *gauge-related*. The degeneracy of the scalar product (9.5.21) expresses its gauge invariance. The set of *gauge equivalence classes* can be given the structure of a vector space on which $\mathcal{P}_+^\uparrow$ operates linearly and on which the expression (9.5.21) defines a *definite* scalar product. In this quotient space unitary representations may thus be realized.

This is also seen in eq. (9.5.24): the effect of a null rotation may be compensated by a gauge transformation, since the latter adds to the Fourier components $\tilde{A}^i(\bar{p})$ a multiple of $\bar{p}^i$, and these contributions are factored out in the formation of the quotient space. Geometrically, the situation is illustrated as follows. The equation $\bar{p}_i \tilde{A}^i = 0$ defines, in the 'little vector space' of the 4-vectors $A^i(\bar{p})$, a lightlike hyperplane—a 3-dimensional subspace; in it, the multiples of $\bar{p}^i$ form a 1-dimensional subspace; the points $\tilde{A}^i + \bar{p}^i \tilde{\Lambda}$ form parallel lightlike straight lines on keeping $\tilde{A}^i$ fixed and varying Λ (the 'generators' of the hyperplane—see Fig. 9.3). These lines are the 'points' of the 2-dimensional 'little' quotient space. Each line remains invariant under null rotations as a whole, the points of it getting permuted, however: there is no possibility to distinguish a point on it in an invariant manner, i.e., there is no possibility to impose a further invariant gauge condition. When the gauge is fixed by some noncovariant condition, Lorentz covariance can only be verified when combined with a suitable gauge transformation. (Our picture is slightly misleading in that the little vector space to be considered is actually complex, but since it is the complexification of a real vector space, one can actually consider real and imaginary parts separately here.)

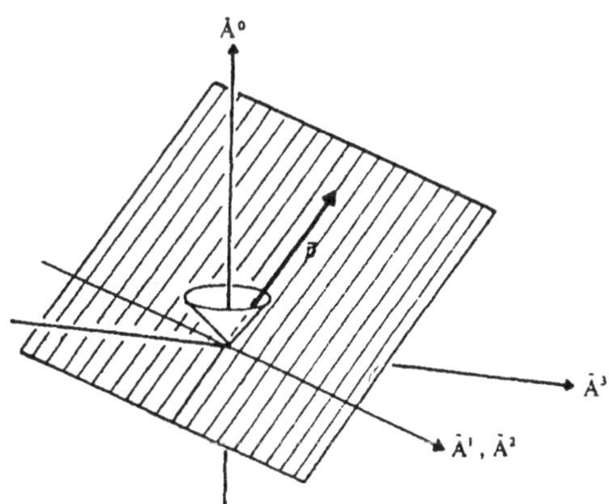

Fig. 9.3. Lightlike hyperplane with lightlike generators

To complete the reduction of the representation in the quotient space of gauge equivalence classes, we remark that for the remaining transformations of $\mathcal{K}_{\bar{p}}$—rotations about the 3-axis—the effect on $\tilde{A}^i(\bar{p})$ is the transformation

$$\tilde{A}^0 \mapsto \tilde{A}^0, \quad \tilde{A}^3 \mapsto \tilde{A}^3, \quad \tilde{A}^1 \pm i\tilde{A}^2 \mapsto \exp(\pm i\alpha)(\tilde{A}^1 \pm i\tilde{A}^2). \tag{9.5.27}$$

Because of $\bar{p}_i \tilde{A}^i(\bar{p}) = 0$, i.e., $\tilde{A}^0 = \tilde{A}^3$, each equivalence class has a representative $\tilde{A}^i(\bar{p}) = (0, \tilde{A}^1, \tilde{A}^2, 0)^\top$; it follows that the quotient space carries helicity $\lambda = \pm 1$. Equation (9.3.11), $W_k = \lambda P_k$, unsolvable in the space of massless and divergenceless vector fields when $\lambda = \pm 1$, does have a solution in the quotient space (exercise). Of course, the solutions are equivalence classes for which the tensor field $F_{ik} := \partial_i A_k - \partial_k A_i$ is selfdual or antiselfdual, and each class may uniquely be decomposed as a sum of two such classes. The physical interpretation has already been talked about. It should be obvious that for scalar fields the transition $m^2 \to 0$ is less problematic than for vector fields.

In a different guise, all peculiarities of vector fields encountered here reappear when these fields are quantized. Also, one may ask whether a theory using gauge equivalence classes can still be regarded as a local field theory. A discussion of this point, which involves the Bohm-Aharanov effect, is given in Feynman (1965).

Let us finally consider *(bi-)spinor fields* $\psi(x)$. For the reasons already known we restrict to fields having $\tilde{\psi}(p) = 0$ for $p^2 < 0$, and we treat the case $p^2 = 0$ separately. After reduction with respect to mass square and sign of p^0, we obtain, in the little spinor space of the $\tilde{\psi}(\bar{p})$, a 4-dimensional representation of $\mathcal{K}_{\bar{p}}$, whose decomposition gives $D^{(1/2,0)} \oplus D^{(0,1/2)} = D^{(1/2)} \oplus D^{(1/2)}$. For each $m^2 > 0$ and sign $p^0 = +1$ (or -1), spin $s = 1/2$ occurs twice: we have an isotypical representation (cf. exercise 6 of sect. 6.6).

In this case, the decomposition into irreducible parts is unique only up to equivalence—at variance with the situation where the complementary irreducible parts are inequivalent. A special way to sort out just one irreducible constituent is given by the Dirac equation[1] $i\gamma^k \partial_k \psi = m\psi$ or $p_k \gamma^k \tilde{\psi}(p) = m\tilde{\psi}(p)$: one easily verifies (exercise) that the matrices

$$\Lambda_\pm(p) := \frac{m \mp p_k \gamma^k}{2m} \tag{9.5.28}$$

yield two complementary projections, one of which projects onto the space of solutions of the Dirac equation in momentum space[2]. To see explicitly that, e.g., $\Lambda_+(p)$ projects onto a spin $1/2$ part, pass to the little spinor space over $\bar{p}$ to obtain there the equation $\gamma^0 \tilde{\psi}(\bar{p}) = \pm\tilde{\psi}(\bar{p})$ and the projection matrices $(1 \pm \gamma^0)/2$. By the equivalence transformation $\tilde{\psi} \mapsto S\tilde{\psi}$, where

$$S := \exp\left(\frac{\pi}{4}\gamma^0\gamma^5\right) = (1 + \gamma^0\gamma^5)/\sqrt{2}, \tag{9.5.29}$$

---

[1] Note that this involves some special choice of the gamma matrices, which is again unique only up to equivalence!

[2] These matrices are also used in electron-positron theory, with a slight change of interpretation; see, e.g., Bjørken and Drell (1964).

they take the form

$$S^{-1} \frac{1 \pm \gamma^0}{2} S = \frac{1 \pm \gamma^5}{2} \tag{9.5.30}$$

(exercise); we have seen already (cf. (9.1.32)) that these effect the splitting of the bispinor space into $D^{(1/2,0)}$ and $D^{(0,1/2)}$ parts.

To compare the scalar product of solutions to the Dirac equation that was already written down in eq. (9.3.20) to the expression (9.4.27) we have to make an expansion analogous to the one given by eqs. (9.5.7,10) and also to pass from the usual p-independent bispinor basis to a Wigner basis, using the bispinor representation $S(\Lambda_p)$ of the Λ_p; finally, an equivalence transformation as above is required. We leave the detailed execution as well as the discussion of the indefinite sesquilinear form

$$\int d^4x \, \bar\psi(x) \, \varphi(x) = \int \frac{d^4p}{(2\pi)^4} \, \tilde{\bar\psi}(p) \, \tilde\varphi(p) \tag{9.5.31}$$

as an exercise.

To treat the case $m^2 = 0$, we can first reduce using $(1 \pm \gamma^5)/2$. Next, it is easy to see that the null rotations get represented trivially only if Weyl equations are satisfied—the converse was seen already in the solution to exercise 8 of the preceding section. A suitable scalar product is (exercise 10)

$$\int_\sigma d\sigma^k \, \varphi^*_{\dot{X}} \, \sigma_k{}^{A\dot{X}} \, \psi_A. \tag{9.5.32}$$

Exercises

1. Verify eq. (9.5.16)!

2. Show that the expression (9.5.21) vanishes modulo condition (9.5.22).

3. Verify eq. (9.5.24)!

4. Discuss how it becomes possible to fulfill the condition $W_k = \pm P_k$ in the space of gauge equivalence classes of vector fields A^i satisfying eqs. (9.5.25).

5. Verify that $\Lambda_\pm(p)$, eq. (9.5.28), are complementary projections, i.e.,

$$\Lambda_\pm^2 = \Lambda_\pm, \qquad \Lambda_\pm \Lambda_\mp = 0, \qquad \Lambda_+ + \Lambda_- = \mathrm{id}. \tag{9.5.33}$$

6. Verify eq. (9.5.30)!

   Remark: S is the spinor representation of a 90° rotation in the (0,5)-plane of a 5-dimensional pseudo-Euclidean space with metric diag $(1, -1, -1, -1, -1)$. The Clifford algebra (9.1.17) for this metric allows for an irreducible representation by the matrices γ^k, γ^5, and with their help the generators of rotations may be formed precisely as in eq. (9.1.28).

7. Carry out the detailed comparison between expressions (9.3.20) and (9.4.27) for $s = 1/2$. (Cf. Fonda and Ghirardi 1970 and Fortschr. Phys. *17*, 727 (1969).)

8. Discuss expression (9.5.31) in the irreducible subspaces of the space of bispinor fields.

9. Show that in the space of bispinor fields with $m^2 = 0$ the null rotations of the little group get represented trivially iff the chiral components satisfy Weyl equations.

10. Show that expression (9.5.32) is hypersurface independent and definite; carry out the comparison with the corresponding $m^2 = 0$ version of eq. (9.4.27).

11. Let $A_{ij\ldots k}(x)$ be totally symmetric tensor fields of degree s that satisfy the Klein-Gordon equation with $m^2 > 0$ as well as the constraint conditions

$$A^i{}_{i\ldots k} = 0, \qquad\qquad \partial^i A_{ij\ldots k} = 0. \qquad\qquad (9.5.34)$$

Which irreducible representations of $\mathcal{P}_+^\uparrow$ are contained in the space of these fields?
Hint: Count the number of independet components of $\tilde{A}_{ij\ldots k}(\bar{p})$! See M. Fierz, Helv. Phys. Acta XII, 3 (1939).

12. Show that $m = 0$, $\lambda = 0$ may be represented in the space of symmetric tensor fields ψ_{ik} having $\Box\psi_{ik} = 0$, $\psi_{i[k,l]} = 0$.

13. $m = 0$, $|\lambda| = 1$ represents not only in the quotient
$\{A_i : \Box A_i = 0, \partial^i A_i = 0\}/\{A_i : \partial^i A_i = 0, \partial_{[i}A_{k]} = 0\}$ but also in the quotient
$\{A_i : (A_{i,k} - A_{k,i})^{,k} = 0\}/\{A_i : \partial_{[i}A_{k]} = 0\}$.

14. $m = 0$, $|\lambda| = 1$ represents in $\mathbf{V}'/\mathbf{V}''$, where
$\mathbf{V}' := \{\psi_{ik} : \psi_{[ik]} = 0, \psi_k{}^k = 0, \psi_{ik}{}^{,k} = 0, \partial_{[j}\psi_{i][k,l]} = 0\}$,
$\mathbf{V}'' := \{$subspace as in exercise 12$\}$.
Hint: $\partial_{[j}\psi_{k]][k,l]} = 0$ is the integrability condition for the existence of some A_{ik} having $A_{ik,j} = \frac{1}{2}(\psi_{ik,j} + \psi_{ij,k} + \psi_{jk,i})$, and for such A_{ik} we have $(A_{ik} + A_{ki})_{,j} = \psi_{ik,j}$, thus $\psi_{ik} = A_{ik} + A_{ki}$ (+const. = 0 by boundary conditions). Since further $A_{i[k,j]} \equiv 0$, there exists A_i having $A_{ik} = A_{i,k}$, so $\psi_{ik} = A_{i,k} + A_{k,i}$. The remaining conditions give $\Box A_i = \partial^i A_i = 0$, one thus returns to eq. (9.5.23).

15. $m = 0$, $|\lambda| = 2$ represents in $\mathbf{V}/\mathbf{V}'$, where $\mathbf{V}'$ is as in exercise 14 and where
$\mathbf{V} := \{\psi_{ik} : \psi_{[ik]} = 0, \psi_k^k = 0, \psi_{ik}^{,k} = 0, \Box\psi_{ik} = 0\}$.
Here the *gauge transformations* given by adding elements of the form $\psi_{ik} = A_{i,k} + A_{k,i} \in \mathbf{V}'$ do not change the tensor field $r_{jikl} := \partial_{[j}\psi_{i][k,l]}$. The latter satisfies
a. $r_{(ji)kl} \equiv r_{ji(kl)} \equiv r_{j[ikl]} \equiv 0$, $r_{ji[kl,m]} \equiv 0$ by its definition, and
b. $r^k{}_{ikl} = 0$ by the conditions in $\mathbf{V}$.

16. $m = 0$, $|\lambda| = 2$ is realized in the space of tensor fields r_{ijkl} satisfying the properties a and b of last exercise; a definite helicity being achieved by imposing selfduality or antiselfduality—either on the first, or, equivalently (proof?) on the second pair of indices. Cf. eqs. (8.4.27) and (9.4.41)!

17. Property a of exercise 15 is sufficient for a tensor field r_{jikl} to be written in terms of some 'tensor potential' ψ_{ik} as $r_{jikl} = \partial_{[j}\psi_{i][k,l]}$, where ψ_{ik} is determined

only up to gauge transformations $\psi_{ik} \mapsto \psi_{ik} + A_{i,k} + A_{k,i}$. (See F. A. E. Pirani in Deser and Ford (1965)). Therefore, in analogy to exercise 14, the case $m = 0$, $|\lambda| = 2$ may also be represented by

$$\{\psi_{ik} : \ \psi_{[ik]} = 0, \ \eta^{ik}\partial_{[j}\psi_{i][k,l]} = 0\}/\{\psi_{ik} : \ \psi_{[ik]} = 0, \ \partial_{[j}\psi_{i][k,l]} = 0\}.$$

18. $m = 0$, $|\lambda| = 3/2$ represents in $\mathbf{V}/\mathbf{V}'$ (*Rarita-Schwinger formalism*), where
$\mathbf{V} := \{\psi_i = \text{vector-(bi)spinor}, \ \gamma^k\partial_k\psi_i = 0, \ \partial^i\psi_i = 0, \ \gamma^i\psi_i = 0\}$,
$\mathbf{V}' := \{\psi_i = \partial_i\psi, \ \psi = \text{bispinor}, \gamma^k\partial_k\psi = 0\}$.
Definite helicity obtains on imposing $\gamma_5\psi_i = \pm\psi_i$, i.e., on effectively considering chiral vector-spinors.

19. $m = 0$, $|\lambda| = 3/2$ may also be realized in the space

$$\{\psi_i : \gamma_{[i}\psi_{j,k]} = 0\}/\{\psi_i : \psi_{[i,j]} = 0\}.$$

How can definite helicity be achieved?

20. $m = 0$, $|\lambda| = 3/2$ may also be realized in a space $\mathbf{V}$ of sixtor-spinor fields ψ_{ik}:

$$\mathbf{V} := \{\psi_{ik} : \psi_{(ik)} = 0, \ \gamma_{[i}\psi_{jk]} = 0, \ \psi_{[ij,k]} = 0\}.$$

How can definite helicity be achieved?

9.6 Irreducible Semiunitary Ray Representations of $\mathcal{P}$

In this section we give a brief discussion of those irreducible semiunitary ray representations of the full Poincaré group $\mathcal{P}$ which upon restriction to $\mathcal{P}_+^\uparrow$ contain only the representations treated in sect. 9.4 as direct summands. We shall stay here on the abstract level, relying on the general theorems formulated in sects. 7.9 and 9.2, which allow to construct the the representations of $\mathcal{P}$ that we want from those of $\mathcal{P}_+^\uparrow$.

The first thing to do is to decide which of the reversals P, T, PT are to be represented linearly or antilinearly. On a purely mathematical basis, both possibilities are open, but since the choice of the subrepresentations admitted for $\mathcal{P}_+^\uparrow$ are physically motivated already, again a physical argument will make the decision.

We start with a formal preparation, which will also show that the values of a cocycle $\omega(.,.)$ belonging to a semiunitary ray representation of $\mathcal{P}$ are already determined, up to equivalence, by its values on the restricted group $\mathcal{P}_+^\uparrow$ and on the four-group $\mathcal{V}_4 = \{E, P, T, PT\}$. Let I be one of the reversals and $g = (a, L) \in \mathcal{P}_+^\uparrow$—if necessary, write $(0, I)$ instead of I for more clarity. Also observe in the following that $I^2 = E$ and $IgI \in \mathcal{P}_+^\uparrow$. In complete analogy to eq. (7.10.7) we then have

$$U_I U_g U_I^{-1} = \gamma(g) U_{IgI} \tag{9.6.1}$$

with some phase factor $\gamma(g)$. With $h \in \mathcal{P}_+^\uparrow$ we form

$$(U_I U_g U_I^{-1})(U_I U_h U_I^{-1}) \equiv U_I(U_g(U_I^{-1} U_I)U_h)U_I^{-1}$$

and evaluate this, using eq. (9.2.1), in the sense of both ways of bracketing. To take into account the possible antiunitaity of U_I, let $\sigma_I(\dots)$ be the identity or complex conjugation, depending on the linearity or antilinearity of U_I, respectively. It then follows that

$$\gamma(g)\gamma(h) = \frac{\sigma_I(\omega(g,h))}{\omega(IgI, IhI)}\, \gamma(gh),$$

i.e., the assignment $g \mapsto \gamma(g)$ is a 1-dimensional unitary ray representation of $\mathcal{P}_+^\uparrow$ with the fraction above as its cocycle. When $\omega(g, h)$ belongs to a single-valued representation, so does the fraction; when $\omega(g, h)$ belongs to a double-valued representation of $\mathcal{P}_+^\uparrow$, numerator and denominator—which each have the cocycle property—belong to a double-valued representation and the fraction thus again belongs to a single-valued representation of $\mathcal{P}_+^\uparrow$. This means that the fraction has the value 1 if the phases of the U_g are chosen such that $g \mapsto \pm U_g$ is a genuine representation of $\widetilde{\mathcal{P}_+^\uparrow}$. Then $g \mapsto \gamma(g)$ is a genuine 1-dimensional representation of $\mathcal{P}_+^\uparrow$, for which we know there is only the trivial possibility $\gamma(g) = 1$. Equation (9.6.1) thus goes over into

$$U_I U_g = U_{IgI} U_I, \tag{9.6.2}$$

entailing

$$\omega(I, g) = \omega(IgI, I).$$

The cocycle relations for $U_I U_g U_h$ and $U_g U_h U_I$ yield

$$\omega(Ig, h) = \frac{\sigma_I(\omega(g,h))\omega(I, gh)}{\omega(I, g)}$$

$$\omega(g, hI) = \frac{\omega(g, h)\omega(gh, I)}{\omega(h, I)}.$$

Multiplying eq.(9.6.2) by $U_{I'} U_h$—where also $I' \in \mathcal{V}_4$—finally gives

$$\omega(Ig, I'h) = \frac{\omega(I, I')\omega(II', h)\omega(IgI, II'h)}{\omega(I, g)\sigma_I(\omega(I', h))},$$

so that indeed ω is known on $\mathcal{P}$ up to equivalence as soon as its values on $\mathcal{P}_+^\uparrow$ and $\mathcal{V}_4$ are specified, as we may achieve $\omega(I, g) = 1$ by a suitable choice of λ_{Ig} (changing the phases of the U_{Ig}; see below).

To decide between linearity vs. antilinearity, we apply eq. (9.6.2) with $g = (Ia, E)$ to the vectors $|p, \dots\rangle$ in (9.4.1). Since $IgI = (0, I)(Ia, E)(0, I) = (a, E)$, we get

$$U_a U_I\,|p, \dots\rangle = U_I U_{Ia}\,|p, \dots\rangle = U_I \exp(i(Ia)^k p_k)\,|p, \dots\rangle =$$

$$= \sigma_I(\exp(ia^k(Ip)_k)\,U_I\,|p, \dots\rangle$$

in analogy to eq. (9.4.2), saying that $U_I\,|p, \dots\rangle$ is an eigenvector of U_a for the eigenvalue $\exp(\sigma_I(i)a^k(Ip)_k)$. Thus, if we want to stay among the 'physical' representations a_+, b_+ of sect. 9.4, we must, putting $I = P$, have $\sigma_P(i) = +i$, as $(Pp)_0 > 0$ for $p_0 > 0$: P therefore must be represented *linearly*. To the contrary, when $I = T$ or $I = PT$, we have $(Ip)_0 < 0$ for $p_0 > 0$, so that we must choose $\sigma_T(i) = \sigma_{PT}(i) = -i$ in order to get a future-directed 4-momentum (positive energy) again: T and PT therefore must be represented *antilinearly*.

We have written down in exercise 4 of sect. 9.2 the four inequivalent cocycles on $\mathcal{V}_4$ that belong to ray representations in which $\{E, P\}$ and $\{T, PT\}$ get represented linearly and antilinearly, respectively. They are characterized by the relations

$$U_T^2 = \alpha \, \mathrm{id}_\mathbf{H}, \qquad (U_{PT})^2 = \beta \, \mathrm{id}_\mathbf{H} \qquad (9.6.3)$$

corresponding to $T^2 = E = (PT)^2$, where α and β independently take the values ± 1 (which cannot be changed by changing the phases of U_T, U_{PT}). We now specialize the phases of U_{Pg}, U_{Tg} by

$$U_{Pg} := U_P \, U_g, \qquad U_{Tg} := U_T \, U_g,$$

$$\text{i.e.,} \quad \omega(P, g) = 1 = \omega(T, g)$$

Independently of the phase conventions made for $\mathcal{V}_4$ in exercise 4 of sect. 9.2, viz.,

$$U_{PT} = U_P \, U_T, \quad \text{i.e.,} \quad \omega(P, T) = 1,$$

we may now check that eq. (9.6.2) is satisfied for $I = PT$ if it holds for $I = P$ and for $I = T$. We may thus finally make the consistent choice

$$U_{PTg} = U_{PT} \, U_g, \qquad \text{i.e.,} \quad \omega(PT, g) = 1.$$

This gives the cocycle on all of $\mathcal{P}$, unique up to equivalence, once its values on $\mathcal{P}_+^\uparrow$ and on $\mathcal{V}_4$ have been chosen. Of course, it is possible, and is actually done in concrete field theory models, to use phase conventions that deviate from the above. The aim here was to find the possible equivalence classes of ray representations: in addition to the values of m^2 and s or λ they are classified by the values of α $(= \pm 1)$ and β $(= \pm 1)$.

At the same time we found it possible to narrow down the range of values for ω from U(1) to $\{1, -1\} \cong \mathcal{Z}_2$, thereby keeping the relevant extension groups as small as possible. Let us emphasize again what we already stressed for O(3) and $\mathcal{L}$: the aim to narrow down to $\mathcal{Z}_2$ may be reached in more than one way; one can go from one of them to the other if changes by factors $\lambda_g \in$ U(1) are admitted; one cannot always do this if only $\lambda_g \in \{1, -1\}$ is admitted. For the latter restriction, there is no (quantum) physical reason at the moment, however.

We now have found the splitting $\mathcal{G} = \mathcal{G}_1 \cup \mathcal{G}_2$—which here is the splitting $\mathcal{P} = \mathcal{P}^\uparrow \cup \mathcal{P}^\downarrow$—and the extension cocycles ω which enter the theorems 1, 2 and their supplements from sect. 9.2; we can thus apply them to get all the irreducible semiunitary ray representations of $\mathcal{P}$ which upon restriction to $\mathcal{P}_+^\uparrow$ decompose into those studied in sect. 9.4. We cannot go into a detailed execution of this program here, but refer to the extensive modern discussion given in the readable account by R. Shaw and J. Lever, Commun. Math. Phys. *38*, 279 (1974); of course, the original work by E. Wigner in Gürzey (1964) is mentioned there together with other relevant work.

The result for $\mathcal{P}^\uparrow$ is easier to obtain, since up to equivalence there is only one nontrivial extension cocycle and no antilinearity: it suffices to apply the theorems and supplements of sect. 7.9. We have given it already at the end of sect. 9.4.

For the group $\mathcal{P}$ one gets, starting from a representation of $\mathcal{P}_+^\uparrow$ having $m^2 > 0$ and spin s, a doubling of the dimension of the 'little' vector spaces (types II, III), except for $\alpha = \beta = (-1)^{2s}$ (type I). Starting from $m^2 = 0$ and helicity λ, one gets a doubling of that dimension if $\lambda = 0$ except for $\alpha = \beta = 1$ (type I), while for $\lambda \neq 0$ there is doubling for $\alpha = (-1)^{2\lambda}$ and quadruplication for $\alpha = -(-1)^{2\lambda}$.

In the usual field theory models one has $\alpha = \beta = (-1)^{2s}$ or $=(-1)^{2\lambda}$. We cannot go into a systematic discussion similar to the one given in sect. 9.5, as far as reversals are concerned. For scalar fields, $\alpha = \beta = 1$ is clear (cf. Appendix D.1). For the electromagnetic field it suggests itself to look at the time reversal behavior of the classical field (see sect. 8.5) and to add a complex conjugation for quantum mechanical (photon) wave functions (which belong to the complexification of the space of classical fields); then again $\alpha = \beta = 1$ is clear. For Dirac-spinor fields the situation is more complicated—it will be explained in Appendix C.2; indeed $\alpha = \beta = -1$ results from it. This suggests that in all field theoretical models one has the values of α and β given above.

At this point we once again stress that the quest for the nonisomorphic double-covering groups of $\mathcal{L}$, and in particular for those that can be described using (bi)spinors (cf. sect. 9.1), is irrelevant for the abstract problem of finding or classifying the semiunitary ray representations of $\mathcal{P}$. In the context of the latter this question becomes just a matter of phase conventions, where, however, such a convention must remain fixed during the course of a concrete theory.

From time to time the question is raised whether there are 'right' phase conventions which may be confirmed as relative phases in interference experiments. For instance, it is known (theory: Y. Aharonov, L. Susskind, Phys. Rev. *158*, 1237 (1967); experiment: H. Rauch et al., Phys. Lett. A *54*, 425 (1975)) that the phase factor (-1) that spinors pick up under 360°-rotations is experimentally seen in neutron interference experiments. However, it is necessary in this experiment to split the neutron beam, to perform the rotation on only one of the two beams, and to unite the beams afterwards. The rotation of the partial beam is done with the help of a magnetic field—thus by a dynamic effect. Contrary to this, our considerations in sect. 9.2 referred to operations on the total system and involved only the most general structures of quantum mechanics, but no dynamical effects in subsystems. The effect considered here is, nevertheless, in full agreement with the spinor nature of the neutron wave function. It appears difficult, however, to dynamically imitate a space reversal on a partial beam, since it is not continuously connected to the unit element in $\mathcal{P}$.

In fact, another effect of the choice for a covering group has been suggested, not relating to interference between split beams but operating with a system as a whole: it is described in the work of DeWitt and DeWitt quoted at the end of sect. 1.5. In this, however, the global structure of Minkowski space is changed, breaking global translation invariance; so this again is outside the assumptions of the present formalism.

Exercise

Show that the time reversal behavior derived in Appendix C.2 for Dirac spinor fields is antiunitary in the sense of the scalar product (9.5.20)!

10 Conservation Laws in Relativistic Field Theory

In sect. 5.9 we derived the conservation laws for energy and momentum of the electromagnetic field, with only a hint at angular momentum. In the present chapter we are going to show quite generally that conservation of energy, momentum, and angular momentum, as well as the law of motion for the center of mass(-energy) is intimately related to Poincaré covariance of the *Lagrangian formulation* of the dynamics. More precisely, we shall be able to associate a divergence-free symmetric energy-momentum tensor with any physical system whose dynamics derives from a 'principle of stationary action' that is Poincaré-covariant: translational covariance produces a divergence-free tensor, and rotational covariance allows to symmetrize it.

There is a general connection between symmetries and conservation laws. This connection is most natural and direct in the formalism of quantum mechanics: since any semilinear operator commuting with the Hamiltonian of a system is conserved in time, this is the case, in particular, for any semi*unitary* operator commuting with the Hamiltonian. For each one-parameter group of such symmetries, the corresponding Hermitian generator is likewise conserved. This latter version, the conservation of the 'infinitesimal' generator, also holds classically in the Hamiltonian formalism. However, the Hamiltonian formalism is less suitable for making the relativistic symmetry manifest; for that purpose, the Lagrangian formulation in terms of an action principle is optimal. Here we have again a relation between symmetries and conservation laws, known as E. Noether's theorem: If the dynamical equations can be written as the Euler equations of an action principle, then to each one-parameter invariance group of the action integral there is a conservation law.

In recent years, it has been (re)discovered that it is possible to set up a 'covariant Hamiltonian formalism', avoiding the usual transition via the Legendre transformation which breaks manifest covariance. Roughly, this is achieved by taking as the phase space the space of solutions of the dynamical equations, rather than the space of canonical initial data: it is possible to describe the important structures of phase space directly in terms of the space of solutions. See, e.g., J. Lee, R. M. Wald, J. Math. Phys. *31*, 725 (1990).

Since the proof of Noether's theorem yields an explicit construction of the conserved quantities whose quantum analogs can, in many cases, be simply guessed, we shall present it here. The conserved quantities so obtained behave additively for composite but noninteracting systems. (Those quantum mechanically conserved quantities stemming from symmetries which cannot be imbedded into connected symmetry groups of the Hamiltonian behave multiplicatively; e.g., parity.)

In this chapter we[1] shall mainly proceed deductively, treating applications in the exercises.

---

[1] 'We' includes the reader.

10.1 Action Principle and Noether's Theorem

The field equations for a field $\phi_\mu(x^k)$ (μ is an index cumulating tensor, spinor, and other indices) in many cases are equivalent to equations of the form

$$\partial_i \frac{\partial \mathcal{L}}{\partial \phi_{\mu,i}} - \frac{\partial \mathcal{L}}{\partial \phi_\mu} = 0 \qquad (10.1.1)$$

(*Euler equations*), where $\mathcal{L}(x^k, \phi_\mu, \phi_{\mu,i})$ is a function of the variables x^k, ϕ_μ, $\phi_{\mu,i}$ (the *Lagrangian density*), and where after carrying out the differentiation with respect to $\phi_{\mu,i}$ the usual identification $\phi_{\mu,i} \equiv \partial\phi_\mu/\partial x^i$ is to be made. As we shall see in a moment, eq. (10.1.1) is related to the following problem. Assume given the *action integral*

$$W = \int_{\mathcal{D}} d^4x \, \mathcal{L}(\phi_\mu, \phi_{\mu,i}, x^k), \qquad (10.1.2)$$

where $\mathcal{D}$ is a 4-dimensional domain in Minkowski space with boundary $\partial\mathcal{D}$. How does W change if ϕ_μ as well as $\mathcal{D}$ are changed infinitesimally? To specify this, we assume that (1) for every x, the value $\phi_\mu(x)$ is changed by an amount $\delta\phi_\mu(x)$, and (2) at every x—at least at those in a neighborhood of $\partial\mathcal{D}$—a displacement vector Δx^k is defined by which the point x is to be shifted (Fig. 10.1).

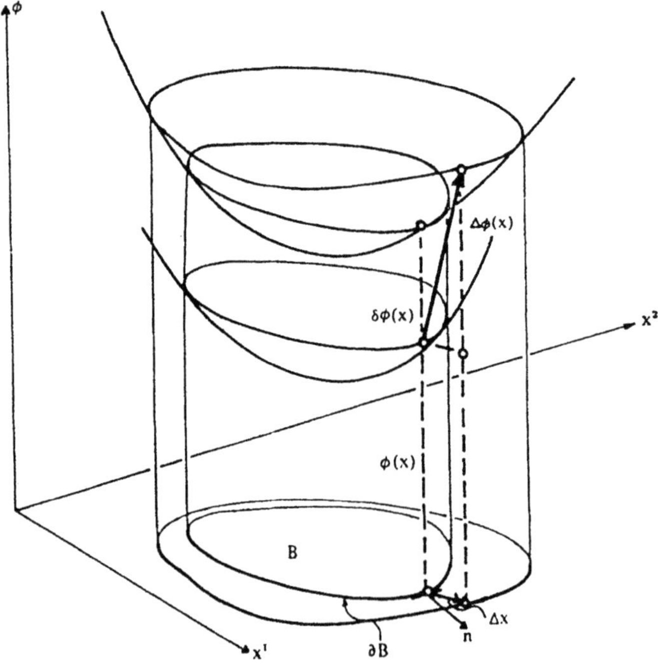

Fig. 10.1. The graph of $\phi_\mu = \phi_\mu(x)$ and its variation in (x, ϕ)-space

Since every element $d\sigma$ of the boundary surface $\partial\mathcal{D}$ in lowest order makes a contribution $\mathcal{L}\,d\sigma\,n_k\,\Delta x^k = \mathcal{L}\,\Delta x^k\,d\sigma_k$ to the change in W—where n_k is the unit normal as in Fig. 10.1—we get for the change in W to lowest order

$$\Delta W = \int_{\mathcal{D}} d^4x \left[\frac{\partial\mathcal{L}}{\partial\phi_\mu}\delta\phi_\mu + \frac{\partial\mathcal{L}}{\partial\phi_{\mu,i}}\delta\phi_{\mu,i}\right] + \int_{\partial\mathcal{D}} d\sigma_i\,\mathcal{L}\,\Delta x^i =$$
$$= \int_{\mathcal{D}} d^4x \left[\frac{\partial\mathcal{L}}{\partial\phi_\mu} - \partial_i\frac{\partial\mathcal{L}}{\partial\phi_{\mu,i}}\right]\delta\phi_\mu + \int_{\partial\mathcal{D}} d\sigma_i\left[\frac{\partial\mathcal{L}}{\partial\phi_{\mu,i}}\delta\phi_\mu + \mathcal{L}\,\Delta x^i\right]. \tag{10.1.3}$$

Here we have used that under the variation above the derivative $\phi_{\mu,i}(x)$ at x changes by $\partial_i\delta\phi_\mu(x)$, i.e.,

$$\delta\phi_{\mu,i} = \partial_i\delta\phi_\mu, \tag{10.1.4}$$

and we have used Gauss' theorem (5.6.11). When eq. (10.1.1) holds, ΔW becomes a boundary integral. We still rewrite ΔW by introducing the difference between $\phi_\mu(x)$ and the value of the varied function $\phi_\mu + \delta\phi_\mu$ at the shifted point $x + \Delta x$:

$$\Delta\phi_\mu := \delta\phi_\mu + \phi_{\mu,k}\Delta x^k. \tag{10.1.5}$$

(See Fig. 10.1, which also shows why $\delta\phi$ is called 'vertical' variation, while $\Delta\phi$ is the 'skew' variation.) With the further definition

$$\Theta^i{}_k := \frac{\partial\mathcal{L}}{\partial\phi_{\mu,i}}\phi_{\mu,k} - \delta^i{}_k\,\mathcal{L} \tag{10.1.6}$$

we get finally

$$\Delta W = \int_{\partial\mathcal{D}} d\sigma_i\left[\frac{\partial\mathcal{L}}{\partial\phi_{\mu,i}}\Delta\phi_\mu - \Theta^i{}_k\,\Delta x^k\right] + \int_{\mathcal{D}} d^4x\left[\frac{\partial\mathcal{L}}{\partial\phi_\mu} - \partial_i\frac{\partial\mathcal{L}}{\partial\phi_{\mu,i}}\right]\delta\phi_\mu. \tag{10.1.7}$$

From eq. (10.1.7) we first draw the following conclusion: the solutions of eq. (10.1.1) have the characteristic property that for them the value of the integral (10.12) is *stationary*, $\Delta W = 0$, against those variations where the boundary $\partial\mathcal{D}$ as well as the values of ϕ_μ on it are kept fixed, i.e., $\Delta x^k = 0$, $\delta\phi_\mu = 0$ on $\partial\mathcal{D}$. The special choice of $\mathcal{D}$ has no influence on the form of the condition (10.1.1) and plays a role only in that the values of ϕ_μ on $\partial\mathcal{D}$ are boundary values for special solutions of eqs. (10.1.1). We therefore say that $\mathcal{L}$ defines an *action principle* or a *variation principle* for eqs. (10.1.1), while postulating stationarity with prescribed $\mathcal{D}$ and boundary values on $\partial\mathcal{D}$ defines a (special) variational *problem*. We shall not dwell on the latter here.

This characterization of solutions to eq. (10.1.1) has a number of useful implications. For instance, given a transformation $(x, \phi) \mapsto (x', \phi')$ of the (x, ϕ) space (which we may again consider as a *fibered space* over spacetime), the solutions of the transformed equations (10.1.1) are stationary for the transformed action integral (10.1.2). It is simpler, however, to transform the action integral first and then to form the Euler equations (10.1.1), instead of transforming the Euler equations directly. In particular, the Euler equations are covariant (form-invariant) under transformations leaving the action integral invariant; the latter is easier to decide in most cases, however.

One may compare the relation between the Lagrange density and the Euler-Lagrange expression (the left-hand side of eq. (10.1.1)) to the relation between potential and force. Indeed, this parallel goes much further than can be described here. E.g., there is an analog to the criterion for the existence of a potential, etc.

To make the concept of form invariance of the action integral precise and at the same time draw an important consequence, we write the transformation explicitly as

$$x^i \mapsto x'^i = X^i(x^k, \phi_\nu) \tag{10.1.8a}$$

$$\phi_\mu \mapsto \phi'_\mu = \Phi_\mu(x^k, \phi_\nu). \tag{10.1.8b}$$

If in eqs. (10.1.8) for ϕ_μ a concrete field $\phi_\mu(x)$ (a cross section of the fibered space) is substituted, one may imagine the variables x to be eliminated to obtain the transformed field in the form $\phi'_\mu(x')$. The domain $\mathcal{D}$ over which one integrates the x in the action integral is transformed into a domain $\mathcal{D}'$ for the x'. The value of the action integral of the transformed field $\phi'_\mu(x')$ over the transformed domain $\mathcal{D}'$ then is

$$W' = \int_{\mathcal{D}'} \mathcal{L}(x', \phi'(x'), \phi'_{,i'}(x')) \, d^4x' = \int_{\mathcal{D}} \mathcal{L}'(x, \phi(x), \phi_{,i}(x)) \, d^4x, \tag{10.1.9}$$

where the last form was achieved by reintroducing the x as integration variables; the function $\mathcal{L}'$ is *defined* by this. Invariance of the action integral now says that $\mathcal{L}$ and $\mathcal{L}'$ as functions of their 3 arguments are identical, $\mathcal{L} \equiv \mathcal{L}'$. If this is assumed, then also $W = W'$.

If instead of a single transformation (10.1.8) we have a 1-parameter group of such transformations

$$x'^i = X^i(x, \phi; \tau), \qquad \phi'_\mu = \Phi_\mu(x, \phi; \tau), \tag{10.1.10}$$

where $\tau = 0$ gives the identity, we can compare the assumed result $W' - W = 0$ for *infinitesimal* $\tau = \Delta\tau$ with the general formula (10.1.7), in which we put

$$\Delta x^k = \left.\frac{\partial X^k}{\partial \tau}\right|_{\tau=0} \Delta\tau, \qquad \Delta\phi_\mu = \left.\frac{\partial \Phi_\mu}{\partial \tau}\right|_{\tau=0} \Delta\tau. \tag{10.1.11}$$

(Observe that the transformation (10.1.10) of the (x, ϕ) space corresponds to a *skew* variation as indicated by the arrow in Fig. 10.1.) If we further assume that $\phi_\mu(x)$ satisfies eq. (10.1.1), that comparison gives

$$\int_{\partial\mathcal{D}} d\sigma_i \, j^i = 0, \qquad \text{where} \qquad j^i \Delta\tau := \frac{\partial \mathcal{L}}{\partial \phi_{\mu,i}} \Delta\phi_\mu - \Theta^i{}_k \Delta x^k. \tag{10.1.12}$$

By Gauss' theorem we also get $\int_{\mathcal{D}} d^4x \, j^i{}_{,i} = 0$, and since $\mathcal{D}$ is arbitrary, it follows that

$$\partial_i j^i = 0. \tag{10.1.13}$$

The *four-current density* j^i thus satisfies an *equation of continuity*, and on assuming suitable fall-off behavior of ϕ_μ at spatial infinity, as in sect. 5.7 we conclude that the '*total charge*'

$$Q := \int_\sigma d\sigma_i\, j^i \qquad (10.1.14)$$

is independent of the spacelike hypersurface σ. Choosing σ as $x^0 = t = const.$, $d\sigma_k = (d^3x, \mathbf{0})$, we get $Q = \int d^3x\, j^0$. The charge $Q_{\mathcal V}$ contained in a finite volume $\mathcal V \subset \sigma$ satisfies

$$\frac{\partial}{\partial t}\, Q_{\mathcal V} = \int_{\mathcal V} d^3x\, \frac{\partial j^0}{\partial x^0} = -\int_{\mathcal V} d^3x\, \boldsymbol{\nabla}\mathbf{j} = -\int_{\partial\mathcal V} d\mathbf{O}\,\mathbf{j}, \qquad (10.1.15)$$

as follows from eq. (10.1.13). This gives the interpretation of $j^0(x)$ as a '*density of charge*' and of $\mathbf{j}(x)$ as the *density of current*.

Therefore we have, for each 1-parameter group of transformations of the (x,ϕ) space that leaves invariant the action integral (10.1.2), a local conservation law (10.1.13) and a conserved quantity (10.1.14) for the solutions of eq. (10.1.1). This is the (first) *Noether theorem* on invariant action principles.

The transformations (10.1.10) are usually not needed in the generality written, but either in the special form

$$x'^i = x^i, \qquad\qquad \phi'_\mu = \phi'_\mu(\phi_\nu; \tau) \qquad (10.1.16)$$

characterizing (global) 'inner symmetries' (e.g., the well-known isospin or SU(3) symmetry of strong interactions), or in the special form

$$x'^i = x'^i(x^k; \tau), \qquad\qquad \phi'_\mu = \phi'_\mu(x^k, \phi_\nu; \tau) \qquad (10.1.17)$$

characterizing spacetime symmetries. The conservation laws associated with the latter are therefore also called *geometric conservation laws*. In the next section we are going to study the geometric conservation laws associated to the various 1-parameter subgroups of the Poincaré group $\mathcal{P}_+^\uparrow$.

The theorem admits a few generalizations. An important one is based on the remark that the Euler equations do no fix the Lagrange density completely, so that eq. (10.1.1) may be covariant even under transformations that do not leave the integrand of the action integral invariant. E.g., if $\mathcal L$ is replaced by $\mathcal L' = const.\mathcal L + \mathcal F$, where $\mathcal F(x,\phi,\phi_{,i})$ has the form of a 'complete divergence'

$$\mathcal F = \frac{\partial f^i}{\partial x^i} + \frac{\partial f^i}{\partial \phi_\mu}\,\phi_{\mu,i}, \qquad\qquad f^i = f^i(x,\phi), \qquad (10.1.18)$$

then $\mathcal L$ and $\mathcal L'$ give equivalent Euler equations, as may be shown as an exercise. For the existence of a conservation law analogous to eq. (10.1.14) it is then sufficient to assume that the function $\mathcal L'$ occurring in eq. (10.1.9) under *infinitesimal* transformations (10.1.10) takes the form $\mathcal L' = \mathcal L + \mathcal F\Delta\tau + +O(\Delta\tau^2)$, where $\mathcal F$ is a complete divergence (10.1.18). The conserved current then differs from eq. (10.1.12) by the term $f^i(x,\phi)$.

Another generalization is to admit transformations (10.1.10) where the right-hand sides also depend on the $\phi_{\mu,i}$ (so-called 'contact transformations'). For a geometrical description one then needs the terminology of fibered manifolds and their 'jet extensions'—cf. Hermann (1970); A. Trautman, Commun. Math. Phys. *6*, 248 (1967). This generalization has, however, not found applications in physics, apart from the case of a single independent variable, where the canonical formalism is usually preferred.

We finally mention that there is a second Noether theorem on action principles that are invariant under transformations depending on 'arbitrary' functions instead of parameters—e.g., gauge transformations

$$A_i \to A_i + \partial_i \Lambda \tag{10.1.19}$$

of the 4-potential of electrodynamics. This will not be considered here. We also omit the explicit formulation of the theorem in the case where the Lagrangian depends on higher-order derivatives, which would lead to field equations of order higher than 2.

Exercises

1. Show that

$$W = \int d^4x \cdot \frac{1}{2}(\Phi_{,i}\,\Phi^{,i} - m^2\Phi^2) \tag{10.1.20}$$

$$W = \int d^4x\, \bar\psi \,(i\gamma^k\partial_k\psi - m\psi) \tag{10.1.21}$$

   are action integrals for the Klein-Gordon and the Dirac equation.

   Hint: Convince yourself that ψ and $\bar\psi$ may be varied independently.

2. If one satisfies the homogeneous Maxwell equations identically by the ansatz $F_{ik} = A_{k,i} - A_{i,k}$, one can obtain the inhomogeneous Maxwell equations $F^{ik}{}_{,k} = -4\pi j^i$ from the action principle

$$W = \int d^4x \left\{ \frac{1}{16\pi}(A^{i,k} - A^{k,i})\,(A_{k,i} - A_{i,k}) - j_i A^i \right\}. \tag{10.1.22}$$

   It has to be observed that even in the source-free case $j^i \equiv 0$, in which $W = -\frac{1}{16\pi}\int d^4x \cdot F^{ik}F_{ik}$, it is nevertheless A_i that has to be regarded as the field variable, and not F_{ik}. Without using A_i one cannot write a local coupling term.

3. Show that

$$W = \int d^4x \left\{ \frac{1}{4}(A_{i,k} - A_{k,i})\,(A^{k,i} - A^{i,k}) + \frac{1}{2}m^2 A_i\,A^i \right\} \tag{10.1.23}$$

   is an action integral for the Proca equations (9.3.22).
   Hint: Form the divergence of the Euler equations that result from this action principle and observe the assumption $m \neq 0$.

4. How does the formalism of action principles look like in the case of only one independent variable? Show that the relativistic equation of motion of a point charge in a given electromagnetic field, eqs. (4.1.10), (5.3.2), has the action integral

$$W = \int \left\{ \frac{m}{2}\eta_{ik}\frac{dx^i}{ds}\frac{dx^k}{ds} + e\,A_i(x)\frac{dx^i}{ds} \right\} ds, \tag{10.1.24}$$

   where $A_i(x)$ is the 4-potential.

5. Show that the solutions of the Euler equations for

$$W = \int ds = \int \sqrt{1 - \left(\frac{d\mathbf{x}}{dt}\right)^2}\, dt \qquad (10.1.25)$$

are straight lines of Minkowski space. Consider now straight lines through a given point $P(t_1, \mathbf{x}_1)$, intersecting a given hypersurface F. Show that the integral W, extended from P to the intersection point Q, is stationary against variation of Q along F if Q is such that the line is orthogonal to F (in the sense of Minkowski geometry).
Hint: Use a formula analogous to eq. (10.1.7) for the change in W when Q is changed, observing that the changed point is also in F.

6. Show that $\mathcal{L}$ and $\mathcal{L}' = const.\mathcal{L} + \mathcal{F}$ lead to the same Euler equations when $\mathcal{F}$ has the form (10.1.18).
Hint: This may be verified directly, or one can perform the variation with fixed values on a fixed boundary, in which $\mathcal{F}$ gives only boundary terms.

7. The action integral

$$W = \int d^4x \left(\Phi^*_{,i}\,\Phi^{,i} - m^2 \Phi^* \Phi\right)$$

for a *complex*-valued scalar field is invariant under $\Phi \mapsto e^{i\tau}\Phi$, just as the action integral (10.1.21) is invariant under $\psi \mapsto e^{i\tau}\psi$. Calculate the conserved current in both cases and compare with eqs. (9.1.47), (9.5.15) or (9.1.46), (9.3.20)!

10.2 Application to Poincaré-Covariant Field Theory

The examples given in the exercises of the last section show that there are action principles for all field equations considered so far. Their Poincaré covariance is immediate since d^4x is invariant and the Lagrange densities $\mathcal{L}$ in all cases are scalars formed from the fields and their derivatives; the absence of an explicit x-dependence of $\mathcal{L}$ guarantees translational invariance. Conversely, one may proceed by writing down an invariant action principle for a given collection of tensor and spinor fields and derive covariant field equations from it. This is important when it comes to the construction of the dynamics of interacting fields. In the simplest cases one forms the sum of the Lagrangians of the free fields and adds an interaction term (coupling term) containing the products of the fields to be coupled. The condition that the result must be a scalar already restricts the number of possibilities. (In terms of representation theory the question is how often the trivial representation is contained in the tensor product of representations.)

The coupling of fields enables the conserved quantities constructed à la Noether to be interpreted physically. Namely, for the free electromagnetic field it turns out that the conserved quantities associated to spacetime translations agree with the expressions given in sect. 5.9 for energy-momentum. Since for coupled systems it is the total energy-momentum that is conserved, the interpretation of the conserved quantities associated with translational invariance as energy-momentum follows. A similar argument applies for angular momentum and the center of mass-energy, which are associated to rotational and boost invariance.

Let us now consider translational, rotational, and boost invariance in turn, restricting to the case where the Lagrange densitiy is constructed from tensor and spinor fields, so that the invariance is manifest. If we put infinitesimally $x'^i = x^i + a^i \Delta \tau$, then by definition of the translation behavior of tensor and spinor fields φ:

$$\varphi'(x') = \varphi(x) \Rightarrow \Delta \varphi = 0. \tag{10.2.1}$$

Observe that here the skew variation $\varphi'(x') - \varphi(x)$ has been calculated, whereas for the determination of the generator for translations in the space of fields the vertical variation $\varphi'(x) - \varphi(x) = \varphi(x - a\Delta\tau) - \varphi(x)$ was used—cf. eq. (9.3.12). In connection with spacetime transformations the negative of the vertical variation—which compares φ and φ' at the *same* point—is also called the *Lie differential* of φ along Δx^k.

Observe also that eq. (10.2.1) does not hold when fields with another kind of translational behavior are involved. One could, e.g., think of replacing the representation $(a, L) \mapsto D(L)$ in eq. (9.1.1) by another finite-dimensional representation of $\mathcal{P}$ in which the translations are represented nontrivially. An example of this obtains in the bispinor representation: here one may, by assigning $(a, E) \mapsto E + a_k \gamma^k (1 - \gamma^5)/2$, extend the representation in a nontrivial fashion to $\mathcal{P}$, as may be verified as an exercise. Objects of this type will not be considered as bispinors, however.

When $\Delta x^k = a^k \Delta \tau$, $\Delta \varphi = 0$ are now inserted into eq. (10.1.12), we find as a divergence-free 4-vector field $j^i = -\Theta^i{}_k a^k$. As the a^k are arbitrary, it follows that

$$\partial_i \Theta^i{}_k = 0$$

$$P_k := \int_\sigma d\sigma_i\, \Theta^i{}_k = \text{conserved}. \tag{10.2.2}$$

$\Theta^i{}_k$ is called the *canonical energy-momentum tensor* field of the system described by $\mathcal{L}$, while P_k is called its *energy-momentum* 4-vector. Because of

$$P^k = \int_{t=const.} d^3x\, \Theta^{0k}, \qquad\qquad \frac{\partial}{\partial t} \int_V d^3x\, \Theta^{0k} = - \int_V dO^\alpha\, \Theta^{\alpha k} \tag{10.2.3}$$

one would interpret Θ^{00} as *energy density*, $\Theta^{0\alpha}$ as *momentum density*, $\Theta^{\alpha 0}$ as *energy current* density and $\Theta^{\alpha\beta}$ as the *stress tensor* density. However, one may object that this involves a conclusion from the integral to the integrand, so that the suggested localization of energy, momentum, etc., contains some arbitrariness. Indeed, let $f^{ji}{}_k$ be an arbitrary tensor field, antisymmetric in i and j and suitably vanishing at infinity; then

$$\partial_i \left(\partial_j f^{ji}{}_k \right) \equiv 0 \tag{10.2.4}$$

and

$$\int_\sigma d\sigma_i\, \partial_j f^{ji}{}_k = \int d^3x\, \partial_j f^{j0}{}_k = \int d^3x\, \partial_\alpha f^{\alpha 0}{}_k = 0, \tag{10.2.5}$$

since the σ-integral is in fact independent of σ, and the d^3x integration can be converted into a surface integral at spatial infinity. Therefore the tensor field

$$T^i{}_k := \Theta^i{}_k + \partial_j f^{ji}{}_k \tag{10.2.6}$$

is divergence-free as well and yields the same value for the total energy-momentum P_k, just localizing it differently: the amount of energy-momentum contained in a finite spatial volume will depend on the choice of $f^{ji}{}_k$. Further arguments are necessary to fix $f^{ji}{}_k$. For the moment, let us just remark that in the case of the free Maxwell field the canonical energy-momentum tensor $\Theta^i{}_k$ as calculated from eq. (10.1.22) does *not* agree with the trace-free, symmetric, gauge-invariant tensor (5.9.13)! The total energy-momentum for both versions is the same, however.

Next, let us consider infinitesimal Lorentz transformations $x'^i = x^i + \omega^i{}_k x^k \Delta\tau$. If φ is a tensor or spinor field, we have

$$\varphi'(x') = \mathrm{D}(L)\,\varphi(x) \Rightarrow \Delta\varphi = \frac{1}{2}\omega_{ab}\Sigma^{ab}\varphi, \tag{10.2.7}$$

where $\Sigma^{ab} = -\Sigma^{ba}$ are the six generators of the finite-dimensional representation $\mathrm{D}(L)$, differing only by factors i from the operators S^{ab} introduced in eqs. (9.3.17–19) for concrete cases.

Note again that this is the skew variation: the vertical variation would also bring in terms corresponding to eq. (9.3.15), i.e., to orbital angular momentum in the wave mechanical interpretation. In line with this occurrence of the operators of *total* angular momentum, the vertical variation under our transformation is sometimes also called *total* variation (or Lie differential, in the case of tensor fields, up to a sign).

Equation (10.1.12) now gives the divergence-free current

$$j^i = \frac{\partial\mathcal{L}}{\partial\varphi_{,i}}\,\frac{1}{2}\,\omega_{ab}\,\Sigma^{ab}\,\varphi - \Theta^i{}_a\,\omega^{ab}\,x_b, \tag{10.2.8}$$

and so, since the ω_{ab} are antisymmetric but arbitrary otherwise:

$$\partial_i\,j^i{}_{ab} = 0, \qquad j^i{}_{ab} := \frac{\partial\mathcal{L}}{\partial\varphi_{,i}}\,\Sigma_{ab}\,\varphi - \left(\Theta^i{}_a\,x_b - \Theta^i{}_b\,x_a\right). \tag{10.2.9}$$

(The component indices on φ and Σ^{ab} have been suppressed.)

To interpret the resulting six conservation laws

$$j_{ab} := \int_\sigma d\sigma_i\,j^i{}_{ab} \tag{10.2.10}$$

consider first the case of a scalar field, for which $\Sigma^{ab} \equiv 0$. The equation $\partial_i\,j^i{}_{ab} = 0$ gives, in this case, because of $\partial_i\Theta^i{}_a = 0$, that $\Theta_{ab} = \Theta_{ba}$: for a *scalar* field, the canonical energy-momentum tensor is *symmetric*.

We shall later find a method of choosing the $f^{ji}{}_k$ in eq. (10.2.6) in such a way as to enable a symmetric energy-momentum tensor for every field. Let us assume now that this has been achieved or that we are in the case of the scalar field. Thus let $T^i{}_k$ be a tensor field with the properties

$$T_{ik} = T_{ki}, \qquad\qquad \partial_i\,T^i{}_k = 0, \tag{10.2.11}$$

which also correctly localizes energy and momentum. From it, we form the analog of the expression that eq. (10.2.9) yields for a scalar field, viz., the moments

$$J^i{}_{ab} := x_a\,T^i{}_b - x_b\,T^i{}_a, \qquad\qquad \partial_i\,J^i{}_{ab} = 0, \tag{10.2.12}$$

which lead to conserved quantities

$$J^{ab} = \int d\sigma_i \left(x^a\, T^{ib} - x^b\, T^{ia} \right) = \int d^3x \left(x^a\, T^{0b} - x^b\, T^{0a} \right).$$ (10.2.13)

Since we are assuming that

$$T^{00} =: \mathcal{E}, \qquad\qquad T^{0\mu} =: \mathcal{P}^{\mu}$$ (10.2.14)

are the densities of energy and momentum, we interpret

$$\mathcal{J}_\gamma := \frac{1}{2}\epsilon_{\gamma\alpha\beta} J^{0\alpha\beta} = \epsilon_{\gamma\alpha\beta} x^\alpha \mathcal{P}^\beta$$ (10.2.15)

as the density of angular momentum and

$$\mathbf{J} = \int d^3x\, \mathbf{x} \times \vec{\mathcal{P}}$$ (10.2.16)

as the total *angular momentum* of the system. Finally, the conservation of $J^{\alpha 0}$,

$$\int d^3x \left(x^\alpha \mathcal{E} - x^0 \mathcal{P}^\alpha \right) = J^{\alpha 0} = const. \,,$$ (10.2.17)

on dividing by the total energy $E := P^0 = \int d^3x\, \mathcal{E}$ may be brought to the form

$$\frac{\int d^3x\, \mathbf{x}\, \mathcal{E}}{\int d^3x\, \mathcal{E}} = \frac{\mathbf{P}}{E}\, t + \mathbf{const.},$$ (10.2.18)

thus expressing a law of motion for the *center of mass-energy (centroid)*: it moves uniformly and rectilinearly with velocity $\mathbf{P}/E$ with respect to the inertial frame in which the space-time splitting involved in eqs. (10.2.14–18) has been made.

It must be emphasized that the position of the world line of such a centroid in spacetime depends on the inertial frame used for its definition; only its direction is uniquely given by the total 4-momentum P^k. When the latter is *timelike*, however, the system described by $T^i{}_k$ itself distinguishes the 4-velocity of an inertial frame—its rest frame. In this case, the centroid with respect to the rest frame is called *relativistic center of mass*, and the total angular momentum with respect to the rest frame is called the total *spin* of the system.

A condition on the energy-momentum tensor to guarantee that the total 4-momentum be time-like, sometimes called *dominant energy condition*, is the following: for every observer with 4-velocity u, the 4-current of energy $\mathcal{E}^i := T^i{}_k u^k$ is timelike and future-directed. This suffices, since a sum or integral of vectors of this kind is also inside the future light cone (cf. exercise 1 of sect. 3.2).

Let us derive 4-dimensional expressions for spin and center of mass in terms of Minkowski geometry, assuming a timelike total 4-momentum. For this we observe that P^k and J^{ik} differ from each other as far as the translational behavior is concerned: while the former is a genuine 4-vector, unchanged under translations, the latter changes under the translation $x \mapsto \tilde{x} = x - a$ as

$$J^{ik} \mapsto \tilde{J}^{ik} = \int d^3\tilde{x} \left(\tilde{x}^i\, T^{0k} - \tilde{x}^k\, T^{0i} \right) = J^{ik} - a^i P^k + a^k P^i,$$ (10.2.19)

i.e., (P^i, J^{ik}) transforms according to the adjoint representation of the Poincaré group (cf. eq. (9.3.4d)). We now explicitly introduce the 4-velocity u of the inertial frame used above to make the space-time split (10.2.14–18). The world line of the centroid for u is then the set of points a for which

$$\tilde{J}^{ik} u_k = 0, \tag{10.2.20}$$

since in the rest frame of u this reads $\tilde{J}^{\alpha 0} = 0$ or

$$a^\alpha = a^0 \frac{P^\alpha}{P^0} + \frac{J^{\alpha 0}}{P^0}, \tag{10.2.21}$$

in agreement with eqs. (10.2.17,18). The 4-dimensional version of eq. (10.2.21) is

$$a^i = \lambda \frac{P^i}{Pu} + \frac{J^{ik} u_k}{Pu}, \tag{10.2.22}$$

since this equation when taken in the rest frame of u gives for $i = 0$ the parameter value $\lambda = a^0$ and for $i = \alpha$ reproduces eq. (10.2.21). When we put $u = P/\sqrt{P^2}$, eq. (10.2.22) goes over into the world line of the relativistic center of mass

$$a^i = \lambda \frac{P^i}{\sqrt{P^2}} + \frac{J^{ik} P_k}{P^2}, \tag{10.2.23}$$

where λ is its proper time.

The tensor of angular momentum with respect to the center of mass world line is called the (classical) *spin tensor* S^{ik} of the system described by T_{ik}. It satisfies eq. (10.2.20) with $u \propto P$, i.e.,

$$S^{ik} P_k = 0. \tag{10.2.24}$$

Because of this relation, S^{ik} does not contain more information than the relativistic *spin vector* $S_i/\sqrt{P^2}$,

$$S_i := \frac{1}{2} \epsilon_{iabc} S^{ab} P^c \equiv \frac{1}{2} \epsilon_{iabc} J^{ab} P^c \tag{10.2.25}$$

(cf. eq. (9.3.8)), from which it may be reconstructed as (use eq. (5.5.9))

$$S_{ab} = -\epsilon_{abik} S^i P^k / P^2 = J_{ab} + (P_a J_{bc} - P_b J_{ac}) P^c / P^2. \tag{10.2.26}$$

As it should be, expression (10.2.26) also results from inserting eq. (10.2.23) into eq. (10.2.19). S_i is orthogonal to P^i,

$$S_i P^i = 0 \; ; \tag{10.2.27}$$

since P^i had to be timelike in all these considerations, S_i is a spacelike vector:

$$S_i S^i \equiv -\frac{1}{2} S_{ab} S^{ab} P^2 < 0. \tag{10.2.28}$$

We emphasize that all considerations here can be made, and all quantities introduced here can be formed, whenever a divergence-free symmetric energy-momentum tensor is available for which P

is timelike—thus in particular if the dominant energy condition is satisfied. This offers the possibility to describe physical systems *phenomenologically* by an energy-momentum tensor without specifying how the latter is constructed from more elementary fields.

An interesting general statement can be derived from the formulae above if a *convex* body is considered, i.e., the spacelike cross sections of the world tube that forms the support of T^i_k are compact and convex. Then every centroid is contained in its interior, as follows from the expression (10.2.18). If $\tilde{J}^{ik}$ defines the centroid for the observer u and if a is the connecting vector to the relativistic center of mass, it follows from $\tilde{J}^{ik} u_k = 0$ and $\tilde{J}^{ik} = S^{ik} - a^i P^k + a^k P^i$ that

$$a^i = \frac{ua}{Pu} P^i + \frac{S^{ik} u_k}{Pu}. \tag{10.2.29}$$

The projection of a^i normal to P^i is the spacelike vector $S^{ik} u_k / Pu$, whose length $r = r(u)$ is given by

$$r^2(u) = -\frac{S^{ik} u_k \, S_{ij} u^j}{(Pu)^2} = \frac{-S^2}{(P^2)^2} - \frac{u^2(-S^2)}{P^2(Pu)^2} - \frac{(Su)^2}{P^2(Pu)^2}. \tag{10.2.30}$$

The first term here is positive by eq. (10.2.28), while the others are negative. On varying u we vary the centroid, and when u tends to a lightlike vector orthogonal to S, $r^2(u)$ tends to

$$r^2 := \frac{-S^2}{(P^2)^2} = \frac{1}{2} \frac{S_{ab} S^{ab}}{P^2}. \tag{10.2.31}$$

Since all centroids are to lie inside the convex body, there results from eq. (10.2.31) a lower bound for the diameter of a convex body with given mass and spin. The order of magnitude corresponds to the argument that the speed of a peripheral point of the body in a stationary state of rigid rotation should not exceed the speed of light.

We finally turn to the question how to construct from the canonical tensor Θ^i_k a tensor T^i_k with the properties (10.2.11). By actually calculating the divergence in eq. (10.2.9), observing eq. (10.2.2), we obtain

$$\Theta_{ba} - \Theta_{ab} = \partial_i \left(\frac{\partial \mathcal{L}}{\partial \varphi_{,i}} \Sigma_{ab} \, \varphi \right). \tag{10.2.32}$$

If for T^i_k we make the ansatz (10.2.6), then for f^{ji}_k besides the antisymmetry condition

$$f^{ji}_k + f^{ij}_k = 0 \tag{10.2.33}$$

we get from $T_{ik} = T_{ki}$ and eq. (10.2.32)

$$\partial_j \left(f^j_{ik} - f^j_{ki} \right) = \partial_j \left(\frac{\partial \mathcal{L}}{\partial \varphi_{,j}} \Sigma_{ik} \, \varphi \right), \tag{10.2.34}$$

i.e.,

$$f^j_{ik} - f^j_{ki} = \frac{\partial \mathcal{L}}{\partial \varphi_{,j}} \Sigma_{ik} \, \varphi + \partial_l \, g^{lj}_{ik} =: g^j_{ik}, \tag{10.2.35}$$

where g^{lj}_{ik} is antisymmetric on l, j as well as on i, k, but arbitrary otherwise. (It suffices, e.g., to choose $g^{lj}_{ik} \equiv 0$.) The unique solution of eq. (10.2.33,35) is

$$f_{jik} = \frac{1}{2}(g_{jik} + g_{ikj} - g_{kji}). \tag{10.2.36}$$

When T_{ik} is formed in this way, then between J^i_{ab} of eq. (10.2.12) and j^i_{ab} of eq. (10.2.9) there is the relation

$$J^i_{ab} = j^i_{ab} + \partial_l \left(x_a f^{li}_{\ b} - x_b f^{li}_{\ a} + g^{li}_{\ ab} \right), \tag{10.2.37}$$

so that the integrals $\int d\sigma_i\, J^i_{ab}$ and $\int d\sigma_i\, j^i_{ab}$ agree, the choice of $g^{li}_{\ ab}$ entering only the localization of angular momentum.

Having settled the formal side of the problem, there remains the question whether by the symmetry postulate (10.2.11) and some specific choice of $g^{lj}_{\ ab}$ a correct localization of energy, momentum, and angular momentum can be achieved. The case of the Maxwell field shows that the above procedure and the choice $g^{lj}_{\ ab} = 0$ indeed lead to the energy-momentum tensor considered in sect. 5.9 (exercise). A further question is: where does a localization of field energy and field momentum play a role? It is Einstein's insight that the stress-energy-momentum tensor acts as the source of the gravitational field in a relativistic theory of gravitation, just as the 4-current density of electric charge acts as the source of the electromagnetic field. Now the standard version of this relativistic theory of gravitation, usually called *General Relativity*, involves a procedure of calculating a symmetric conserved energy-momentum tensor from the Lagrangian of the (nongravitational) field in question, which is entirely different conceptually from the procedure presented here—due to Belinfante[1]—but which demonstrably agrees with the one constructed here[2] on taking $g^{lj}_{\ ab} = 0$.

It is interesting to note that already in 1914— thus well before Noether's theorem, Belinfante's procedure and the publication of General Relativity—M. Abraham writes (Jahrb. Radioakt. Elektron. *11*, 470) after discussing the energy-momentum tensor of electromagnetism: "If all forces of Nature can be fitted into the scheme of the symmetric world tensor, then the theorem about the momentum of the energy current and the theorem about the inertia of energy that follows from it have universal validity ...".

While the arguments in favor of the Belinfante tensor given above appear to be convincing, it must be admitted that there are alternatives to General Relativity (e.g., the Einstein-Cartan theory[3]) that prefer the canonical tensor. A final decision has not yet been made, and it seems difficult to use experiments for this purpose.

For the experimental demonstration of momentum and angular momentum of electromagnetic radiation we refer the reader to the laser experiments on the pressure of light described in *Scientific American 226*, Nr. 2, 62 (1972), and to the movie "The Angular Momentum of Circularly Polarized Radiation", EDC College Physics Film Series.

Our considerations in this section were purely classical. Although the only known fundamental fields that are macroscopically observable are the electromagnetic field and the gravitational field (which, however, had to remain excluded from our considerations for other reasons), the domain of applicability of our considerations is larger: firstly, with only slight modifications, they can also be adapted to apply to quantized fields; and secondly, as stressed before, the conserved quantities can be formed whenever a physical system is described phenomenologically and a symmetric

---

[1] F. Belinfante, Physica *6*, 887 (1939).
[2] L. Rosenfeld, Mem. Acad. R. Belg. *6*, 30 (1940).
[3] See, e.g., F. Hehl, Rev. Mod. Phys. 48, 393 (1976).

divergenceless energy-momentum tensor is associated with it. In the next section we shall illustrate this procedure by the example of relativistic hydrodynamics.

Exercises

1. Determine the canonical and the Belinfante-symmetrized energy-momentum tensor for the scalar Klein-Gordon field, eq. (10.1.20), for the electromagnetic field, eq. (10.1.22) with $j^i = 0$, for the Proca field, eq. (10.1.23), for the Dirac and the Weyl field!

2. For which of these fields (considered as classical!) does the dominant energy condition hold—perhaps in the *weak* form where for every observer u the energy 4-current is to be non-spacelike and future-directed:

$$\mathcal{E}^i := T^{ik} u_k \quad \text{satisfies} \quad \mathcal{E}^i u_i > 0, \quad \mathcal{E}^i \mathcal{E}_i \geq 0. \tag{10.2.38}$$

Which are the fields having $\mathcal{E}^i \mathcal{E}_i = 0$?

3. Show that the weakly dominant energy condition is equivalent to the statement

$$T^{00} \geq |T^{ab}| \quad \text{for every inertial frame.} \tag{10.2.39}$$

4. Verify eq. (10.2.26) and the equality (10.2.28).

5. Verify eqs. (10.2.30).

6. Verify eqs. (10.2.36,37).

7. The vector S_i may be formed even in the case where $P_i P^i = 0$, since we avoided in definition (10.2.25) to divide by $(P_i P^i)^{1/2}$. If one tries to define, by analogy to eq. (10.2.20), a spin tensor via $S^{ik} P_k = 0$ even in the null case $P_i P^i = 0$, one finds as a necessary condition $J^{ik} P_k \propto P^i$.

   a. Show that this condition is translationally invariant and is equivalent to $S_i \propto P_i$. The factor of proportionality corresponds to the *helicity* λ (cf. eqs. (9.3.11), (9.4.37)).

   b. The lightlike vector P^i determines, according to eq. (8.4.23), a spinor π^A up to a phase factor $e^{i\varphi}$. Show that the condition $J^{ik} P_k \propto P^i$ requires one of the two principal spinors (cf. eq. (8.3.20)) of J^{ik} to be proportional to π^A. If α^A is the other one, then the pair (P^j, J^{ik}) determines the pair $(\alpha^A, \pi^*_{\dot{X}})$ uniquely up to the phase factor $e^{-i\varphi}$ ('*twistor*').

   c. What is the translation behavior (up to a phase factor) that results for the pair $(\alpha^A, \pi^*_{\dot{X}})$?

   d. Show that $\alpha^A \pi_A - \alpha^{*\dot{X}} \pi^*_{\dot{X}}$ is also translation invariant! How does it relate to λ?

8. Show that the assignment

$$(a, L) \to \left[1 + \frac{1}{2} a_i \gamma^i (1 - \gamma^5)\right] S(L) \qquad (10.2.40)$$

mentioned in the text—where $S(L)$ is as in eq. (9.1.21)—furnishes a 4-dimensional reducible (but not fully reducible) representation of $\mathcal{P}_+^\uparrow$, under which from the $\mathcal{L}_+^\uparrow$-invariant forms (9.1.33,34,38,39) only the form (9.1.39) remains invariant. Show the equivalence of this representation to the one that results from c of the previous exercise, and give the relation between the invariant in d of that exercise and eq. (9.1.39).

Remark: We stress that in contradistinction to this '*twistor representation*' of $\mathcal{P}_+^\uparrow$ the bispinor representation of $\mathcal{P}_+^\uparrow$ is given by $(a, L) \mapsto S(L)$. Together with the 10-dimensional adjoint representation and the 5-dimensional representation (9.3.6) we have here a further finite-dimensional example in which the translations are represented nontrivially.

Remark: For the exercises 7 and 8 cf. Penrose and Rindler (1986) and the references therein.

10.3 Relativistic Hydrodynamics

Relativistic hydrodynamics was, for a long time, a branch of Relativity Theory which seemed to be particularly far from any application. Where does one encounter fluids or gases that would stream with velocities comparable to the speed of light? The numerous theoretical investigations in this subject thus only served conceptual clarifications, involving some unexpected problems, to be mentioned later.

This situation has changed drastically now. Relativistic hydrodynamics forms an important part of cosmology as well as of the theory of processes going on in the neighborhood of neutron stars and Black Holes. The fluid flows under the influence of the strong gravitational forces prevailing there reach relativistic speeds, leading to enormous heating and X-ray emission. These concrete applications are, however, outside the framework of special-relativistic hydrodynamics and cannot be treated here.

N.R. hydrodynamics of ideal fluids and gases is governed by the continuity and Euler equation

$$\frac{\partial \rho}{\partial t} + \operatorname{div}(\rho \mathbf{v}) = 0 \qquad (10.3.1)$$

$$\rho \frac{d\mathbf{v}}{dt} + \operatorname{grad} p = \mathbf{f}, \qquad (10.3.2)$$

where ρ and p are the *mass density* and *pressure* of the fluid, respectively, and $\mathbf{f}$ is an external volume force. To close the system, these equations still have to be supplemented by the *equation of state*

$$p = p(\rho). \qquad (10.3.3)$$

It is wrong to conjecture that eq. (10.3.1) can be brought to a covariant form by defining a 4-vector of mass current $j^i = \rho(x) u^i$ and rewriting it as $j^i{}_{,i} = 0$: it is a characteristic feature of relativity that the mass density $\rho(x)$ does not satisfy such law—in fact, we shall obtain a modified form of this law.

To arrive at the correct equations we proceed by analogy to sect. 5.9 where we found the physical interpretation for the components of the stress-energy-momentum tensor of the electromagnetic field. Here we start from this interpretation and write for the *stress-energy-momentum tensor* of an *ideal fluid*

$$
T_{ik} = \begin{pmatrix} \rho & 0 & 0 & 0 \\ 0 & p & 0 & 0 \\ 0 & 0 & p & 0 \\ 0 & 0 & 0 & p \end{pmatrix}
\tag{10.3.4}
$$

in the rest system of a fluid element at x: ideal fluids are characterized by the feature that their stress tensor $T_{\alpha\beta}$ contains no shear stresses and is thus proportional to $\delta_{\alpha\beta}$.

The generalization of expression (10.3.4) to an arbitrary frame, in which the fluid element moves with 4-velocity components u^i, is then obviously

$$
T_{ik} = (\rho + p)\, u_i u_k - \eta_{ik}\, p.
\tag{10.3.5}
$$

Notice that again $\rho(x)$ and $p(x)$ denote energy density and pressure as measured in the *rest* system of the fluid element.

The equations of motion result, in the absence of external volume forces, from the conservation laws $T^{ik}{}_{,k} = 0$ as

$$
T^{ik}{}_{,k} = [(\rho + p)\, u^i u^k]_{,k} - p^{,i} = 0.
\tag{10.3.6}
$$

To compare them with their N.R. approximations above, we first multiply eq. (10.3.6) by u_i; a short calculation, using $u^i u_i = 1$, $u^i{}_{,k}\, u_i = 0$, gives

$$
(\rho u^k)_{,k} + p\, u^k{}_{,k} = 0.
\tag{10.3.7}
$$

This shows indeed that the mass current ρu^k is not conserved. Before entering into the significance of this result, we first consider the space part of eq. (10.3.6):

$$
[(\rho + p)\, \mathbf{u}\, u^k]_{,k} + \operatorname{grad} p = 0.
\tag{10.3.8}
$$

If we define the *convective* or *comoving derivative* of an arbitrary tensor field T as

$$
\dot{T} = T_{,k}\, u^k,
\tag{10.3.9}
$$

we can rewrite eq. (10.3.8), using eq. (10.3.7), as

$$
\dot{p}\, \mathbf{u} + (\rho + p)\, \dot{\mathbf{u}} + \operatorname{grad} p = 0.
\tag{10.3.10}
$$

This is the relativistic version of the *Euler equation of hydrodynamics*, as can be seen immediately by going to the comoving frame $u^i = (1, \mathbf{0})^\top$. Its main difference to the N.R. version is the addition of the pressure (p/c^2 in usual units!) to the mass-energy density in the inertial term.

For electromagnetic radiation we have $p = \rho/3$, since the tracelessness of the electromagnetic stress-energy-momentum tensor survives averaging. This introduces a factor $4/3$ which is analogous to the factors $4/3$ that occurred in our investigations on charged particles, and it supplements our earlier historical remarks insofar as the factor here corresponds exactly to the calculations of F. Hasenöhrl from 1904 (cf. sect. 5.10).

We now return to eq. (10.3.7). The time component of eq. (10.3.6) has not turned out to be the relativistic version of eq. (10.3.1); the latter rather has to be postulated separately, using additional insights from elementary particle physics. In contradistinction to mass density $\rho(x)$, the *baryon density*[1] $n(x)$ does satisfy an *equation of continuity*

$$(n\,u^k)_{,k} = 0, \tag{10.3.11}$$

expressing the conservation of baryon number (conservation in time, independence of inertial frame). Here $n(x)$ is defined such that for some 'normalizing state'—e.g., rarefied hydrogen gas—one has $n = \rho$.

For an electron gas the baryon density has to be replaced by the *lepton density* in the equation of continuity. For photons and mesons there is *no* continuity equation since they can be created and annihilated arbitrarily.

The relation between n, p and ρ follows from the equation of state (10.3.3) and the thermodynamical definition of pressure

$$p = -\frac{d\,(\text{energy per baryon})}{d\,(\text{volume per baryon})} = -\frac{d\,(\rho/n)}{d\,(1/n)} = n\frac{d\rho}{dn} - \rho \tag{10.3.12}$$

or

$$\int \frac{d\rho}{p(\rho) + \rho} = \int \frac{dn}{n}. \tag{10.3.13}$$

With this, also $n(\rho)$ is known.

Mass density ρ and baryon density n differ by the density $n\varepsilon$ of inner energy ($\varepsilon = $ *specific inner energy* $=$ inner energy per baryon):

$$\rho = n(1 + \varepsilon). \tag{10.3.14}$$

The inner energy is negative if energy is released at the formation of the state ρ (e.g., binding energy of nuclei), and is positive if energy has to be spent (e.g., compressional work).

Specific *entropy* s ($=$entropy per baryon) and *temperature* T are defined by postulating $1/T$ to be an integrating factor for the equation

$$ds = \frac{1}{T}\left(d\varepsilon + p\,d\left(\frac{1}{n}\right)\right), \tag{10.3.15}$$

since $v = 1/n$ is the specific volume. The constancy of entropy along a stream line of an ideal fluid follows directly from eq. (10.3.7):

$$\dot{p} = [(\rho + p)\,u^k]_{,k} = [(n + \varepsilon n + p)\,u^k]_{,k} = n\dot{\varepsilon} + p\,u^k{}_{,k} + \dot{p}. \tag{10.3.16}$$

---

[1]See textbooks on particle physics for the concept of baryon and lepton number.

Dividing by n yields, using eq. (10.3.11),

$$T\dot{s} = \dot{\varepsilon} + p\left(\frac{1}{n}\right) = 0. \qquad (10.3.17)$$

The time component of the conservation law (10.3.6) thus tells us that in the case of an ideal fluid no energy is converted into heat, entropy remaining constant.

For *nonideal fluids* the ansatz (10.3.5) must be generalized as

$$T_{ik} = (\rho + p)\,u_i u_k + (q_i u_k + q_k u_i) - \eta_{ik} p - \pi_{ik}. \qquad (10.3.18)$$

Here q_i describes an energy flow relative to u_i, and π^{ik} is the anisotropy in pressure. (Cf. G. F. R. Ellis in Sachs (1971), where the general-relativistic version is found together with cosmological applications; the relativistic theory of viscous fluids is found in Weinberg (1972).)

The approach to the equations of relativistic hydrodynamics chosen here has the advantage of being simple and immediate. It is also possible to derive the Euler equations from a *variational principle*—see, e.g., Yourgrau and Mandelstam (1968). Relativistic hydrodynamics can also be generalized to *charged* fluids and plasmas (relativistic magneto-hydrodynamics): one adds the energy-momentum tensor (5.9.12) of the electromagnetic field to the tensor (10.3.5) of the fluid and sets the divergence of the sum equal to zero, obtaining a Lorentz force term in the Euler equations.

In the phenomenological approach above we automatically also obtained some equations belonging to *thermodynamics*. We thus could think of basing our considerations on some relativistic kinetic theory or some relativistic statistical mechanics. We make a few comments on these.

In the older literature on relativistic thermodynamics one finds mainly discussions of the behavior of thermodynamic quantities under change of the reference frame: what will the temperature, the entropy, etc., be as measured by a moving observer? The first considerations stem from Einstein and Planck (M. Planck, Berl. Ber. 1907, p. 152; Ann. Phys. (Leipzig) *26*, 1 (1908); A. Einstein, Jahrb. Radioakt. Elektron. *4*, 411 (1907)). They obtained the result that entropy is relativistically invariant, while the temperature has to be transformed as $T = T_0\sqrt{1 - v^2}$. Later it was mainly R. C. Tolman (1934) who took up these ideas, creating a general-relativistic thermodynamics. This seemed to settle the matter, until H. Ott (Z. Phys. *175*, 70 (1963)) derived a transformation law for the temperature which deviated from the Einstein-Planck one: $T = T_0/\sqrt{1 - v^2}$. Subsequently a number of other papers dedicated to this problem appeared, e.g., D. ter Haar, H. Wergeland, Phys. Rep. C *1*, 31 (1971); P. T. Landsberg in: Conn and Fowler (1970); O. Grøn, Nuovo Cimento B *17*, 141 (1973); D. Eimerl, Ann. Phys. (N.Y.) *91*, 481 (1975); G. Horwitz, J. Katz, Ann. Phys. (N.Y.) *76*, 301 (1973).

It seems, however, that the discussion was on a pseudo-problem. For instance, the difference between Einstein-Planck and Ott comes about because addition of heat also means addition of energy, whereby the mass of the thermodynamic system is increased. One may now postulate the heat transfer to be at constant velocity or at constant momentum of the system: depending on the choice, one gets one or the other of the transformation laws. The discussions are characterized by the absence of hints as to how the different behaviors for the temperature could be distinguished experimentally. How does one measure the temperature of a moving system? Imagine a cavity filled with radiation: if it moves relative to us, we see a loss of isotropy, due to aberration, and the Doppler effect will change the wavelength in a direction-dependent way. The result thus is not an isotropic radiation with a definite transformation rule for its temperature. (An effect of this kind is, e.g., seen in the 3 K cosmic background radiation due to the motion of the earth; see G. Smoot et al., Phys. Rev. Lett. *39*, 898 (1977); Astrophys. J. *371*, L1, (1991).)

Since the measurement of thermodynamic quantities presupposes thermodynamic equilibrium, it makes more sense to formulate the laws of thermodynamics in the rest system, regarding them as form-invariant. There is a special case, however, that needs an extra treatment. Landau and Lifshitz (1958b) showed that the equilibrium conditions of statistical mechanics can be satisfied only if the total system is in constant translational motion or in a state of rigid rotation about a fixed axis. Now while the treatment of systems in uniform rectilinear relative motion with scalar thermodynamic laws is easy, the thermodynamics of rotating systems requires a more detailed consideration. The issue here is the relation between global quantities (total energy, total entropy, ...) and local ones (pressure, density, temperature, ...). This problem was treated by Horwitz and Katz, loc. cit. They show that equilibrium requires $T_G = T_L\sqrt{1 - v^2}$, where T_G and T_L are the global and the local temperature, respectively, and v is the velocity of the volume element considered relative to the axis of rotation. Thus a rotationg body has, in a state of thermodynamic equilibrium, not a constant but a spatially variable local temperature.

The relativistic formulation of *statistical mechanics* of noninteracting particles was first given by F. Jüttner (Ann. Phys. (Leipzig) *34*, 856 (1911)); it offers no basic difficulties. Theory and applications are found, e.g., in Huang (1963) or Landau and Lifshitz (1958b). In recent years, fields of application for relativistic thermodynamics have been the early universe (cf. E. R. Harrison, Annu. Rev. Astron. Astrophys. (1973)) and the theory of neutron stars and collapsed objects (cf. V. Canuto, Annu. Rev. Astron. Astrophys. (1974)).

The problem of statistical mechanics of *interacting* particles has been treated in recent decades from two points of view. On the one hand, the *Boltzmann equation* has been written within the framework of a relativistic kinetic theory of interacting particles (cf. J. Ehlers in Sachs (1971) and Stewart (1971)). On the other hand, mainly Balescu (J. Phys. Soc. Japan *26*, Suppl. 313-315; article in Stuart and Brainard (1970)) has tried to attack the problem of a genuine statistical mechanics of interacting relativistic particles directly. The difficulty comes from the 'no-interaction theorems' mentioned in sect. 5.1 that seem to exclude the description of interaction by retarded action at a distance. If, as an alternative, one describes the interaction between particles as mediated by fields[1] (as in relativistic electrodynamics), problems caused by the infinite number of degrees of freedom of dynamical fields enter the stage, posing delicate mathematical questions concerning the concept of phase space and a measure on it, etc., and extreme care is necessary here when it comes to make reliable predictions about phenomena like phase transitions.

---

[1]See, e.g., M. LeBellac (1996), A. Das (1997).

Appendix A
Basic Concepts from Group Theory

For the convenience of the reader, we include here the basic definitions, concepts, and facts from group theory. This is done in a quite dogmatic fashion—it is hoped that the main text provides ample motivation to study and use these concepts. Only algebraic features are presented.

A.1 Definition of Groups

A nonempty set $\mathcal{G}$ is called a *group* if there is given a composition, or *multiplication*, rule assigning to every pair $(g, h) \in \mathcal{G} \times \mathcal{G}$ an element $gh \in \mathcal{G}$ (called the *product* of g and h) such that the following hold:

1. *associativity*: $(g_1 g_2) g_3 = g_1 (g_2 g_3) =: g_1 g_2 g_3$ for every triple $(g_1, g_2, g_3) \in$
   $\in \mathcal{G} \times \mathcal{G} \times \mathcal{G}$;

2. *unit element*: there exists an element $e \in \mathcal{G}$ such that $eg = g$ for all $g \in \mathcal{G}$;

3. *inverse element*: for every $g \in \mathcal{G}$ there exists an inverse g^{-1} such that $g^{-1} g = e$.

   The group is called *commutative* or *Abelian* iff in addition to the above one has

4. $gh = hg$ for all $(g, h) \in \mathcal{G} \times \mathcal{G}$.

It follows from 1, 2, 3 that also $ge = g$, $gg^{-1} = e$ for all $g \in \mathcal{G}$, and that e, g^{-1} are uniquely determined by the above properties. One further has

$$(g_1 g_2 g_3 \ldots)^{-1} = \ldots g_3^{-1} g_2^{-1} g_1^{-1}.$$

A.2 Subgroups and Factor Groups

A nonempty subset $\mathcal{G}_1 \subset \mathcal{G}$ is called a *subgroup* of $\mathcal{G}$ if it forms a group with respect to the multiplication given in $\mathcal{G}$; it suffices for this that for all $(g, h) \in \mathcal{G}_1 \times \mathcal{G}_1$ one has $gh^{-1} \in \mathcal{G}_1$. The intersection of two subgroups is a subgroup. The union of two subgroups is not—but just as any nonempty subset $\mathcal{M} \subset \mathcal{G}$, it *generates* a subgroup: the subgroup generated by $\mathcal{M}$ is the smallest subgroup $\prec \mathcal{M} \succ$ containing it, which is the same as the intersection of all subgroups containing it.

In what follows we shall write $g\mathcal{M}$ and $\mathcal{M}g$ (where $g \in \mathcal{G}$) for the set of all products gm and mg, respectively, when m runs through the subset $\mathcal{M}$. Similarly, for two nonempty subsets $\mathcal{M}, \mathcal{N}$ we shall mean by the products $\mathcal{M}\mathcal{N}$ and $\mathcal{N}\mathcal{M}$ the subsets formed by all products mn and nm, respectively, where (m, n) runs through $\mathcal{M} \times \mathcal{N}$.

If $\mathcal{G}_1 \subset \mathcal{G}$ is a subgroup, subsets of the form $g\mathcal{G}_1$ and $\mathcal{G}_1 g$ are called left and right *cosets* of $\mathcal{G}_1$, respectively. If $g \in \mathcal{G}_1$, then $g\mathcal{G}_1 = \mathcal{G}_1$. Two different (say, left) cosets are disjoint, and the whole group $\mathcal{G}$ is a disjoint union

$$\mathcal{G} = \mathcal{G}_1 \cup g\mathcal{G}_1 \cup h\mathcal{G}_1 \cup \ldots$$

or

$$\mathcal{G} = \mathcal{G}_1 \cup \mathcal{G}_1 g \cup \mathcal{G}_1 h \cup \ldots$$

of left, or also of right, cosets. For a given subgroup $\mathcal{G}_1$, the set of all (say, left) cosets is written $\mathcal{G}/\mathcal{G}_1$. Since $g_1 \mapsto gg_1$ is a bijective assignment, all cosets have the same cardinality; it follows that for a finite group the cardinality ($=order$) of a subgroup divides the order of the group. The cardinality of $\mathcal{G}/\mathcal{G}_1$—which may be finite even in infinite groups—is called the *index* of $\mathcal{G}_1$ in $\mathcal{G}$.

If for a subgroup $\mathcal{G}_1 \subset \mathcal{G}$ each right coset $\mathcal{G}_1 g$ coincides with the corresponding left coset $g\mathcal{G}_1$, it is called a *normal*, or *invariant*, subgroup ('invariance' referring to inner automorphisms, see below). Every subgroup of index 2 is an invariant subgroup. In an Abelian group, all subgroups are invariant. The intersection of two invariant subgroups is invariant.

An element commuting with all group elements is called *central* in $\mathcal{G}$. The set $\mathcal{Z}(\mathcal{G})$ of all central elements is called the *center* of $\mathcal{G}$; it is a special Abelian invariant subgroup. A central subset is one consisting of central elements.

With the multiplication of subsets introduced above, the (left=right) cosets with respect to an invariant subgroup $\mathcal{G}_1$ of $\mathcal{G}$ form a group, called the *factor*, or *quotient*, group $\mathcal{G}/\mathcal{G}_1$.

A.3 Homomorphisms, Extensions, Products

A map φ of a group $\mathcal{G}$ *into* another group $\mathcal{G}'$ is called a *homomorphism* if the image of the product of two elements equals the product (in $\mathcal{G}'$) of the images. When the map is surjective (*onto*), the homomorphism is called epimorphism; if it is injective (one-to-one) it is called monomorphism; if it is bijective (one-to-one and onto) it is called *isomorphism* and the groups are called then *isomorphic*, in symbols: $\mathcal{G} \cong \mathcal{G}'$. Isomorphisms of a group onto itself are called *automorphisms*; they form a group $Aut(\mathcal{G})$ under composition of maps. *Inner* automorphisms are given by the operation of *conjugating* with group elements $h \in \mathcal{G}$, consisting in the assignment $g \mapsto hgh^{-1}$; they are trivial iff h is central. Normal (=invariant) subgroups are invariant under all inner automorphisms—hence the name. The inner automorphisms form an invariant subgroup $Int(\mathcal{G}) \cong \mathcal{G}/\mathcal{Z}(\mathcal{G})$ of $Aut(\mathcal{G})$.

In general, a homomorphism $\varphi\colon \mathcal{G} \to \mathcal{G}'$ will not be injective, i.e., several elements in $\mathcal{G}$ will have the same image in $\mathcal{G}'$. The set $\ker\varphi$ of elements in $\mathcal{G}$ that get mapped to the unit element in $\mathcal{G}'$ is called the *kernel* of φ; it is an invariant subgroup, whose cosets are collections of elements with the same image. The image of $\mathcal{G}$ under φ, written $\varphi(\mathcal{G}) =: \mathrm{im}\varphi \subset \mathcal{G}'$, is then *isomorphic* to the factor group $\mathcal{G}/\ker\varphi$, to which $\mathcal{G}$ is mapped by the *sur*jective homomorphism $g \mapsto g\ker\varphi$ (called the *canonical projection* to the factor group).

If $\mathcal{G}_1$ is an invariant subgroup and $\mathcal{G}_2$ is a subgroup of $\mathcal{G}$, then $\mathcal{G}_1\mathcal{G}_2 = \mathcal{G}_2\mathcal{G}_1$ is a subgroup of $\mathcal{G}$; when the canonical projection $\mathcal{G} \to \mathcal{G}/\mathcal{G}_1$ is restricted to $\mathcal{G}_2$, then the kernel and image of that restriction are $\mathcal{G}_1 \cap \mathcal{G}_2$ and $\mathcal{G}_1\mathcal{G}_2/\mathcal{G}_1$, respectively, so that $\mathcal{G}_2/\mathcal{G}_1 \cap \mathcal{G}_2 \cong \mathcal{G}_1\mathcal{G}_2/\mathcal{G}_1$ are isomorphic. If here $\mathcal{G}_1 \cap \mathcal{G}_2 = \{e\}$ and $\mathcal{G}_1\mathcal{G}_2 = \mathcal{G}$, it follows that $\mathcal{G}/\mathcal{G}_1 \cong \mathcal{G}_2$, and every $g \in \mathcal{G}$ has a unique decomposition $g = g_1 g_2$,

$g_i \in \mathcal{G}_i$. Conversely, if to $\mathcal{G}/\mathcal{G}_1 =: \mathcal{G}_0$ there is an isomorphic subgroup $\mathcal{G}_2 \subset \mathcal{G}$ that canonically projects onto $\mathcal{G}_0$, we are in the situation just described, which is referred to by saying that $\mathcal{G}$ has the structure of an *(internal) semidirect product* of the invariant subgroup $\mathcal{G}_1$ and the subgroup $\mathcal{G}_2$. The still more special case where the elements of the two subgroups commute (so that both subgroups are invariant) is referred to as an *(internal) direct product* structure.

A group $\mathcal{G}$ is called an *extension* of a group $\mathcal{G}_0$ *by* a group $\mathcal{G}_1$ (or of $\mathcal{G}_1$ by $\mathcal{G}_0$, according to some authors) if it contains an invariant subgroup isomorphic to $\mathcal{G}_1$ such that the factor group is isomorphic to $\mathcal{G}_0$. (These data do not fix the extension; on the other hand, data that would allow to make the extension unique in general do not allow for the existence of the extension[1]). Note that the term 'extension' should not make us assume that $\mathcal{G}$ contains a subgroup $\mathcal{G}_2$ isomorphic to $\mathcal{G}_0$ under the projection. If the latter is indeed the case, the extension is called *inessential*, and we are just in the situation of a semidirect product. Another important special case obtains when the invariant subgroup of $\mathcal{G}$ that corresponds to $\mathcal{G}_1$ (now to be assumed Abelian) is central: the extension is then called *central*. For more on the latter, see sec. 7.10.

A group $\mathcal{G}$ is called *simple* if it is nonabelian and contains no nontrivial invariant subgroup. Homomorphisms of simple groups are trivial or injective, thus representations are trivial or faithful.

The *external direct product* of two groups was defined in exercise 6 of sect. 3.1, and its relation to the internal direct product structure was given in exercise 4 of sect. 7.9. Similarly, one can define an *external semidirect product* $\mathcal{G}_1 \times_\Sigma \mathcal{G}_2$ of two groups $\mathcal{G}_1$, $\mathcal{G}_2$ with respect to a homomorphism Σ of $\mathcal{G}_2$ into the group $Aut(\mathcal{G}_1)$. Here the product *set* $\mathcal{G}_1 \times \mathcal{G}_2$ is made into a group by the multiplication rule

$$(g_1, g_2) \circledS (h_1, h_2) := (g_1 \Sigma_{g_2} h_1, g_2 h_2).$$

In it, $\mathcal{G}_1$, $\mathcal{G}_2$ are imbedded isomorphically by $g_1 \mapsto (g_1, e_2)$, $g_2 \mapsto (e_1, g_2)$, $\mathcal{G}_1$ being isomorphic to an invariant subgroup and $\mathcal{G}_2$ being isomorphic to the factor group; the action of the automorphism Σ_{g_2} on g_1 corresponds to conjugating (g_1, e_2) with (e_1, g_2). We thus have the structure of an internal semidirect product. The special case where $\Sigma_{g_2} = \mathrm{id}_{\mathcal{G}_1}$ for all $g_2 \in \mathcal{G}_2$ leads back to the direct product.

To the semidirect product of two Lie groups corresponds a notion of *semidirect sum of Lie algebras* $\mathbf{L}_1$, $\mathbf{L}_2$, which is abstractly defined as follows. A *derivation* D in a (Lie) algebra $\mathbf{L}$ is a linear map $D : \mathbf{L} \to \mathbf{L}$ which satisfies, with respect to the product $\circ$, the *Leibniz rule*

$$D(A \circ B) = D(A) \circ B + A \circ D(B).$$

The derivations of $\mathbf{L}$ form a Lie algebra $Der(\mathbf{L})$ under the ordinary commutator $[D, D']$. Given now a Lie algebra homomorphism $\sigma : \mathbf{L}_2 \to Der(\mathbf{L}_1)$, one can define a Lie algebra structure on the direct sum of vector spaces $\mathbf{L}_1 \oplus \mathbf{L}_2$ by putting

$$(A_1, A_2) \circ (B_1, B_2) = (A_1 \circ B_1 + \sigma_{A_2}(B_1) - \sigma_{B_2}(A_1), A_2 \circ B_2).$$

The special case where $\sigma_{A_2} = 0$ for all $A_2 \in \mathbf{L}_2$ is just the direct sum as introduced in sect. 8.1.

---

[1] Cf. Kirillov (1976), but beware of some misprints!

A.4 Transformation Groups

One says that a group $\mathcal{G}$ *acts*, or *operates*, as a transformation group on a set $\mathbf{M}$, or is *realized* as a transformation group, and $\mathbf{M}$ is called a $\mathcal{G}$-space, if to every $g \in \mathcal{G}$ and to every $m \in M$ there is assigned a transformed element $\rho(g, m) \equiv \rho_g(m) \in M$, such that in the case of a *left* or a *right* action one has $\rho_{gh}(m) = \rho_g(\rho_h(m))$ and $= \rho_h(\rho_g(m))$, respectively, as well as $\rho_e(m) = m$ for all $m \in M$. The latter implies (and is implied by) the fact that the ρ_g are self-bijections (permutations) of $\mathbf{M}$, and we have a homomorphism of $\mathcal{G}$ into the group of all permutations of $\mathbf{M}$. One says that $\mathcal{G}$ acts *effectively*, or *freely*, on $\mathbf{M}$ if the statement "$\rho_g(m) = m$ for all, or some, m" implies $g = e$. The action is called *transitive* if for every pair m, m' there is a $g \in \mathcal{G}$ such that $\rho_g(m) = m'$. In this case, $\mathbf{M}$ is called a *homogeneous* $\mathcal{G}$-space; if any $m_0 \in \mathbf{M}$ is selected as an 'origin', there is a bijection between this 'pointed' homogeneous $\mathcal{G}$-space and the coset space $\mathcal{G}/\mathcal{G}_0$, where $\mathcal{G}_0$ is the subgroup of elements that leave m_0 fixed—its *isotropy*, or *stable*, subgroup. (These are left or right cosets when the action is from the left or right; different choices for the origin lead to conjugate isotropy subgroups.) If the action is transitive as well as free—also called *simply transitive*—the space $\mathbf{M}$ is in bijection with the group, and is sometimes called a *principal* $\mathcal{G}$-space.

Realizations on a vector space by linear or semilinear (see Appendix B.1) operators are called (linear) *representations* or semilinear representations. A projective, or ray, representation is a realization on a projective space $\mathbf{M} = \mathrm{P}(\mathbf{V})$ (the set of 1-dimensional subspaces of a vector space $\mathbf{V}$) by projective transformations (induced on the projective space by semilinear transformations of the vector space). See sects. 7.10 and 9.2 for more on ray representations.

Appendix B
Abstract Multilinear Algebra

Linear and *multilinear algebra* is one of the most elementary as well as most often used branches of mathematics, and there are numerous texts presenting it in its abstract—i.e., basis-independent—form, originally invented for the purpose of treating infinite-dimensional spaces, but then also found to be useful in finite dimension. Its presently most general form (modules over noncommutative rings) is found, e.g., in Bourbaki (1970), and almost everything found there is 'in immediate danger of being applied' in fundamental theoretical physics of the day. We bring only a few portions of it, adapted to our purposes, abstract versions as a background to the component versions used in the main text as well as to the appendices to follow.

We will consider *vector spaces* (=linear spaces) over a commutative field $\mathbf{F}$—the field of *scalars*—which will always be specialized to be $\mathbf{R}$ or $\mathbf{C}$. We shall assume $\mathbf{V}$ to be *finite-dimensional* in order to avoid an even higher degree of abstraction in the definition of the tensor product, as well as to escape the necessity of additional concepts from functional analysis. In a trivial way, $\mathbf{F}$ is a 1-dimensional vector space over $\mathbf{F}$, but it is also a vector space over each of its subfields. Extension fields of $\mathbf{F}$ are vector spaces over $\mathbf{F}$; in concrete terms: $\mathbf{C}$ is 2-dimensional over $\mathbf{R}$ but 1-dimensional over $\mathbf{C}$. If for a vector space over $\mathbf{F}$ the field of scalars is restricted to a subfield, the dimension is increased accordingly: in particular, when a complex vector space is looked at as a vector space over the reals—i.e., when we look at its *realification*—the dimension over the reals is twice the complex dimension.

We should like to point out that the conceptual distinction between an abstract vector space $\mathbf{V}$ over $\mathbf{F}$ and the 'numerical' vector space $\mathbf{F}^n$ (where $n = \dim_{\mathbf{F}} \mathbf{V}$), to which it is isomorphic in many ways corresponding to choices of bases, is physically relevant not only in Relativity, where we set out for frame-independent concepts. The simplest case of a 1-dimensional vector space over $\mathbf{R}$ should make this clear: after choosing a zero point for the time axis, the latter is a 1-dimensional vector space, and choosing the basis vector 'second' gives a definite isomorphism to $\mathbf{R}$; its dual space of frequencies is also 1-dimensional, and the standard isomorphism to $\mathbf{R}$ is obtained by taking the dual basis 'Hertz'; ...

B.1 Semilinear Maps

Let $\mathbf{V}$, $\mathbf{W}$ vector spaces over $\mathbf{F}$. A map $A\colon \mathbf{V} \to \mathbf{W}$ is called $\mathbf{F}$-*semilinear* if it is additive, $A(v + w) = A(v) + A(w)$, and maps rays (=1-dimensional subspaces) to rays: $A(\alpha v) = \alpha' A(v)$ for all $v, w \in \mathbf{V}$, $\alpha \in \mathbf{F}$. It follows that the assignment $\alpha \mapsto \alpha'$ is an automorphism σ of $\mathbf{F}$ associated with A. For $\mathbf{F} = \mathbf{R}$ there is only the trivial automorphism $\sigma = \mathrm{id}$. Generally, if $\sigma = \mathrm{id}$, A is called *linear*. For $\mathbf{F} = \mathbf{C}$ there is the additional possibility[1] of σ being complex conjugation, in which case A is called

---

[1] According to some rumor, there is an indenumerable set of other automorphisms for $\mathbf{C}$ whose existence has been proved without constructing one. They are all discontinuous and do not preserve the subfield $\mathbf{R}$, and we shall disregard them.

antilinear. When $\mathbf{W} = \mathbf{F}$, a semilinear map is also called a semilinear *functional*, or *form*, on $\mathbf{V}$. Images and inverse images of linear subspaces are linear subspaces; in particular, $\operatorname{im} A = A(\mathbf{V}) \subset \mathbf{W}$ and $\ker A = \{v \in \mathbf{V} : Av = 0\}$ are subspaces.

When $\mathbf{W} = \mathbf{V}$, the invertible semilinear maps (=transformations, operators) form a group under composition, the semilinear group of the space $\mathbf{V}$; the linear ones forming a subgroup of index 2. (Note that the associated automorphisms also form a group and that, in particular, linear $\circ$ linear = antilinear $\circ$ antilinear = linear, linear $\circ$ antilinear = antilinear $\circ$ linear = antilinear.)

The semilinear maps $\mathbf{V} \to \mathbf{W}$ associated with a *fixed* automorphism σ may also be added and multiplied with scalars, defining $\alpha A + B$ by $(\alpha A + B)v = \alpha Av + Bv$. They thus form a vector space over $\mathbf{F}$ which in the linear case will be written $L(\mathbf{V}, \mathbf{W})$.

B.2 Dual Space[1]

The vector space $\widetilde{\mathbf{V}}$ over $\mathbf{F}$ formed by the $\mathbf{F}$-linear functionals is called the *dual space* of $\mathbf{V}$, and its elements are called *covectors* or covariant vectors. (The elements of the original space are then called contravariant.) If $\{b_i\}$ is a basis in $\mathbf{V}$, then every vector $v \in \mathbf{V}$ has a unique decomposition $v = v^i b_i$, and the linear functionals $\tilde{b}^i \colon v \mapsto v^i$ form the *cobasis* $\{\tilde{b}^i\}$ in $\widetilde{\mathbf{V}}$ which is dual to $\{b_i\}$: each $a \in \widetilde{\mathbf{V}}$ may be written $a = a_i \tilde{b}^i$ with $a_i = a(b_i)$, as can be seen by applying the functional to $v = v^i b_i$.

Each $v \in \mathbf{V}$ defines on $\widetilde{\mathbf{V}}$ a linear functional $\tilde{v}$ by $\tilde{v}(a) := a(v) \ \forall a \in \widetilde{\mathbf{V}}$. This imbeds $\mathbf{V}$ linearly in a natural way into its double dual $\widetilde{\widetilde{\mathbf{V}}}$ (i.e., in a way that does not require any new structure). In finite dimension, $\mathbf{V}$ and $\widetilde{\widetilde{\mathbf{V}}}$ may be identified this way, while such would not be possible without a further structure (e.g., an inner product) among $\mathbf{V}$, $\widetilde{\mathbf{V}}$. The bilinear map $\mathbf{V} \times \widetilde{\mathbf{V}} \to \mathbf{F}$ that to the pair (v, a) assigns the value $a(v) =: (a \,|\, v) =: (v \,|\, a)$ is called the canonical *inner* or *scalar product between* $\mathbf{V}$ and $\widetilde{\mathbf{V}}$.

B.3 Complex-Conjugate Space[1]

The vector space $\widetilde{\mathbf{V}}^*$ formed by the *anti*linear functionals on $\mathbf{V}$ is called the *complex conjugate dual* space. To each covector $a \in \widetilde{\mathbf{V}}$ there is the *complex-conjugate* covector $a^* \in \widetilde{\mathbf{V}}^*$, which is given by $a^*(v) := (a \,|\, v)^* \ \forall v \in \mathbf{V}$. This gives an antilinear map $\widetilde{\mathbf{V}} \to \widetilde{\mathbf{V}}^*$ called complex conjugation.

Similarly one forms the space of antilinear functionals on $\widetilde{\mathbf{V}}$; the *complex conjugation* $\mathcal{K}$ antilinearly imbeds $\mathbf{V}$ into it, sending $\mathbf{V} \ni v \mapsto v^*$, where the latter functional is defined by $v^*(a) = (a \,|\, v)^* \ \forall a \in \widetilde{\mathbf{V}}$. The image $\mathcal{K}(\mathbf{V}) = \mathbf{V}^*$ is called the vector space *complex-conjugate* to $\mathbf{V}$. (It is canonically isomorphic with $\mathbf{V}$ equipped with the new multiplication with scalars given by $\alpha \circ v := \alpha^* v$, where on the right one has the original multiplication; however, sometimes it is more convenient to keep the two apart even as sets.) In the same vein there are natural isomorphisms $\mathbf{V} \cong \mathbf{V}^{**}$, $\widetilde{\mathbf{V}}^* \cong \widetilde{\mathbf{V}}^*$, ... , and in this sense we have $(v^*)^* = v$, $(a^*)^* = a$ for $v \in \mathbf{V}$, $a \in \widetilde{\mathbf{V}}$.

---

[1] In today's mathematics, the symbols commonly used for the dual and the complex conjugate space are $\mathbf{V}^*$, or $\mathbf{V}'$, and $\overline{\mathbf{V}}$, respectively.

To each basis $\{b_i\}$ in $\mathbf{V}$ there is the dual basis $\{\tilde{b}^i\}$ in $\widetilde{\mathbf{V}}$, the complex-conjugate basis $\{b_i^*\}$ in $\mathbf{V}^*$ and the complex-conjugate dual basis $\{\tilde{b}^{*i}\}$ in $\widetilde{\mathbf{V}}^*$. Note again that one can dualize a basis only as a whole, while complex conjugation can be applied to each vector individually.

To every antilinear map $A\colon \mathbf{V} \to \mathbf{W}$ there is the *linear* map $\mathcal{K}_{\mathbf{W}} \circ A$ from $\mathbf{V}$ to $\mathbf{W}^*$, and conversely. We stress again that for an abstract complex vector space, the operation of complex conjugation maps into *another* space! Complex conjugation by conjugating the numerical components would stay in the same space but is a basis-dependent concept—actually depending on a basis up to basis changes with real coefficients, called a real(ity) structure in $\mathbf{V}$ (see B.6).

B.4 Transposition, Complex and Hermitian Conjugation

A semilinear map $A\colon \mathbf{V} \to \mathbf{W}$ defines for every $b \in \widetilde{\mathbf{W}}$ a linear functional on $\mathbf{V}$ by $v \mapsto \sigma(b\,|\,Av)$; it will be written $A^\top b$. This defines the (semilinear) *transposed map* $A^\top\colon \widetilde{\mathbf{W}} \to \widetilde{\mathbf{V}}$. When A is invertible, the map $\tilde{A}:= (A^\top)^{-1} = (A^{-1})^\top$ is called the *contragredient* of A.

Similarly, the *Hermitian conjugate* map $A^\dagger\colon \widetilde{\mathbf{W}}^* \to \widetilde{\mathbf{V}}^*$ is defined by $(A^\dagger b\,|\,v):= \sigma(b\,|\,Av)$ for $b \in \widetilde{\mathbf{W}}^*$, $v \in \mathbf{V}$, and the complex-conjugate map $A^*\colon \mathbf{V}^* \to \mathbf{W}^*$ is defined by $(A^*v^*\,|\,b):= \sigma(v^*\,|\,A^\top b)$ for $v^* \in \mathbf{V}^*$, $b \in \widetilde{\mathbf{W}}$, i.e., $A^* = \mathcal{K}_{\mathbf{W}} \circ A \circ \mathcal{K}_{\mathbf{V}}^{-1}$. The operations $*$, $^\top$, † commute with each other (in the sense of natural isomorphisms between the target spaces as mentioned before), and we have $(A^\top)^\top = (A^*)^* = (A^\dagger)^\dagger = A$, $(A^*)^\top = A^\dagger$, ... $(Av)^* = A^*v^*$; the composition behavior is $(B \circ A)^* = B^* \circ A^*$, $(B \circ A)^\top = A^\top \circ B^\top$, $(B \circ A)^\dagger = A^\dagger \circ B^\dagger$.

If A is linear and a matrix (A_i^α) is assigned to it as usual by choosing a basis $\{b_i\}$ in $\mathbf{V}$, a basis $\{e_\mu\}$ in $\mathbf{W}$, and decomposing $Ab_i = A_i^\alpha e_\alpha$, then to the maps $A^\top$, A^*, $A^\dagger$ there belong the transposed, the complex-conjugate and the Hermitian-conjugate matrices, respectively, iff in the corresponding spaces one uses the dual, the complex-conjugate, and the complex-conjugate dual bases.

B.5 Bi- and Sesquilinear Forms

A linear map $g\colon \mathbf{V} \to \widetilde{\mathbf{V}}$ determines by $g(v,v'):= (gv\,|\,v')$ a *bilinear form* on $\mathbf{V}$ (cf. eqs. (7.5.11,13a)) denoted by the same letter, and conversely, a bilinear form g determines such map by $v \mapsto g(v, .)$. To the transposed map $g^\top\colon \mathbf{V} \to \widetilde{\mathbf{V}}$ corresponds the transposed bilinear form. Similarly, a linear map $g^\#\colon \widetilde{\mathbf{V}} \to \mathbf{V}$ determines a bilinear form on $\widetilde{\mathbf{V}}$. If for a semilinear operator $S\colon \mathbf{V} \to \mathbf{V}$ we have $S^\top g S = g$, the corresponding bilinear form is invariant in the sense that $g(Sv, Sv') = \sigma(g(v, v'))$, and conversely; if S is invertible, the condition rewrites as the intertwining property $gS = \tilde{S}g$. If the maps g, $g^\#$ are invertible, the corresponding bilinear forms are nondegenerate (cf. eq. (7.5.12)), and conversely. In particular, one may then choose $g^\# = g^{-1}$, or also $g^\# = (g^\top)^{-1} = \tilde{g}$; however, it is only the latter version which is natural in the sense that the tensors defined by these bilinear forms—see below— get transformed into each other by the maps induced in tensor spaces by the maps corresponding to the forms. (This observation leads to the choice of sign in eq. (8.5.3).)

Similarly, a linear map $h\colon \mathbf{V} \to \widetilde{\mathbf{V}}^*$ determines by $v \mapsto (h\,v)^*$ an antilinear map $\mathbf{V} \to \widetilde{\mathbf{V}}$ and a sesquilinear form h on $\mathbf{V}$ (cf. eqs. (7.5.11,13*b*)), and conversely. To the Hermitian-conjugate map $h^\dagger\colon \mathbf{V} \to \widetilde{\mathbf{V}}^*$ corresponds the Hermitian-conjugate sesquilinear form. Regarding invariance and nondegeneracy we have the same observations as before. If we write components with respect to the basis $\{\tilde{b}^{k*}\}$ in $\widetilde{\mathbf{V}}^*$ with dotted indices, we have $h(b_i) = h_{\dot{k}i}\tilde{b}^{k*}$.

For the relations between nondegenerate bilinear and sesquilinear forms with special symmetry properties ($g^\top \propto g$, $h^\dagger \propto h$) to pseudo-Euclidean, symplectic, and pseudo-unitary geometries in $\mathbf{V}$ cf. sect. 7.5, and Porteous (1981).

B.6 Real and Complex Structures

Linear maps $C\colon \mathbf{V} \to \mathbf{V}^*$ determine antilinear maps $\mathcal{C}\colon \mathbf{V} \to \mathbf{V}$ via $v \mapsto (Cv)^*$, and conversely. The vectors v invariant under $\mathcal{C}$ form a subset $\mathbf{V}'$ which is a real vector space only; however, it is nontrivial only if the linear map $\mathcal{C}^2 = C^*C$ possesses the eigenvalue 1. The maximal dimension of that real vector space—equal to the complex dimension of $\mathbf{V}$—is reached when $\mathcal{C}^2 = C^*C = \mathrm{id}_{\mathbf{V}}$. The operator $\mathcal{C}$ is then called an *anti-involution of first kind*, or a complex conjugation *in* $\mathbf{V}$, or a *real(ity) structure* for $\mathbf{V}$. Vectors v having $\mathcal{C}v = v$ and linear operators $S\colon \mathbf{V} \to \mathbf{V}$ having $CS = S^*C^*$ ($\Leftrightarrow \mathcal{C}S = S\mathcal{C}$; invariance of the real structure under S) are called *real with respect to* $\mathcal{C}$. A real basis for $\mathbf{V}'$ is also a basis for $\mathbf{V}$; with respect to it, real vectors have real components and real operators have real matrices, while to C corresponds the unit matrix when the conjugate basis in $\mathbf{V}^*$ is used. $\mathbf{V}$ is isomorphic to the complexification (see below) of $\mathbf{V}'$. For a subspace $\mathbf{W} \subset \mathbf{V}$, the number $\dim(\mathbf{W} \cap \mathcal{C}\mathbf{W})$ is called its *real index* with respect to $\mathcal{C}$.

Each, pseudo-Euclidean, symplectic, and pseudo-unitary structures in a vector space, may be brought to well-known numerically simple normal forms by a suitable choice of basis; however, this will not be possible for two of them simultaneously, in general. This, among other things, justifies the abstract characterization of these structures given here. For an application we refer the reader to Appendix C.

In a *real* vector space $\mathbf{V}$—where there are no anti-involutions—one may distinguish *involutions* of first and second kind, i.e., linear transformations J with squares $J^2 = +\mathrm{id}_{\mathbf{V}}$ and $J^2 = -\mathrm{id}_{\mathbf{V}}$, respectively. The first kind defines pairs of complementary projections $P_\pm := \frac{1}{2}(\mathrm{id}_{\mathbf{V}} \pm J)$ and thus decompositions of $\mathbf{V}$. An involution of second kind is also called *complex structure* in $\mathbf{V}$, allowing to consider $\mathbf{V}$ as a complex vector space of half dimension by defining the multiplication by complex numbers α as $\alpha v := (\mathrm{Re}\,\alpha)v + (\mathrm{Im}\,\alpha)Jv$. (All axioms of a complex vector space may be verified; the real dimension of $\mathbf{V}$ must be even for J to exist: $\dim_{\mathbf{R}}\mathbf{V} = 2m$.)

The last mentioned situation has to be strictly distinguished from the concept of complexification where one passes from the set $\mathbf{V}$ to a larger one (see below) but does not need an involution of second kind.

In complex spaces the distinction between (linear) involutions of first and second kind does not make much sense since one may pass from one to the other by $J \mapsto iJ$. *Anti-involutions of second kind* $\mathcal{J}$, by definition satisfying $\mathcal{J}^2 = -\mathrm{id}_{\mathbf{V}}$, would allow to view a complex vector space as a vector space over the skew field of quaternions of half the complex dimension (which must be even for $\mathcal{J}$

to exist). This is why $\mathcal{J}$ is also called a *quaternionic structure* (one also encounters the adjectives pseudo-real and anti-real); in this book, no use of this possibility will be made, however. Examples would be (i) $(u_1, u_2)^\top \mapsto (-u_2^*, u_1^*)^\top$ in $\mathbf{C}^2$, commuting with the action of SU(2); (ii) $\mathcal{C}'$ as given in Appendix C.2.

B.7 Direct Sums

The *direct sum* $\sum \oplus \mathbf{V}_i$ of vector spaces $\mathbf{V}_1$, $\mathbf{V}_2$, ... was introduced in sect. 6.6 in the case of two summands; the definition in the case of a finite number of summands is analogous; for a denumerable infinity of summands one defines an analogous structure on the set of all sequences $(v_1, v_2, \ldots) = v_1 \oplus v_2 \oplus \ldots$, $v_i \in \mathbf{V}_i$, where there are only finitely many nonzero members in each sequence. (In the Hilbert space framework this set has still to be completed in norm, thus adding those sequences with infinitely many nonzero members for which the sum of norm squares converges.) From bases $\{b_{(i)\mu_i} \mid \mu_i = 1, \ldots, \dim \mathbf{V}_i\}$ for the $\mathbf{V}_i$ one constructs vectors of the form

$$0 \oplus \ldots \oplus 0 \oplus b_{(i)\mu_i} \oplus 0 \oplus \ldots ,$$

which together form a basis for $\sum \oplus \mathbf{V}_i$. The dimension of the latter is therefore $\sum \dim \mathbf{V}_i$.

Given semilinear maps $A_i \colon \mathbf{V}_i \to \mathbf{W}_i$ which all have the *same* associated automorphism σ of $\mathbf{F}$, one can form their direct sum $A = \sum \oplus A_i$, a semilinear map $\sum \oplus \mathbf{V}_i \to \sum \oplus \mathbf{W}_i$ with associated automorphism σ, defined as $A(\sum \oplus v_i) := \sum \oplus A_i v_i$.

The formation of direct sums commutes with dualization and complex conjugation in the sense of the existence of natural isomorphisms $(\mathbf{V}_1 \oplus \mathbf{V}_2)^\sim \cong \widetilde{\mathbf{V}}_1 \oplus \widetilde{\mathbf{V}}_2$, etc., and of relations $(A_1 \oplus A_2)^\top = A_1^\top \oplus A_2^\top$, etc. (There is also a natural isomorphism among $\mathbf{V}_1 \oplus \mathbf{V}_2$ and $\mathbf{V}_2 \oplus \mathbf{V}_1$, etc.) There result, as a consequence, inner products, complex and real structures on $\sum \oplus \mathbf{V}_i$ if such are provided for the summands.

B.8 Tensor Products

The *tensor product* $\prod \otimes \mathbf{V}_i$ of vector spaces $\mathbf{V}_1$, $\mathbf{V}_2, \ldots$ (finitely many factors) was introduced in the main text in a basis-dependent way. Here we choose to define it as the vector space formed by the multilinear functionals $f \colon \widetilde{\mathbf{V}}_1 \times \widetilde{\mathbf{V}}_2 \times \ldots \to \mathbf{F}$ on the Cartesian product of the dual spaces. Thus, for every such f, we have $f(a_1, a_2, \ldots) \in \mathbf{F}$, where $a_i \in \widetilde{\mathbf{V}}_i$ and where f is separately linear in each argument.

If $v_i \in \mathbf{V}_i$, the tensor product $f = v_1 \otimes v_2 \otimes \ldots = \prod \otimes v_i$ of these vectors is defined to be the multilinear functional for which

$$f(a_1, a_2, \ldots) = a_1(v_1)\, a_2(v_2)\ \ldots , \qquad a_i \in \widetilde{\mathbf{V}}_i.$$

Given bases in the $\mathbf{V}_i$, all possible tensor products $\prod_i \otimes b_{(i)\mu_i}$ taken together form a basis for $\prod \otimes \mathbf{V}_i$, whose dimension therefore is $\prod \dim \mathbf{V}_i$. Each $f \in \prod \otimes \mathbf{V}_i$ has, with respect to the product basis, the expansion

$$f = f^{\mu_1 \mu_2 \cdots}\, b_{(1)\mu_1} \otimes b_{(2)\mu_2} \otimes \ldots \quad \text{with} \quad f^{\mu_1 \mu_2 \cdots} := f\left(\tilde{b}_{(1)}^{\mu_1}, \tilde{b}_{(2)}^{\mu_2}, \ldots\right),$$

from which one easily reads off the transformation behavior of components under a change of bases.

Given semilinear maps $A_i\colon \mathbf{V}_i \to \mathbf{W}_i$ all associated with the *same* automorphism σ of $\mathbf{F}$, their tensor product $\prod \otimes A_i$ is defined to be the semilinear map $A\colon \prod \otimes \mathbf{V}_i \to \prod \otimes \mathbf{W}_i$ with associated automorphism σ that has

$$(Af)(b_1, b_2, \ldots) = \sigma\, f(A_1^\top b_1, A_2^\top b_2, \ldots), \quad b_i \in \widetilde{\mathbf{W}}_i.$$

It follows for $v_i \in \mathbf{V}_i$

$$A \prod \otimes v_i = \prod \otimes A_i v_i,$$

and we have the multiplication rule (cf. eq. (6.5.5))

$$\left(\prod \otimes B_i\right) \circ \left(\prod \otimes A_i\right) = \prod \otimes (B_i \circ A_i).$$

When this definition is applied to the product basis, one obtains, in the linear case, the component form of the definition as used in the text (Kronecker product of matrices). *Warning*: There is no meaningful definition of tensor product between semilinear maps whose associated automorphisms are different!

Again there are some more or less obvious isomorphisms concerning relations between the tensor product and earlier constructions, such as $\mathbf{V}_1 \otimes \mathbf{V}_2 \cong \mathbf{V}_2 \otimes \mathbf{V}_1, \ldots,$ $(\mathbf{V}_1 \otimes \mathbf{V}_2) \otimes \mathbf{V}_3 \cong \mathbf{V}_1 \otimes \mathbf{V}_2 \otimes \mathbf{V}_3, \ldots, (\mathbf{V}_1 \otimes \mathbf{V}_2)^\sim \cong \widetilde{\mathbf{V}}_1 \otimes \widetilde{\mathbf{V}}_2, \ldots, (\mathbf{V}_1 \otimes \mathbf{V}_2)^* \cong \mathbf{V}_1^* \otimes \mathbf{V}_2^*,$ $(\mathbf{V}_1 \oplus \mathbf{V}_2) \otimes \mathbf{V}_3 \cong (\mathbf{V}_1 \otimes \mathbf{V}_3) \oplus (\mathbf{V}_2 \otimes \mathbf{V}_3), \ldots, \mathbf{V} \otimes \mathbf{F} \cong \mathbf{V}$. In the sense of these relations, there are then analogous relations for maps, such as $(A_1 \otimes A_2)^\top = A_1^\top \otimes A_2^\top,$ $(A_1 \oplus A_2) \otimes A_3 = (A_1 \otimes A_3) \oplus (A_2 \otimes A_3)$ etc. Also, $L(\mathbf{V}, \mathbf{W})$, the space of linear maps $\mathbf{V} \to \mathbf{W}$, is naturally isomorphic to $\widetilde{\mathbf{V}} \otimes \mathbf{W}$: we can assign to $f \in L(\mathbf{V}, \mathbf{W})$ the bilinear functional on $\mathbf{V} \times \widetilde{\mathbf{W}}$ whose value for the arguments $v \in \mathbf{V}$, $b \in \widetilde{\mathbf{W}}$ equals $(b \mid fv)$.

In component language, all these isomorphisms simply become identities. One can, therefore, try and introduce an 'abstract index' formalism (Penrose and Rindler 1984) that keeps the advantages of the component-index notation without actually referring to bases and components. In this, indices do not take numerical values but symbolize quantities that are identical once the natural isomorphisms are made, and operations on them.

According to these constructions, inner products, complex and real structures on the spaces $\mathbf{V}_i$ allow to define corresponding ones on the product spaces $\prod \otimes \mathbf{V}_i$. Observe, however, that the type of structure might change. For instance, from a symplectic type bilinear form on $\mathbf{V}$ one derives a symmetric type bilinear form on $\mathbf{V} \otimes \mathbf{V}$ or $\mathbf{V} \otimes \mathbf{V}^*$ and on other even tensorial powers; complex conjugation $\mathbf{V} \to \mathbf{V}^*$ yields a real structure on $\mathbf{V} \otimes \mathbf{V}^*$. (When $\dim \mathbf{V} = 2$, this is the basis of the relation between spinors and tensors!) The tensor product of two anti-involutions of second kind is of first kind; ....

B.9 Complexification

An elegant application of the tensor product is the abstract definition of the *complexification* $\mathbf{V}^c$ of a real vector space $\mathbf{V}$. If we think of the extension field $\mathbf{C} \supset \mathbf{R}$ as a (2-dimensional) vector space over $\mathbf{R}$ we can form the tensor product of real vector spaces $\mathbf{V}^c := \mathbf{C} \otimes \mathbf{V}$. In this space we can define the product with complex numbers α by requiring it to be distributive and by putting $\alpha(\beta \otimes v) := (\alpha\beta) \otimes v$: one verifies the axioms of a complex vector space. This method is—because of $\mathbf{C} = \mathbf{R} \oplus \mathbf{R}$—equivalent with setting $\mathbf{V}^c = \mathbf{V} \oplus \mathbf{V}$ and defining $\alpha(v \oplus v') = (\operatorname{Re}\alpha v - \operatorname{Im}\alpha v') \oplus (\operatorname{Re}\alpha v' + \operatorname{Im}\alpha v)$. Also, because of natural isomorphisms mentioned above, we can look at $\mathbf{V}^c$ as sitting inside the space of complex-valued linear functionals on $\widetilde{\mathbf{V}}$, the multiplication with complex numbers taking place in the target space of the functionals. $\mathbf{V}^c$ has a canonical real structure $\mathcal{C}\colon \alpha \otimes v \mapsto \alpha^* \otimes v$ with real subspace $\mathbf{R} \otimes \mathbf{V} \cong \mathbf{V}$. We have canonical isomorphisms $(\mathbf{V}^c)^c \cong \mathbf{V}^c$, $(\mathbf{V} \oplus \mathbf{W})^c \cong \mathbf{V}^c \oplus \mathbf{W}^c$, $(\mathbf{V} \otimes_{\mathbf{R}} \mathbf{W})^c \cong \mathbf{V}^c \otimes_{\mathbf{C}} \mathbf{W}^c$, $(\widetilde{\mathbf{V}})^c \cong \widetilde{\mathbf{V}^c}, \ldots$

B.10 The Tensor Algebra over a Vector Space

Starting from a vector space $\mathbf{V}$ we can form the *tensorial powers* $\mathbf{V}^2 = \mathbf{V} \otimes \mathbf{V}, \ldots,$ $\mathbf{V}^p, \ldots$ and we further put $\mathbf{V}^1 = \widetilde{\widetilde{\mathbf{V}}} \cong \mathbf{V}$, $\mathbf{V}^0 = \mathbf{F}$. Taking into account natural isomorphisms mentioned above, the direct sum $\sum \oplus \mathbf{V}^p$ becomes an associative algebra with respect to the tensor product as the algebra's multiplication: this is the *contravariant tensor algebra* over $\mathbf{V}$. Starting from $\widetilde{\mathbf{V}}$ instead, we construct the *covariant tensor algebra*, and finally $(\sum \oplus \mathbf{V}^p) \otimes (\sum \oplus \widetilde{\mathbf{V}}^q)$ is the *mixed tensor algebra* over $\mathbf{V}$. Elements possessing only a component in the subspace $\mathbf{V}^p_q := \mathbf{V}^p \otimes \widetilde{\mathbf{V}}^q$ are called *homogeneous of bidegree*, or type, (p, q). (Up to now, we only considered homogeneous tensors, but in second quantization one also uses inhomogeneous tensors.)

New operations becoming possible here are the various *contractions*, i.e., linear maps $C^i_j\colon \mathbf{V}^p_q \to \mathbf{V}^{p-1}_{q-1}$, which map elements of product form

$$v_1 \otimes \ldots \otimes v_i \otimes \ldots \otimes v_p \otimes a_1 \otimes \ldots \otimes a_j \otimes \ldots \otimes a_q$$

—where $v_1, \ldots \in \mathbf{V}$, $a_1, \ldots \in \widetilde{\mathbf{V}}$—to

$$(a_j|v_i)v_1 \otimes \ldots \not{v}_i \ldots \otimes v_p \otimes a_1 \otimes \ldots \not{a}_j \ldots \otimes a_q$$

(with the indicated omissions). By combining tensorial multiplication and contraction, i.e., by *transvection*, elements from $\mathbf{V}^p_q$ may be used in many ways to map $\mathbf{V}^a_b$ linearly into $\mathbf{V}^{a+p-m}_{b+q-m}$ (m = number of contractions). Conversely, each map of this kind defines an element from $\mathbf{V}^p_q$ (cf. $L(\mathbf{V}, \mathbf{V}) \cong \mathbf{V} \otimes \widetilde{\mathbf{V}}$ and also the 'quotient theorem'). For complex vector spaces, one constructs an analogous algebra also over $\mathbf{V}^*$, and one can still tensor it with the algebra just considered. Note, however, that there are no contractions between $\mathbf{V}^*$ and $\widetilde{\mathbf{V}}$!

For a semilinear map $A\colon \mathbf{V} \to \mathbf{W}$ there are the *tensorial powers* $A^{\otimes p}$, semilinear maps $\mathbf{V}^p \to \mathbf{W}^p$, and the tensorial powers

$$A^{\top \otimes q}\colon \widetilde{\mathbf{W}}^q \equiv \mathbf{W}_q \to \widetilde{\mathbf{V}}^q \equiv \mathbf{V}_q.$$

(We put $A^{\otimes 1} = A$, $A^{\otimes 0} = \mathrm{id}_{\mathbf{V}}$, etc.) If A is invertible one can form

$$\tilde{A}^{\otimes q}\colon \mathbf{V}_q \to \mathbf{W}_q, \qquad A^{\otimes p} \otimes \tilde{A}^{\otimes q}\colon \mathbf{V}^p_q \to \mathbf{W}^p_q,$$

i.e., A may be extended in a natural manner as a *type-preserving map* $A^\otimes$ to the whole mixed tensor algebra, commuting with tensorial multiplication and contraction. Conversely, it may be shown that all type-preserving invertible semilinear maps of the tensor algebras possessing these commutation properties arise in the way described.

If $\mathbf{W} \equiv \mathbf{V}$, also an arbitrary linear operator $A \colon \mathbf{V} \to \mathbf{V}$, not necessarily invertible, may be extended to the whole mixed tensor algebra in a natural but different manner. This arises when A is interpreted as the generator of a 1-parameter group of isomorphisms $U(\tau) = \exp(\tau A) \colon \mathbf{V} \to \mathbf{V}$. The latter induce, in each $\mathbf{V}^p_q$, the group $U^{\otimes p}(\tau) \otimes \widetilde{U}^{\otimes q}(\tau)$, which in turn has a generator D (omitting type indices). On differentiating at $\tau = 0$ it is seen that D commutes with contractions and satisfies, for arbitrary tensors T', T'' of the algebra, the *Leibniz rule*

$$D(T' \otimes T'') = D(T') \otimes T'' + T' \otimes D(T'').$$

A type-preserving linear map D of the tensor algebra, in the mixed case commuting with contractions, satisfying this purely algebraic condition, is called a *derivation* of the algebra. Derivations form a Lie algebra with respect to the commutator $[D, D'] = DD' - D'D$. Every linear map $A \colon \mathbf{V} \to \mathbf{V}$ may, by the definitions: $D = 0$ on $\mathbf{V}^0 = \mathbf{F}$, $D = A$ on $\mathbf{V}^1$, $D = -A^T$ on $\mathbf{V}_1$ and by postulating the Leibniz rule, be extended uniquely to the tensor algebra as a derivation, and every derivation of the algebra may be shown to arise in this way.

The constructions just considered, as well as the ones to follow in the next section, will be important in the formal structure of second quantization.

B.11 Symmetric and Exterior Algebra

To every permutation π of p elements we can associate a linear map $A_\pi \colon \mathbf{V}^p \to \mathbf{V}^p$ by

$$(A_\pi f)(a_1, \dots, a_p) := f(a_{\pi^{-1}(1)}, \dots, a_{\pi^{-1}(p)}), \qquad (a_1, \dots \in \widetilde{\mathbf{V}}),$$

which for each $A \colon \mathbf{V} \to \mathbf{V}$ commutes with $A^{\otimes p}$. We have $A_\rho A_\pi = A_{\rho\pi}$ for the product of the permutations ρ, π; $\pi \mapsto A_\pi$ is thus a representation of the symmetric permutation group G_p of p elements in the space $\mathbf{V}^p$. This representation is reducible, the reduction yielding the various *symmetry classes of tensors* (cf. Boerner 1955; Fulton and Harris 1991). Of particular importance are the 1-dimensional representations $\pi \mapsto \mathrm{id}$ and $\pi \mapsto \mathrm{sign}(\pi)\,\mathrm{id}$, being carried by tensors $T \in \mathbf{V}^p$ having $A_\pi T = T$ and $A_\pi T = \mathrm{sign}(\pi)\,T$, called totally symmetric and antisymmetric ($=$ skew), respectively. These tensors form subspaces $\bigvee^p(\mathbf{V})$ and $\bigwedge^p(\mathbf{V})$ of $\mathbf{V}^p$, invariant under all maps $A^{\otimes p}$, onto which the operators

$$\mathrm{Sym} := \frac{1}{p!} \sum_{\pi \in G_p} A_\pi \quad \text{and} \quad \mathrm{Alt} := \frac{1}{p!} \sum_{\pi \in G_p} \mathrm{sign}(\pi)\, A_\pi$$

project, respectively. The direct sums $\sum_{p=0}^{\infty} \oplus \bigvee^p(\mathbf{V}) =: \bigvee(\mathbf{V})$ and $\sum_{p=0}^{\infty} \oplus \bigwedge^p(\mathbf{V}) = : \bigwedge(\mathbf{V})$ become associative algebras—called the *symmetric* and the *exterior* algebra over **V**—with respect to the *symmetric* and *exterior* product, defined by

$$T \vee D := \frac{(p+q)!}{p!\,q!}\,Sym(T \otimes D) \in \vee^{p+q}(\mathbf{V}) \ \text{ for } \ T \in \vee^{p}(\mathbf{V}), \ D \in \vee^{q}(\mathbf{V})$$

and

$$T \wedge D := \frac{(p+q)!}{p!\,q!}\,Alt(T \otimes D) \in \wedge^{p+q}(\mathbf{V}) \ \text{ for } \ T \in \wedge^{p}(\mathbf{V}), \ D \in \wedge^{q}(\mathbf{V})$$

and distributivity for direct sums. We have the properties

$$T \vee D = D \vee T \ \text{ and } \ T \wedge D = (-1)^{pq} D \wedge T \ \text{ for } \ T \in \wedge^{p}(\mathbf{V}), \ D \in \wedge^{q}(\mathbf{V}).$$

Depending on purposes, one finds in the literature varying conventions concerning the combinatorial factors in the definitions of $\wedge$, $\vee$: for measuring volumes, the factor written above for $\wedge$ is convenient; for the isomorphism with polynomial algebra (cf. sect. 7.6) for $\vee$ the factor 1 is convenient. It is essential to guarantee associativity; one may verify that the above choice is o.k. in this respect.

To each semilinear map $A:\mathbf{V} \to \mathbf{W}$ there are the *symmetric* and the *exterior* powers $A^{\vee p}$ and $A^{\wedge p}$ that $A^{\otimes p}$ induces from $\vee^{p}(\mathbf{V})$ to $\vee^{p}(\mathbf{W})$ and from $\wedge^{p}(\mathbf{V})$ to $\wedge^{p}(\mathbf{W})$. This also yields semilinear actions $A^{\vee}$ and $A^{\wedge}$ of A on all of $\vee(\mathbf{V})$ and $\wedge(\mathbf{V})$ by forming the direct sums of the powers. In this way, scalar products and other structures are transferred to these spaces.

Just as in the case of the total tensor algebra, linear operators $\mathbf{V} \to \mathbf{V}$ extend also in a second way, namely as *derivations* of the algebras $\vee(\mathbf{V})$ and $\wedge(\mathbf{V})$, where now the Leibniz rule refers to the $\vee$ and the $\wedge$ product. We may mention that in the formalism of second quantization the one-particle observables are extended in this way to the whole Fock space.

The dual spaces $(\vee^{p}(\mathbf{V}))^{\sim}$ and $(\wedge^{p}(\mathbf{V}))^{\sim}$ are isomorphic to $\vee^{p}(\widetilde{\mathbf{V}})$ and $\wedge^{p}(\widetilde{\mathbf{V}})$, respectively, in a natural way. This is analogous to the isomorphism $(\mathbf{V} \otimes \mathbf{W})^{\sim} \cong \widetilde{\mathbf{V}} \otimes \widetilde{\mathbf{W}}$, but we want to be more explicit here because of the occurrence of combinatorial factors that are subject to conventions. In the last-mentioned case the isomorphism is given by associating to a linear functional f on $\mathbf{V} \otimes \mathbf{W}$ a bilinear functional f' on $\mathbf{V} \times \mathbf{W}$, defined by $f'(v,w) = f(v \otimes w)$. Conversely, f is known once we know f', and a possible constant factor in the definition of f' has been set equal to 1 in order to achieve that the inverse image of $a \otimes b \in \widetilde{\mathbf{V}} \otimes \widetilde{\mathbf{W}}$ evaluated on $v \otimes w$ yield $(a \otimes b)(v,w) = a(v)b(w)$. One proceeds in an analogous fashion in the case of several factors. However, if $\mathbf{W} = \ldots = \mathbf{V}$, a simple restriction of the above assignment $(\mathbf{V}^{p})^{\sim} \cong (\widetilde{\mathbf{V}})^{p}$ to the symmetric and the antisymmetric subspaces leads to combinatorial factors which are sometimes unwanted and may be avoided by choosing another value for the numerical factor mentioned before (namely $= 1/p!$ in the case of our definition of the $\vee$ and the $\wedge$ product). This means that if $a_1 \vee \ldots \vee a_p$ and $a_1 \wedge \ldots \wedge a_p$ are thought of as elements of $(\vee^{p}(\mathbf{V}))^{\sim}$ and $(\wedge^{p}(\mathbf{V}))^{\sim}$, we have as the definitions of the inner products

$$(a_1 \vee \ldots \vee a_p \mid v_1 \vee \ldots \vee v_p) = \operatorname{perm} a_i(v_j) = (a_1 \vee \ldots \vee a_p)(v_1, \ldots, v_p)$$

and

$$(a_1 \wedge \ldots \wedge a_p \mid v_1 \wedge \ldots \wedge v_p) = \det a_i(v_j) = (a_1 \wedge \ldots \wedge a_p)(v_1, \ldots, v_p).$$

Here the *permanent* of a matrix arises from the determinant by writing the full expansion of the latter and converting all minuses into pluses.

The scalar products induced in these spaces that originate from linear or antilinear maps $\mathbf{V} \to \widetilde{\mathbf{V}}$ are also to be understood in this sense, and the basis of $\bigvee^p(\widetilde{\mathbf{V}})$ formed by the products $\tilde{b}^{i_1} \vee \ldots \vee \tilde{b}^{i_p}$ is then dual, in the sense of $(\bigvee^p(\mathbf{V}))^\sim$, to the symmetric product basis $b_{i_1} \vee \ldots \vee b_{i_p}$. (A corresponding statement holds for $\wedge$.) If we have a symmetric or Hermitian scalar product in $\mathbf{V}$ for which the basis $\{b_i\}$ is orthonormal, then the product basis is orthonormal in the sense of the induced scalar product discussed. We then also have a scalar product on the full algebra by declaring tensors of different degree to be orthogonal, and by requiring bilinearity or sesquilinearity also with respect to the direct sum operation. Similarly, one defines the scalar product between $\bigvee^p(\mathbf{V})$ and $\bigvee^q(\widetilde{\mathbf{V}})$ to be zero when $p \neq q$, and the same in the antisymmetric case. This then fixes the duality of the total algebras over dual spaces.

$\wedge(\widetilde{\mathbf{V}})$ is sometimes called the *Grassmann algebra* over $\mathbf{V}$; $\bigvee(\widetilde{\mathbf{V}})$—with a suitable convention in the definition of the symmetric product—is isomorphic to the algebra of *polynomials* in $\dim \mathbf{V}$ variables.

B.12 Inner Product. Creation and Annihilation Operators

An element T from $\bigvee(\mathbf{V})$ or $\wedge(\mathbf{V})$ determines a linear operator $\mu(T)$ on that space sending $T' \mapsto T \vee T'$ or $T' \mapsto T \wedge T'$. (Generally, for an (associative) algebra, $T \mapsto \mu(T)$ is called its (left) regular representation; we encountered it in dual form in the group context in sect. 7.7. Observe that it is now essential to consider all inhomogeneous elements of the symmetric or exterior algebra in order to stay within one single space, although many definitions of semilinear operators that follow are written only for homogeneous elements: they are to be extended to inhomogeneous elements by distributivity and semilinearity with respect to the direct sum.)

An element $\widetilde{T}$ from $\bigvee(\widetilde{\mathbf{V}})$ or $\wedge(\widetilde{\mathbf{V}})$ determines an operator in these spaces in the same way, whose transpose $\iota(\widetilde{T})$ is an operator in $\bigvee(\mathbf{V})$ or $\wedge(\mathbf{V})$. The image $\iota(\widetilde{T})\,T'$ of a tensor T' is called the *(left) inner product* with $\widetilde{T}$, sometimes also denoted by $\widetilde{T} \lrcorner T'$. For us, the most important case will be the one where $T = v \in \mathbf{V}, \widetilde{T} = a \in \widetilde{\mathbf{V}}$. Explicitly, $\iota(a)$ is just transvection with a: $\iota(a)\,T' = C_1^1(a \otimes T')$, or

$$(\iota(a)\,T')(a_1, \ldots, a_{p-1}) = T'(a, a_1, \ldots, a_{p-1})$$

$$\text{for} \quad T' \in \textstyle\bigvee^p(\mathbf{V}) \quad \text{or} \quad T' \in \textstyle\wedge^p(\mathbf{V}) \quad \text{and} \quad a_1, \ldots \in \widetilde{\mathbf{V}}.$$

For $v, v' \in \mathbf{V}$ it follows from associativity and the commutation laws for $\vee, \wedge$ written above that

$$\mu(v)\,\mu(v') = \mu(v')\,\mu(v) \quad \text{or} \quad = -\mu(v')\,\mu(v)\,;$$

for $a, a' \in \widetilde{\mathbf{V}}$ we have the same relations in dual space, and therefore we find, by taking transposes, that

$$\iota(a)\,\iota(a') = \iota(a')\,\iota(a) \quad \text{or} \quad = -\iota(a')\,\iota(a).$$

With slightly more labor one can also verify that $\iota(a)$ is a derivation or an antiderivation (of degree -1), i.e.,

$$\iota(a)(T' \vee T'') = (\iota(a)\,T') \vee T'' + T' \vee \iota(a)\,T''$$

or

$$\iota(a)(T' \wedge T'') = (\iota(a)\,T') \wedge T'' + (-1)^p\, T' \wedge \iota(a)\,T'' \quad \text{for} \quad T' \in \textstyle\bigwedge^p(\mathbf{V}).$$

By taking $T' = v \in \mathbf{V}$ we obtain the further (anti)commutation rule

$$\iota(a)\,\mu(v) - \mu(v)\,\iota(a) = a(v)\,\mathrm{id} \quad \text{or} \quad \iota(a)\,\mu(v) + \mu(v)\,\iota(a) = a(v)\,\mathrm{id}.$$

The *commutation rules* found here are essentially those between *creation* $(\mu(v))$ and *annihilation* $(\iota(a))$ *operators* in the formalism of second quantization, which thus reveals itself algebraically as a part of tensor algebra. We emphasize that there would be combinatorial factors in the definitions of μ, ι if other combinatorial factors had been chosen in the definition of $\vee$, $\wedge$, to guarantee the above form of the commutation relations, and that under the present conventions the mentioned normalization of scalar products has to be observed.

For the discussion of relativistic covariance in the formalism of second quantization we still derive the relation that exists between $\mu(Av)$, $\iota(\tilde{A}v)$ and $\mu(v)$, $\iota(a)$, where A is a semilinear map $\mathbf{V} \to \mathbf{W}$, inducing maps $A^{\vee}\colon \bigvee(\mathbf{V}) \to \bigvee(\mathbf{W})$ and $A^{\wedge}\colon \bigwedge(\mathbf{V}) \to \bigwedge(\mathbf{W})$ as explained. It follows from the definitions that

$$A^{\vee}\mu(v) = \mu(Av)\,A^{\vee} \quad \text{and} \quad A^{\wedge}\mu(v) = \mu(Av)\,A^{\wedge},$$

and by transposing the analogous relations for the dual spaces—$A^{\top}$, a replacing A, v—we get

$$\iota(a)\,A^{\vee} = A^{\vee}\,\iota(A^{\top}a) \quad \text{and} \quad \iota(a)\,A^{\wedge} = A^{\wedge}\,\iota(A^{\top}a).$$

If A is invertible, it follows that

$$\mu(Av) = A^{\vee}\mu(v)(A^{\vee})^{-1}, \qquad \iota(\tilde{A}a) = A^{\vee}\iota(a)(A^{\vee})^{-1},$$

and

$$\mu(Av) = A^{\wedge}\mu(v)(A^{\wedge})^{-1}, \qquad \iota(\tilde{A}a) = A^{\wedge}\iota(a)(A^{\wedge})^{-1}.$$

(We were using obvious relations such as $(A^{\top})^{\vee} = (A^{\vee})^{\top}$, etc.)

B.13 Duality in Exterior Algebra

Writing $\dim \mathbf{V} = n$, we have from elementary combinatorics that

$$\dim \overset{p}{\textstyle\bigvee}(\mathbf{V}) = \dim \overset{p}{\textstyle\bigvee}(\widetilde{\mathbf{V}}) = \binom{n + p - 1}{p}$$

and

$$\dim \overset{p}{\textstyle\bigwedge}(\mathbf{V}) = \dim \overset{p}{\textstyle\bigwedge}(\widetilde{\mathbf{V}}) = \binom{n}{p}.$$

From a basis $\{b_i\}$ for $\mathbf{V}$ we get bases for $\bigvee^p(\mathbf{V})$ and $\bigwedge^p(\mathbf{V})$ by taking all products

$$b_I := \underbrace{b_{i_1} \vee \ldots \vee b_{i_1}}_{p_{i_1}} \vee \underbrace{b_{i_2} \vee \ldots \vee b_{i_2}}_{p_{i_2}} \vee \ldots \vee \underbrace{b_{i_s} \vee \ldots \vee b_{i_s}}_{p_{i_s}} = \bigvee_{i=1}^{n} b_i^{\vee p_i},$$

$$\text{where} \quad 1 \le i_1 < i_2 < \ldots < i_s \le n, \quad \textstyle\sum_{k=1}^{s} p_{i_k} = p = \sum_{i=1}^{n} p_i,$$

and

$$b_I := b_{i_1} \wedge b_{i_2} \wedge \ldots \wedge b_{i_p} = \bigwedge_{i=1}^{n} b_i^{\wedge p_i}, \quad \text{where} \quad 1 \le i_1 < i_2 < \ldots < i_p \le n.$$

In the second version of writing these products, the integer exponents p_i satisfy $0 \le p_i \le p$ and $p_i \in \{0,1\}$, respectively, where $b_i^{\vee 0} = b_i^{\wedge 0} := 1$; similarly for bases of $\bigvee^p(\widetilde{\mathbf{V}})$ and $\bigwedge^p(\widetilde{\mathbf{V}})$.

In the language of second quantization, the p_i are the *occupation numbers* of the 'one-particle states' b_i. According to an observation of P. Ehrenfest and R. Kamerlingh-Onnes one gets

$$\dim\bigvee^{p}(\mathbf{V}) = (p + n - 1)!/p!(n - 1)!$$

as the number of permutations of $p + n - 1$ symbols, among which there are p equal symbols $\flat$ and $n - 1$ equal symbols $\vee$, by remarking that the basis vectors listed are in bijection with 'distribution symbols' $\flat \ldots \flat \vee \flat \ldots \flat \vee \vee \ldots$, upon the convention that the consecutive occurrence of two symbols $\vee$ means the nonoccurrence of one of the b_i in b_I.

Our point is now that while $\dim\bigvee^p(\mathbf{V})$ steadily increases with p—so that the symmetric algebra over a finite-dimensional vector space is infinite-dimensional— the dimension of $\bigwedge^p(\mathbf{V})$ and of $\bigwedge^p(\widetilde{\mathbf{V}})$ first increases but then decreases again, as we have $\binom{n}{p} = \binom{n}{n-p}$. Indeed, the exterior algebra over a finite-dimensional vector space has the finite dimension $\sum \binom{n}{p} = (1+1)^n = 2^n$. Despite the former relation, one needs some extra structure to specify an isomorphism between $\bigwedge^p(\mathbf{V})$ (or $\bigwedge^p(\widetilde{\mathbf{V}})$) and $\bigwedge^{n-p}(\mathbf{V})$ (or $\bigwedge^{n-p}(\widetilde{\mathbf{V}})$). Least expensive is the specification of an isomorphism between the 1-dimensional spaces $\bigwedge^0(\mathbf{V}) = \mathbf{F}$ and $\bigwedge^n(\widetilde{\mathbf{V}})$ by giving the image of 1: $1 \mapsto \tilde{e} \in \bigwedge^n(\widetilde{\mathbf{V}})$. In a real vector space, $\tilde{e}$ is called an *oriented volume element*, since it assigns to every parallelopiped spanned by n vectors $v_1, \ldots, v_n$ a real number $\tilde{e}(v_1, \ldots, v_n)$, depending linearly on each edge and vanishing iff the vectors are dependent, making the parallopiped degenerate. In the complex case one rather speaks of a *determinant function* or of a *unimodular structure*. The point is now that this structure yields linear maps—observe $\mu(1) = \mathrm{id} \Rightarrow \iota(1) = \mathrm{id}$—

$$\bigwedge^p(\mathbf{V}) \xrightarrow[*]{} \bigwedge^{n-p}(\widetilde{\mathbf{V}}), \, T \mapsto {}_*T := \iota(T)\tilde{e},$$

and, using the basis $\{e\}$ of $\bigwedge^n(\mathbf{V})$ dual to the basis $\{\tilde{e}\}$ of $\bigwedge^n(\widetilde{\mathbf{V}})$, linear maps

$$\bigwedge^p(\widetilde{\mathbf{V}}) \xrightarrow{*} \bigwedge^{n-p}(\mathbf{V}), \, \widetilde{T} \mapsto {}^*\widetilde{T} := \iota(\widetilde{T}) \, e.$$

As we shall not prove in detail, these maps are essentially inverses of each other: for $T \in \bigwedge^p(\mathbf{V})$, $\tilde{T} \in \bigwedge^p(\widetilde{\mathbf{V}})$ we have

$$*(_*T) = (-1)^{p(n-p)}\, T, \qquad\qquad {}_*(^*\tilde{T}) = (-1)^{p(n-p)}\, \tilde{T};$$

whence it follows that they all are isomorphisms. Also, the inner product between $\bigwedge^p(\widetilde{\mathbf{V}})$ and $\bigwedge^p(\mathbf{V})$ is preserved in the sense that

$$(\tilde{T} \mid D) = (_*D \mid {}^*\tilde{T}).$$

The reader may check that the development in sect. 5.5 is just a component version of this ('Poincaré') duality $\bigwedge^p(\mathbf{V}) \xleftrightarrow[*]{*} \bigwedge^{n-p}(\widetilde{\mathbf{V}})$, referred to *unimodular* bases $\{b_i\}$ in $\mathbf{V}$—bases having $b_1 \wedge \ldots \wedge b_n = e \Leftrightarrow \tilde{e}(b_1, \ldots, b_n) = 1$—which are related among each other by substitutions $b'_i = S^k{}_i b_k$ having $\det(S^k{}_i) = 1$. (Observe that the *determinant* of a linear map $S\colon \mathbf{V} \to \mathbf{V}$ may be defined by $S^{\wedge n} e =: (\det S)\, e$, but is actually independent of the special determinant function chosen.)

   If a nonsingular linear map $g\colon \mathbf{V} \to \widetilde{\mathbf{V}}$ is specified—e.g., by a scalar product on $\mathbf{V}$ ($g^{\mathsf{T}} = \gamma g$, $\gamma = \pm 1$)—one also gets maps $\bigwedge^p(\mathbf{V}) \to \bigwedge^{n-p}(\mathbf{V})$ by composing

$$\tilde{g} \circ {}_* \equiv \frac{1}{g(e,e)}\, {}^* \circ g.$$

(To simplify notation, the induced maps and bilinear forms are here all denoted by the same letter g.) Under this map, $g(\cdot,\cdot) =: \langle\,\cdot\mid\cdot\,\rangle$ is 'almost' invariant: it follows from the conservation of the inner product between $\bigwedge^p(\mathbf{V})$ and $\bigwedge^p(\widetilde{\mathbf{V}})$ written above that

$$\langle T \mid D \rangle = \langle \tilde{g}_* D \mid {}^*g T \rangle = \langle e \mid e \rangle \langle \tilde{g}_* D \mid \tilde{g}_* T \rangle.$$

These formulae simplify slightly if the determinant function is chosen *compatible with* g, i.e., if $\langle e \mid e \rangle = 1$. This is always possible over $\mathbf{C}$, while over $\mathbf{R}$ this possibility depends on the signature of the quadratic form induced by $\langle\,\cdot\mid\cdot\,\rangle$ in $\bigwedge^p(\mathbf{V})$, and only $|\langle e \mid e \rangle| = 1$ may always be achieved. The operation $\tilde{g} \circ {}_* =: *$, perhaps with a conventional sign factor, is called ('Hodge') *star* operation (duality). (In the applications of the present text, where g is the Minkowski metric η, one can achieve $\langle e \mid e \rangle = -1$.) If we further assume—as we had for the inner products on $\mathbf{V}$—that $g^{\mathsf{T}} = \gamma g$, the formulae above yield for the iterated star operator

$$* * = \gamma \frac{(-1)^{p(n-p)}}{\langle e \mid e \rangle}\ \mathrm{id}.$$

   A simple geometric interpretation of the operations presented purely algebraically in this section is obtained when the tensors considered are *simple* (or *decomposable*, also sometimes called (Plücker-Grassmann) *extensors*), i.e., have the product form $v_1 \wedge \ldots \wedge v_p$. One may think of p-dimensional subspaces $\mathbf{V}'$ of $\mathbf{V}$ either as spanned by p independent vectors v_i, in which case $v_1 \wedge \ldots \wedge v_p$ is, up to a numerical factor, independent of their special choice in $\mathbf{V}'$. Or one can think of them as being given by $n - p$ independent linear homogeneous equations $(a_{p+1} \mid v) = 0, \ldots, (a_n \mid v) = 0$; what matters here is only the *annihilating* space $\widetilde{\mathbf{V}}' \subset \widetilde{\mathbf{V}}$ spanned by the covectors a_i, similarly associated with the exterior product $a_{p+1} \wedge \ldots \wedge a_n$. For the same subspace, the two extensors—the spanning and the annihilating one—are just related by $v_1 \wedge \ldots \wedge v_p \propto {}^*(a_{p+1} \wedge \ldots \wedge a_n)$; because of the free numerical factor, the normalization of the determinant function involved is unimportant. Also, if a scalar product $\langle \cdot \mid \cdot \rangle$ is given, the extensors $v_1 \wedge \ldots \wedge v_p$ and $* (v_1 \wedge \ldots \wedge v_p)$ are associated with orthogonal subspaces. Much more could be said about this 'geometric algebra' of Grassmann and Clifford, but this is not the place to do so.

B.14 $\mathcal{G}$-Geometries and Quantities of Type $(\mathcal{G}, \sigma)$

In the abstract version of linear algebra, vectors and tensors and the operations with them were introduced without the use of components and their transformation behavior; rather, the latter comes as a consequence of the abstract definitions. Although there is no doubt about the usefulness of them, there are cases where the component version cannot be avoided: this happens whenever the component version involves nonrational functions.

To illustrate this point, consider, for an n-dimensional space $\mathbf{V}$, the 1-dimensional spaces $\mathbf{W}^{(p)} := (\wedge^n(\mathbf{V}))^{\otimes p}$. A linear transformation $S : \mathbf{V} \to \mathbf{V}$ induces in $\mathbf{W}^{(p)}$ a linear transformation that simply consists in multiplying by $(\det S)^p$. This gives, for every integer $p \geq 0$, a 1-dimensional representation of the group $\mathrm{GL}(\mathbf{V})$ of all nonsingular S. This can still be extended to negative integers by considering $\tilde{S}$ and the transformations it induces on the spaces $(\wedge^n(\widetilde{\mathbf{V}}))^{\otimes p}$. If we work over $\mathbf{C}$, then also $S \mapsto |\det S|^p$ with arbitrary real p gives a representation of the group; over $\mathbf{R}$, even the discontinuous function $S \mapsto \mathrm{sign} \det S$ gives a representation. Spaces carrying these representations cannot be obtained from $\mathbf{V}$ using the tensorial constructions available up to now, since in the tensorial case the elements of the representing matrices involve only *rational* functions of the matrix elements of S with respect to any basis in $\mathbf{V}$. Nevertheless these irrational representations and their tensor products with tensor representations are needed in physics—in particular when $p = \pm 1/2$. One speaks of *relative tensors of weight p*.

When these 'new' representations are restricted to the unimodular subgroup $\mathrm{SL}(\mathbf{V})$ they collapse to the usual tensor representations. However, upon further restriction, again new nontensorial representations may come up. For instance, if $\mathbf{V}$ is real and 4-dimensional with a symmetric bilinear form $\eta(\cdot \mid \cdot)$ of signature $(+ - - -)$, the subgroup of $\mathrm{SL}(\mathbf{V})$ leaving η invariant is the proper Lorentz group $\mathcal{L}_+$. The subset $\{v \in \mathbf{V} : \eta(v, v) \geq 0,\ v \neq 0\}$ of $\mathbf{V}$ is invariant under $\mathcal{L}_+$ and consists, in the sense of the standard topology of a real vector space, of two connected components (in physical language: future and past light cone and interior of each). The subgroup of $\mathcal{L}_+$ leaving these components invariant is $\mathcal{L}_+^\uparrow$. We then have a nontrivial 1-dimensional representation of $\mathcal{L}_+ = \mathcal{L}_+^\uparrow \cup \mathcal{L}_+^\downarrow$ by $S \mapsto 1$ or $S \mapsto -1$ for $S \in \mathcal{L}_+^\uparrow$ or $S \in \mathcal{L}_+^\downarrow$; however, the representation space for it cannot be obtained from $\mathbf{V}$ by any of the abstract tensor constructions treated up to this point. At the same time we observe that there enter, in these examples, special properties of the base fields $\mathbf{C}$, $\mathbf{R}$ used ($|\ldots|$ is defined on $\mathbf{C}$ and $\mathbf{R}$ and is multiplicative; $>$ and $\alpha^p > 0$ for $\alpha > 0$ are well-defined in $\mathbf{R}$), and also topological considerations came in.

Since in physics one cannot afford a puristically algebraic point of view, we now give a modernized version of the component definition of tensors which allows for the generalizations just discussed. We shall have to use the standard numerical vector space $\mathbf{F}^n$ over $\mathbf{F}$ with its canonical basis vectors $(1,0,0,\ldots)^{\mathsf{T}}$, $(0,1,0,\ldots)^{\mathsf{T}}$, $\ldots$, which we want somehow to get rid of again (e.g., to satisfy the principle of relativity).

Let $\mathbf{B}(\mathbf{V})$ be the set of all bases, or frames, in $\mathbf{V}$: then an element $\mathbf{b} = \{b_i\}$ defines an isomorphism (which we denote by the same letter) $\mathbf{b}\colon \mathbf{F}^n \to \mathbf{V}$ by assigning to every column vector $\mathbf{v} = (v^i) \in \mathbf{F}^n$ the vector $\mathbf{bv} = b_i v^i \in \mathbf{V}$. (We have written here multiplication by scalars as right multiplication, just to be able to interpret $\mathbf{bv}$ symbolically as well as in the sense of matrix multiplication, reading $\mathbf{b}$ as a row matrix whose entries are the basis vectors.) Similarly, the elements $\tilde{\mathbf{b}} = \{\tilde{b}^i\}$ of $\tilde{\mathbf{B}}(\mathbf{V}) := \mathbf{B}(\tilde{\mathbf{V}})$ give us maps $\tilde{\mathbf{b}}\colon \mathbf{V} \to \mathbf{F}^n$, $\tilde{\mathbf{b}}v := \mathbf{v} = (v^i) = (\tilde{b}^i(v))$, and in this sense we have $\tilde{\mathbf{b}} = \mathbf{b}^{-1} : \mathbf{b} \circ \tilde{\mathbf{b}} = \mathrm{id}_{\mathbf{V}}$, $\tilde{\mathbf{b}} \circ \mathbf{b} = \mathbf{1}$. (Again, these equations also have a matrix interpretation when $\tilde{\mathbf{b}}$ is taken as a column of covectors and $\circ$ is taken to mean $\otimes$ and $(\,|\,)$ in the first and second of these equations, respectively.)

Distinguishing some arbitrary basis $\mathbf{b}$ yields a bijection $\mathbf{B}(\mathbf{V}){\leftrightarrow}\mathrm{GL}(n,\mathbf{F})$ (nonsingular $n \times n$ matrices with elements from $\mathbf{F}$) which associates to every $\mathbf{b}' \in \mathbf{B}(\mathbf{V})$ the matrix $\mathbf{S} = \tilde{\mathbf{b}}\mathbf{b}'$ (i.e., $S^k{}_i = (\tilde{b}^k\,|\,b'_i)$) and to every $\mathbf{S}$ the basis $\mathbf{b}' = \mathbf{bS}$ (i.e., $b'_i = b_k S^k{}_i$). Since $\mathbf{b}$ was chosen at will, this bijection is not canonical, and as a consequence only a small part of the group structure of $\mathrm{GL}(n,\mathbf{F})$ passes on to $\mathbf{B}(\mathbf{V})$: there is no neutral basis, there are no inverse bases and no multiplication of bases *in* $\mathbf{B}(\mathbf{V})$ which would be naturally defined. Just to right multiplication $\mathbf{S} \mapsto \mathbf{SS}'$ in $\mathrm{GL}(n,\mathbf{F})$ there corresponds a *right action* of $\mathrm{GL}(n,\mathbf{F})$ on $\mathbf{B}(\mathbf{V})$: $\mathbf{b} \mapsto \mathbf{bS}'$. To left multiplication in $\mathrm{GL}(n,\mathbf{F})$, $\mathbf{S} \mapsto \mathbf{S}'\mathbf{S}$, nothing immediate corresponds; however, a left action of $\mathrm{GL}(n,\mathbf{F})$ on $\mathbf{B}(\mathbf{V})$ is formally given by $\mathbf{b} \mapsto \mathbf{bS}'^{-1}$. The right action $\mathbf{b} \mapsto \mathbf{bS}$ of $\mathrm{GL}(n,\mathbf{F})$ on $\mathbf{B}(\mathbf{V})$ is *simply transitive*, making $\mathbf{B}(\mathbf{V})$ into a principal $\mathrm{GL}(n,\mathbf{F})$-space. (Cf. Appendix A.)

The group $\mathrm{GL}(n,\mathbf{F})$ acts on $\mathbf{F}^n$ in the usual manner: $\mathbf{v} \mapsto \mathbf{Sv}$, which is a *left* action. The group then acts also on the Cartesian product $\mathbf{B}(\mathbf{V}) \times \mathbf{F}^n$, the group element $\mathbf{S}$ sending the pair $(\mathbf{b},\mathbf{v})$ into the pair $(\mathbf{bS}^{-1}, \mathbf{Sv})$. If we call two pairs equivalent if one can be carried to the other by some $\mathbf{S} \in \mathrm{GL}(n,\mathbf{F})$, we can form the quotient $(\mathbf{B}(\mathbf{V}) \times \mathbf{F}^n)/\mathrm{GL}(n,\mathbf{F})$ with respect to this equivalence relation. It is then clear that the equivalence classes are in bijection with the elements of $\mathbf{V}$: $v = \mathbf{bv} = \mathbf{bS}^{-1}\mathbf{Sv} \leftrightarrow$ class of $(\mathbf{b},\mathbf{v})$. This precisely corresponds to the component definition of vectors: $\mathbf{v}$ and $\mathbf{Sv}$ are the columns of components of v with respect to the frames $\mathbf{b}$ and $\mathbf{bS}^{-1}$, respectively.

In our 'reconstruction' of $\mathbf{V}$ from $\mathbf{B}(\mathbf{V})$ and $\mathbf{F}^n$, whose aim was the removal of the preferred role of the canonical basis of $\mathbf{F}^n$, $\mathbf{B}(\mathbf{V})$ and $\mathrm{GL}(n,\mathbf{F})$ play the primary role. Thus while earlier the $\mathbf{b} \in \mathbf{B}(\mathbf{V})$ were looked at as maps $\mathbf{F}^n \to \mathbf{V}$, it is more appropriate now to look at the $v \in \mathbf{V}$ as maps $\mathbf{B}(\mathbf{V}) \to \mathbf{F}^n$, $v\colon \mathbf{b} \mapsto \tilde{\mathbf{b}}v$ which are *equivariant* with respect to the action of $\mathrm{GL}(n,\mathbf{F})$ on $\mathbf{B}(\mathbf{V})$ and on $\mathbf{F}^n$, i.e., maps satisfying $v(\mathbf{bS}^{-1}) = \mathbf{S}v(\mathbf{b})$. It is then also appropriate to write $\mathbf{B}$ instead of $\mathbf{B}(\mathbf{V})$, and to just consider it as some principal $\mathrm{GL}(n,\mathbf{F})$-space.

Two small changes in this construction now allow to include the nontensorial quantities mentioned (relative tensors with noninteger weights, space- and time-pseudotensors for $\mathcal{L}$, $\ldots$) into our scheme. In the most abstract version one considers, instead of $\mathrm{GL}(n, \mathbf{F})$, just some group $\mathcal{G}$, and instead of $\mathbf{F}^n$ just any $\mathcal{G}$-space $\mathbf{M}$, i.e., we have a homomorphism σ from $\mathcal{G}$ into the group of all permutations of $\mathbf{M}$; $\mathbf{B}$ is replaced by some principal $\mathcal{G}$-space $\mathbf{B}_\mathcal{G}$. (The latter may be interpreted as a set of 'reference frames', all on the same footing with respect to the group, but nothing being said about their detailed nature.) One then forms the quotient $(\mathbf{B} \times \mathbf{M})/\mathcal{G}$ with respect to the equivalence relation '$(\mathbf{b}, \mathbf{m}) \sim (\mathbf{b}', \mathbf{m}')$ iff $\mathbf{m}' = \sigma(g)\mathbf{m} \in \mathbf{M}$ and $\mathbf{b}' = \mathbf{b}g^{-1}$ for some $g \in \mathcal{G}$'—the right action of g^{-1} on $\mathbf{b}$ having been written simply as $\mathbf{b}g^{-1}$). Again, these equivalence classes also correspond bijectively to the $\mathcal{G}$-equivariant maps $\mathbf{B} \to \mathbf{M}$ (i.e., maps where $\mathbf{b} \mapsto \mathbf{m}$ implies $\mathbf{b}g^{-1} \mapsto \sigma(g)\mathbf{m}$). If $\mathbf{M} = \mathbf{F}^m$ and $\sigma : \mathcal{G} \to \mathrm{GL}(m, \mathbf{F})$ is a matrix representation of $\mathcal{G}$ in $\mathbf{F}^m$, then the quotient inherits from $\mathbf{F}^m$ an isomorphic vector space structure: each equivalence class has a standard representative of the form $(\mathbf{b}_0, \mathbf{v})$, where $\mathbf{b}_0$ is arbitrary but the same for all classes; if $[\ldots]$ denotes the projection to the quotient, one can define the vector space structure by $\alpha[\mathbf{b}_0, \mathbf{v}] + \beta[\mathbf{b}_0, \mathbf{w}] := [\mathbf{b}_0, \alpha\mathbf{v} + \beta\mathbf{w}]$, which is independent of the special $\mathbf{b}_0$ chosen.

Now let $\mathbf{V}$ be an n-dimensional vector space over $\mathbf{F}$, and assume $\mathbf{B} = \mathbf{B}_\mathcal{G}(\mathbf{V}) \subset \mathbf{B}(\mathbf{V})$ to be a subset of $\mathbf{B}(\mathbf{V})$, whereby $\mathcal{G}$ becomes isomorphic to a subgroup of $\mathrm{GL}(n, \mathbf{F})$: then one says that one has a $\mathcal{G}$-*geometry*, or $\mathcal{G}$-*structure*, defined on $\mathbf{V}$, which is nothing but a subset of bases on which $\mathcal{G}$ acts in a simply transitive manner. The bases $\mathbf{b} \in \mathbf{B}_\mathcal{G}(\mathbf{V})$ are called $\mathcal{G}$-*bases*. For any two of them, $\mathbf{b}$ and $\mathbf{b}'$, the matrix $\tilde{\mathbf{b}}\mathbf{b}' = \mathbf{S}$ belongs to $\mathcal{G} \subset \mathrm{GL}(n, \mathbf{F})$, and no $\mathcal{G}$-basis is preferred over any other. If σ is a representation of $\mathcal{G}$ in $\mathbf{F}^m$, the elements of the vector space $(\mathbf{B}_\mathcal{G}(\mathbf{V}) \times \mathbf{F}^m)/\mathcal{G}$ are called *quantities of type* $(\mathcal{G}, \sigma)$ over $\mathbf{V}$. Again, they may also be thought of as equivariant maps $\mathbf{B}_\mathcal{G}(\mathbf{V}) \to \mathbf{F}^m$. When σ is a tensor representation of $\mathcal{G} \subset \mathrm{GL}(n, \mathbf{F})$, the vector space just considered may be identified with a tensor space over $\mathbf{V}$; but the essence of the apparently circumstantial construction is that one may have on $\mathbf{V}$ many—albeit isomorphic—$\mathcal{G}$-geometries for a fixed group $\mathcal{G}$: they correspond bijectively to the cosets in $\mathrm{GL}(n, \mathbf{F})/\mathcal{G}$.

The definitions of (pseudo)Euclidean, symplectic, and (pseudo)unitary geometries given in sect. 7.5 fall into the present scheme: by a choice of basis, the matrix of scalar products $\langle b_i \mid b_k \rangle$ may be brought to some known standard form that remains invariant only under some subgroup $\mathcal{G} \subset \mathrm{GL}(n, \mathbf{F})$—the (pseudo)orthogonal, symplectic, or (pseudo)unitary group, as the case may be, transforming inside the subset $\mathbf{B}_\mathcal{G}$ of (pseudo)orthonormal or symplectic bases. But also real, complex, and quaternionic structures fall into the scheme. While one can—as actually done before—describe these structures purely tensorially, there are indeed subgroups $\mathcal{G}$ for which this is impossible (so-called nonalgebraic linear Lie groups), so that the definition of $\mathcal{G}$-structure in their case cannot avoid the explicit use of $\mathbf{B}_\mathcal{G}(\mathbf{V})$. Also, there are groups where the tensorial characterization of $\mathcal{G}$-geometries is possible in principle but rather involved for practical purposes (as happens for the so-called exceptional simple Lie groups).

Let us also remember that the use of $\mathbf{B}_\mathcal{G}$ for $\mathcal{G} = \mathcal{L}_+^\uparrow$ was the physical starting point for the theory of special relativity: we considered the set $\mathbf{B} = \mathcal{I}$ of all inertial frames and found, on the basis of two principles, that it is a principal $\mathcal{P}_+^\uparrow$-space. The special structure of $\mathcal{P}_+^\uparrow$ as a semidirect product of $\mathcal{L}_+^\uparrow$ with the 4-dimensional translation group $\mathcal{T}$ allowed to construct from $\mathbf{B}_{\mathcal{P}_+^\uparrow}$ and the set $\mathbf{R}^4$ of event coordinates the Minkowski space $\mathbf{X}_4 = (\mathbf{B}_{\mathcal{P}_+^\uparrow} \times \mathbf{R}^4)/\mathcal{P}_+^\uparrow$, to give it the structure of an affine space with a pseudo-metric, and to study the associated Minkowski vector space $\mathbf{V}_4$ with scalar product η. We indeed confined ourselves to use only $\mathcal{L}$-bases in $\mathbf{V}_4$, or rather even only $\mathcal{L}^\uparrow$-bases—and thus an $\mathcal{L}^\uparrow$-structure—for reasons of practical realizability as well as because of the existence in nature of an *arrow of time*.

If $\mathcal{G} \subset \mathcal{G}'$, a $\mathcal{G}'$-structure compatible with a $\mathcal{G}$-structure—i.e. $\mathbf{B}_\mathcal{G} \subset \mathbf{B}_{\mathcal{G}'}$ —is already uniquely determined by $\mathbf{B}_\mathcal{G}$. This does not mean, however, that for all quantities of type $(\mathcal{G}, \sigma)$ there exist quantities of type $(\mathcal{G}', \sigma)$, since the representation σ in general will not extend to $\mathcal{G}'$.

Appendix C
Majorana Spinors, Charge Conjugation and Time Reversal in Dirac Theory

C.1 Dirac Algebra Reconsidered

Although the developments of the appendix to sect. 9.1 were given in matrix language, we intentionally avoided almost everywhere any use of a special matrix realization of the Dirac matrices, basing all considerations on the anticommutation relations (9.1.14,17) alone. We thus can adopt an abstract vector space language here to repeat and extend some of the results obtained there. To be specific, we restrict our consideration to spinors associated to 4-dimensional Minkowski vector space. The basic theorem was that a complex spinor space Σ together with a quadruple of linear operators γ_i acting irreducibly on Σ and satisfying the anticommutation relations $\gamma_{(i}\gamma_{k)} = \eta_{ik}\,\mathrm{id}_\Sigma$ is uniquely determined up to equivalence and that $\dim \Sigma = 4$.

Since the quadruples $-\gamma_i^\top$, $\gamma_i^\dagger$, $-\gamma_i^*$ act irreducibly on the complex spaces $\tilde{\Sigma}$, $\tilde{\Sigma}^*$, Σ^*, satisfying formally the same anticommutation relations there, it follows that there exist equivalence maps $B\colon \Sigma \to \tilde{\Sigma}$, $A\colon \Sigma \to \tilde{\Sigma}^*$, $C\colon \Sigma \to \Sigma^*$, such that

$$-\gamma_i^\top = B\,\gamma_i\,B^{-1}, \qquad \gamma_i^\dagger = A\,\gamma_i\,A^{-1}, \qquad -\gamma_i^* = C\,\gamma_i\,C^{-1}.$$

A and B were found to be determined uniquely up to a positive and a complex factor, respectively, by imposing the condition that the former be Hermitian with $A\gamma_0$ positive-definite, while the latter was found to be antisymmetric (sect. 9.1, exercise 9).

By the arguments sketched in exercise 6 of sect. 7.5 (ABC-theorem), it also follows that $C^*C = c\,\mathrm{id}_\Sigma$ with c real, C becoming unique up to a phase factor by requiring $c = \pm 1$, and that it needs only an adjustment of the free factors just mentioned to have the relation $B = A^\top C$, entailing $A(\psi,\psi) = -cA(\varphi,\varphi)$ when $\psi := (C\varphi)^*$. The last relation serves to couple the signature of the Hermitian form determined by A and the actual value of $c = \pm 1$. At the moment, both are still unknown, however; but by a completely analogous computation one finds that

$$A(\psi, \gamma_i\psi) = cA(\varphi, \gamma_i\varphi),$$

and by invoking the definiteness of $A\gamma_0$ we can now conclude that $c = +1$. (The previous relation now says that the Hermitian form associated with A itself has neutral signature.) Therefore, C gives rise to a real structure, or complex conjugation, $\mathcal{C} := \mathcal{K}^{-1}C$ in Σ (cf. Appendix B.6), which will be studied in more detail in the next section. (Hence the mnemonic C for conjugation.)

Of course, also the quadruples $+\gamma_i^\top$, $-\gamma_i^\dagger$, $+\gamma_i^*$ give representations of the Clifford algebra, intertwined to γ_i by $A' \propto A\gamma$, $B' \propto B\gamma$, $C' \propto C\gamma$. Our choices of signs in the definitions of A and C are dictated by the purposes they are to serve in the Dirac theory—see eqs. (9.1.41,42), (9.3.20) and sect. C.2—together with our chosen signature convention $(+ - - -)$; the choice for B then follows if we wish to have $B \propto A^\top C$. For the convention $(+ + + -)$, the physical definitions for A, C would require the opposite signs. To complete the number of sources for possible confusion, we remark that for arbitrary signature (p,q) with p pluses and q minuses, the signs in the definitions of A,

B, C *must* be taken as $(-1)^q$, $(-1)^m$, $(-1)^{\tilde{m}}$, respectively, when $p + q = 2m + 1$, $p - q = 2\tilde{m} + 1$ are *odd*—otherwise the intertwiners A, B, C do not exist for an irreducible representation; when $p + q = 2m$, $p - q = 2\tilde{m}$ are even, the opposite signs are possible, the corresponding intertwiners being $A' \propto A\gamma$, etc. So from the point of view of n-dimensional systematics (as required in some attempts to construct unified theories of all interactions), our A, B, C for $(+ - - -)$ Dirac theory should have been written A', B', C; but we omitted the primes for notational convenience. We add that in the general case one has $B^\top = (-1)^{m(m+1)/2}B$, $C^2 = (-1)^{\tilde{m}(\tilde{m}+1)/2}\mathrm{id}_\Sigma$ and definiteness for $A\gamma_H$ with $H = \{p+1,\ldots,p+q\}$. Note that in certain signatures both C and $C' := C\gamma$ are of second kind.

We can summarize our findings and conventions by saying that A, B, C define, on the spinor space Σ, a Hermitian form (of neutral signature but such that $A\gamma_i n^i$ is positive-definite for all timelike future-directed n), a symplectic form, and a real structure, such that the following diagram ('Carter's diamond') is commutative:

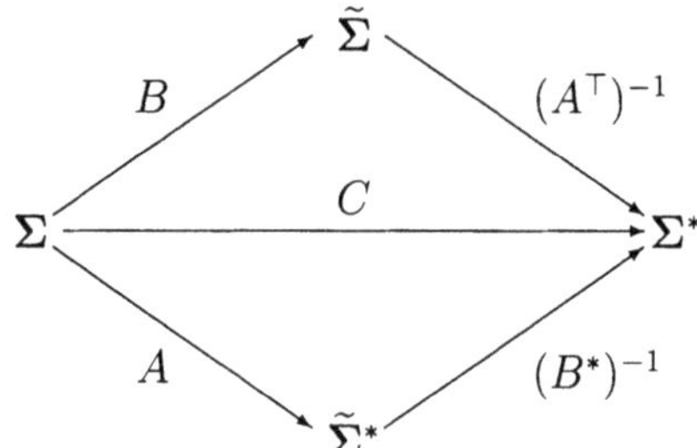

We add the chiral properties of A, B, C which follow from the definition of the chiral projectors $P_\pm := (\mathrm{id} \pm i\gamma)/2$ and the defining relations for A, B, C. One derives

$$AP_\pm = P_\mp^\dagger A, \quad BP_\pm = P_\pm^\top B, \quad CP_\pm = P_\mp^* C,$$

from which

$$A(P_+\psi, P_+\varphi) = 0 = A(P_-\psi, P_-\varphi)$$

$$B(P_+\psi, P_-\varphi) = 0 = B(P_-\psi, P_+\varphi)$$

$$CP_\pm = P_\mp^* C \Leftrightarrow CP_\pm = P_\mp C.$$

A and C are thus chirality mixing (chirality odd), while B is chirality splitting (chirality even), and so is the antisymmetric bilinear form associated with $B\gamma$: the antisymmetric bilinear forms defined by the latter in the chiral subspaces are essentially the (semi-)spinor 'metrics' ϵ, $\dot\epsilon$ considered in chap. 8. If we use a basis for Σ adapted to the chiral decomposition $\Sigma = P_+\Sigma \oplus P_-\Sigma$ (these subspaces were denoted $\mathbf{S}$, $\dot{\mathbf{S}}$ in sect. 8.4), we can still normalize the subbases to get the following matrices for A, B, C:

$$B = \begin{pmatrix} -\epsilon & 0 \\ 0 & \epsilon \end{pmatrix}, \quad C = \begin{pmatrix} 0 & \epsilon \\ -\epsilon & 0 \end{pmatrix}, \quad A = \begin{pmatrix} 0 & 1 \\ 1 & 0 \end{pmatrix}, \quad \text{where} \quad \epsilon := \begin{pmatrix} 0 & 1 \\ -1 & 0 \end{pmatrix}.$$

This takes into account all conventions made and fixes the matrices up to a common sign in B and C. Note again that statements like "$A = \gamma_0$, $A^2 = -C^2 = E$ (unit matrix), $\gamma_0 = $ Hermitian, $\gamma_\mu = $ anti-Hermitian, ..." make sense only as matrix relations with respect to a certain class of bases in Σ but not as relations between maps, simply because the source and target spaces of the maps do not match appropriately. Due to the possibility of taking $A = \gamma_0$ in a restricted class of matrix representations, in many texts there is no symbol for the geometrical structure A at all, while B is written C since it serves to define $\mathcal{C}$ via our $\mathcal{K}^{-1}(A^\top)^{-1}B$. The geometrical, or covariance, properties are not brought out by such a formulation, creating a possibility for confusion when other dimensions and signatures—notably the Euclidean one— or the curved Riemannian spacetime of General Relativity get involved.

C.2 Majorana Spinors, Charge Conjugation, Time Reversal

The antilinear operator $\mathcal{C} := \mathcal{K}^{-1}C$ on Σ satisfies (from Appendices B.4,6 and eqs. (9.1.27,36))

$$\mathcal{C}\,\gamma_i = -\gamma_i\,\mathcal{C}, \quad \mathcal{C}^2 = \mathrm{id}_\Sigma, \quad S^{-1}\mathcal{C}\,S = d(L)\,\mathrm{sign}L^0{}_0\,\mathcal{C},$$

so that the real structure defined by it on Σ is invariant under $\mathcal{L}_+^\uparrow$—and for the choice $d(L) = \mathrm{sign}L^0{}_0$ even under $\mathcal{L}$. The real spinors with respect to $\mathcal{C}$ (i.e., spinors invariant under $\mathcal{C}$) are also called *Majorana spinors*; with respect to a Majorana basis they have real components, and the $S(L)$ have real matrices (at least for $\mathcal{L}_+^\uparrow$). On the other hand, the γ_i anticommute with $\mathcal{C}$, and so with respect to a Majorana basis have *purely imaginary* matrices (*Majorana representation*).

For the signature convention $\eta_{ik} = (-+++)$ the γ_i matrices would be *real* in a Majorana representation, since here one would have to replace $\mathcal{C}$ by $\mathcal{C}' := \mathcal{C}\,\gamma$, satisfying

$$\mathcal{C}'\gamma_i\,\mathcal{C}'^{-1} = \gamma_i,$$

to achieve $\mathcal{C}'^2 = \mathrm{id}_\Sigma$. The anti-involution $\mathcal{C}'$ defined in the same manner in *our* signature $(+---)$ would be of *second* kind, defining a quaternionic structure. This means that one could work with spinors having only two quaternionic components, and with 2×2 matrices having quaternionic elements, but this is practically never done in physics.

The *physical* interest in the antilinear operator $\mathcal{C}$ lies in the following fact: If a spinor field ψ satisfies the Dirac equation with 'minimal' coupling to an electromagnetic field as described by a 4-potential A_k ($\hbar = c = 1$, $e =$ charge of the particle):

$$\gamma^k(i\partial_k - eA_k(x))\,\psi(x) = m\,\psi(x),$$

then $\psi^c := \mathcal{C}\psi = (C\psi)^* = C^*\psi^*$ satisfies

$$\gamma^k(i\partial_k + eA_k(x))\,\psi^c(x) = m\,\psi^c(x),$$

where the sign of charge appears reversed. Therefore, the operator $\mathcal{C}$ is called *charge conjugation* (in its first-quantized version; in second quantization, charge conjugation becomes linear again—cf. textbooks on elementary particle physics for a more detailed and generalized treatment of this concept).

It should be noted that the decisive feature of the charge conjugation operator is its anticommuting with the γ_k (for our signature, or its commuting with the γ_k for the opposite signature—the point being that the 4-momentum of a free Dirac particle must be timelike, which dictates the occurrence or nonoccurrence of a factor i in the Dirac equation). Its property of being an anti-involution of first rather than second kind is 'accidental' here in that in higher Lorentzian dimensions it indeed may be of second kind, so that Majorana particles (charge self-conjugate) then do not exist. On the other hand, the concept of Lorentz invariant real structure primarily needs an anti-involution of the first kind, anticommuting or commuting with the γ_k; in certain higher Lorentzian dimensions neither a commuting nor an anticommuting one exists, in others both exist.

The same operator also allows to formulate the correct version—in the sense of sect. 9.6—of Lorentz transformations containing a reversal of time, $L \in \mathcal{L}^\downarrow$, for Dirac fields, which has to be antilinear. Indeed, the linear transformation $S(L)$ constructed in eq. (9.1.21) (sometimes called Racah time reversal)—while granting the correct covariance behavior, eq. (9.1.26)—is linear and also does not correspond to the time-pseudovector nature (sect. 8.5) of the electromagnetic 4-potential if the latter is included into the Dirac equation as written above. Namely, from this equation we get, applying $S(L)$,

$$\gamma^k(iL_k{}^j\partial_j - eL_k{}^j A_j)S(L)\psi = mS(L)\psi,$$

and for $x^{i'} = L^i{}_k x^k$ we have $L_k{}^j\partial_j = \partial'_k$; but for physical reasons we have $A'_k = \operatorname{sign} L^0{}_0 \cdot L_k{}^j A_j$. Both, this mismatch as well as the lack of antilinearity, are removed simultaneously by taking, for $L \in \mathcal{L}^\downarrow$, as the transformed spinor

$$\psi'(x) = \mathcal{C}\, S(L)\, \psi(L^{-1}x),$$

as the application of $\mathcal{C}$ to the foregoing equation shows, taking antilinearity and $\mathcal{C}\gamma_i\mathcal{C}^{-1} = -\gamma_i$ into account (*Wigner time reversal*).

It is of interest, in view of sect. 9.6, to determine the square of the operators for T and PT. Up to phase factors which are unimportant here, we had—cf. eq. (9.1.29)— $S(PT) = \gamma$, $S(P) = \gamma^0$, so $S(T) = \gamma^1\gamma^2\gamma^3$, or more generally, $S(n) = \gamma\gamma_i n^i$ for a reflection in a hyperplane with timelike unit normal n. Using the relations of the previous section one then verifies

$$(\mathcal{C}S(n))^2 = (\mathcal{C}S(T))^2 = (\mathcal{C}S(PT))^2 = -\operatorname{id}_\Sigma.$$

Let us carry out here explicitly the transformation leading from a basis (b_1, b_2, b_3, b_4), to which the matrix representation (9.1.13) of the γ_i may refer, to a Majorana basis (m_1, m_2, m_3, m_4). (Other Majorana bases are then obtained by arbitrary (regular) *real* substitutions.) For the matrix representation (9.1.13) the matrix for C was written down in sect. C.1. We now write $\binom{u}{v}$ for the column of components of a Majorana spinor in the present basis, u, v being 2-rowed columns. The Majorana reality condition $C\psi = \psi^*$ then simply requires $v = -\epsilon u^*$, $u = $ arbitrary. To get a basis of Majorana spinors, pick for u the columns $\binom{1}{0}$, $\binom{0}{1}$, $\binom{i}{0}$, $\binom{0}{i}$. (Note that these columns are linearly dependent over $\mathbf{C}$; but the reality condition is not $\mathbf{C}$-linear and leads to independent Majorana basis spinors!) The columns $\binom{u}{v}$ obtained in this way also form the columns of the transformation matrix S between the bases:

$$(m_1, m_2, m_3, m_4) = (b_1, b_2, b_3, b_4)\begin{pmatrix} 1 & 0 & i & 0 \\ 0 & 1 & 0 & i \\ 0 & -1 & 0 & i \\ 1 & 0 & -i & 0 \end{pmatrix} = (b_1 + b_4, b_2 - b_3, i(b_1 - b_4), i(b_2 + b_3))$$

$$\Rightarrow (b_1, b_2, b_3, b_4) = \frac{1}{2}(m_1 - im_3, m_2 - im_4, -m_2 - im_4, m_1 + im_3).$$

We can read off S^{-1} from the second line to find a Majorana version $\left(\gamma^k\right)_{\mathrm{Majorana}} = S^{-1} \cdot (9.1.13) \cdot S$ for the γ-matrices which is purely imaginary as expected:

$$\gamma^0_{\mathrm{Maj}} = \begin{pmatrix} 0 & 0 & 0 & i \\ 0 & 0 & -i & 0 \\ 0 & i & 0 & 0 \\ -i & 0 & 0 & 0 \end{pmatrix} = \begin{pmatrix} \mathbf{0} & -\sigma_2 \\ -\sigma_2 & \mathbf{0} \end{pmatrix}, \quad \gamma^1_{\mathrm{Maj}} = \begin{pmatrix} 0 & 0 & i & 0 \\ 0 & 0 & 0 & -i \\ i & 0 & 0 & 0 \\ 0 & -i & 0 & 0 \end{pmatrix} = \begin{pmatrix} \mathbf{0} & i\sigma_3 \\ i\sigma_3 & \mathbf{0} \end{pmatrix},$$

$$\gamma^2_{\mathrm{Maj}} = \begin{pmatrix} i & 0 & 0 & 0 \\ 0 & i & 0 & 0 \\ 0 & 0 & -i & 0 \\ 0 & 0 & 0 & -i \end{pmatrix} = \begin{pmatrix} i\mathbf{1} & \mathbf{0} \\ \mathbf{0} & -i\mathbf{1} \end{pmatrix}, \quad \gamma^3_{\mathrm{Maj}} = \begin{pmatrix} 0 & 0 & 0 & -i \\ 0 & 0 & -i & 0 \\ 0 & -i & 0 & 0 \\ -i & 0 & 0 & 0 \end{pmatrix} = \begin{pmatrix} \mathbf{0} & -i\sigma_1 \\ -i\sigma_1 & \mathbf{0} \end{pmatrix}.$$

Appendix D
Poincaré Covariance in Second Quantization

In sect. 9.5, we mentioned the difficulties of relativistic wave mechanics in the narrow sense of a one-particle formalism, perfectly consistent in N.R. theory, difficulties caused by the necessary occurrence of negative energy solutions. The antiparticle interpretation of the negative energy states in general requires, in the presence of interactions, to work with a many-particle formalism, and it is useful to study this formalism already in the noninteracting situation (Fock space). In the interacting case, then, at least two such Fock spaces are needed ('ingoing' and 'outgoing' Fock spaces), which are mapped into each other by the 'S-operator' characterizing the interaction and describing scattering, annihilation and creation processes. (See, e.g., Henley and Thirring 1962; H. Rumpf, H. Urbantke, Ann. Phys. (N.Y.) *114*, 332 (1978)).

In the present appendix we restrict our considerations to the simplest case—free neutral (particle = antiparticle) spinless particles, described by scalar wave functions. We 'construct' the associated quantum field and discuss the Poincaré covariance of the formalism. A more detailed treatment including the functional analysis necessary due to the occurrence of infinite-dimensional spaces is given, e.g., by Kastler (1961), who dedicates more than half of his text to a careful treatment of free quantum fields, in particular of the Maxwell and the Dirac field. A physical discussion of the observables and states and the nontrivial features of relativistic free quantum fields is found in Henley and Thirring (1962); it might be useful, however, to make clear which aspects of the theory are just 'trivial linear algebra'.

We should add here that our presentation has, in addition to its lack of mathematical rigor concerning functional analysis, another defect. Namely, the modern concept of a quantum field is such that our semi-historical approach, motivated from the desire to surmount the difficulties encountered in the construction of quantum mechanics of a relativistic particle, yields just a very special—although important—operator representation of an object that might be loosely called the abstract field algebra. From the point of view of trying to set up the quantum mechanics of a relativistic *field*—rather than *particle*—the field algebra is the primary object, and usually one nowadays approaches the subject from this side, considering the many-particle quantum mechanics as just one possibility to represent the quantum field. However, our modest aim is just to define the field operators in one representation only, and to write down the Poincaré covariance properties of them.

D.1 The One-Particle Space

We consider the space $\mathbf{H}_m$ of *complex-valued* solutions φ of the Klein-Gordon equation $(\Box + m^2)\varphi = 0$, normalizable in the sense of the scalar product $\langle\!\langle \ | \ \rangle\!\rangle_m$ obtained from eq. (9.5.11) by omitting the factor $2\pi\delta(m'^2 - m^2)$. $\mathbf{H}_m$ is the orthogonal direct sum $\mathbf{H}_m^+ \oplus \mathbf{H}_m^-$, where $\mathbf{H}_m^+$, or $\mathbf{H}_m^-$, contains solutions with only positive, or negative, frequencies ($A_-(\mathbf{p}) \equiv 0$ or $A_+(\mathbf{p}) \equiv 0$). The scalar product induced by $\langle\!\langle \ | \ \rangle\!\rangle$ on $\mathbf{H}^\pm$ may also be written $\pm\langle \ | \ \rangle$, where $\langle \ | \ \rangle$ is the Hermitian sesquilinear form induced from the form (9.5.15). (Cf. eq. (9.5.16); we will omit the mass parameter m as an index in most formulae to follow.)

The *one-particle states* for free, neutral spinless particles are then described by 1-dimensional subspaces (rays) of $\mathbf{H}^+$, to avoid negative energies. On this space we have the irreducible unitary action $\varphi \mapsto U(a, L)\varphi$ of $\mathcal{P}^\uparrow$ given by

$$(U(a, L)\varphi)(x) = \varphi(L^{-1}(x - a)) \quad \text{for} \quad L \in \mathcal{L}^\uparrow.$$

However, for $L \in \mathcal{L}^\downarrow$ this definition would give an operator that leads from $\mathbf{H}^+$ to $\mathbf{H}^-$. We can remedy this by introducing an additional complex conjugation, which leads from $\mathbf{H}^\pm$ to $\mathbf{H}^\mp$ as is immediate from looking at the Fourier transform. The operator thus defined is then *antiunitary*, carrying $\langle \varphi \,|\, \psi \rangle$ into $\langle \varphi \,|\, \psi \rangle^*$ (cf. sect. 9.2).

Note that while we were preaching in Appendix B.3 that complex conjugation for an abstract complex vector space $\mathbf{H}$ would lead to another space, we have here the situation of a complex conjugation *in* $\mathbf{H}$, since this space, being the complexification of the space of real-valued solutions of the Klein-Gordon equation, has a natural real structure in the sense of Appendix B.6. On the other hand, the direct sum decomposition of $\mathbf{H}$ into a pair of complex-conjugate subspaces corresponds in the real subspace to the presence of a complex structure J such that its complex-linear extension to $\mathbf{H}$ defines the above decomposition via the projection operators $P^\pm = (\mathrm{id} \pm iJ)/2$.

The projection operators $P^\pm$ associated with our $\mathcal{P}$-invariant decomposition of $\mathbf{H}$ may be represented by integral kernels $\triangle^\pm(x; x')$ with respect to $\langle \,|\, \rangle$, the latter being given by an integral: let $\{^+\varphi_k \in \mathbf{H}^+,\ ^-\varphi_l \in \mathbf{H}^-\}$ be any complete orthonormal system in $\mathbf{H}$ adapted to $\mathbf{H}^\pm$, so that

$$\langle\, ^+\varphi_k \,|\, ^+\varphi_{k'} \,\rangle = \delta_{kk'}, \qquad \langle\, ^+\varphi_k \,|\, ^-\varphi_l \,\rangle = 0, \qquad \langle\, ^-\varphi_l \,|\, ^-\varphi_{l'} \,\rangle = -\delta_{ll'}$$

(the $^-\varphi_l$ may be taken, e.g., as $(^+\varphi_l)^*$, but need not be), then we have the expansion

$$\varphi = \sum_k \langle\, ^+\varphi_k \,|\, \varphi \,\rangle\, ^+\varphi_k - \sum_l \langle\, ^-\varphi_l \,|\, \varphi \,\rangle\, ^-\varphi_l = P^+\varphi + P^-\varphi,$$

which is, more explicitly,

$$(P^\pm\varphi)(x) = \int_\sigma d\sigma'^j\, \triangle^\pm(x; x')\, \overset{\leftrightarrow}{\partial_j'}\, \varphi(x'),$$

where

$$\triangle^+(x; x') := +i \sum_k {}^+\varphi_k(x)\, {}^+\varphi_k^*(x'), \qquad \triangle^-(x; x') := -i \sum_l {}^-\varphi_l(x)\, {}^-\varphi_l^*(x').$$

Since $P^+ + P^- = \mathrm{id}_{\mathbf{H}}$, the integral kernel

$$\triangle(x; x') := \triangle^+(x; x') + \triangle^-(x; x')$$

yields the solution of the Cauchy problem for the Klein-Gordon equation with initial values on σ:

$$\varphi(x) = \int_\sigma d\sigma'^j \triangle(x; x')\, \overset{\leftrightarrow}{\partial_j'}\, \varphi(x').$$

Taking for σ the hypersurface $t' = t$, where $x = (\mathbf{x}, t)$ in some inertial system, we see that we must have

$$\Delta(x; x')\big|_{t = t'} = 0, \qquad \partial_{t'}\Delta(x; x')\big|_{t = t'} = -\delta(\mathbf{x} - \mathbf{x}').$$

Since "$t = t'$ in some inertial system" just means that x and x' are spacelike to each other, we conclude that $\Delta(x; x') = 0$ whenever $(x - x')^2 < 0$.

From their definitions, the integral kernels $\Delta^\pm(x; x')$, $\Delta(x; x')$ satisfy the Klein-Gordon equation in each of their arguments, and we also have the relations

$$\Delta^\pm(x; x')^* = -\Delta^\pm(x'; x) = \Delta^\mp(x; x'),$$

$$\Delta(x; x')^* = -\Delta(x'; x) = \Delta(x; x').$$

Defining now for each of these kernels $K(x; x')$ a (distributional) wave function K_x, depending on x as a parameter, by

$$K_x(x') = K(x'; x),$$

we can rewrite the projections $P^\pm$ as

$$(P^\pm\varphi)(x) = \left\langle \frac{1}{i}\Delta_x^\pm \,\bigg|\, \varphi \right\rangle.$$

This way of writing them will be useful in the following sections.

D.2 Fock Space and Field Operator

As is well known and as is plausible from eq. (7.8.5), the Hilbert space of a quantum mechanical multiparticle system is the tensor product of the one-particle spaces. If the particles are all of the same kind, the multiparticle space will be a tensorial power of the one-particle space; on it, we have the action of the permutation group as in Appendix B.11. The principle of indistinguishability further requires that only the trivial or the alternating representation is actually allowed to occur (cf. Landau and Lifshitz 1958a), leading to *Bose* or *Fermi statistics*, respectively. We shall impose Bose statistics—we already mentioned that in relativistic theory Fermi statistics for integer spin fields leads to difficulties—and so describe (pure) p-particle states by rays from the subspace $\bigvee^p(\mathbf{H}^+)$.

To allow for changes in the number of particles under the influence of external actions, one now takes as the state vector space the direct sum $\bigvee(\mathbf{H}^+)$, called the (Bosonic) *Fock space over* $\mathbf{H}^+$. The ray given by the 1-dimensional subspace $\bigvee^0(\mathbf{H}^+) = \mathbf{C}$ is called the *vacuum state*. From it, a basis for the whole Fock space is obtained by applying polynomials formed from the *creation operators* $a^\dagger(\varphi) := \mu(\varphi)$, $\varphi \in \mathbf{H}_m^+$ (see Appendix B.12). The states corresponding to inhomogeneous tensors do not have a well-defined number of particles but only well-defined probabilities to find p particles in them. Thus, the particle number now is a quantum mechanical observable, i.e., we have the *particle number operator* N, defined by the property that the $\mathsf{V}^p(\mathbf{H}^+)$ be its eigenspaces for the eigenvalues p. The scalar product $\langle\,|\,\rangle$ in $\mathbf{H}^+$ assigns, in an *anti*linear fashion, to every $\varphi \in \mathbf{H}^+$ the element $\langle\varphi|\cdot\rangle$ of the dual space, and the operator on $\mathsf{V}(\mathbf{H}^+)$ given by the inner product with that element, $\iota(\langle\varphi|\cdot\rangle)$, is called the *annihilation operator* $a(\varphi)$ associated with φ. By the definition of ι (Appendix B.12), $a(\varphi)$ and $a^\dagger(\varphi)$ are are Hermitian conjugates in the sense of the scalar product defined by $\langle\,|\,\rangle$ on $\mathsf{V}(\mathbf{H}^+)$, and we have from Appendix B.12 the *commutation relations*

$$[a(\varphi), a(\psi)] = 0 = [a^\dagger(\varphi), a^\dagger(\psi)],$$

$$[a(\varphi), a^\dagger(\psi)] = \langle\,\varphi\,|\,\psi\,\rangle\,\mathrm{id}_{\mathsf{V}(\mathbf{H}^+)}.$$

The operator

$$\Phi(x) := \underbrace{a^\dagger\left(\frac{1}{i}\triangle_x^+\right)}_{=:\ ^-\Phi(x)} + \underbrace{a\left(\frac{1}{i}\triangle_x^+\right)}_{=:\ ^+\Phi(x)}$$

on $\mathsf{V}(\mathbf{H}^+)$ is called the *field operator*. $\Phi(x)$ satisfies the Klein-Gordon equation with respect to x, and since it is Hermitian one says that it represents a real, or neutral, quantum field. (We emphasize that this terminology does *not* mean that the wave functions $\varphi \in \mathbf{H}^+$ representing one-particle *states* are real!) If we insert the expansion of $\triangle_x^+$ we get the usual expansion of $\Phi(x) = {}^+\Phi(x) + {}^-\Phi(x)$ with respect to a complete system,

$$^+\Phi(x) = \sum_k a_k\,{}^+\varphi_k(x), \quad {}^-\Phi(x) = \sum_k a_k^\dagger\,{}^+\varphi_k^*(x), \quad a_k := a({}^+\varphi_k), \quad a_k^\dagger := a^\dagger({}^+\varphi_k).$$

From the general commutation relations above we get

$$[a_k, a_{k'}] = 0, \qquad [a_k, a_{k'}^\dagger] = \delta_{kk'}, \qquad [a_k^\dagger, a_{k'}^\dagger] = 0,$$

and for the commutators between the field operators at different points x, y we have $[\Phi(x), \Phi(y)] = \langle -i\triangle_x^+\,|\,-i\triangle_y^+\rangle - \langle -i\triangle_y^+\,|\,-i\triangle_x^+\rangle$. Since P^+ is idempotent and $\triangle_y^+$ contains only positive frequencies we have $\langle -i\triangle_x^+\,|\,\triangle_y^+\rangle = \triangle_y^+(x) = \triangle^+(x,y)$, so that we finally have the *commutation relation*[1]

$$[\Phi(x), \Phi(y)] = -i\triangle(x,y).$$

From the relations given above for $\triangle(x,y)$ at equal time arguments we get the *equal time commutation relations* (with respect to any inertial system)

$$[\Phi(x), \Phi(x')]_{t\,=\,t'} = 0, \qquad\qquad [\dot\Phi(x), \dot\Phi(x')]_{t\,=\,t'} = 0,$$

$$[\Phi(x), \dot\Phi(x')]_{t\,=\,t'} = i\delta(\mathbf{x} - \mathbf{x}').$$

---

[1]With Fermi statistics, we would have obtained here the anticommutator relation $\Phi(x)\,\Phi(y) + \Phi(y)\,\Phi(x) = -i(\triangle^+(x,y) - \triangle^-(x,y)) =: -i\triangle_1(x,y)$; contrary to $\triangle(x,y)$, this does *not* vanish for spacelike separation of the arguments, thus creating causality problems.

When the quantum dynamics for a classical real scalar field is set up by the procedure of canonical quantization, one brings the dynamics given by the Klein-Gordon equation to Hamiltonian form, the momentum canonically conjugate to $\Phi(\mathbf{x}, t)$ being $\Pi(\mathbf{x}, t)$, and then 'quantizes' by regarding the algebra generated by the $\Phi(\mathbf{x}, t)$, $\Pi(\mathbf{x}, t)$ (Hermitian) as noncommutative, satisfying the equal time commutation relations by fiat. One then tries to represent this algebra by operators on some Hilbert space. One important (irreducible) operator realization is given by the Fock space construction above (Fock representation).

D.3 Poincaré Covariance and Conserved Quantities

The physical interpretation of the formalism uses the absolute values of expressions of the type $\langle Z \mid \Phi(x)\,\Phi(y)\ldots Z'\rangle$, where Z, Z' are vectors of the Fock space $\vee(\mathbf{H}^{+})$ and $\langle \mid \rangle$ is the scalar product there. Writing U instead of $U(a, L)$, the Poincaré covariance of the formalism requires that

$$|\langle U^{\vee}Z \mid \Phi(Lx + a)\,\Phi(Ly + a)\ldots U^{\vee}Z'\rangle| = |\langle Z \mid \Phi(x)\,\Phi(y)\ldots Z'\rangle|.$$

This will certainly be satisfied if $(U^{\vee})^{-1}\,\Phi(Lx + a)\,U^{\vee} = \Phi(x)$, or

$$(U^{\vee})^{-1}\,\Phi(x)\,U^{\vee} = \Phi(L^{-1}(x - a)).$$

To demonstrate this fundamental *transformation property of field operators*, we go back to its definition as well as to the transformation properties of the creation and annihilation operators obtained at the end of Appendix B.12. We have

$$(U^{\wedge})^{-1}\,a^{\dagger}(-i\triangle_{x}^{+})\,U^{\wedge} = a^{\dagger}(-iU^{-1}\triangle_{x}^{+})$$

and an analogous relation for the annihilation operator, taking into account the unitarity of U. (To save space, we restrict to the group $\mathcal{P}^{\uparrow}$, but there are similar manipulations with interpolated complex conjugations that do the job for $\mathcal{P}^{\downarrow}$.) The proof will be completed when we will have shown that

$$U^{-1}\triangle_{x}^{\pm} = \triangle_{L^{-1}(x-a)}^{\pm},$$

which also yields the $\mathcal{P}^{\uparrow}$-invariance of the integral kernels

$$\triangle^{\pm}(x, x') = \triangle^{\pm}\big(L^{-1}(x - a), L^{-1}(x' - a)\big).$$

But this follows from the $\mathcal{P}^{\uparrow}$-invariance of the spaces $\mathbf{H}^{\pm}$, expressible as $P^{\pm}U = UP^{\pm}$: for any $\varphi \in \mathbf{H}$ we have

$$\langle U^{-1}(-i\triangle_{x}^{\pm}) \mid \varphi\rangle = \langle -i\triangle_{x}^{\pm} \mid U\varphi\rangle = (P^{\pm}U\varphi)(x) =$$

$$= (UP^{\pm}\varphi)(x) = (P^{\pm}\varphi)(L^{-1}(x - a)) = \langle -i\triangle_{L^{-1}(x-a)}^{\pm} \mid \varphi\rangle.$$

We have reached the aim of this appendix, namely to show the relation between the action of $\mathcal{P}$ on wave functions and on quantum field operators. We just add a few complements.

Since we have shown at the same time that $\triangle(x, x')$ and, with it, the commutation relations for $\Phi(x)$ are $\mathcal{P}^\uparrow$-invariant, we can state that $\Phi(x) \mapsto \Phi(Lx + a)$ is an *automorphism of the field algebra*. We now show that this automorphism (we restrict to $\mathcal{P}_+^\uparrow$, but space reversals could be included) is *inner* (in the sense of Appendix A), meaning that the operator $U^\vee$ above on the Fock space may be expressed in terms of the field operator, thus being the representative of an element of the 'abstract' field algebra. This we do by explicitly expressing the (Hermitian) generators of the action of $\mathcal{P}_+^\uparrow$ in terms of the field operator. (It should be underlined again at this point that not only the 'proofs' sketched in this appendix but also the statements themselves need qualifications and refinements of a functional analytic nature to become mathematically acceptable!)

The generators in question are, according to our remarks in Appendix B.12, certain derivations whose action is known once they are defined on $\mathbf{H}^+$. The (Hermitian) generators of $\mathcal{P}_+^\uparrow$ on $\mathbf{H}^+$ are the differential operators given in eqs. (9.3.12,15); we will write them here in a unified fashion as $\varphi \mapsto X\varphi$, $(X\varphi)(x) = i\xi^k(x)\,\varphi_{,k}(x)$, where the vector field $\xi^k(x)$ satisfies the *Killing equation* (5.9.29). Now for $\varphi \in \mathbf{H}^+$ we have $X\varphi \in \mathbf{H}^+$ and thus $\varphi = P^+\varphi$ and $X\varphi = P^+X\varphi$, which we write explicitly as

$$\mathrm{id}_{\mathbf{H}^+}\varphi = i\int_\sigma d\sigma'^j(-i\triangle_{x'}^+)\,\overset{\leftrightarrow}{\partial}_j'\,\varphi(x') = i\int_\sigma d\sigma'^j(-i\triangle_{x'}^+)\,\overset{\leftrightarrow}{\partial}_j'\,\langle\,-i\triangle_{x'}^+\,|\,\varphi\,\rangle$$

and as

$$X|_{\mathbf{H}^+}\varphi = i\int_\sigma d\sigma'^j(-i\triangle_{x'}^+)\,\overset{\leftrightarrow}{\partial}_j'\,i\xi^k(x')\,\partial_k'\varphi(x') =$$

$$= i\int_\sigma d\sigma'^j(-i\triangle_{x'}^+)\,\overset{\leftrightarrow}{\partial}_j'\,\left\langle\,i\xi^k(x')\,\partial_k'\frac{1}{i}\triangle_{x'}^+\,\Big|\,\varphi\,\right\rangle.$$

When we remember the definition of $^\pm\Phi(x)$ as well as that of the annihilation and creation operators $a(\ldots)$, $a^\dagger(\ldots)$, we see that the action on $\varphi \in \mathbf{H}^+$ of the Fock space operators

$$N := i\int_\sigma d\sigma'^j\,{}^-\Phi(x')\,\overset{\leftrightarrow}{\partial}_j'\,{}^+\Phi(x') \quad \text{and} \quad N_\xi := i\int_\sigma d\sigma'^j\,{}^-\Phi(x')\,\overset{\leftrightarrow}{\partial}_j'\,(i\xi^k(x')\,\partial_k'\,{}^+\Phi(x'))$$

is identical with the expressions for φ and $X\varphi$ just written. On the other hand, these Fock space operators are derivations on $\vee(\mathbf{H}^+)$, since it follows from Appendix B.12 that any product of a creation and an annihilation operator, in this order, is a derivation in $\vee(\mathbf{V})$ and also in $\wedge(\mathbf{V})$ (of degree 0, i.e., type-preserving). More explicitly, the derivation property implies for $\varphi \in \vee^p(\mathbf{H}^+)$ that

$$N|_{\vee^p(\mathbf{H}^+)} = p\,\mathrm{id}_{\vee^p(\mathbf{H}^+)}$$

$$\left(N_\xi|_{\vee^p(\mathbf{H}^+)}\,\varphi\right)(x_1, \ldots, x_p) = \left(i\xi^k(x_1)\frac{\partial}{\partial x_1^k} + \ldots + i\xi^k(x_p)\frac{\partial}{\partial x_p^k}\right)\varphi(x_1, \ldots, x_p).$$

Thus N is nothing but the particle number operator defined in Appendix D.2 (another expression for it will be written below), and N_ξ is the desired generator as expressed in terms of Φ.

The Hermitian generators of the action of $\mathcal{P}_{+}^{\uparrow}$ on Fock space just obtained are closely related to the conserved quantities $\int d\sigma_i \Theta^i{}_k \xi^k$ found in sect. 10.2 for a classical free real scalar field. Indeed, for the classical field we have

$$\int_\sigma d\sigma_i \Theta^i{}_k \xi^k = -\frac{1}{2} \int_\sigma d\sigma_i\, \Phi\, \overset{\leftrightarrow}{\partial}{}^i (\xi^k \partial_k \Phi),$$

since the difference of the integrands may be converted into $(\Phi\, \Phi^{[i} \xi^{k]})_{,k}$ using the field equation and Killing's equation[1]; thus its hypersurface integral $\int d\sigma_i$ converts into a 2-surface integral at infinity which vanishes under the usual boundary conditions. (This is most easily seen on taking σ: $x^0 = const.$ as in sect. 10.2.) Now the same manipulations are possible (without using commutativity) for the operator field $\Phi(x)$ upon the understanding that $\Theta^i{}_k$ denote the 'Hermiticized' expression, i.e., $\Phi^{,i}\Phi_{,k}$ is replaced by $\frac{1}{2}(\Phi^{,i}\Phi_{,k} + \Phi_{,k}\Phi^{,i})$. The generator N_ξ is then the *normally ordered* form

$$N_\xi =: \int d\sigma_i \Theta^i{}_k \xi^k :$$

of that expression, arising when $\Phi = {}^+\Phi + {}^-\Phi$ is substituted and the factors ${}^+\Phi$ are everywhere written to the right of the factors ${}^-\Phi$. (Observe the orthogonality between positive and negative frequency solutions.) The normally ordered form differs from the original expression by a (somewhat ill-defined) multiple of the identity operator: this is most clearly seen when one expands ${}^+\Phi$ with respect to an orthonormal basis of eigenfunctions φ_k of $X_{|\mathbf{H}+}$ (and ${}^-\Phi$ with respect to the complex-conjugate functions). Denoting the eigenvalues of X by X_k (assumed discrete just for the ease of writing), we get

$$N = \sum_k a_k^\dagger a_k, \qquad\qquad N_\xi = \sum_k X_k\, a_k^\dagger a_k,$$

$$-\frac{1}{2}\int_\sigma d\sigma_i\, \Phi\, \overset{\leftrightarrow}{\partial}{}^i(\xi^k \partial_k \Phi) = \frac{1}{2}\sum_k X_k\left(a_k^\dagger a_k + a_k a_k^\dagger\right) = N_\xi + \frac{1}{2}\operatorname{Tr} X_{|\mathbf{H}+}\mathrm{id},$$

using the commutation relations. The formally infinite vacuum expectation term $\frac{1}{2}\operatorname{Tr} X_{|\mathbf{H}+}$, usually neglected, does not seem to have received its final correct treatment, in particular in view of mass-energy as a source of gravitation.

---

[1] This identity actually comes from a general formula for the *variation* of the geometrical conserved quantities (10.1.12,17) as specialized to the case of a quadratic Lagrangian (linear homogeneous field equations). See, e.g., J. Lee, R. M. Wald, J. Math. Phys. *31*, 725 (1990).

Notation and Conventions

1. General Mathematical Symbols

$\Rightarrow$ implies

$\Leftrightarrow$ implies and is implied by

$\rightarrow$ tends to; is replaced by; mapping between sets

$\mapsto$ mapping of elements

$\approx$ approximately equal to (within the accuracy considered)

$\propto$ is proportional to

$\begin{matrix} A := B \\ B =: A \end{matrix}$ A is defined by B

$\equiv$ identically equal to

$\mathbf{R}$ the set of real numbers (as a real vector space also: $\mathbf{R}^1$)

$\mathbf{C}$ the set of complex numbers (as a complex vector space also: $\mathbf{C}^1$)

* complex conjugation

$*$ in prefix position: formation of duals according to eq. (5.5.10)

$m \in M,\ M \ni m \ldots$ m is an element of the set M

$N \subset M,\ M \supset N \ldots$ N is a subset of M

$M = \{m\,|\ldots\} = \{m : \ldots\}$ M is the set of all m specified by $\ldots$

$M \cap N$ intersection of the sets M, N

$M \cup N$ union of the sets M, N

$\emptyset$ empty set

$M \times N$ Cartesian product of the sets M, N ($=$ set of all ordered pairs (m, n), where $m \in M$, $n \in N$)

$\circ$ composition of maps; binary composition law

iff   if and only if

2. Differentiation and Integration

$$\frac{\partial f}{\partial x} =: \partial_x f =: f_{,x}$$

$$\frac{\partial f}{\partial x^i} =: \partial_i f =: f_{,i} =: \nabla_i f$$

$d^3 x$ volume element in $\mathbf{R}^3$, $= dx^1\, dx^2\, dx^3 = dx\, dy\, dz$

$d^4 x$ volume element in $\mathbf{R}^4$ or in Minkowski space

$d\mathbf{O}$ vectorial surface element of a surface in $\mathbf{R}^3$

$d\sigma_i$ vectorial normal surface element of a hypersurface in Minkowski space according to eq. (5.6.8)

$d\sigma_{ik}$ tensorial normal surface element of a 2-surface in Minkowski space according to eq. (5.7.9)

$\partial \mathcal{D}$ boundary of the domain $\mathcal{D}$ of integration

3. Dirac Function

1-dimensional: $\delta(x)$, $\int \delta(x)\, f(x)\, dx = f(0)$
3-dimensional: $\delta^3(x)$
4-dimensional: $\delta^4(x)$
For a function $g(x)$ of one variable with simple zeroes x_A one has

$$\delta(g(x)) = \sum_A \frac{\delta(x - x_A)}{|g'(x_A)|} \ .$$

4. Linear Spaces, Operators, Matrices

$\mathbf{R}^n$	space of n-tuples of real numbers
$\mathbf{C}^n$	space of n-tuples of complex numbers
$\mathbf{S}_n$	unit sphere in $\mathbf{R}^{n+1}$
$\mathbf{V}, \mathbf{W}$	abstract vector spaces (over $\mathbf{C}$ or $\mathbf{R}$)
$\mathbf{H}$	Hilbert space
$\dim_{\mathbf{F}} \mathbf{V}$	dimension of the vector space $\mathbf{V}$ over the field $\mathbf{F}$
$\mathbf{V} \oplus \mathbf{W}$	direct sum of the vector spaces $\mathbf{V}$, $\mathbf{W}$, see secs. 6.6 and B.7
$\mathbf{V} \otimes \mathbf{W}$	tensor product of the vector spaces $\mathbf{V}$, $\mathbf{W}$, see secs. 6.6 and B.8
$T: \mathbf{V} \to \mathbf{W}$	map from $\mathbf{V}$ to $\mathbf{W}$
$\mathrm{id}_{\mathbf{V}}$	unit operator (identical map) of $\mathbf{V}$ onto itself (frequently written simply id)
$\mathbf{1}$	unit operator (unit matrix) in $\mathbf{C}^2$ or $\mathbf{R}^3$, $\mathbf{C}^3$
E	unit operator (unit matrix) in $\mathbf{R}^4$
$T \otimes T'$	Kronecker or tensor product of the matrices or linear maps T, T'
$T \oplus T'$	direct sum of the matrices or linear maps T, T'

$M^{\top}, M^{-1}, M^*, M^{\dagger}, \widetilde{M}$ transposed, inverse, complex-conjugate, Hermitian conjugate, contragredient of matrix M, respectively
$(M^{\dagger} = (M^{\top})^* = (M^*)^{\top}, \ \widetilde{M} = (M^{\top})^{-1} = (M^{-1})^{\top})$

diag $(a, b, \dots)$ diagonal matrix with diagonal elements a, b, $\dots$

5. Groups

$\mathcal{G}, \mathcal{H},$	abstract groups
e	unit element of $\mathcal{G}$
g^{-1}	inverse element of $g \in \mathcal{G}$
$\mathcal{G} \cong \mathcal{H}$	isomorphic groups
$\mathcal{P}$	Poincaré group
$\mathcal{L}$	Lorentz group
$\mathcal{T}$	translation group

The subgroups $\mathcal{L}_+^{\uparrow}$, $\mathcal{L}_+$, $\mathcal{L}^{\uparrow}$, $\mathcal{L}_0$ of $\mathcal{L}$ are defined in sect. 6.3;
the correseponding subgroups of $\mathcal{P}$ are written $\mathcal{P}_+^{\uparrow}$, $\mathcal{P}_+$, $\mathcal{P}^{\uparrow}$, $\mathcal{P}_0$

$\mathrm{O}(n)$ or $\mathrm{O}(n, \mathbf{C})$ orthogonal group of $\mathbf{R}^n$ or $\mathbf{C}^n$

$\mathrm{O}(p, q)$ pseudo-orthogonal group of $\mathbf{R}^{p+q}$ (leaving invariant the quadratic form $x_1^2 + \ldots + x_p^2 - x_{p+1}^2 - \ldots - x_{p+q}^2$)

$\mathrm{U}(p, q)$ pseudo-unitary group of $\mathbf{C}^{p+q}$ (leaving invariant the Hermitian form $|x_1|^2 + \ldots + |x_p|^2 - |x_{p+1}|^2 - \ldots - |x_{p+q}|^2$)

$\mathrm{U}(n) := \mathrm{U}(n, 0)$

$\mathrm{GL}(n)$ or $\mathrm{GL}(n, \mathbf{C})$ group of all nonsingular $n \times n$ matrices (real or complex)

$\mathrm{SL}(n)$ or $\mathrm{SL}(n, \mathbf{C})$ unimodular $(\det = 1)$ subgroup of $\mathrm{GL}(n)$ or $\mathrm{GL}(n, \mathbf{C})$

$\mathrm{SO}(p, q)$ or $\mathrm{SU}(p, q)$ $\mathrm{O}(p, q) \cap \mathrm{SL}(p + q)$ or $\mathrm{U}(p, q) \cap \mathrm{SL}(p + q, \mathbf{C})$

6. Vectors, Tensors, Spinors

$x^0 = t$ (inertial) time coordinate

$(x^1, x^2, x^3) = \mathbf{x}$ orthogonal Cartesian coordinates in 3-space

Index notation: $x^i, \quad i = 0, 1, 2, 3$

$\qquad\qquad\qquad x^\mu, \quad \mu = 1, 2, 3$

Summation convention: $x^i a_i := \sum_i x^i a_i$, etc., i.e., unless the contrary is explicitly stated, a sum over its range is understood whenever an index occurs twice within the same monomial

Spinor indices: $\varphi^A, \psi^{\dot X}, \ldots : A = 1, 2, \ \dot X = 1, 2$

Kronecker symbol: $\delta^i{}_k, \delta^\mu{}_\nu, \delta_{\mu\nu}, \delta_A{}^B, \ldots = \begin{cases} 1 & \text{if indices are equal} \\ 0 & \text{if indices are unequal} \end{cases}$

Generalized Kronecker symbol: $\delta^{ikm\cdots}_{abc\cdots}$, see eq. (5.5.4)

Total symmetrization and antisymmetrization: $T_{(ik\ldots m)}$ and $T_{[ik\ldots m]}$, ibidem

Permutation symbol: $\epsilon(ikmn\ldots) := \begin{cases} 0 & \text{if there are any two equal indices} \\ +1 & \text{even permutation of natural order} \\ -1 & \text{odd permutation of natural order} \end{cases}$

ϵ-tensors: $\quad \epsilon_{ikmn} := \epsilon(ikmn)$ in Minkowski space; see eqs. (5.5.8ff.)

$\qquad\qquad \epsilon_{\mu\nu\lambda} := \epsilon(\mu\nu\lambda) \quad$ for $\mathbf{R}^3$

$\qquad\qquad \epsilon_{AB} := \epsilon(AB) \quad$ for 2-spinor space

Metric tensor: $\quad \eta_{ik} := \mathrm{diag}\,(1, -1, -1, -1) \quad$ (sign convention)

Remark: This convention is convenient for the 2-component spinor formalism; it also makes it easy to memorize positivity properties for time components which some important physical quantities have. These advantages are counteracted by disadvantages when space-time splittings are made, as the following circumstantial rule for index transport shows.

Index transport: For quantities with 4-tensor character use η_{ik}, η^{ik} as in eqs. (3.4.1,5). For a quantity that is 3-tensorial but is not part of a 4-tensorial quantity of the *same* type, as well as in all of chap. 7: use $\delta_{\mu\nu}$, $\delta^{\mu\nu}$. Examples: 3-velocity $\mathbf{v} = (v^\mu) = (v_\mu)$, field strenghts $\mathbf{E} = (E_\mu) = (E^\mu)$, $\mathbf{B} = (B_\mu) = (B^\mu)$, $\epsilon_{\mu\nu\lambda} = \epsilon^{\mu\nu\lambda} = \epsilon_\mu{}^{\nu\lambda} = \ldots$.

Spinorial quantities: $\Phi^A = \epsilon^{AB}\Phi_B$, $\Phi_B = \Phi^C\epsilon_{CB}$

Space-time splitting: $(x^i) = (x^0, x^\mu) = (t, \mathbf{x})$
$$(x_i) = (t, -\mathbf{x}) \quad \text{etc.}$$

Exception: $\partial_i := \left(\dfrac{\partial}{\partial x^0}, \dfrac{\partial}{\partial x^\mu} \right) = (\partial_t, +\boldsymbol{\nabla})$

Abstract, index, and matrix notation:
v, v' 4-vectors
v^i, v'^i components of v, v' in some frame I
$v^{i'}$ components of v in a 'primed reference frame' I'
$\mathrm{v} = (v^i)$ column matrix of components v^i (in the text also as a row)
4-scalar product: $uv := u^i v^k \eta_{ik} =: \mathrm{u}\,\mathrm{v}$

3-vectors: $\mathbf{v}$, $\vec{\mathcal{P}}$ for the column of components, but sometimes also for the abstract, or geometric, object
$\boldsymbol{\nabla}$ nabla operator
Scalar product: $\mathbf{uv}$
Vector product: $\mathbf{u} \times \mathbf{v}$
Tensor product: $\mathbf{u} \otimes \mathbf{v} =$ tensor with components $u^\mu v^\nu$

Scalar product in Hilbert space: $\langle x, y \rangle$, or $\langle x \,|\, y \rangle$ in Dirac notation

7. Physical Conventions

c speed of light, made equal to 1 by suitable choice of units
h Planck's quantum of action
$\hbar$ $h/2\pi$, made equal to 1 by suitable choice of units
$\mathbf{v}$ relative velocity between inertial systems

$$\gamma := \frac{1}{\sqrt{1 - v^2/c^2}}, \quad \text{often written as} =: \gamma_\mathbf{v} \text{ for clarity}$$

$$s := \int \sqrt{1 - v^2/c^2}\, dt \ \ldots \ \text{proper time}$$

4-potential: $(A^i) = (V, \mathbf{A})$, where V is the scalar and $\mathbf{A}$ is the vector potential
Convention for the tensor of electromagnetic field strengths: eq. (5.2.18)
Convention for electromagnetic stress-energy-momentum tensor: eq. (5.9.12)

Bibliography

1. Books Quoted

Alexandrow, P.S. (ed.) (1971) Die Hilbertschen Probleme (Ostwalds Klassiker). Leipzig: Geest und Portig.

Altmann, S.L. (1986) Rotations, Quaternions, and Double Groups. Oxford: Clarendon Press.

Anderson, J. (1967) Principles of Relativity Physics. New York: Academic Press.

Bacry, H. (1967) Leçons sur la théorie des groupes et les symétries des particules élémentaires. Paris: Dunod.

Barut, A.O. (ed.) (1973) Studies in Mathematical Physics. (Nato Advanced Study Institute Series, series C, vol. 1). Dordrecht: Reidel.

Biedenharn, L.C., Louck, J.D. (1981) Angular Momentum in Quantum Physics. Reading, Mass.: Addison-Wesley.

Bjørken, J.D., Drell, S.D. (1964) Relativistic Quantum Mechanics. New York: McGraw-Hill.

Bjørken, J.D., Drell, S.D. (1965) Relativistic Quantum Fields. New York: McGraw-Hill.

Blatt, J.M., Weisskopf, V.F. (1952) Theoretical Nuclear Physics. New York: Wiley.

Boerner, H. (1970) Representations of Groups. Amsterdam: North-Holland.

Bogolyubov, N.N., Logunov. A.A., Oksak, A.I., Todorov, I.T. (1990) General Principles of Quantum Field Theory. Dordrecht: Kluwer.

Bogolyubov, N.N., Shirkov, D.V. (1980) Introduction to the Theory of Quantized Fields. New York: Wiley.

Borel, É. (1914) Introduction géométrique à quelques théories physiques. Paris: Gauthier-Villars.

Born, M., Wolf, E. (1970) Principles of Optics. Oxford: Pergamon Press.

Bourbaki, N. (1970) Algèbre. Paris: Hermann et Cie. (English translation: Reading, Mass.: Addison-Wesley, 1974.)

Brillouin, L. (1960) Wave Propagation and Group Velocity. New York: Academic Press.

Browder, F. (ed.) (1976) Mathematical Developments Arising from Hilbert Problems. Proc. Symp. Pure Math. No. 28. Providence, R.I.: American Mathematical Society

Budinich, P., Trautman, A. (1988) The Spinorial Chessboard. Berlin Heidelberg New York Tokyo: Springer.

Byckling, E., Kajantie, K. (1973) Particle Kinematics. New York: Wiley.

Cartan, É. (1966) The Theory of Spinors. Cambridge, Mass.: MIT Press.

Chevalley, C. (1946) The Theory of Lie Groups. Princeton, N.J.: Princeton University Press.

Chevalley, C. (1956) Fundamental Concepts of Algebra. New York: Academic Press.

Conn, G.K.T., Fowler, G.N. (1970) Essays in Physics, vol. 2. London, New York: Academic Press.

Cornwell, J.F. (1985) Group Theory in Physics, vol. 2. London: Academic Press.

Dadhich, N., Narlikar, J. (eds.) (1998) Gravitation and Relativity: At the Turn of the Millennium. Pune, India: Inter-University Center for Astronomy and Astrophysics.

Davies, P.C.W. (1974) The Physics of Time Asymmetry. London: Surrey University Press.

DeWitt, C., DeWitt, B.S. (eds.) (1973) Black Holes. New York: Gordon and Breach.

Dieudonné, J.A. (1972) Treatise on Analysis, vol. III. New York: Academic Press.

Dieudonné, J.A. (1977) Treatise on Analysis, vol. V. New York: Academic Press.

Dieudonné, J.A. (1980) Special Functions and Linear Representations of Lie Groups. Providence, R.I.: American Mathematical Society.

Dieudonné, J.A., Carrell, J.B. (1971) Invariant Theory, Old and New. New York: Academic Press.

Dingle, H. (1961) The Special Theory of Relativity. London: Methuen.

Dym, H., McKean, H. P. (1972) Fourier Series and Integrals. New York: Academic Press.

Edmonds, A.R. (1960) Angular Momentum in Quantum Mechanics. Princeton, N.J.: University Press.

Feynman, R.P. (1965) Lectures on Physics, vol. 3. Reading, Mass.: Addison-Wesley.

Flügge, S. (1964) Quantentheorie I. Berlin Göttingen Heidelberg: Springer.

Fock, W.A. (1960) The Theory of Space, Time, and Gravitation. New York: Pergamon Press.

Fonda, L., Ghirardi, G.C. (1970) Symmetry Principles in Quantum Physics. New York: Dekker Inc.

French, A.P. (1968) Special Relativity. Cambridge, Mass.: MIT Press.

Fulton, W., Harris, J. (1991) Representation Theory. Berlin Heidelberg New York Tokyo: Springer.

Gasiorowicz, S. (1966) Elementary Particle Physics. New York: Wiley.

Gelfand, I.M., Minlos, R.A., Shapiro, Z.Ya. (1963) Representations of the Rotation and Lorentz Group and Their Applications. Oxford: Pergamon Press.

Goldstein, H. (1959) Classical Mechanics. Reading, Mass.: Addison-Wesley.

Greub, W. (1975) Linear Algebra, 4th edn. New York Heidelberg Berlin: Springer.

Greub, W. (1978) Multilinear Algebra, 2nd edn. New York Heidelberg Berlin: Springer.

Grünbaum, A. (1973) Philosophical Problems of Space and Time. Dordrecht: Reidel.

Gürzey, F. (1964) Group Theoretical Concepts and Methods in Elementary Particle Physics. New York: Gordon and Breach.

Hagedorn, R. (1963) Relativistic Kinematics. New York: Benjamin.

Halmos, P. (1974) Finite Dimensional Vector Spaces. New York Berlin Heidelberg: Springer.

Halpern, F. (1968) Special Relativity and Quantum Mechanics. Englewood Cliffs, N.J.: Prentice-Hall.

Hamermesh, M. (1962) Group Theory and Its Application to Physical Problems. Reading, Mass.: Addison-Wesley.

Hawking, S.W., Ellis, G.F.R. (1973) The Large Scale Structure of Space Time. Cambridge: Cambridge University Press.

Helgason, S. (1962) Differential Geometry and Symmetric Spaces. New York: Academic Press.

Henley, E.M., Thirring, W. (1962) Elementary Quantum Field Theory. New York: McGraw-Hill.

Hermann, R. (1966) Lie Groups for Physicists. New York: Benjamin.

Hermann, R. (1970) Vector Bundles in Physics. New York: Benjamin.

Ho-Kim, Q., Pham, X.Y. (1998) Elementary Particles and Their Interactions. Berlin Heidelberg New York Tokyo: Springer.

Holton, G. (1973) Thematic Origins of Scientific Thought: Kepler to Einstein. Cambridge, Mass.: Harvard University Press.

Huang, K. (1963) Statistical Mechanics, vol. 2. New York: Wiley.

Itzykson, C., Zuber, J.-B. (1980) Quantum Field Theory. New York: McGraw-Hill.

Jackson, J.D. (1999) Classical Electrodynamics. New York: Wiley.

Jacobson, N. (1962) Lie Algebras. New York: Interscience.

Janoschek, R. (ed.) (1991) Chirality from Weak Bosons to the α-Helix. Berlin Heidelberg New York Tokyo: Springer.

Jost, R. (1965) The General Theory of Quantized Fields. Providence, R.I.: American Mathematical Society.

Kabir, P.K. (1968) The CP Puzzle. London, New York: Academic Press.

Kacser, C. (1970) Introduction to the Special Theory of Relativity. Englewood Cliffs, N.J.: Prentice-Hall.

Källén, G. (1964) Elementary Particle Physics. Reading, Mass.: Addison-Wesley.

Kastler, D. (1961) Introduction à l'électrodynamique quantique. Paris: Dunod.

Kerner, E.H. (1972) The Theory of Action-at-a-distance in Relativistic Particle Dynamics. New York: Gordon and Breach.

Kilmister, C.W. (1970) Special Theory of Relativity. New York: Pergamon Press.

Kirillov, A.A. (1976) Elements of the Theory of Representations. Berlin Heidelberg New York: Springer.

Klauder, J. (ed.) (1972) Magic without Magic: John Archibald Wheeler. San Francisco: Freeman.

Kuhn, T. S. (1970) The Structure of Scientific Revolutions. Chicago: University of Chicago Press.

Landau, L.D., Lifshitz, E.M. (1958a) Quantum Mechanics. London: Pergamon Press.

Landau, L.D., Lifshitz, E.M. (1958b) Statistical Physics. London: Pergamon Press.

Landau, L.D., Lifshitz, E.M. (1961) Classical Field Theory. London: Pergamon Press.

Lang, S. (1966) Linear Algebra. Reading, Mass.: Addison-Wesley.

Larmor, J.J. (1900) Aether and Matter. Cambridge: Cambridge University Press.

Leinfellner, W. (1965) Einführung in die Erkenntnis- und Wissenschaftstheorie. Mannheim: Bibliographisches Institut A.G.

Loebl, E.M. (ed.) (1968) Group Theory and Its Applications. New York: Academic Press.

Lorentz, H.A. (1909) The Theory of Electrons. Leipzig: Teubner.

Lorentz, H.A., Einstein, A., Minkowski, H. (1958) Das Relativitätsprinzip. Stuttgart: Teubner

Mackey, G. (1968) Induced Representations and Quantum Mechanics. New York: Benjamin.

Marder, L. (1971) Time and the Space Traveler. London: Allen and Unwin.

Mittelstaedt, P. (1989) Der Zeitbegriff in der Physik. Mannheim: Bibliographisches Institut A.G.

Moszkowski, A. (1922)Einstein—Einblicke in seine Gedankenwelt. Berlin: F. Fontane.

Naimark, M.A. (1960) Normed Rings. Groningen: Noordhoff.

Naimark, M.A. (1964) Linear Representations of the Lorentz Group. New York: Pergamon Press.

Penrose, R., Rindler, W. (1984) Spinors and Space-Time, vol. 1: Two-Spinor Calculus and Relativistic Fields. Cambridge: Cambridge University Press.

Penrose, R., Rindler, W. (1986) Spinors and Space-Time, vol. 2: Spinor and Twistor Methods in Space-Time Geometry. Cambridge: Cambridge University Press.

Petrov, A.Z. (1969) Einstein Spaces. Oxford: Pergamon Press.

Pickert, G. (1961) Analytische Geometrie. Leipzig: Geest und Portig.

Pietschmann, H.V.R. (1983) Formulae and Results in Weak Interactions and Derivations. Wien New York: Springer.

Pontryagin, L.S. (1966) Topological Groups. New York: Gordon and Breach.

Popper, K. (1982) The Logic of Scientific Discovery. London: Hutchinson.

Post, E.J. (1962) Formal Structure of Electromagnetics. Amsterdam: North-Holland.

Reed, M., Simon, B. (1972–78) Methods of Modern Mathematical Physics, vol. I–IV. New York: Academic Press.

Reich, K. (1994) Die Entwicklung des Tensorkalküls: Vom absoluten Differentialkalkül zur Relativitätstheorie. Basel: Birkhäuser.

Rindler, W. (1960) Special Relativity. Edinburgh, London: Oliver and Boyd.

Rindler, W. (1982) Introduction to Special Relativity. Oxford: Clarendon Press.

Rindler, W. (1977) Essential Relativity: Special, General and Cosmological. New York Berlin Heidelberg: Springer.

Robertson, H.P., Noonan, T.W. (1968) Relativity and Cosmology. Philadelphia: W.B. Saunders.

Rohrlich, F. (1965) Classical Charged Particles. Reading, Mass.: Addison-Wesley.

Roman, P. (1960) Theory of Elementary Particles. Amsterdam: North-Holland.

Roman, P. (1969) Introduction to Quantum Field Theory. New York: Wiley.

Ryan, M.R., Shepley, L.C. (1975) Homogeneous Relativistic Cosmologies. Princeton, N.J.: Princeton University Press.

Sachs, R.K. (1971) General Relativity and Cosmology. New York: Academic Press.

Samelson, H. (1990) Notes on Lie-Algebras. Berlin Heidelberg New York Tokyo: Springer.

Schiff, L.I. (1968) Quantum Mechanics. New York: McGraw-Hill.

Schmutzer, E. (1968) Relativistische Physik. Leipzig: Teubner.

Schwartz, H.M. (1968) Introduction to Special Relativity. New York: McGraw-Hill.

Schweber, S. (1961) An Introduction to Relativistic Quantum Field Theory. Evanston, Ill.: Row and Peterson.

Segrè, E. (ed.) (1953) Experimental Nuclear Physics, vol. I. New York: Wiley.

Sexl, R.U., Sexl, H. (1973) Weiße Zwerge—Schwarze Löcher. Reinbek und Braunschweig: Rowohlt und Vieweg.

Sexl, R.U., Urbantke, H.K. (1995) Gravitation und Kosmologie. Heidelberg: Spektrum Akademischer Verlag.

Shaw, R. (1982) Linear Algebra and Group Representations, vol. I: Linear Algebra and Introduction to Group Representations. London: Academic Press.

Shaw, R. (1983) Linear Algebra and Group Representations, vol. II: Multilinear Algebra and Group Representations. London: Academic Press.

Silberstein, L. (1914) The Theory of Relativity. London: Macmillan.

Simms, D.J. (1968) Lie Groups and Quantum Mechanics. Berlin Heidelberg New York: Springer. (Lecture Notes im Mathematics 52)

Smirnov, V.I. (1964) A Course of Higher Mathematics, vol. III, part 2. New York: Pergamon Press.

Spivak, M. (1965) Calculus on Manifolds. New York: Benjamin.

Stewart, J. (1971) Non Equilibrium Relativistic Kinetic Theory. Berlin Heidelberg New York: Springer.

Streater, R.F., Wightman, A.S. (1964) PCT, Spin and Statistics, and All That. New York: Benjamin.

Stuart, E.B., Brainard, A.J. (1970) A Critical Review of Thermodynamics. Baltimore: Mono Book Corp.

Talman, J.D., Wigner, E.P. (1968) Special Functions: A Group Theoretic Approach. New York: Benjamin.

Terletskii, Y.P. (1968) Paradoxes in the Theory of Relativity. New York: Plenum Press.

Thomson, J.J. (1904) Electricity and Matter. (Silliman Lectures, Yale University, May 1903.) Westminster: Archibald Constable and Co., Ltd.

Tits, J. (1983) Liesche Gruppen und Algebren. Berlin Heidelberg New York: Springer.

Tolman, R.C. (1934) Relativity, Thermodynamics, and Cosmology. Oxford: Clarendon Press.

Trump, M.A., Schieve, W.C. (1999) Classical Relativistic Many-Body Dynamics. Dordrecht: Kluwer.

Urban, P. (ed.) (1964) Acta Physica Austriaca 1964, Suppl. I. Wien, New York: Springer.

Van der Waerden, B.L. (1966) Algebra I. Berlin Heidelberg New York: Springer.

Van der Waerden, B.L.(1967) Algebra II. Berlin Heidelberg New York: Springer.

Varadarajan, V.S. (1985) Geometry of Quantum Theory, 2nd edn. New York Berlin Heidelberg: Springer.

Warner, F. (1983) Foundations of Differentiable Manifolds and Lie Groups. New York Berlin Heidelberg: Springer.

Weber, J., Karade, T.M. (eds.) (1985) Gravitational Radiation and Relativity. Singapore: World Scientific.

Weinberg, S. (1972) Gravitation and Cosmology. New York: Wiley.

Weinberg, S. (1995) The Quantum Theory of Fields. Cambridge: Cambridge University Press.

Weitzenböck, R. (1923) Invariantentheorie. Groningen: Noordhoff.

Wentzel, G. (1949) Quantum Theory of Fields. New York: Interscience.

Weyl, H. (1923) Mathematische Analyse des Raumproblems. Berlin: Springer.

Weyl, H. (1946) The Classical Groups. Princeton, N.J.: Princeton University Press.

Weyl, H. (1955) Group Theory and Quantum Mechanics. New York: Dover.

Whittaker, E. (1960) A History of the Theories of Aether and Electricity, 2 vols. New York: Harper Torchbooks.

Wigner, E.P. (1959) Group Theory and Its Applications to the Quantum Mechanics of Atomic Spectra. New York: Academic Press.

Yourgrau, W., Mandelstam, S. (1968) Variational Principles in Dynamics and Quantum Theory. London: Pitman.

Zhang, Y. Zh. (1997) Special Relativity and Its Experimental Foundations. Singapore: World Scientific.

2. Other Selected and Advanced Books

Aharoni, J. (1965) The Special Theory of Relativity. Oxford: Clarendon Press.

Barut, A.O. (1964) Electrodynamics and Classical Theory of Fields and Particles. New York: Macmillan.

Corson, E.M. (1953) Introduction to Tensors, Spinors, and Relativistic Wave Equations. London: Blackie and Sons.

Deser, S., Ford, W. (eds.) (1965) Brandeis Summer Institute 1964. Englewood Cliffs, N.J.: Prentice-Hall.

DeWitt, B.S., Stora, R. (eds.) (1984) Relativity, Groups, and Topology II. Amsterdam: North-Holland.

DeWitt, C., DeWitt, B.S. (eds.) (1964) Relativity, Groups, and Topology. New York: Gordon and Breach.

DeWitt, C., Omnès, R. (eds.) (1960) Relations de dispersion et particules élémentaires. Paris: Hermann et Cie. (Article by A. S. Wightman).

Dyson, F.J. (ed.) (1966) Symmetry Groups. New York: Benjamin.

Kahan, Th. (1965) Theory of Groups in Classical and Quantum Physics. Edinburgh: Oliver and Boyd.

Lipkin, H.J. (1965) Lie Groups for Pedestrians. Amsterdam: North-Holland.

Møller, C. (1952) The Theory of Relativity. Oxford: Clarendon Press.

Pauli, W. (1958) Theory of Relativity. New York: Pergamon Press.

Perkins, D.H. (1987) Introduction to High Energy Physics. Reading, Mass.: Addison-Wesley.

Segal, I. (1963) Mathematical Problems of Relativistic Physics. Providence, R.I.: American Mathematical Society.

Smirnov, V.I. (1964) A Course of Higher Mathematics, vol. III, part I. New York: Pergamon Press.

Synge, J.L. (1965) Relativity: The Special Theory. Amsterdam: North-Holland.

Subject Index